Morphological Design and Synthesis of Nanoparticles

Morphological Design and Synthesis of Nanoparticles

Editors

Andrei Honciuc
Mirela Honciuc

MDPI

Basel • Beijing • Wuhan • Barcelona • Belgrade • Novi Sad • Cluj • Manchester

Editors

Andrei Honciuc
Electroactive Polymers and Plasmochemistry
"Petru Poni" Institute of Macromolecular Chemistry
Iasi
Romania

Mirela Honciuc
Natural Polymers, Bioactive and Biocompatible Materials
"Petru Poni" Institute of Macromolecular Chemistry
Iasi
Romania

Editorial Office
MDPI
St. Alban-Anlage 66
4052 Basel, Switzerland

This is a reprint of articles from the Special Issue published online in the open access journal *Nanomaterials* (ISSN 2079-4991) (available at: www.mdpi.com/journal/nanomaterials/special_issues/YQ0YE5QQV5).

For citation purposes, cite each article independently as indicated on the article page online and as indicated below:

Lastname, A.A.; Lastname, B.B. Article Title. *Journal Name* **Year**, *Volume Number*, Page Range.

ISBN 978-3-7258-0396-5 (Hbk)
ISBN 978-3-7258-0395-8 (PDF)
doi.org/10.3390/books978-3-7258-0395-8

Contents

About the Editors . . . . **vii**

Preface . . . . **ix**

Mirela Honciuc and Andrei Honciuc
Morphological Design and Synthesis of Nanoparticles
Reprinted from: *Nanomaterials* **2024**, *14*, 360, doi:10.3390/nano14040360 . . . . **1**

Stepan Sysak, Beata Czarczynska-Goslinska, Piotr Szyk, Tomasz Koczorowski, Dariusz T. Mlynarczyk and Wojciech Szczolko et al.
Metal Nanoparticle-Flavonoid Connections: Synthesis, Physicochemical and Biological Properties, as Well as Potential Applications in Medicine
Reprinted from: *Nanomaterials* **2023**, *13*, 1531, doi:10.3390/nano13091531 . . . . **6**

Matheus A. Chaves, Letícia S. Ferreira, Lucia Baldino, Samantha C. Pinho and Ernesto Reverchon
Current Applications of Liposomes for the Delivery of Vitamins: A Systematic Review
Reprinted from: *Nanomaterials* **2023**, *13*, 1557, doi:10.3390/nano13091557 . . . . **53**

Sedigheh Cheraghali, Ghasem Dini, Isabella Caligiuri, Michele Back and Flavio Rizzolio
PEG-Coated MnZn Ferrite Nanoparticles with Hierarchical Structure as MRI Contrast Agent
Reprinted from: *Nanomaterials* **2023**, *13*, 452, doi:10.3390/nano13030452 . . . . **90**

Atia Selmani, Ramona Jeitler, Michael Auinger, Carolin Tetyczka, Peter Banzer and Brian Kantor et al.
Investigation of the Influence of Wound-Treatment-Relevant Buffer Systems on the Colloidal and Optical Properties of Gold Nanoparticles
Reprinted from: *Nanomaterials* **2023**, *13*, 1878, doi:10.3390/nano13121878 . . . . **107**

Natalia Domenikou, Ioannis Thanopulos, Vassilios Yannopapas and Emmanuel Paspalakis
Nonlinear Optical Rectification in an Inversion-Symmetry-Broken Molecule Near a Metallic Nanoparticle
Reprinted from: *Nanomaterials* **2022**, *12*, 1020, doi:10.3390/nano12061020 . . . . **128**

Xingwang Shen, Junjie Li and Shuang Xi
High Strength Die-Attach Joint Formation by Pressureless Sintering of Organic Amine Modified Ag Nanoparticle Paste
Reprinted from: *Nanomaterials* **2022**, *12*, 3351, doi:10.3390/nano12193351 . . . . **140**

Erik Biehler, Qui Quach and Tarek M. Abdel-Fattah
Synthesis of Platinum Nanoparticles Supported on Fused Nanosized Carbon Spheres Derived from Sustainable Source for Application in a Hydrogen Generation Reaction
Reprinted from: *Nanomaterials* **2023**, *13*, 1994, doi:10.3390/nano13131994 . . . . **160**

Guanyu Cai, Luidgi Giordano, Cyrille Richard and Bruno Viana
Effect of the Elaboration Method on Structural and Optical Properties of $Zn_{1.33}Ga_{1.335}Sn_{0.33}O_4$:0.5%$Cr^{3+}$ Persistent Luminescent Nanomaterials
Reprinted from: *Nanomaterials* **2023**, *13*, 2175, doi:10.3390/nano13152175 . . . . **173**

Oliver Pauli and Andrei Honciuc
Extraction of Metal Ions by Interfacially Active Janus Nanoparticles Supported by Wax Colloidosomes Obtained from Pickering Emulsions
Reprinted from: *Nanomaterials* **2022**, *12*, 3738, doi:10.3390/nano12213738 . . . . **188**

Andrei Honciuc, Ana-Maria Solonaru and Mirela Honciuc
Water-Floating Hydrogel Polymer Microsphere Composites for Application in Hydrological Mining of Cu(II) Ions
Reprinted from: *Nanomaterials* **2023**, *13*, 2619, doi:10.3390/nano13192619 . . . . **202**

Mun'delanji C. Vestergaard, Yuki Nishida, Lihn T. T. Tran, Neha Sharma, Xiaoxiao Zhang and Masayuki Nakamura et al.
Antifungal Activity and Molecular Mechanisms of Copper Nanoforms against *Colletotrichum gloeosporioides*
Reprinted from: *Nanomaterials* **2023**, *13*, 2990, doi:10.3390/nano13232990 . . . . **220**

Anastasiya A. Shesterkina, Olga A. Kirichenko, Olga P. Tkachenko, Alexander L. Kustov and Leonid M. Kustov
Liquid-Phase Partial Hydrogenation of Phenylacetylene at Ambient Conditions Catalyzed by Pd-Fe-O Nanoparticles Supported on Silica
Reprinted from: *Nanomaterials* **2023**, *13*, 2247, doi:10.3390/nano13152247 . . . . **237**

Marco Sanna Angotzi, Valentina Mameli, Dominika Zákutná, Fausto Secci, Huolin L. Xin and Carla Cannas
Hard–Soft Core–Shell Architecture Formation from Cubic Cobalt Ferrite Nanoparticles
Reprinted from: *Nanomaterials* **2023**, *13*, 1679, doi:10.3390/nano13101679 . . . . **251**

Inna Y. Khairani, Qiyuan Lin, Joachim Landers, Soma Salamon, Carlos Doñate-Buendía and Evguenia Karapetrova et al.
Solvent Influence on the Magnetization and Phase of Fe-Ni Alloy Nanoparticles Generated by Laser Ablation in Liquids
Reprinted from: *Nanomaterials* **2023**, *13*, 227, doi:10.3390/nano13020227 . . . . **266**

Ahmad Hamdan and Luc Stafford
A Versatile Route for Synthesis of Metal Nanoalloys by Discharges at the Interface of Two Immiscible Liquids
Reprinted from: *Nanomaterials* **2022**, *12*, 3603, doi:10.3390/nano12203603 . . . . **284**

Qi Yuan, Wencai Hu, Tao Wang, Sen Wang, Gaobin Liu and Xueyan Han et al.
Electrochemical Synthesis of Nb-Doped $BaTiO_3$ Nanoparticles with Titanium-Niobium Alloy as Electrode
Reprinted from: *Nanomaterials* **2023**, *13*, 252, doi:10.3390/nano13020252 . . . . **294**

Bojana Milićević, Jovana Periša, Zoran Ristić, Katarina Milenković, Željka Antić and Krisjanis Smits et al.
Hydrothermal Synthesis and Properties of Yb^{3+}/Tm^{3+} Doped Sr_2LaF_7 Upconversion Nanoparticles
Reprinted from: *Nanomaterials* **2022**, *13*, 30, doi:10.3390/nano13010030 . . . . **304**

Andrei Honciuc and Oana-Iuliana Negru
Monitoring the Surface Energy Change of Nanoparticles in Functionalization Reactions with the NanoTraPPED Method
Reprinted from: *Nanomaterials* **2023**, *13*, 1246, doi:10.3390/nano13071246 . . . . **314**

About the Editors

Andrei Honciuc

Dr. Andrei Honciuc is currently a senior scientific researcher at the "Petru Poni" Institute of Macromolecular Chemistry in Romania. Before, he was a Metrohm-endowed Professor at the Zurich University of Applied Science. There, he led a research group with a primary focus on nanomaterials and interfacial science. Currently, Dr. Andrei Honciuc's research activities revolve around the synthesis of Janus nanoparticles and their applications in Pickering emulsions and self-assembled suprastructures. Prior to this, he was employed at BASF as a chemist, where he developed materials for photolithographic processes used in the production of the most recent generation of computer chips and processors. He earned his PhD from the University of Alabama, USA, in 2006 and pursued his postdoctoral studies, delving into various aspects of surfaces, interfaces, and nanotechnology at the University of Colorado at Boulder and as an Alexander von Humboldt Fellow at the Friedrich-Alexander University in Erlangen.

Mirela Honciuc

Dr. Mirela Honciuc's interests cover the broad topics of synthesis and the application of microstructured and hydrogel-based composites.

Her current work is focused on advanced synthetic strategies for nanoparticles with unique morphologies, the synthesis of composite materials by combining hydrogels, nano- and microparticles with unique properties, especially in the realms of stimuli-responsive materials, conductive materials, and programmable materials.

Dr. Mirela Honciuc is interested in deploying advanced materials in various fields, such as hydro mining, biomedical applications, the water treatment and food industries, and gas and humidity sensors.

Preface

The significance of nanoparticles in the realms of science and technology is growing, leading to breakthroughs across diverse fields such as medicine, electronics, energy, and environmental science. In the medical field, nanoparticles are employed in targeted drug delivery systems, enabling the direct treatment of diseased cells while reducing harm to healthy ones, thereby enhancing therapeutic results. In electronics, they facilitate the creation of smaller, more efficient, and more potent devices. In the energy sector, nanoparticles aid in the development of more effective solar cells and batteries with improved storage capacities. Furthermore, in environmental applications, they are utilized for pollution control and water purification, demonstrating their adaptability and usefulness in tackling some of the world's most urgent challenges. In this reprint, we meticulously chose contributions that cover all these crucial aspects of advanced nanoparticle applications.

The aim of this reprint is to capture some of the current trends in nanoparticle research focus on the correlation between nanoparticle morphology and function. For instance, the asymmetry of Janus nanoparticles gives them amphiphilic properties and the ability to partition at interfaces, self-assemble into superstructures, emulsify, or act as unidirectional nanomotors, among other things. However, there are countless other examples of functionality that stem from the morphological design of the nanoparticles, such as in nanoparticle catalysis, in drug delivery systems, or for nanoparticles used as technology facilitators for designing nanostructured materials, interfaces, and composites. This compilation of articles, bound in the book titled "Morphological Design and Synthesis of Nanoparticles," aims to provide a snapshot of the latest developments in the synthetic strategies of nanoparticles with unique morphologies that grant them special functions, covering a wide range of applications, from biology to catalysis, optoelectronics, and beyond. This collection of articles is committed to advancing the synthetic strategies of uniquely shaped nanoparticles, the design of materials derived from the use of functional nanoparticles, the physicochemical investigation of phenomena resulting from such nanoparticles, devices incorporating these nanoparticles as active components, and new applications.

This reprint is addressed to a broad audience that includes chemists, environmental scientists, and biologist but also students of science curricula interested in the latest developments of nanoparticles and their potential uses.

The editors express their gratitude to all the authors that contributed their work to this collection and for the great help and support from the assistant editors and staff of MDPI.

Andrei Honciuc and Mirela Honciuc
Editors

nanomaterials

Editorial

Morphological Design and Synthesis of Nanoparticles

Mirela Honciuc * and Andrei Honciuc *

"Petru Poni" Institute of Macromolecular Chemistry, Gr. Ghica Voda Alley 41A, 700487 Iasi, Romania
* Correspondence: teodorescu.mirela@icmpp.ro (M.H.); honciuc.andrei@icmpp.ro (A.H.)

Nanoparticles are particles with dimensions measured in nanometers, and exist at a scale where the physical, chemical, and biological properties of materials can differ significantly from those at a larger scale. Their unique characteristics are not merely due to their small size, but also arise from their high surface area to volume ratio, quantum effects, and the specific arrangements of their atoms.

The importance of nanoparticles in science and technology cannot be overstated, with numerous advancements being made across a wide array of disciplines, including medicine, electronics, energy, and environmental science. In medicine, nanoparticles are used for targeted drug delivery systems, which allow for the direct treatment of diseased cells while minimizing damage to healthy ones, significantly improving therapeutic outcomes. In electronics, they enable the development of smaller, more efficient, and more powerful devices. In the realm of energy, nanoparticles contribute to the creation of more efficient solar cells and batteries. Moreover, in environmental applications, they are used for pollution remediation and water purification, showcasing their versatility and utility in addressing some of the most pressing global challenges. In this Special Issue, we have carefully selected contributions that address all of these key aspects of the advanced applications of nanoparticles.

Nanoparticles are at the forefront of biomedical innovations and drug delivery systems. This is demonstrated by Sysak et al. [1] in their review article, which explores the subject of the synthesis and characterization of metal nanoparticle–flavonoid conjugates, emphasizing their potential to enhance bioavailability and target specificity in medical applications. The combination addresses the limitations of flavonoids, such as poor solubility and rapid metabolism, by leveraging the unique physicochemical properties of metal nanoparticles. The discussion spans various synthetic strategies, physicochemical properties like size and surface charge, and the biological implications of these hybrid materials, particularly their applications in cancer therapy, immune modulation, and as potent antioxidants. Yet another contribution from the forefront of nanoparticle applications is the review by Chaves et al. [2] that highlights the potential use of liposomes, which are soft nanoparticles/biomimetic particles, as versatile drug delivery vehicles for encapsulating bioactive compounds for improved stability and efficacy. Liposomes are intricate vesicular structures formed by one or more phospholipid bilayers separating the exterior aqueous medium from the interior one, with diameters ranging from 20 nm to several microns. This systematic review highlights the burgeoning application of liposomes in both cosmetics and food industries, focusing on the encapsulation of vitamins for enhancing their bio-accessibility and bioavailability, thus bridging the gap between pharmaceutical and food science applications. The research of Cheraghali et al. [3] further highlights the application potential of nanoparticles in medicine as MRI contrasting agents by performing an in vitro evaluation of MnZn ferrite nanoparticles coated with polyethylene glycol (PEG). The hierarchical morphology of these nanoparticles, resembling dandelion structures, offers an increased specific surface area, potentially improving the efficacy of contrast enhancement in MRI applications. The study not only compares the morphological and surface characteristics of these nanoparticles to those with a normal structure, but also assesses their cytotoxicity and hemocompatibility.

check for updates

Citation: Honciuc, M.; Honciuc, A. Morphological Design and Synthesis of Nanoparticles. *Nanomaterials* **2024**, *14*, 360. https://doi.org/10.3390/nano14040360

Received: 7 February 2024
Accepted: 9 February 2024
Published: 15 February 2024

Improving the synthesis methods to generate biocompatible nanoparticles for use in wound healing applications is also a field of interest. This is demonstrated by the work of Selmani et al. [4], which focuses on exploring improved synthetic methods of Au nanoparticles via the Turkevich method. This research explores the potential of Au nanoparticles to enhance wound healing through their radical scavenging activity, contributing to the development of innovative treatments.

The use of nanoparticles in optics, electronics, and energy applications is a subject of heightened interest for fundamental and applied research. The computational study conducted by Domenikou et al. [5] explores the enhancement of nonlinear optical properties near metallic nanoparticles using a polar zinc–phthalocyanine molecule near a gold nanosphere as a model system. The research underscores the impact of nanoparticle proximity on optical rectification coefficients under various external field conditions, providing insights into the design and optimization of photonic materials and devices that exploit the unique electromagnetic interactions between nanoparticles and molecular systems. The experimental work of Shen et al. [6] introduces a novel approach to enhancing the sintering performance of silver nanoparticle pastes through surface modification with organic amines. This method significantly improves the electrical conductivity and mechanical strength of sintered joints, providing a promising solution for high-performance electronic packaging applications.

High-performance materials are in high demand for energy applications. In this case, nanoparticles can make enormous contributions, as highlighted by the experimental work of Biehler et al. [7]. The authors of this work focused on the use of platinum nanoparticles supported on carbon spheres derived from sugar, a sustainable source, for the efficient generation of hydrogen—a clean energy carrier.

In addition to bioimaging applications, the design of luminescent materials is also of paramount importance for civil applications; for example, luminescent nanoparticles in pigments for road signs, markings, and lines enhance nighttime visibility and traffic guidance. The study of Cai et al. [8] focuses on improved luminescence persistence by improving the parameters of the synthesis method.

The application of nanoparticles in environmental applications and sustainability represents an emerging field with significant advantages over traditional materials. Pauli and Honciuc [9] explored an innovative approach to water purification: their study describes the use of Janus nanoparticles supported by wax colloidosomes for the extraction of metal ions from wastewater. The research highlights the potential of these nanostructured materials to float on water surfaces, efficiently adsorb and recover ions, and withstand multiple extraction cycles, offering a novel and sustainable alternative to traditional ion-exchange technologies. Further, the work of Honciuc et al. [10] proposes the concept of the hydrological mining of Cu(II) ions, with the help of specially designed nanoparticles–hydrogel polymer–microsphere composites capable of floating on water surfaces. The composite material combines nanostructured polymer microspheres with a polyvinyl alcohol (PVA) matrix, demonstrating the ability to adsorb and recover metal ions efficiently. This innovative approach suggests a more environmentally friendly method for extracting valuable metals from water bodies, reducing the need for energy-intensive pumping and processing.

The realm of nanoparticle applications extends to their potential use in agriculture as effective antifungal agents. For example, the antifungal properties of synthesized copper nanoforms against Colletotrichum gloeosporioides, a plant pathogen, are demonstrated in the experimental study of Vestergaard et al. [11], which reveals the potential of copper nanoparticles to serve as effective antifungal agents in agriculture. This research examines the impact of nanoparticle size, distribution, and oxidation state on their antifungal efficacy, providing valuable insights into the mechanisms of action and suggesting strategies for the sustainable management of plant diseases.

The use of nanoparticles as advanced catalysts represents the traditional realm of nanoparticle applications, which is ever-expanding. The work of Shesterkina et al. [12] demonstrates that bimetallic Pd-Fe/SiO2 catalysts have wide potential practical implications as new non-toxic alternative to the Lindlar catalyst for the selective hydrogenation of triple C≡C bonds in the liquid phase at room temperature. The selective hydrogenation of alkynes is important for the synthesis of pharmaceuticals, vitamins, nutraceuticals, fragrances, etc.

Creating advanced nanoparticle composites with unique magnetic properties is a widely pursued application area, as these nanoparticles could be valuable as active components in miniaturized transformers and reactive electronic components, electromagnetic shielding, actuators, spintronics, and beyond. The study of Angotzi et al. [13] examines the formation mechanisms of bi-magnetic core–shell nanoarchitectures, using cobalt ferrite nanoparticles as seeds to grow a manganese ferrite shell, shedding light on the competitive nucleation processes and the impact on magnetic properties. The findings contribute to the development of materials with tailored magnetic behaviors, offering potential applications in data storage and medical imaging. Khairani et al. [14] study the solvent influence on the magnetization and phase of the Fe-Ni alloy nanoparticles. The research provides a detailed analysis of the phases, magnetization, and oxidation levels of nanoparticles synthesized in various solvents, offering insights into the control of nanoparticle characteristics through solvent manipulation.

Further, controlling the morphology, or the shape and structure, of nanoparticles is crucial for optimizing their properties and functionalities. The morphology determines how nanoparticles interact with their environment and, by extension, their effectiveness in a given application. For instance, the shape of a nanoparticle can influence how it is absorbed by cells, its catalytic activity, and its optical properties. This control over morphology allows scientists and engineers to tailor nanoparticles for specific purposes, enhancing their performance and opening up new application possibilities.

Designing nanoparticles with specific morphologies and developing physical synthetic methods to create them are fundamental aspects of nanotechnology research. The ability to design nanoparticles deliberately involves understanding the relationship between the structure of nanoparticles and their properties. This knowledge guides the development of synthetic methods that are not only capable of producing nanoparticles with the desired characteristics, but also do so in a reliable, scalable, and environmentally friendly manner with the help of physical methods. Advancements in synthetic methods are essential for the practical application of nanoparticles, enabling the mass production of nanomaterials with controlled properties and ensuring their widespread use in various industries. Hamdan and Stafford [15] introduce us to a versatile route for the synthesis of metal nanoalloys using a novel spark discharge method for producing metal nanoalloys in a liquid environment, exploiting the interface between two immiscible liquids. The technique enables the synthesis of nanoparticles with controlled composition and embedded in a carbon matrix, opening up new avenues for the creation of nanoalloys with tailored properties for catalysis, plasmonics, and energy conversion applications. The electrochemical method is yet another synthesis method for nanoparticles, as demonstrated by the work of Yuan et al. [16]. This work reports an innovative electrochemical approach for the synthesis of Nb-doped $BaTiO_3$ nanoparticles and highlights the ability to control dopant concentrations and achieve high crystallinity under mild conditions. The research emphasizes the significance of alkalinity in the synthesis process, detailing the impact on crystal grain size, distribution, and the potential applications of the resulting nanoparticles in electronic materials, showcasing a novel route for doping and tailoring the properties of nanoceramics. The hydrothermal synthesis of nanoparticles is represented by the work of Milićević et al. [17] that focuses on the hydrothermal synthesis of upconversion nanoparticles doped with ytterbium (Yb^{3+}) and thulium (Tm^{3+}), aiming to optimize their luminescent properties for applications in bioimaging and security. The study systematically analyzes the effect of dopant concentrations on emission properties, providing insights into the structural and photoluminescent characteristics of these nanoparticles, and suggesting optimal doping levels for enhanced near-infrared emission.

Accompanying the discussion of synthesis and functionalization methods, this Special Issue also highlights the importance of the development of metrology tools for nanoparticles, especially the development of new methods of measuring the nanoparticle surface properties. While monitoring the physicochemical transformation of macroscopic surfaces is trivial, monitoring such changes at the nanoscale is extremely challenging due to the lack of methods and tools. Among these changes in the physicochemical properties of nanoparticles are changes in surface wettability from the liquid of nanoparticles, following a chemical surface modification via physical treatment. Monitoring the change in the wettability of nanoparticles via the contact angle and surface energy is extremely important in predicting the nanoparticles' behavior in terms of their dispersibility in water and air, pelleting ability, and interaction with solvents or other molecules, and could predict their potential risks toward the environment and living organisms. In this context, a new method for measuring the surface energy of nanoparticles was developed by Honciuc and Negru [18], namely, the NanoTraPPED method. This research offers a novel approach to monitoring the surface energy changes of nanoparticles during functionalization reactions. The study details the application of this method to silica nanoparticles undergoing various surface reactions, providing insights into the physicochemical transformations and the impact of molecular complexity on surface energy. This contribution advances the understanding of nanoparticle functionalization, offering a valuable tool for characterizing surface modifications.

These detailed descriptions offer a comprehensive overview of the contributions to the Special Issue, highlighting the multifaceted nature of nanoparticle research and its potential to address complex challenges across a wide spectrum of scientific and technological domains.

Funding: This work was supported by a grant from the Ministry of Research, Innovation and Digitization of Romania, CNCS/CCCDI-UEFISCDI, project number PN-III-P4-PCE-2021-0306 (Contract Nr. PCE62/2022).

Conflicts of Interest: The authors declare no conflicts of interest.

References

1. Sysak, S.; Czarczynska-Goslinska, B.; Szyk, P.; Koczorowski, T.; Mlynarczyk, D.T.; Szczolko, W.; Lesyk, R.; Goslinski, T. Metal Nanoparticle-Flavonoid Connections: Synthesis, Physicochemical and Biological Properties, as Well as Potential Applications in Medicine. *Nanomaterials* **2023**, *13*, 1531. [CrossRef] [PubMed]
2. Chaves, M.A.; Ferreira, L.S.; Baldino, L.; Pinho, S.C.; Reverchon, E. Current Applications of Liposomes for the Delivery of Vitamins: A Systematic Review. *Nanomaterials* **2023**, *13*, 1557. [CrossRef] [PubMed]
3. Cheraghali, S.; Dini, G.; Caligiuri, I.; Back, M.; Rizzolio, F. PEG-Coated MnZn Ferrite Nanoparticles with Hierarchical Structure as MRI Contrast Agent. *Nanomaterials* **2023**, *13*, 452. [CrossRef] [PubMed]
4. Selmani, A.; Jeitler, R.; Auinger, M.; Tetyczka, C.; Banzer, P.; Kantor, B.; Leitinger, G.; Roblegg, E. Investigation of the Influence of Wound-Treatment-Relevant Buffer Systems on the Colloidal and Optical Properties of Gold Nanoparticles. *Nanomaterials* **2023**, *13*, 1878. [CrossRef] [PubMed]
5. Domenikou, N.; Thanopulos, I.; Yannopapas, V.; Paspalakis, E. Nonlinear Optical Rectification in an Inversion-Symmetry-Broken Molecule near a Metallic Nanoparticle. *Nanomaterials* **2022**, *12*, 1020. [CrossRef] [PubMed]
6. Shen, X.; Li, J.; Xi, S. High Strength Die-Attach Joint Formation by Pressureless Sintering of Organic Amine Modified Ag Nanoparticle Paste. *Nanomaterials* **2022**, *12*, 3351. [CrossRef] [PubMed]
7. Biehler, E.; Quach, Q.; Abdel-Fattah, T.M. Synthesis of Platinum Nanoparticles Supported on Fused Nanosized Carbon Spheres Derived from Sustainable Source for Application in a Hydrogen Generation Reaction. *Nanomaterials* **2023**, *13*, 1994. [CrossRef] [PubMed]
8. Cai, G.; Giordano, L.; Richard, C.; Viana, B. Effect of the Elaboration Method on Structural and Optical Properties of Zn1.33Ga1.335Sn0.33O4:0.5%Cr3+ Persistent Luminescent Nanomaterials. *Nanomaterials* **2023**, *13*, 2175. [CrossRef] [PubMed]
9. Pauli, O.; Honciuc, A. Extraction of Metal Ions by Interfacially Active Janus Nanoparticles Supported by Wax Colloidosomes Obtained from Pickering Emulsions. *Nanomaterials* **2022**, *12*, 3738. [CrossRef] [PubMed]
10. Honciuc, A.; Solonaru, A.-M.; Honciuc, M. Water-Floating Hydrogel Polymer Microsphere Composites for Application in Hydrological Mining of Cu(II) Ions. *Nanomaterials* **2023**, *13*, 2619. [CrossRef] [PubMed]

11. Vestergaard, M.C.; Nishida, Y.; Tran, L.T.T.; Sharma, N.; Zhang, X.; Nakamura, M.; Oussou-Azo, A.F.; Nakama, T. Antifungal Activity and Molecular Mechanisms of Copper Nanoforms against Colletotrichum Gloeosporioides. *Nanomaterials* **2023**, *13*, 2990. [CrossRef] [PubMed]
12. Shesterkina, A.A.; Kirichenko, O.A.; Tkachenko, O.P.; Kustov, A.L.; Kustov, L.M. Liquid-Phase Partial Hydrogenation of Phenylacetylene at Ambient Conditions Catalyzed by Pd-Fe-O Nanoparticles Supported on Silica. *Nanomaterials* **2023**, *13*, 2247. [CrossRef] [PubMed]
13. Sanna Angotzi, M.; Mameli, V.; Zákutná, D.; Secci, F.; Xin, H.L.; Cannas, C. Hard–Soft Core–Shell Architecture Formation from Cubic Cobalt Ferrite Nanoparticles. *Nanomaterials* **2023**, *13*, 1679. [CrossRef] [PubMed]
14. Khairani, I.Y.; Lin, Q.; Landers, J.; Salamon, S.; Doñate-Buendía, C.; Karapetrova, E.; Wende, H.; Zangari, G.; Gökce, B. Solvent Influence on the Magnetization and Phase of Fe-Ni Alloy Nanoparticles Generated by Laser Ablation in Liquids. *Nanomaterials* **2023**, *13*, 227. [CrossRef] [PubMed]
15. Hamdan, A.; Stafford, L. A Versatile Route for Synthesis of Metal Nanoalloys by Discharges at the Interface of Two Immiscible Liquids. *Nanomaterials* **2022**, *12*, 3603. [CrossRef] [PubMed]
16. Yuan, Q.; Hu, W.; Wang, T.; Wang, S.; Liu, G.; Han, X.; Guo, F.; Fan, Y. Electrochemical Synthesis of Nb-Doped BaTiO3 Nanoparticles with Titanium-Niobium Alloy as Electrode. *Nanomaterials* **2023**, *13*, 252. [CrossRef]
17. Milićević, B.; Periša, J.; Ristić, Z.; Milenković, K.; Antić, Ž.; Smits, K.; Kemere, M.; Vitols, K.; Sarakovskis, A.; Dramićanin, M. Hydrothermal Synthesis and Properties of Yb3+/Tm3+ Doped Sr2LaF7 Upconversion Nanoparticles. *Nanomaterials* **2022**, *13*, 30. [CrossRef] [PubMed]
18. Honciuc, A.; Negru, O.-I. Monitoring the Surface Energy Change of Nanoparticles in Functionalization Reactions with the NanoTraPPED Method. *Nanomaterials* **2023**, *13*, 1246. [CrossRef]

nanomaterials

Review

Metal Nanoparticle-Flavonoid Connections: Synthesis, Physicochemical and Biological Properties, as Well as Potential Applications in Medicine

Stepan Sysak [1,2], Beata Czarczynska-Goslinska [3], Piotr Szyk [1], Tomasz Koczorowski [1], Dariusz T. Mlynarczyk [1], Wojciech Szczolko [1], Roman Lesyk [4,5] and Tomasz Goslinski [1,*]

1 Chair and Department of Chemical Technology of Drugs, Poznan University of Medical Sciences, Grunwaldzka 6, 60-780 Poznań, Poland
2 Doctoral School, Poznan University of Medical Sciences, Bukowska 70, 60-812 Poznań, Poland
3 Chair and Department of Pharmaceutical Technology, Poznan University of Medical Sciences, Grunwaldzka 6, 60-780 Poznań, Poland
4 Department of Biotechnology and Cell Biology, Medical College, University of Information Technology and Management in Rzeszów, Sucharskiego 2, 35-225 Rzeszow, Poland
5 Department of Pharmaceutical, Organic and Bioorganic Chemistry, Danylo Halytsky Lviv National Medical University, Pekarska 69, 79010 Lviv, Ukraine
* Correspondence: tomasz.goslinski@ump.edu.pl

Abstract: Flavonoids are polyphenolic compounds widely occurring throughout the plant kingdom. They are biologically active and have many medical applications. Flavonoids reveal chemopreventive, anticarcinogenic, and antioxidant properties, as well as being able to modulate the immune system response and inhibit inflammation, angiogenesis, and metastasis. Polyphenols are also believed to reverse multidrug resistance via various mechanisms, induce apoptosis, and activate cell death signals in tumor cells by modulating cell signaling pathways. The main limitation to the broader usage of flavonoids is their low solubility, poor absorption, and rapid metabolism. To tackle this, the combining of flavonoids with nanocarriers could improve their bioavailability and create systems of wider functionalities. Recently, interest in hybrid materials based on combinations of metal nanoparticles with flavonoids has increased due to their unique physicochemical and biological properties, including improved selectivity toward target sites. In addition, flavonoids have further utilities, even in the initial step of preparation of metal nanomaterials. The review offers knowledge on multiple possibilities of the synthesis of flavonoid-metal nanoparticle conjugates, as well as presents some of their features such as size, shape, surface charge, and stability. The flavonoid-metal nanoparticles are also discussed regarding their biological properties and potential medical applications.

Keywords: cancer; flavonoids; metal nanoparticles; polyphenols; reactive oxygen species

Citation: Sysak, S.; Czarczynska-Goslinska, B.; Szyk, P.; Koczorowski, T.; Mlynarczyk, D.T.; Szczolko, W.; Lesyk, R.; Goslinski, T. Metal Nanoparticle-Flavonoid Connections: Synthesis, Physicochemical and Biological Properties, as Well as Potential Applications in Medicine. *Nanomaterials* **2023**, *13*, 1531. https://doi.org/10.3390/nano13091531

Academic Editors: Andrei Honciuc and Mirela Teodorescu

Received: 3 April 2023
Revised: 26 April 2023
Accepted: 27 April 2023
Published: 2 May 2023

1. Introduction

Flavonoids represent a category of polyphenolic compounds found throughout the plant kingdom that has received much interest due to their numerous biological activities [1]. They have chemopreventive, anticarcinogenic properties, and demonstrate antiproliferative activity on tumor cells. In addition, they inhibit inflammation, angiogenesis, and metastasis [2]. Polyphenols are also believed to reverse multidrug resistance via various mechanisms as well as induce apoptosis and activate cell death signals in tumor cells by modulating cell signaling pathways, such as activator protein-1, nuclear factor NF-kappa-B (NF-κB) or mitogen-activated protein kinases [1,3]. Of importance is the modulation of the immune system response by flavonoids as well as their antioxidant activity, which manifests itself through the ability of free radical scavenging [2,4]. Despite their promising health-promoting potential, numerous flavonoids, including (−)-epigallocatechin-3-*O*-gallate (EGCG), quercetin (QUR), genistein, apigenin (AP), naringenin, silibinin, and kaempferol, reveal low solubility, poor absorption, and rapid metabolism. Therefore, combining flavonoids with nanocarriers could improve their bioavailability and create systems of wider functionalities [5].

In recent years, nanoparticles have attracted the broad attention of researchers dealing with various scientific disciplines, mainly due to their interesting physicochemical properties and potential applicability in medicine. Therefore, interest in combining flavonoids with metal nanoparticles (NPs) has increased due to the unique physical, chemical, and biological properties of the resulting connections or hybrid materials. Chemical reduction of the noble metal precursor is one of the most popular methods of preparation of noble metal nanoparticles. Metal nanomaterials formulated with active substances of plant origin can be synthesized using physical, chemical, or biological methods and characterized with analytical techniques involving microscopic and spectroscopic studies. Other benefits of NPs are low cost, simple synthesis, and the possibility of controlling both their shapes and their sizes. It is also interesting that flavonoids themselves are useful for the preparation of metal nanomaterials, with these being reducing and electrostatic agents for the 'Green' synthesis of NPs from their metal salt precursors [6–9]. In metal nanoparticles, features of metals are exploited, such as optical polarizability, electrical conductivity, chemical properties, antibacterial effectiveness, and biocompatibility [10].

Metal nanoparticles are becoming more and more widely considered as perspective pharmaceutical carriers because of their numerous advantages. Metal nanoparticles have been developed as a superior alternative to conventional cancer therapy treatment due to theranostic properties that provide both diagnosis and drug delivery designed to monitor the therapy [11]. Noble metal NPs enable the tracking of nano-complex therapeutic carriers within the body owing to their unique plasmonic properties, which makes such therapy more efficient and safer. Regarding non-noble metal NPs, they are cost-effective, can convert electromagnetic energy into heat (hyperthermia), and possess magnetic properties [12,13]. Metal nanoparticles present increased stability and half-life in circulation, as well as appropriate biodistribution [14]. Due to the surface modification and incorporation of different ligands, they can target specific tissues and cells [4]. Functionalizing the nanoparticles with flavonoid ligands was found to improve their selectivity and allow them to reach the target sites [8].

Overall, this review summarizes the current state of flavonoid-metal nanoparticle conjugates and hybrids, including multiple possibilities for their synthesis and modifying features such as size, shape, surface charge, and stability (Figure 1). The flavonoid-metal nanoparticles are also discussed regarding their biological properties and medical applications, including potential utility as drug delivery vehicles, modulators of cellular responses, applications in cancer therapy, antibacterial treatment, and tissue engineering.

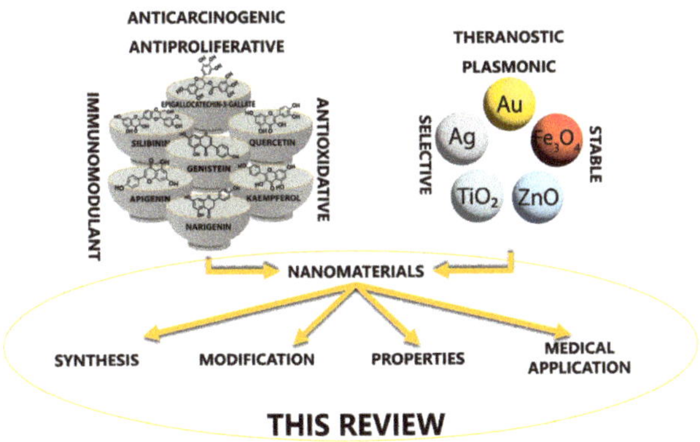

Figure 1. Flavonoid-metal nanoparticle conjugates and hybrids discussed in the review in terms of their synthesis, physicochemical, and biological features.

2. Physicochemical and Biological Properties of Flavonoids

Flavonoids are secondary plant metabolites, often acting as pigments responsible for the colors of fruit, flowers, and vegetables. They are characterized by a 2-phenylchromane scaffold (C6-C3-C6) with a heterocyclic pyran ring (C) fused with the benzene ring (A) and linked to the phenyl ring (B) (Figure 2). Flavonoid chemical structures have various substituents, including multiple hydroxyls (-OH), methoxyl (-OCH_3), and glycoside groups, along with an oxo group at position 4 of the C-ring. Flavonoids can be classified into various subclasses based on the oxidation level, unsaturation and substitution pattern of the C-ring, and the bonding position of the B-ring to C2/C3/C4 carbons of the C-ring. The subclasses of flavonoids include flavones, flavonols, flavanones, flavanonols, flavan-3-ols, isoflavones, neoflavonoids, anthocyanidins, chalcones, dihydrochalcones, and aurones. Flavonoids may exist as aglycones or as their derivatives. The structural features and configurations of flavonoids determine their bioavailability, metabolism, biochemical, and pharmacological activities [15].

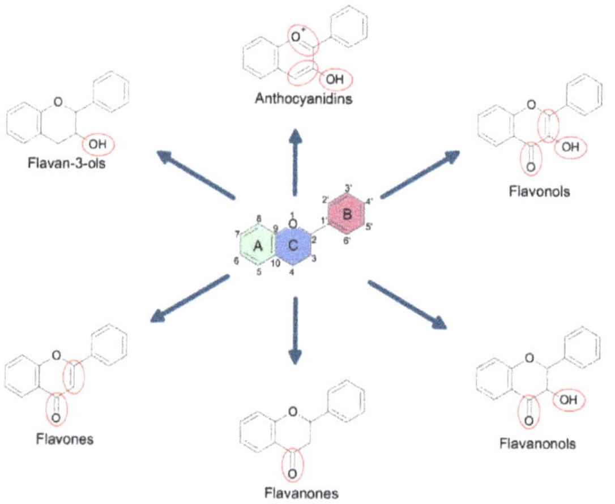

Figure 2. A 2-phenylchromane scaffold of flavonoids with a heterocyclic pyran ring (C) fused with the benzene ring (A) and linked to the phenyl ring (B). Selected subclasses of flavonoids.

Flavonoids exhibit unique physicochemical properties that influence their solubility in various solvents. Several factors affect their solubility in water, including the presence of a double bond in the C-ring and the number of hydroxyl (-OH) substituents in the B-ring. The solubility of flavonoids in 1-octanol is adversely affected by the number of hydroxyl groups, whereas the solubility in water generally increases with a rise in the number of hydroxyl groups. The addition of an -OCH_3 substituent to the B-ring reduces their solubility in both solvents. Additionally, the linkage between C2 and C3 plays a significant role in determining solubility, as the presence of a double bond results in lower solubility in both water and 1-octanol. An -OH substituent at C3 decreases aqueous solubility but increases 1-octanol solubility when the linkage between C2 and C3 is a double bond. The dissolution of flavonoids in 1-octanol can be either entropy-driven (chrysin, apigenin, kaempferol, morin) or enthalpy-determined. Overall, the solubility in water is a critical factor in the spontaneous transfer of flavonoids from water to 1-octanol [16].

Upon spectroscopic examination, flavonoids display two primary absorbance bands. Band I, in the range of 320–385 nm, results from the B-ring absorption, while Band II, within 250–285 nm, originates from the A-ring absorption. Any changes introduced to the structure cause differences in shifts of absorbance bands, e.g., kaempferol at 367 nm, quercetin at 371 nm, and myricetin at 374 nm. Flavanones, featuring a C-ring without a double bond bear different spectral profiles from other flavonoid subclasses. Typically, the spectra of flavanones exhibit a major Band II peak within the 270–295 nm range (288 nm for naringenin or 285 nm for taxifolin) along with a smaller Band I peak (326–327 nm). When a highly substituted B-ring is present, Band II often appears as a multitude of peaks. Anthocyanins, containing cinnamoyl/benzoyl units, display distinct absorbance properties, exhibiting a bandwidth ranging between 450 and 560 nm and another between 240 and 280 nm, subject to modification by hydroxyl groups connected to the B-ring [17]. Flavonoid compounds are sensitive to elevated temperatures encountered during microwave-assisted (MAE) and ultrasonic-assisted solvent extractions (UAE). These molecules may undergo degradation when subjected to excessive thermal stress. Many factors contribute to this phenomenon, including the quantity and positioning of hydroxyl moieties within their structures and the existence of accompanying substituents such as carbohydrates. The extent to which glycosidic bonds impede flavonoid stability appears relatively inconsequential compared to these other influences. Therefore, careful consideration is needed prior to employing MAE or UAE techniques to prevent undue deterioration of target analytes. Although conventional heating approaches have been shown to provide comparable extraction efficacy, they require significantly lengthier durations per unit of mass processed [18].

The composition and arrangement of hydroxyl groups within the A-, B-, and C-rings contribute to flavonoid antioxidant potential. Hydroxyls on the B-ring, particularly those in the *o*-dihydroxy configuration, increase the stability of the radical formed while enhancing electron delocalization. Pyrogallol groups augment antioxidant capacity, whereas an additional C2–C3 double bond and oxo-functionality in the pyran ring promotes further electron delocalization, leading to improved antioxidant performance. Flavonoids with a single 3-OH group and paired hydroxyls at positions 5 or 8 in the A-ring display favorable antioxidant characteristics. The glycosylation of flavonoids typically reduces antioxidant capacity relative to their aglycone forms. The type, location, and structural features of attached sugars play crucial roles in determining antioxidant efficacy. Modifications such as esterification, acylation, methylation, sulfation, and other substitutions commonly diminish antioxidant capacity, with sensitivity towards alterations appearing more prominent in the B-ring than in other regions. Researchers utilize different experimental methods, each yielding unique results depending on the specific testing conditions employed, so as to better understand flavonoids' antioxidative capabilities [19].

Flavonoid stability varies significantly when subjected to external stimuli such as heat and light. Studies have demonstrated that flavonoids undergo oxidative transformations upon exposure to these factors. These reactions lead to diverse outcomes regarding the preservation of their antioxidant potential. While some flavonoids, such as rutin and luteolin 7-*O*-glucoside, show reduced antioxidant capacity following heat treatment, others, such as quercetin, display remarkable resilience. Eriodictyol, another representative flavonoid, exhibits enhanced antioxidant activity upon thermal degradation due to the generation of novel compounds. Mesquitol appears to be highly responsive to both heat and light stress, resulting in improved antioxidant activity post-decomposition. Investigations into the impact of environmental conditions on flavonoids' antioxidant behavior typically involve studying their concentrated forms (model solutions). This approach allows for greater control over experimental variables and facilitates a better understanding of how heat treatment affects flavonoids [20].

Flavonoids, originally utilized for their dyeing and preservation properties, have emerged as promising candidates for medicinal chemistry due to their remarkable chemical diversity and recognized biological properties. Flavonoids exhibit variations in different aspects of their absorption, metabolism, and bioavailability within living systems. In the body, these compounds occur in different forms (commonly found as polar conjugates, glucuronides, and sulfates in plasma) and are attached primarily to serum albumin. Thus, these factors limit their distribution throughout the body, including entry into the central nervous system (CNS). During digestion, flavonoids undergo alterations involving changes such as glucuronidation in the intestine lumen, enterohepatic recirculation, hydroxylation, and dehydrogenation, predominantly in the liver. Some flavonoids also undergo transformations via bacterial action in the colon. Finally, certain flavonoid metabolites display considerable pharmacological activity both in vitro and potentially in vivo, yet further research remains necessary to fully comprehend how these combinations contribute to overall health effects [21]. The intricate structural framework of flavonoids permits them to interact with various biological macromolecules such as proteins [22], DNA [23], RNA [24], receptors [25], and bacterial cell walls [26], exhibiting diverse biological activities. The pharmacological versatility of flavonoids has broadened their potential utility in different fields of medicine, including preventive care and palliative treatment for life-threatening disorders. Consequently, extensive research endeavors have been focused on unraveling the underlying mechanisms of flavonoid action, paving the way for novel therapeutic interventions based on their multifaceted properties.

Flavonoids reveal a range of biological effects, including antioxidant [27–29], anti-inflammatory [30], antiviral [31], anticancer [1,32], and neuroprotective activities [33,34]. These properties can be attributed to the specific chemical structure and functional groups present within each flavonoid molecule. For instance, the hydroxyl groups present in flavonoids can donate hydrogen atoms, leading to the scavenging of free radicals and inhibition of oxidative stress. Moreover, flavonoids can modulate several signaling pathways by interacting with cell surface receptors and enzymes, ultimately regulating cell growth, proliferation, and apoptosis. Furthermore, the ability of flavonoids to chelate metal ions can disrupt the formation of reactive oxygen species and prevent cellular damage.

According to recent studies, flavonoids may have characteristics that aid in wound-healing processes, which were thoroughly discussed in an excellent review by Zulkefli et al. [35]. This is due to their ability to reduce inflammation, promote the formation of new blood vessels, facilitate skin cell renewal, and protect against the damage caused by toxic chemicals. Flavonoids can influence wound healing by modulating the expression of specific markers linked with pathways such as Angiopoietin-1/Tie-2 (Ang-1/Tie-2) [36], Focal Adhesion Kinase (FAK)/Src [37], c-Jun N-Terminal Kinase (JNK) [38], Mitogen-Activated Protein Kinase/Extracellular Signal-Regulated Kinase (MAPK/ERK) [39], Nuclear Factor Erythroid 2-Related Factor 2/Antioxidant Response Element (Nrf2/ARE) [40], Nuclear Factor Kappa B (NF-κB) [41], p38 Mitogen-Activated Kinase (MAPK) [38], Phosphatidylinositol 3-Kinase/Protein Kinase B (PI3K/AKT) [39], Transforming Growth Factor-beta (TGF-β) [36], Transforming Growth Factor/Suppressor of Mothers against Decapentaplegic (TGF-β/Smads) [36], and Wnt/β-catenin [42].

In conclusion, the intricate and diverse chemical structure of flavonoids enables them to interact with a wide range of biological targets, showcasing their potential as a vital tool in medicinal chemistry. The elucidation of the mechanisms behind flavonoid activity will aid in developing novel therapeutic interventions, which may prove beneficial in preventing and treating various diseases.

3. Nanoparticle-Flavonoid Connections

Many studies present connections of flavonoids with metal nanoparticles, mainly Ag, Au, oxides of Fe, Zn, and Ti, which resulted in materials of interesting physicochemical and biological properties.

3.1. Silver Nanoparticles

Of great interest are the potential antioxidant, antibacterial, antifungal, antiparasitic, antiviral, and anticancer properties of silver nanoparticles. To study the perspective medical applications, silver nanoparticles were functionalized with apigenin, catechin, EGCG, kaempferol, myricetin, 4′,7-dihydroxyflavone, dihydromyricetin, hesperidin, or quercetin. In addition, silver@quercetin nanoparticles were researched as a biocompatible and photostable aggregation-induced emission luminogen for in situ and real-time monitoring of biomolecules and biological processes. In the biological study, silver nanoparticles with curcumin and quercetin were analyzed in terms of the potential anti-inflammatory effect. Additionally, isoorientin-loaded silver nanoparticles were investigated in terms of their potential toxicity and activity on enzymes related to type II diabetes and obesity (Figure 3).

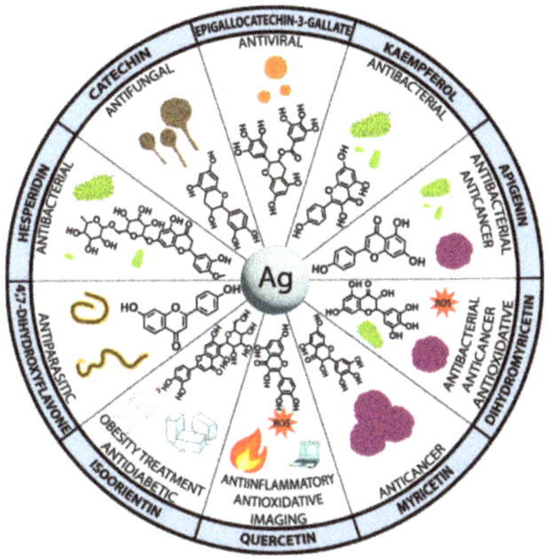

Figure 3. Spectrum of flavonoid-silver nanoparticle conjugates and hybrids with their most striking biological features.

3.1.1. Silver Nanoparticles in Therapy of Infectious Diseases

In 2022, Zhao et al. published a study on the synthesis of highly bactericidal silver nanoparticles coated with different ratios of hesperidin and pectin (HP-AgNPs) through a microwave-assisted process (Figure 4) [43]. The prepared AgNPs of 11.93–17.34 nm in size were named according to the ratio of hesperidin to pectin used in the synthesis—P-AgNPs for 0:1, HP-AgNPs1 for 1:3, HP-AgNPs2 for 3:1, and H-AgNPs for 1:0. The proportion of both ingredients revealed a tremendous effect on the morphology of the nanomaterials, as AgNPs prepared with either hesperidin or pectin alone had irregular shapes, while those made with both showed uniform morphology. HP-AgNPs2 demonstrated the most potent antibacterial activity, with a minimum inhibitory concentration (MIC) at 66.7 μg/mL against *Escherichia coli*. This value was significantly lower than the MIC of P-AgNPs, which reached around 8 times higher values at 533.3 μg/mL. The MICs of HP-AgNPs1 and H-AgNPs were also lower than that noted for P-AgNPs, at approximately 266.7 μg/mL and 133.3 μg/mL, respectively. In the study, the combination of hesperidin and pectin with AgNPs significantly enhanced the antimicrobial activity of the nanoparticles against *E. coli*. The adsorbance of HP-AgNPs2 on the cell wall caused significant morphological changes, including depression and damage to their cell wall, which induced oxidative stress and led to bacterial death. The direct contact with bacteria, AgNPs caused cell damage and cytotoxicity related to the release of Ag^+ ions. The fact that the MIC of HP-AgNPs2 was much lower than that of P-AgNPs, despite the latter releasing 25% more Ag^+ after 48 h, hints at the possibility of the combined action of particles and Ag^+ ions contributing to the enhanced antibacterial effect of HP-AgNPs2. Among the AgNPs tested, HP-AgNPs2 showed the highest increase in the amount of reactive oxygen species (up to 262.6%), indicating enhanced antibacterial properties.

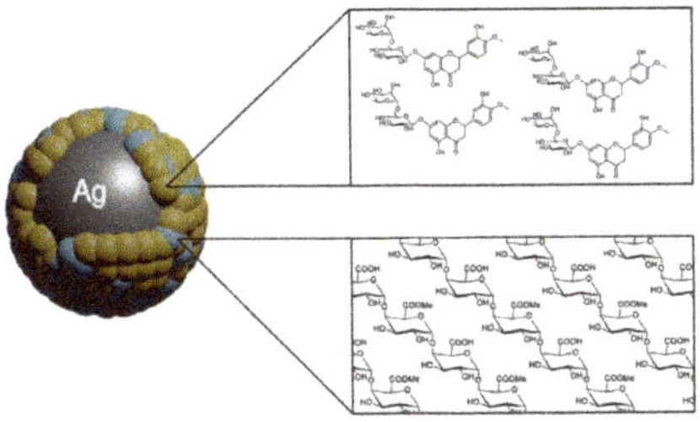

Figure 4. Schematic representation of silver nanoparticles coated with hesperidin and pectin (HP-AgNPs) developed by Zhao et al. [43].

In a similar study, ultra-uniform and colloidally stable hesperidin-capped silver nanoparticles (Ag-Hes NPs) were explored in terms of potential use for the treatment of infected wounds [44]. To prepare Ag-Hes NPs, $AgNO_3$, and hesperidin solutions were mixed with silver nanoparticles and stirred at room temperature. The resulting Ag-Hes NPs were combined with poly(vinyl alcohol)-sodium alginate (PVA-Alg) and electrospun to form Ag-Hes@H nano hydrogels. A variety of techniques was used to characterize their physicochemical properties, as well as to observe the expression of related proteins in cells and detect apoptosis-related proteins. The Ag-Hes NPs were highly uniform and colloidally stable, with a diameter of around 20 nm and a core-shell structure comprising a hesperidin shell surrounding a Ag core. The antibacterial activity evaluation showed that Ag-Hes NPs reveal a higher reduction in turbidity of *S. aureus* suspension compared to bare Ag nanoparticles. The viability of *S. aureus* decreased significantly in the presence of both Ag and Ag-Hes NPs, with a higher inhibition rate of 94.5% for Ag-Hes compared to 62% for Ag NPs. This may be ascribed to the uniform size and antioxidant activity of Ag-Hes NPs, which was confirmed experimentally. Unlike AgNPs, Ag-Hes NPs were found to be non-toxic as evaluated on human umbilical vein endothelial cells (HUVECs), possibly due to the protective effect of the hesperidin shell. As mentioned, Ag-Hes NPs were used to prepare electro-spun nanofibers and a hydrogel (Ag-Hes@H). The latter was found to significantly enhance the migration of HUVECs cells in a cell scratch assay, as shown by the significant reduction in the size of the scratch gap and the higher migration rate compared to control and other treatment groups. Moreover, Ag-Hes@H improved the closure rate of infected wounds in male rats (97% for the Ag-Hes@H group, and 83% for the Ag@H group after treatment). The most complete process of re-epithelialization and the strongest collagen fibers in the wound healing process of rats, as well as an increase in collagen deposition and proliferation of skin cells at the wound surface, were observed for the Ag-Hes@H group. In further in-depth tests, an increased expression of bFGF (basic fibroblast growth factor, a protein involved in skin regeneration) and SIRT1 was noted, while the expression of NF-κB was decreased. Additionally, the levels of the inflammatory factors MMP9, TNF-α, and IL-6 were suppressed in the Ag-Hes@H group. The breakdown of the gel network and the release of silver core and hesperidin molecules upon exposure to reducing substances in the body may contribute to the inhibitory effects on bacterial growth and inflammation observed with Ag-Hes@H.

In their study from 2021, Kannanoor et al. investigated silver nanoparticles (Ag-NPs) conjugated with kaempferol and hydrocortisone to formulate KH-AgNPs, which showed strong antibacterial properties against various bacterial strains [45]. KH-AgNPs were synthesized by mixing kaempferol, hydrocortisone, and $AgNO_3$ solutions with subsequent heating. Next, the mixture was enriched with $NaBH_4$, alkalized with NaOH to pH 12, stirred, cooled, and centrifuged, yielding uniform KH-AgNPs with a diameter of 10–30 nm and a face-centered cubic crystal structure. KH-AgNPs decreased the bacterial growth of *E. coli* with an MIC of 62.5 μg/mL and an MBC of 125 μg/mL. The results of the ROS assay showed that KH-AgNPs dramatically increased oxidative stress in *E. coli* in a concentration-dependent manner, with all KH-AgNP-treated cells producing more ROS than untreated cells. The authors also investigated the effect of KH-AgNPs on *E. coli* biofilms and discovered that KH-AgNPs effectively reduced and dissolved the biofilms by decreasing the colonization of bacteria on the surface. In a separate test, the effect of KH-AgNPs on static biofilm formation was tested at different concentrations. It turned out that treatment with KH-AgNPs for 48 h resulted in a 58% decrease in biofilm formation at a concentration of 110 μg/mL. Treatment with KH-AgNPs led to an increase in the production of free radicals in *E. coli* cells, resulting in lipid peroxidation, which compromises the integrity of the bacterial cell membrane.

Scroccarello et al. recently evaluated the antifungal effectiveness of silver nanoparticles made with several types of PCs with variable antioxidant capabilities, including flavonoids, such as CT and MY [46]. The AgNPs were synthesized by rapid and simple mixing of $AgNO_3$ with PCs in an alkaline environment at room temperature; along with CT and

MY, caffeic (CF) and gallic (GA) acids were employed. AgNPs@PCs were characterized by UV-vis spectroscopy, DLS, and TEM. The 2,2′-azinobis-(3-ethylbenzothiazoline-6-sulfonic acid) (ABTS) assay results revealed that PCs had greater antioxidant activity than AgNPs@PCs, CT was more active than MY, and dihydroxylic PCs were more active than tri-hydroxylic PCs. The same pattern was observed when AgNPs were generated with relative PCs, demonstrating that the PCs were preserved in the AgNPs shell even after purification. The Folin–Ciocalteu reagent did not react with all of the AgNPs@PCs, indicating that the phenols were bound to the AgNPs and that free PCs were not present. The effect of AgNPs@PCs on *Aspergillus niger* spore germination was investigated on the separate isotropic growth and vegetative tube germination phases. AgNPs@CT at the concentration of 30 mg/L inhibited isotropic growth in the most significant way, reaching 33%, followed by AgNPs@GA and AgNPs@CF with 24% and 23%, respectively, and AgNPs@MY with 17%. At the same concentration, AgNPs@CT revealed the most significant inhibitory impact on vegetative tube germination at 36%, followed by AgNPs@MY at 25%, AgNPs@GA at 20%, and AgNPs@CF at 15%. The effect of AgNPs@PCs on the vegetative growth of *A. niger* was studied in relation to the mycelial growth inhibition induced by applying AgNPs@PCs at five concentrations (5–60 mg/L) after 10 days of incubation. All AgNPs@PCs inhibited mycelial growth significantly, depending on the type of PCs employed in the synthesis and the concentration evaluated, while free PCs did not influence mycelial development. AgNPs@PCs inhibitory activity followed the antioxidant activity trend: AgNPs@CT > AgNPs@CF > AgNPs@MY > AgNPs@GA. The inhibition of hyphal growth and cell damage/viability was used to assess the efficacy of AgNPs@PCs on the growth of *A. niger*, and the effects on *A. niger* mycelium were studied using SEM and confocal imaging. AgNPs@PCs attached to and clad around hyphae caused a decrease in length and quantity as well as an aberrant growth pattern indicative of cell death. Cell viability experiments revealed that AgNPs@PCs led to significant cell death.

In 2021, Saadh and Aldalaen investigated the potential use of EGCG in combination with silver nanoparticles (AgNPs) and in co-administration with zinc(II) ions as a novel topical therapeutic with multiple effects against H5N1 influenza [47]. AgNPs were conjugated with EGCG by mixing various concentrations of EGCG solution with an aqueous solution of $AgNO_3$, followed by the addition of $NaBH_4$. The optimal inhibitory concentration of EGCG, both free and conjugated, was deduced to be 60 μM, whereas for zinc sulfate, the optimal concentration concluded 1.5 mg/mL. The combination of EGCG and zinc sulfate showed considerable antiviral activity against the H5N1 virus in embryonated SPF eggs, reducing the $logEID_{50}$/mL value 4.2–5.6 times. Similarly, the combination of EGCG-AgNPs and zinc sulfate demonstrated exceptional antiviral efficacy, with a 7.1 reduction in the $logEID_{50}$/mL value compared with the control group. The MTS proliferation assay with the use of different combinations of EGCG, zinc sulfate, and EGCG-AgNPs did not reveal any significant difference between the applied formulations and the control after 72 h. As a result, at the doses utilized in the study, none of the materials were determined to be cytotoxic.

The potential for the antileishmanial activity of 47DHF-functionalized gold and silver nanoparticles (Au-47DHF and Ag-47DHF) was studied by Sasidharan and Saudagar [48]. Nanoparticles Ag-47DHF were synthesized by adding 47DHF in varying ratios to solutions of $AgNO_3$ and NaOH. They ranged in size from 10 to 30 nm, with an average size of 25.1 nm, and exhibited a zeta potential of 40 mV. The Ag-47DHF nanoparticles had drug loading (DL) efficiencies of 64.13% and revealed cumulative drug release of 14.55% at pH 7.4 and 51.29% at pH 5.8, with a burst in the first 10 h, followed by a slow release. Ag-47DHF was more effective against axenic amastigotes than promastigotes, with the IC_{50} values against promastigotes at 0.8483 μg/mL, and against axenic amastigotes at 0.262 μg/mL. A 50% cytotoxic concentration (CC_{50}) value at 4.95 μg/mL of Ag-47DHF was noted in macrophages. Ag-47DHF with IC_{50} values at 0.215 μg/mL demonstrated antileishmanial action and targeted intracellular amastigotes. In addition, the infectivity index of 30 and the selectivity index (SI) of 24 were noted for Ag-47DHF. Compared to the infected control, the nanoparticle-administered macrophages generated reduced nitrite concentrations, indicating a decrease in parasite burden. The quantity of ROS (the key trigger of apoptosis induction for 47DHF) formed by Ag-47DHF nanoparticles was higher than that in control H_2O_2.

3.1.2. Silver Nanoparticles in Anticancer Therapy

Zarei et al. compared the biological potential of sodium citrate-based (SC-SNPs) and AP-based (AP-SNPs) silver nanoparticles in vitro and in vivo [49]. Either sodium citrate (SC) or AP was dissolved in water, filtered, and added to a $AgNO_3$ solution to create corresponding SNP colloids. The nanoparticles were subjected to UV-vis spectroscopy, field emission scanning electron microscopy (FESEM), and particle size analysis, which revealed for both types of nanoparticles homogeneous dispersions, stable zeta potentials, and pseudospherical forms. SNPs were tested for anticancer effects against the MCF-7 breast cancer cell line. Both nanoparticles suppressed MCF-7 cell growth in a dose-dependent manner, although the pro-apoptotic activity of AP-SNPs was the most significant. The morphological alterations seen in cancer cells treated with nanoparticles included cell shrinkage, plasma membrane blebbing, detachment, and destruction. Caspase-3, a fundamental apoptotic-related biomarker, was up-regulated 3.17-fold with AP-SNPs and 1.75-fold with SC-SNPs treatment of MCF-7 when compared to the untreated reference. In the research on mice, SC-SNPs substantially elevated liver enzymes alkaline phosphatase, aspartate transaminase, and alanine transaminase at 50 mg/kg/day, whereas AP-SNPs also significantly increased the same enzymes at 100 mg/kg/day. AP-SNPs cytoprotected mouse hepatocytes and dramatically decreased lipid peroxidation in the mouse liver. In addition, as compared to SC-SNPs, they boosted the expression of the antioxidant enzymes SOD and GPx in the liver of mice. The histopathology results revealed no significant cellular damage in the livers of mice treated for 30 days with either kind of nanoparticle.

Anwer et al. explored silver nanoparticles labeled with myricetin as a possible treatment for colorectal cancer [50]. The synthesis was carried out via microwave assisted (AgNPs-mw) or aging (AgNPs-aging) processes. The AgNPs-mw revealed a higher maximum absorbance than AgNPs-aging, indicating that the bioreduction process was faster for AgNPs-mw. TEM results showed that the synthesized nanoparticles had a spherical shape with a size range of 12–20 nm and a crystallite size of approximately 18 and 17 nm for AgNPs-mw and AgNPs-aging, respectively. The mean particle size, as determined by DLS analysis, was approximately 61 nm. The functional groups of the nanoparticles were identified using FTIR spectra, which showed shifts in the peaks for O-H, C-O, and NAH vibrations, indicating reduction resulting from the synthesis process. Myricetin was found to be effective in decreasing the viability of human colorectal cancer cells (HCT116) in a dose-dependent manner when tested in a cell viability assay with the IC_{50} value for myricetin at 106.87 µg/mL, while the IC_{50} value for silver nanoparticles labeled with myricetin (mAgNPs) was 34.04 µg/mL. mAgNPs showed some cytotoxicity towards HCT116 cells at concentrations below 200 µg/mL, although they were biocompatible with normal cells (HEK-293) up to a concentration of 400 µg/mL. There was no significant effect on the viability of healthy cells when treated with myricetin or mAgNPs. The researchers also conducted gene expression and pathway enrichment analysis using a colorectal cancer gene expression dataset and found several genes differentially expressed in healthy versus primary adenocarcinoma cells, healthy versus adjacent cells, and adjacent versus primary adenocarcinoma cells. They also noted that certain pathways were activated in each of these comparisons.

3.1.3. Silver Nanoparticles in Multimodal Action Materials

In 2021, Li et al. synthesized and characterized silver nanoparticles (AgNPs) functionalized by DMY with strong antioxidant, antibacterial, and anticancer properties, making them potential candidates for use as antimicrobial materials in the food and pharmaceutical industries [51]. To formulate the desired nanoparticles, the solutions of DMY and $AgNO_3$

were mixed and heated, leading to a grass-green suspension with black DMY-AgNPs. The nanoparticles were mainly spherical in shape and 114.76 nm in size; their zeta potential indicated high stability and dispersibility in water. The presence of characteristic functional groups in the FTIR spectra as well as the detection of C, O, and Ag elements in the XPS full-spectrum scan pattern, confirmed the superficial DMY. According to the XRD pattern, DMY-AgNPs had a face-centered cubic lattice structure of silver corresponding to the Miller index of (111) as the main orientation. The DPPH method was used to investigate the ability of DMY-AgNPs to scavenge free radicals, which was found to be comparable to or better than butylated hydroxytoluene or free DMY, with a scavenging rate of 56–92% at concentrations of 0.01–0.1 mg/mL. DMY-AgNPs inhibited *E. coli* and *Salmonella* well, with MICs of 10^{-6} g/L and 10^{-4} g/L, respectively. The antibacterial activity of DMY-AgNPs was shown to be superior and significant to that of DMY and tetracycline, with the inhibition rate increasing with sample concentration. MTT tests were used to assess the anticancer potential of DMY-AgNPs. When compared to DMY, the nanoparticles had a considerably stronger inhibitory impact on HeLa cells and a superior and consistent inhibitory effect on HepG2 cells. Both DMY and DMY-AgNPs inhibited MDA-MB-231 cells quite well, with the latter demonstrating exceptional inhibition (81.44%) at high concentrations.

Silver@quercetin nanoparticles (Ag@QCNPs) were prepared by adding quercetin (QUR) to an ammoniacal silver nitrate solution, resulting in the development of a biocompatible and photostable AIEgen for in situ and real-time monitoring of biomolecules and biological processes [52]. According to the TEM imaging, the nanoparticles were composed of a 35-nm silver core and a QUR shell. The silver core revealed well-organized crystal grains. The fluorescence emission of Ag@QCNPs in different THF/water ratios shifted from 480 nm to 550 nm as the water content increased, with a decrease in fluorescence intensity of less than 10% after irradiation with strong UV light showing resistance to photobleaching. The particle size and fluorescence intensity of Ag@QCNPs were found to be adjustable by varying the amount of QUR used in their preparation, as raising the amount of QUR resulted in an increase in fluorescence intensity and QUR shell thickness. Ag@QCNPs exhibited both aggregation-induced luminescence and the distinct plasma scattering of silver nanoparticles. Ag@QCNPs showed minimal cytotoxicity in HT-29 cells (>95% cell viability). During 60-min co-incubations at 100 μg/mL, the cytoplasm of HeLa cells retained Ag@QCNPs, as confirmed by CLSM images. In the course of in vivo computerized tomography imaging study in mice bearing S180 sarcoma cells, the researchers noted that the Ag@QCNPs accumulated at the highest level (128.6 HU) in the tumor site 2 h after injection before slowly declining over the following 10 h.

In the 2022 paper, Kumawat et al. reported on the synthesis of silver nanoparticles double functionalized with curcumin and, among others, quercetin (Cur-Ag^{QUR}) to produce a potential anti-inflammatory agent (Figure 5) [53]. The formulation and modification process involved heating and stirring aqueous solutions of curcumin and $AgNO_3$, followed by dialysis to remove unreacted molecules and ions. The emerged Cur-Ag was subjected to further surface functionalization with quercetin to produce Cur-Ag^{QUR}, which was subsequently dialyzed. The functionalization was achieved through binding interactions mediated by the unique properties of the functionalizing molecules. In quercetin, the reduced form of the polyphenol electrostatically stabilized the metal nanoparticles and attached to the surface of Cur-Ag nanoparticles to form Cur-Ag^{QUR} nanoparticles. The success of the functionalization was indicated by a change in color intensity and the excitation of metal nanoparticle surface plasmon vibrations. The stability and surface charge of these nanoparticles were evaluated using UV-vis spectroscopy, TEM, zeta potential, and DLS measurements. The results showed that Cur-Ag^{QUR} was stable, monodisperse, and had a negative surface charge. The surface functionalization of these nanoparticles resulted in an increase in their hydrodynamic size but did not affect their stability or surface charge. The Cur-Ag^{QUR} nanoparticles were then tested for their potential biological properties, including radical scavenging capacity (RSC), haemocompatibility, and anti-inflammatory effects. The results showed that Cur-Ag^{QUR} had the highest RSC among other tested probes (including bare Cur-Ag, conjugates with isoniazid and tyrosine). The haemocompatibility of the nanoparticles was also found to be good, as the percentage of hemolysis was less than 5% in all tested concentrations. In terms of their anti-inflammatory effects, Cur-Ag^{QUR} inhibited the secretion of pro-inflammatory cytokines from macrophages stimulated by lipopolysaccharide (LPS). The researchers also conducted cell viability studies using the MTT assay on mouse macrophages and found that the nanoparticles demonstrated good cell viability at all evaluated doses, with the Cur-Ag^{QUR} nanoparticles showing the highest viability. In addition, the researchers used the 2′,7′-dichlorofluorescin diacetate assay to assess the potential toxicity of nanoparticles on different organelles via reactive oxygen species (ROS) production and showed that Cur-Ag^{QUR} significantly reduced ROS in macrophages. As far as the anti-inflammatory effects of nanoparticles were concerned, it turned out that they significantly reduced the production of pro-inflammatory cytokines in macrophages. The expression of pro-inflammatory cytokines (such as TNF-α, IL-6, and IL-1β) was raised when treated with LPS. However, when co-treated with nanoparticles, the expression of TNF-α was significantly reduced, which, obviously, could be attributed to the presence of curcumin or quercetin biomolecules on Cur-Ag^{QUR}. Silver nanoparticles are known for antibacterial, antifungal, and immunomodulatory activities. Curcumin reveals anti-inflammatory effects due to its structure, which influences transcription factors, cytokines, and protein kinases and inhibits the synthesis of inflammatory molecules such as TNF-α. QUR also inhibits the production of TNF-α and nitric oxide (NO) in murine macrophages. Thus, summing up, double-functionalized nanoparticles can down-regulate pro-inflammatory genes while their biomedical potential can be improved by choosing proper biomolecules and nanoparticles.

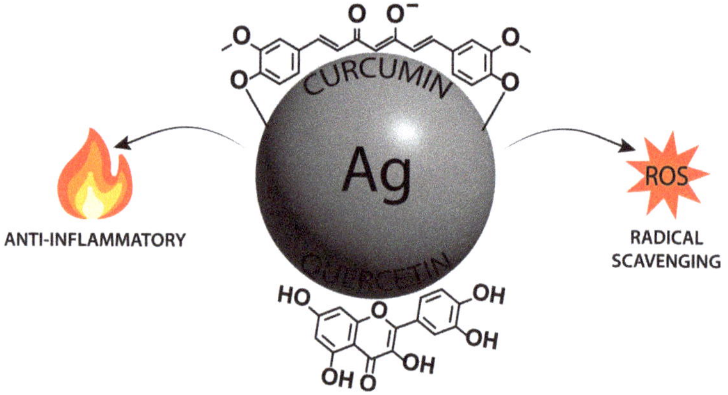

Figure 5. Silver nanoparticles functionalized with curcumin and quercetin and its action.

In their study from 2021, Wang et al. explored the production and characterization of isoorientin-loaded silver nanoparticles (AgNPs-Iso) as well as their stability and possible therapeutic uses [54]. A green synthesis approach was employed to create AgNPs from corn starch. The addition of SC increased AgNP production, as evidenced by the distinctive surface plasmon resonance peak at 403 nm and the color shift from white to brown. Following that, AgNPs were combined with isoorientin (Iso) and centrifuged to produce AgNPs-Iso precipitate with 76.60% loading efficiency, as confirmed by spectrophotometry. AgNPs had a mean size of 65 nm, while AgNPs-Iso had a size of 117 nm. For the characterization of nanoparticles, TEM was used to examine the morphology, whereas FTIR spectroscopy was used to evaluate the functional groups on the surface of nanoparticles. Stability testing revealed that AgNPs and AgNPs-Iso were stable in pH 5–9 and 0–0.30 M NaCl solutions. At lower pH values and higher concentrations of NaCl, aggregation was observed. In the simulated gastrointestinal tract, AgNPs-Iso was shown to be more stable than AgNPs. Moreover, the more stable nanoparticles displayed no changes in Iso retention even after 1 h of UV irradiation. The hemolysis study revealed that while AgNPs and AgNPs-Iso concentrations were reduced, the number of erythrocytes with complete morphology grew, and the erythrocyte hemolysis ratio decreased. The hemolysis ratio of AgNPs-Iso (29%) was much lower than that of AgNPs (about 60%) when the nanoparticles were studied at a concentration of 60 μg/mL. Observations of erythrocyte morphology using an inverted microscope revealed that the AgNPs-Iso nanoparticles are safer than unloaded ones. According to the MTT experiment, AgNPs-Iso had minimal cytotoxicity in HL-7702 human liver cells, suggesting that AgNPs-Iso might mitigate moderate cytotoxicity produced by AgNPs. AgNPs-Iso also inhibited alpha-glucosidase and pancreatic lipase, implicated in the development of type II diabetes and obesity, respectively. These data imply that AgNPs-Iso might constitute a promising therapeutic agent for these disorders.

To sum up

Pros

- The loading of AgNPs with flavonoids induces the biocompatibility of nanoparti-cles and reduces their toxicity.
- The connections of silver nanoparticles with particular flavonoids revealed a wide spectrum of potential antioxidant, antibacterial, antifungal, antiparasitic, antivi-ral, and anticancer properties.

Cons

- Functionalization of nanoparticles with hesperidin was found to cause stability problems, as the zeta potentials of the nanoparticles were closer to neutral [43]. Some of the nanoparticles reveal strong positive zeta potentials, which may cause toxicity [55].
- The nanoparticles discussed in the presented studies were of relatively small size, <20 nm, which is a poor indicator for biocompatibility [56].

AgNPs conjugated with flavonoids exerted satisfying antibacterial [43–45] and antifungal activity [46]. Moreover, when in co-administration with zinc(II) ions they exhibit antiviral activity [47]; furthermore, when combined with gold, antiparasitic activity was observed [48]. Silver@quercetin nanoparticles were proposed as luminogens for the monitoring of biomolecules [52] due to their ability to accumulate in the tumor site, whereas silver@myricetin [50] might be useful in the treatment of human colorectal cancer. Double-functionalized AgNPs with curcumin and quercetin seem to offer a promising anti-inflammatory nanocure [53]. AP-based silver nanoparticles revealed cytoprotective activity on mouse hepatocytes and increased the expression of liver antioxidant enzymes SOD and GPx [49]. Isoorientin-loaded silver nanoparticles (AgNPs-Iso) appeared to inhibit enzymes implicated in the development of type II diabetes and obesity [54], whereas silver nanoparticles (AgNPs) functionalized by DMY demonstrated antioxidant, antibacterial, and anticancer properties [51]. The summary of the data related to connections of silver nanoparticles with flavonoids, which were presented in this section, is included in Table 1.

Table 1. The summary of the data presented in Section 3.1.

Flavonoid/ Compound	Size of NPs	Synthesis Method	Activity	Target	Ref.
Hesperidin, Pectin	11.93–17.34 nm	microwave-assisted reduction	MIC 66.7 μg/mL	*E. coli*	[43]
Hesperidin	~20 nm	chemical reduction of $AgNO_3$	inhibition rate of 94.5%	*S. aureus*	[44]

Table 1. *Cont.*

Flavonoid/ Compound	Size of NPs	Synthesis Method	Activity	Target	Ref.
Kaempferol, Hydrocortisone	10–30 nm	chemical reduction of $AgNO_3$ with $NaBH_4$	MIC 62.5 μg/mL	*E. coli*	[45]
Catechin Myricetin	5 nm (AgNPs@MY) and 8 nm (AgNPs@CT)	chemical reduction of $AgNO_3$ in alkaline environment	33% of growth inhibition at conc. 30 mg/L of AgNPs@CT	*Aspergillus niger*	[46]
EGCG	Not specified	chemical reduction of $AgNO_3$ with $NaBH_4$	7.1 reduction in the $logEID_{50}$/mL value	H5N1 influenza	[47]
4′,7-Dihydroxyflavone	25.1 nm	chemical reduction of $AgNO_3$	0.8483 μg/mL 0.262 μg/mL	promastigotes amastigotes	[48]
Apigenin	93.94 nm	chemical reduction of $AgNO_3$	MIC 0.06–3.75 mg/mL	Gram-positive and Gram-negative bacteria	[49]
Myricetin	12–20 nm (TEM)	microwave-assisted reduction and aging	IC_{50} 34.04 μg/mL	Human colorectal cancer cells (HCT116)	[50]
Dihydromyricetin	114.76 nm	solvothermal method	MIC 10^{-6} g/L (*E. coli*) and 10^{-4} g/L (*Salmonella*) Cell viability = approx. 18.5%	*E. coli* and *Salmonella* MDA-MB-231 breast cancer cells	[51]
Quercetin	35 nm	chemical reduction of $AgNO_3$	Cancer cells imaging	HeLa cells	[52]
Curcumin, Quercetin	32.71 nm (hydrodynamic size)	chemical reduction of $AgNO_3$	Anti-inflammatory activity	Mouse macrophages	[53]
Isoorientin	117 nm	green synthesis approach from corn starch	Not specified	HL-7702 human liver cells	[54]

3.2. Gold Nanoparticles

Gold nanoparticles (AuNPs), similarly to AgNPs, have been eagerly studied since their discovery. This is mainly due to the ease of their preparation, simple surface functionalization, and a wide array of potential uses, such as, but not exclusively, in the medical sciences [57]. Additional application of flavonoid compounds improves the biological effects exerted by gold metallic particles, providing potential therapeutics with a range of properties (Figure 6). Of great interest are the potential antibacterial, antioxidant, antiparasitic, antiangiogenic, and anticancer properties of such nanoparticles. In this regard, the nanoparticles were functionalized with chrysin, kaempferol, quercetin, epigallocatechin gallate, procyanidins, 4′,7-dihydroxyflavone, metal-phenolic networks, and green tea polyphenols. In addition, a gold nanocluster was applied for the sensitive and selective detection of dopamine, whereas hesperidin isolated from orange peel was used in the synthesis of gold nanoparticles, which was applied both as an antioxidant and a photocatalyst for the treatment of industrial wastewater.

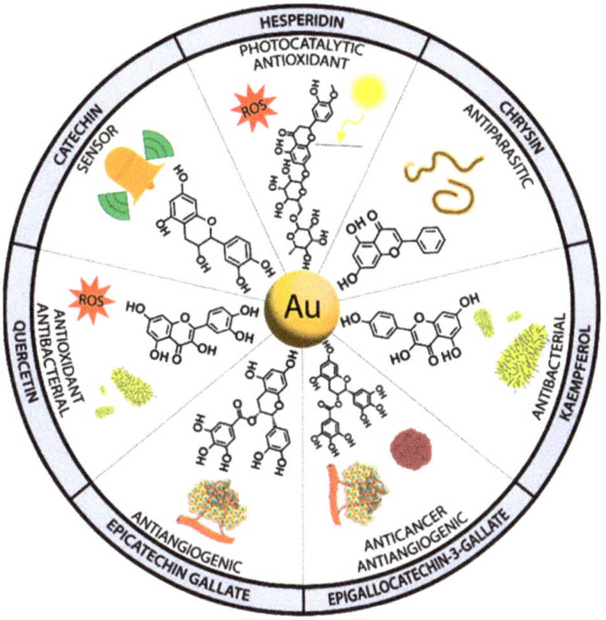

Figure 6. Spectrum of flavonoid-gold nanoparticle conjugates and hybrids with their most striking biological features.

3.2.1. Gold Nanoparticles in Therapy of Infectious Diseases

In 2021, Alhadrami et al. composed gold nanoparticles (GNPs) coated with chrysin (CHY), kaempferol, and QUR, to boost their antibacterial action against Gram-negative bacteria [58]. GNPs were made by combining an aqueous tetrachloroauric acid ($HAuCl_4$) solution with an aqueous GSH solution, adjusting the pH to 8 with NaOH, and then adding $NaBH_4$ until a ruby-red hue emerged. After centrifugation, the freshly formed GSH-GNPs were incubated with various flavonoids, allowing them to attach to the carboxylate groups of GSH and overlay the GSH-coated GNPs. The binding effectiveness of flavonoids to GSH-coated GNPs was 80%, 71%, and 41% for quercetin, kaempferol, and CHY, respectively, indicating a correlation between the level of hydroxylation of flavonoids and their binding efficiencies. Flavonoids were effectively coated on the GNPs, which was confirmed by UV-vis spectroscopic evaluations with characteristic changes in their plasmon resonance absorption band, FTIR spectra, as well as powder XRD patterns with characteristic peaks for Au, GSH, and flavonoids. GNPs revealed an average particle size of 4.1–35 nm, with a monodisperse, homogenous spherical, and hexagonal prism-like form. EDX proved the presence of Au in various percentages in each sample, up to 24.71% in GNP-quercetin, and the presence of C and O in adequate proportions to corroborate the loading of flavonoids. In the in vitro evaluations, GNP-quercetin was the most effective against all Gram-negative bacteria tested, with particularly strong action against *E. coli*, *P. aeruginosa*, and *P. vulgaris* (MIC 30 μg/mL) and *K. pneumonia* (MIC 60 μg/mL). GNP-kaempferol showed a strong antibacterial effect against *E. coli* and *P. vulgaris* (MICs of 60 and 30 μg/mL, respectively), but not against *P. aeruginosa* and *K. pneumonia* (MICs of 240 and 120 μg/mL, respectively). GNP-CHY inhibited *E. coli* well (MIC 60 μg/mL) but was less effective against the other microorganisms tested (MIC > 240 μg/mL). GNP-quercetin conjugate was chosen for the study of the antibacterial action. The TEM pictures evidently demonstrated disrupted membranes and the presence of GNP-quercetin nanoparticles inside *E. coli* and *P. aeruginosa* cells. Such an effect was not noted for uncoated GNPs. When flavonoids were docked to the binding site of DNA gyrase, they acquired greater docking scores with the subunit Gyr-B than with Gyr-A. Quercetin was the most potent (IC_{50} 0.89 μM), and CHY was the least potent (IC_{50} 3.91 μM), which indicates that the degree of hydroxylation of the flavonoids appears to be associated with their inhibitory activity on Gyr-B. In silico experiments were performed to understand better the antibacterial activity of flavonoids. The authors built a model of a 5 nm GNP coupled to GSH molecules and docked each flavonoid against it. According to the study, quercetin and kaempferol presented more stable interactions with GSH than CHY. The authors also investigated the effect of each flavonoid on the bacterial outer membrane, discovering a strong correlation between the flavonoids' growth inhibitory action and the local increase in membrane fluidity. Overall, quercetin was found to be the most effective of the three flavonoids studied.

In another study, Zhang et al. managed to functionalize gold nanorods (GNRs) with metal-phenolic networks (MPNs), rendering photothermal bactericidal nanoparticles (GNRs@MPNs) (Figure 7) [59]. As phenolic motifs, epigallocatechin gallate (EGCG), procyanidins (OPC), and tannic acid (TA) were chosen. To generate starting nanorods, the authors employed the seed-mediated growth method delineated by Nikoobakht and El-Sayed [60] as well as the conformed procedure of Fu et al. [61]. According to the technique, the solutions of $HAuCl_4$ and cetyltrimethylammonium bromide (CTAB) were stirred together and then injected with ice-cold sodium borohydride ($NaBH_4$) solution to provide the brownish seed solution that further was subjected to incubation. The growth mixture was prepared on the basis of a CTAB-sodium oleate binary surfactant mixture with $AgNO_3$ and consecutive addition of $HAuCl_4$. The resulting solution was adjusted with HCl to an appropriate pH level, injected with the ascorbic acid solution, combined with a small amount of the seed solution, and then incubated to undergo centrifugal purification of GNRs later. The obtained supernatant was concentrated and incubated in polyphenolic solutions to replace capped CTAB. To encapsulate MPNs, obtained GNRs were mixed with ultrapure water, sequentially injected with polyphenol solution of a specific kind (EGCG, OPC, or TA) and with iron chloride hexahydrate ($FeCl_3 \cdot 6H_2O$) solution, then combined with Tris-HCl buffer (pH 8). The forthcoming centrifugation and redispersion with water resulted in monolayered GNRs@MPNs (GNRs@MPN1)–GNRs@Fe-EGCG1, GNRs@Fe-OPC1, and GNRs@Fe-TA1 for each particular polyphenolic substrate. Later, the encapsulation process was repeated on GNRs@MPN1 as a precursor leading to GNRs@MPN2/3/4. The 808 nm laser thermography of new assemblies revealed a good synergy between GNRs and MPNs components in terms of temperature increment compared to GNRs alone, which could be reflected in the following manner: @Fe-OPC > @Fe-TA > @Fe-EGCG > GNRs. The in vitro antibacterial studies were conducted with and without NIR irradiation on *S. aureus* and *E. coli* O157:H7 cultures involving unclad GNRs as well as their three-layered polyphenolic compositions. The 808 nm light solely could not induce any apparent damage to bacterial samples. Treatment with bare GNRs caused negligible suppression to *S. aureus* (about 6.4% according to fluorescence imaging) and moderate detriment to *E. coli* O157:H7 (16.23%). However, the NIR laser application increased the antibacterial rate of GNRs to 30.07% and 69.8%, respectively. The irradiated @Fe-OPC3, @Fe-TA3, and @Fe-EGCG3 expressed 98.6%, 88.6%, and 85.4% respective suppression towards *S. aureus*, along with 99.8%, 90.5%, and 85.07% suppression in the case of *E. coli* O157:H7. What is interesting, the authors studied the antibacterial behavior of goldless MPNs towards *S. aureus* colonies, revealing the suppressing activity of 52.1%, 40.5%, and 33.2% for OPC-MPN, TA-MPN, and EGCG-MPN respectively. Further, a mice model with methicillin-resistant *S. aureus* (MRSA)-infected wounds was utilized, allowing for the investigation of the effectiveness of GNRs@MPNs in a more realistic and relevant setting. Treatment of an artificially inflicted wound with @Fe-OPC3 resulted in a slower healing process, with a 32.6% reduction in wound area after 8 days, mostly due to the chemical sterilization properties. However, the use of GNRs and NIR irradiation led to the formation of a scab and a 46.5% reduction in the wound area, indicating that the GNRs had a moderate ability to convert light into heat, which aided in the healing process. Treatment with NIR-irradiated @Fe-OPC3 significantly accelerated wound healing, resulting in an 88.24% reduction in the wound area after 8 days.

This bactericidal outcome likely occurred due to the synergistic photothermal effect of @Fe-OPC3. A plate counting assay also showed a significant reduction in the presence of MRSA bacteria in the wound tissue after treatment with irradiated @Fe-OPC3. The H&E staining histological analysis was used to study the effects of GNRs and @Fe-OPC3 on bacteria and wound healing. GNRs without NIR irradiation revealed weaker performance in wound healing, while NIR-irradiated GNRs partially regenerated epidermis tissue. In contrast, @Fe-OPC3 treated with NIR irradiation was successful in promoting the generation of intact epidermal layers in wound tissue, demonstrating the distinguishable photothermal potential for both antibacterial activity and wound healing as a promising candidate for use in these applications. The biosafety of GNRs@MPNs was extensively tested both in vitro and in vivo. The cytotoxicity was assessed on 3T3 cells, and the nanostructure appeared not to cause a significant decrease in cellular viability at concentrations of up to 100 ppm, indicating its non-toxic nature. According to blood compatibility evaluations with murine red blood cells, GNRs had high hemolysis at 89.79%, indicating poor blood compatibility. @Fe-EGCG3 and @Fe-OPC3 had a low hemolysis ratio of approximately 0% and 1.50%, respectively, while @Fe-TA3 did not present satisfactory blood compatibility at all. Moreover, no significant changes were observed during mice body weight assessment after the injection of GNRs@MPNs (with and without NIR), indicating their favorable biosafety. H&E staining on main organ tissues (including heart, liver, spleen, lung, and kidney) showed that GNRs caused toxic effects, while GNRs@MPNs had no notable abnormalities or damage and were safe for wound healing, even when irradiated. These results demonstrated the effectiveness and biocompatibility of GNRs@MPNs as bactericidal agents.

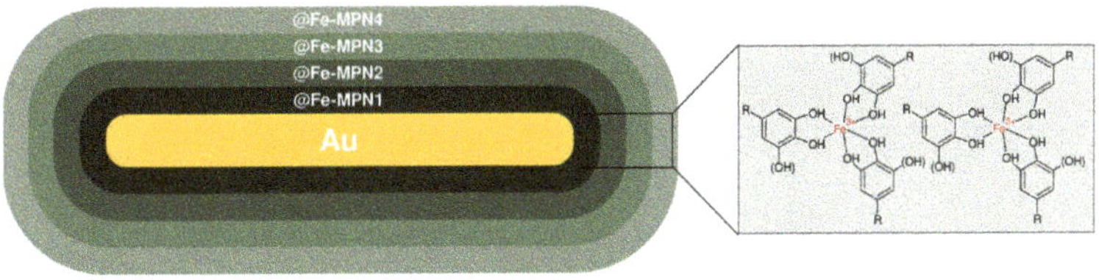

Figure 7. Metal-phenolic networks rendering photothermal bactericidal nanoparticles (GNRs@MPNs) with phenolic fragments such as epigallocatechin gallate, procyanidins, and tannic acid.

In 2021, Das et al. developed quercetin-conjugated gold nanoparticles (QuAunps) as a promising agent of antibacterial and antioxidant features [62]. To obtain nanoparticles, a chloroauric acid solution was mixed with a Tween 80 solution, then mixed with a cyclomixer and combined with a methanolic QUR solution. The mixture was then sonicated, and the progress of the reaction was monitored using UV spectrophotometry. The nano-gold was then separated by centrifugation, washed, and stored for further use. The characterization of the nanomaterial was performed using DLS (a hydrodynamic diameter of 62 nm), HRTEM (spherical shapes with a narrow particle size distribution and an average size of 30 nm), energy dispersive X-ray (EDX), and FTIR. The antioxidant activities of QuAunps were investigated using the measurement of the ability to inhibit oxidation of brilliant cresyl blue. QuAunps presented a concentration-dependent activity manifested by 42% and 67% antioxidant activities at 8 μg/mL and 45.8 μg/mL, respectively. Antibacterial assays showed that QuAunps had a lower MIC (7.6 μg/mL) against *E. coli* than free QUR (24.7 μg/mL) and control citrate-capped Aunps (50.8 μg/mL). The authors also examined the cytotoxicity of QuAunps, QUR, and ciprofloxacin (a control drug) in Vero cells and concluded that QuAunps had the highest biocompatibility with a CC_{50} value at 956.8 μM. The MBC of QuAunps was 10.5 g/mL. On agar plates treated with QuAunps, bacterial growth was noted below the MBC after 24 h, but not at or above the MBC. In addition, the dye uptake efficiency of QuAunps was measured at 48.37%, which was significantly higher than those noted for the ciprofloxacin group (19.39%) and the bacterial control group

(7.33%). Dye inclusion assays showed that QuAunps damaged the bacterial membrane, resulting in increased dye uptake. The TEM micrographs were applied for the study of QuAunps interaction with the bacterial surface.

The potential of selected functionalized gold nanoparticles has also been studied against selected parasites. In the previous chapter, the antileishmanial activity of 47DHF-functionalized silver nanoparticles (Ag-47DHF) was discussed [48]. In the same paper, the authors also studied the potential of Au-47DHF nanoparticles, which were synthesized by adding 47DHF in varying ratios to $HAuCl_4{\cdot}3H_2O$ solutions while stirring the mixture at a constant temperature and allowing the reaction to continue for 10 min after the color change had occurred. The Au-47DHF nanoparticles were oval and spherical in form, with a zeta potential of 60 mV, and had a size range of 6–7 nm, with an average size of 5.8 nm. UV spectroscopy, the high-resolution transmission electron microscope (HRTEM) pictures, DLS investigations, selected area electron diffraction (SAED) patterns, and FTIR analyses confirmed the structures of both nanoparticles. The nanoparticles had drug loading (DL) efficiencies of 80.79% and 64.13%, respectively. The release of 47DHF from the gold nanoparticles was gradual; at pH 7.4 and 5.8, its cumulative drug release was 14.95% and 57.97%, respectively. Au-47DHF was very effective against both developmental stages of the parasite with IC_{50} values against promastigotes and axenic amastigotes at 0.1226 µg/mL and 0.115 µg/mL, respectively. With IC_{50} values of 0.121 µg/mL, Au-47DHF demonstrated antileishmanial action and targeted intracellular amastigotes. When compared to the infected control, the nanoparticle-administered macrophages generated reduced nitrite concentrations, indicating a decrease in parasite burden. The quantity of ROS (the key trigger of apoptosis induction for 47DHF) formed in Au-47DHF was more than that noted in control H_2O_2. In another study related to the antiparasitic action of gold nanoparticles, Raj et al. developed and tested novel chrysin-gold nanoparticles (CHY-AuNPs) and studied their potential against leishmaniasis [63]. CHY-AuNPs were synthesized by mixing CHY and $HAuCl_4{\cdot}3H_2O$ solution in a 1:2 ratio. The synthesis of nanoparticles was optimized through parameters such as temperature, pH, and concentration of metal ions, resulting in smaller and more stable nanoparticles as indicated by UV spectra. An acidic pH and high temperature were found to be favorable for the synthesis of CHY-AuNPs. Biosorption, mediated by the hydroxyl group of CHY, occurred at the surface of the nanoparticles due to the interaction between positively charged functional groups and anionic gold ions. The characterization of nanoparticles was performed using UV-vis, and HRTEM micrographs (an average size of 20 nm and a spherical or oval shape), whereas the high crystallinity was confirmed by the clear dotted lattice fringes observed in selected area diffraction SAED patterns. The nanoformulation was found to have a high DL efficiency of approximately 91% for CHY. During the drug release study, it was found that a higher percentage of the drug was released in neutral pH conditions (40–90%) compared to acidic pH (0.9–30%). When tested for oral efficiency, the release of the drug was minimal in highly acidic conditions, but increased significantly at pH 6.8 and 7.4, with approximately 30% and 80% of the drug released, respectively. The in vitro antileishmanial activity of CHY and CHY-AuNPs was assessed by measuring the reduction in *L. donovani* burden in infected macrophages after 48 h of treatment. The results showed that CHY-AuNPs were more effective in the reduction in the parasite viability with an IC_{50} value of 0.8 µg/mL, compared to CHY with an IC_{50} value of 2.19 µg/mL. This suggests that the enhanced efficacy of the drug conjugated with AuNPs may result from targeted delivery.

3.2.2. Gold Nanoparticles in Anticancer Therapy

In a recent work from 2022, Cunha et al. reported the improvement of the EGCG's efficacy through its conjugation with gold nanoparticles (AuNPs), which led to two nanosystems based on AuNPs [64]. The chitosan (CHI) mediated nanoparticles (EGCG-ChAuNPs) were synthesized by mixing an aqueous solution of chloroauric acid with CHI and heating until a red suspension was obtained and subsequent conjugation with EGCG via carbodiimide-mediated cross-linking (EDC/NHSS). Meanwhile, to prepare the cysteamine-mediated nanoparticles (EGCG-CystAuNPs), the Turkevich method was employed, followed by subsequent functionalization with cysteamine. The resulting CystAuNPs were conjugated with EGCG through EDC/NHSS coupling and stabilized by tween 80. The obtained nanoparticles were characterized using techniques such as UV-vis spectroscopy, DLS, zeta potential study, TEM, and Attenuated Total Reflectance Fourier Transform Infrared spectroscopy (ATR-FTIR). The nanoparticles turned out to exert antioxidant and cytotoxic effects in pancreatic cancer cells, with EGCG-ChAuNPs and EGCG-CystAuNPs inducing 50% cell growth inhibition at EGCG concentrations of 2.2 and 3.7 μM, respectively. The EGCG alone required a concentration of 23 μM to achieve the same level of cytotoxicity. Caspase-3 activity assay also showed that the conjugation of EGCG with AuNPs enhanced apoptosis in the cancer cells compared to EGCG alone. The authors concluded that AuNP complexes could be used as delivery carriers to increase EGCG antioxidant activity in cancer tissues. In the study from 2021, Panda et al. examined the antiangiogenic properties of green tea polyphenols (GTP) combined with gold nanoparticles (GTP-AuNPs) [65]. The nanoparticles were prepared by adding a solution of EGCG or (−)-epicatechin gallate (ECG) to $HAuCl_4$ in deionized water and stirring. The obtained red mixture was then centrifuged, washed with DI water, and stored for future use. AFM, UV-vis spectroscopy, and DLS were used to characterize the size, shape, and stability of GTP-AuNPs, which appeared to be influenced by the solubility of the polyphenols in the aqueous medium, with ECG-AuNPs revealing a larger diameter due to the formation of agglomerates and EGCG-AuNPs being more stable due to their higher solubility. Unlike the wild angiogenin (Ang), the protein used in this study had a 6His-tag at the N-terminal, increasing the count of amino acid residues to 144 (from 123) and the molecular weight to 16 kDa (from 14 kDa). Using an agarose gel-based assay, the inhibitory efficacy of ECG and EGCG against the ribonucleolytic activity of Ang was investigated, and it was discovered that EGCG was a significantly more potent inhibitor than ECG, with a relatively higher band intensity. The binding of ECG and EGCG was similar to binding to ribonuclease A, and both flavonoids decreased Ang catalytic activity through interaction with Lys 40 but did not directly alter the ribonucleolytic site. ECG and EGCG bound to Ang noncompetitively, with the gallate moiety (ring D) playing an essential role. The docking results were supported by the in vitro and in vivo effects of ECG and EGCG on Ang, indicating that the polyphenols inhibit Ang via an allosteric manner of inhibition. Kinetic experiments showed that all polyphenols bound to an allosteric site on Ang in a noncompetitive manner, whereas capped AuNPs demonstrated competitive binding with very low inhibition constants (Ki~4 μg/mL). Fluorescence quenching tests and binding constant estimates further confirmed the capacity of GTP-AuNPs to bind to the active site on Ang. The GTP-AuNPs also suppressed the in vivo angiogenic response in the chick chorioallantoic membrane (CAM) experiment. These findings show that ECG- and EGCG-capped AuNPs possess better antiangiogenic properties than their free analogs.

3.2.3. Other Types of Gold Nanoparticles Activity

In 2022, Liu et al. developed a nanoprobe using a CT-functionalized gold nanocluster (C-Au NC) for sensitive and selective detection of dopamine (DA) [66]. To synthesize C-Au NCs, glutathione (GSH)-Au NCs were first prepared by mixing hydrochloride gold and GSH in ultrapure water, heating the mixture, and purifying the resulting GSH-Au NCs through ethanol precipitation. Catechin was then modified with the chemical linker 4-((2,5-dioxopyrrolidin-1-yloxy)carbonyl) phenylboronic acid (BE) to create a BE-CT complex. This complex was mixed with GSH-Au NCs solution and phosphate buffer (PB), and the mixture was stirred at room temperature for 12 h. The resulting C-Au NCs were obtained through ultrafiltration and stored at 4 °C for further testing and analysis. Characterization techniques, including UV-vis absorption spectra, FTIR, XPS, steady-state and time-resolved fluorescence, DLS, and HRTEM, were used to confirm the successful synthesis of the C-Au NCs. The resulting C-Au NCs were monodisperse with a spherical size of 1.7 nm and a hydrodynamic diameter of 3.6 nm. The surface functionalization of CT did not significantly affect the maximum emission wavelength or fluorescence lifetime of the Au NCs but did lead to an increase in the hydrodynamic diameter. The DA detection mechanism was based on the formation of azamonardine (proved by electrospray ionization-mass spectrometry spectra) via the selective chemical reaction between resorcinol fragment of CT and DA, which leads to enhanced fluorescence emission. The chemical linker used to construct the C-Au NC nanoprobe did not affect the CT-DA reaction, as demonstrated by the fluorescence emission spectra of CT-DA and BE-CT-DA solutions. The optimal pH for the reaction between CT and DA to generate azamonardine compound was found to be 9. The surface CT density of C-Au NCs affected the DA sensing performance, with the reaction kinetics increasing with a higher GSH/BE-CT ratio. The reaction rate was slower than previous reports due to the low reactant concentrations used in this work, as well as the hindering effect of the surface GSH ligand. To ensure the completion of the reaction, the chosen conditions for constructing the ratiometric DA sensing platform were a GSH/BE-CT ratio of 15, pH 9, and a reaction time of 1 h. The sensitivity of the nanoprobes was evaluated by recording the fluorescence emission spectra after adding different concentrations of DA. The test showed a good linear relationship with the DA concentration in the range of 0 to 500 nM, with a detection limit of 1.0 nM. The reproducibility of the nanoprobes was also tested, resulting in a low relative standard deviation of 2.5%. The accuracy of the DA detection was verified through HPLC characterization with a UV detector. The selectivity of the nanoprobes was evaluated in the presence of various interferents and found to only enhance the I461/I560 ratio in the presence of DA. The practicality of the platform for DA detection was demonstrated through testing in urine and cell lysate samples. C-Au NC was not able to detect DA in urine samples but had satisfactory recoveries and low relative standard deviation values when tested by a standard addition method. When tested in cell lysate samples, the platform showed good accuracy, as indicated by favorable recoveries and low relative standard deviation values. According to the results, the platform seems adequate for detecting DA in complex biological media.

In another study, Pradhan et al. investigated the potential of hesperidin isolated from orange peels in the synthesis of gold nanoparticles and their use as both an antioxidant and a photocatalyst for the treatment of industrial wastewater [67]. The synthesis of hesperidin gold nanoparticles was achieved by implementing isolated and recrystallized hesperidin into the gold structure using the chemical reduction method with trisodium citrate as a reducing and stabilizing agent while optimizing the conditions. The nanoparticles were characterized by HRTEM, SAED, and FTIR. The photocatalytic activity of hesperidin gold nanoparticles was evaluated on organic dyes and pollutants using sodium borohydride as a reducing agent. The tested chemicals were methyl orange (MO), methylene blue (MB), bromocresol green (BCG), and 4-nitrophenol (4NP). Hes-Au NPs were found to be efficient in degrading these substances in the presence of visible light, with more than 90% degradation observed in all cases. The authors declared that the optimal volume of Hes-Au NPs for degradation of these dyes was 1 mL for MO, MB, and 4NP, and 1.2 mL for BCG, which allowed to degrade of these substances at a level of 88%, 95%, 90%, and 98%, respectively. The optimal concentration of $NaBH_4$ was 3 mM for 4NP and 5 mM for the dyes, and the kinetics of the degradation followed first-order kinetics. The dyes degraded the most at 1 mM/0.1 mL, whereas 4NP degraded the most at 2 mM/0.1 mL. In the DPPH and ABTS radical scavenging studies, Hes-Au NPs demonstrated strong antioxidant activity, with IC_{50} values of 37.16 μg/mL for DPPH and 53.57 μg/mL for ABTS. The antioxidant activity of Hes-Au NPs increased with concentration, showing inhibition in the range of 34.8–79% for DPPH and 13.8–74% for ABTS at concentrations ranging from 20 to 100 μg/mL, which is lower than that of referential ascorbic acid with respect to both cases. The nanoparticles outperformed ascorbic acid during the hydroxyl radical scavenging analysis, demonstrating an increase in the scavenging of hydroxyl free radicals from 7.3% to 54% in the range of 20–100 μg/mL (30–52.1% for ascorbic acid). The IC_{50} value in this experiment concluded at 94 μg/mL. Scroccarello et al. developed an adhesive film made of nanostructured Ag/Au and functionalized with CT for reagentless H_2O_2 and glucose detection (Figure 8) [68]. The film (PDA@Au-CT@Ag) was created on an enzyme-linked immunosorbent assay (ELISA) polystyrene microplate using a layer-by-layer decoration method that included DA polymerization (to form a PDA layer), the addition of gold nanoparticles (AuNPs), the functionalization of the AuNPs with CT as a reducing agent, and the formation of a silver network by adding $AgNO_3$ and NaOH. The resulting film had a reproducible LSPR peak intensity at 405 nm due to the creation of a nano-Ag network and a shoulder at 500–520 nm due to the gold component. The existence of Ag and Au was confirmed by energy-dispersive X-ray (EDX) analysis, which yielded a ratio of 1.15. During the storing assessment, the most promising results were achieved when the films were preserved for one month in MeOH:H_2O (80:20) and in sodium citrate (10 mM), resulting in 96.5% and 94.2% retention of the original LSPR signal, respectively. In contrast, the PDA@AuNPs film without Ag can be stored at ambient temperature in the dark for one year without significantly changing the LSPR maximum. It was discovered that when PDA@Au-CT@Ag was exposed to H_2O_2, its LSPR absorbance changed with the most severe etching occurring at ambient temperature (25 °C), pH 7, within the first 40 min of H_2O_2 contact as validated per SEM and EDX. The resultant film was subsequently utilized to quantitatively detect H_2O_2 in the range of 1–200 μM, with a detection limit of 0.2 μM. Using GOx as an enzyme, the suggested nanocomposite film was tested for its capacity toward the detection of glucose. The film was discovered to be capable of detecting glucose in the range of 2–250 μM, with a detection limit of 0.4 μM, in a repeatable manner and with a nice linear response. The experiment was performed in a single step and was unaffected by the presence of the enzyme, in contrast to approaches using colloidal dispersions, which frequently require additional stages or enzyme removal procedures to prevent aggregation, precipitation, and enzyme inhibition. The suggested PDA@Au-CT@Ag-based assay was utilized to detect H_2O_2 and glucose in several commercial soft drinks, with no substantial signal loss due to any of the evaluated interferents. The assay was shown to be repeatable and accurate, with recoveries ranging from 84–111% for H_2O_2 and 83–105% for glucose.

To sum up
Pros
• Gold nanoparticles can be easily prepared and functionalized, and they present significant potential for medical sciences, including antibacterial, antioxidant, an-tiparasitic, antiangiogenic potential, and anticancer properties. • The functionalization of gold nanoparticles with flavonoid compounds improves their biological effects. These effects were presented for gold nanoparticles func-tionalized with chrysin, kaempferol, quercetin, epigallocatechin gallate, procya-nidins, 4′,7-dihydroxyflavone, metal-phenolic networks, and green tea polyphe-nols. Due to the photothermal effect, flavonoid-AuNP conjugates reveal significant potential for both antibacterial activity and wound healing. • Depending on the conjugation between proper flavonoid and gold nanoparticles, diverse applicability of fabricated systems can be obtained—from antibacterial agents to biosensors. A gold nanocluster was also applied for sensitive and selec-tive detection of dopamine, whereas hesperidin isolated from orange peels was used in the synthesis of gold nanoparticles, which was applied both as an antiox-idant and a photocatalyst for the treatment of industrial wastewater. • Despite using structurally diverse flavonoids, no influence on the maximum emis-sion wavelength or fluorescence lifetime of the gold nanoparticles was observed.
Cons
• Limited information is known about the long-term toxicity of gold nanoparticles. Some organic functionalizations, such as citrate, might increase toxicity rather than decrease it [69]. Therefore, the toxicity of conjugates should be carefully evaluated. • The strong positive potential of some reported herein nanoparticles might raise safety issues, as the feature was correlated with membrane damage [55]. • In contrast to spherical nanoparticles, gold nanorods are reported to disturb pro-tein structures and reveal reduced mobility - they remain at the injection site, where they might cause damage [69]. • The competitiveness of the currently developed systems is unclear as there is a lack of studies undertaking this topic. • The glucose sensors based on flavonoids [68] still require optimization for better functioning, as commercially used systems are more advantageous [70].

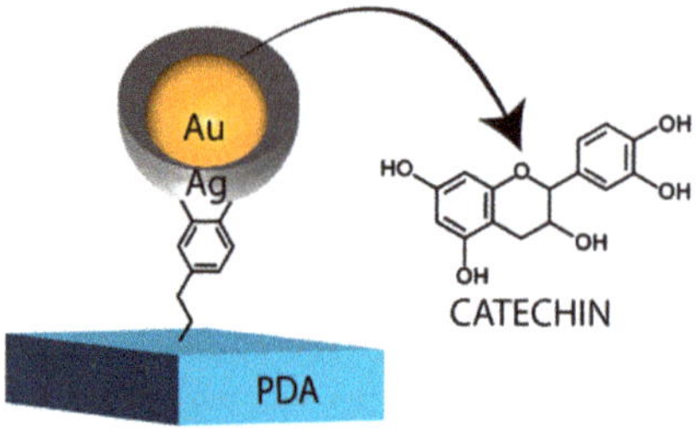

Figure 8. Simplified nanostructured Ag/Au adhesive film.

Gold nanoparticles have also been often functionalized with flavonoid compounds to improve their properties and broaden their applications. Gold nanorods with metal-phenolic networks (containing EGCG, procyanidins (OPC), and tannic acid (TA)) [59] used with NIR-irradiation significantly accelerated the infected wound healing and proved their bactericidal effectiveness. Similarly, quercetin-conjugated nano-gold particles exerted antibacterial and antioxidant activity [62] as well as gold nanoparticles, after coating with CHY, kaempferol, and quercetin, boosted their antibacterial action against Gram-negative bacteria [58]. Novel CHY-gold nanoparticles [63] were able to combat leishmaniasis, probably as a result of targeted delivery, whereas GNRs combined with RSV [71] revealed promising but less effective than antileishmanial amphotericin B activity, potentially due to the ROS generation. The CHI- and cysteine-mediated nanoparticles [64] conjugated with EGCG (EGCG-ChAuNPs and EGCG-CystAuNPs) induced cell growth inhibition of pancreatic cancer cells and as delivery carriers increased EGCG antioxidant activity in cancer tissues. Similarly, the delivery of RVS to PANC-1 cancer cells of PDAC was improved by attaching it to gold nanoparticles (GNPs) [72]. Moreover, GTP combined with gold nanoparticles exerted the antiangiogenic properties of (GTP-AuNPs) [65], useful in cancer treatment. C-Au NC can also be used for the detection of DA [66] in complex biological media. Interestingly, gold nanoparticles combined with hesperidin showed antioxidant and photocatalytic potential for the treatment of industrial wastewater [67]. A film composed of nanostructured Ag/Au functionalized with CT can be useful for reagentless H_2O_2 and glucose detection. The summary of the data related to connections of gold nanoparticles with flavonoids, which were presented in this section, is included in Table 2.

Table 2. The summary of the data presented in Section 3.2.

Flavonoid/ Compound	Size of NPs	Synthesis Method	MIC/ID$_{50}$	Action	Ref.
4′,7-Dihydroxyflavone	8 nm	chemical reduction of $HAuCl_4$	IC_{50prom} 0.1226 μg/mL IC_{50ama} 0.115 μg/mL	promastigotes amastigotes	[48]
Chrysin, kaempferol, and quercetin	4.1–35 nm	chemical reduction of $HAuCl_4$ with $NaBH_4$ in the presence of glutathione	MIC 30 μg/mL for quercetin-AuNPs	*E. coli*, *P. aeruginosa*, and *P. vulgaris*	[58]
EGCG, procyanidins, tannic acid	Not specified	seed-mediated growth method	98.6% *S. aureus* growth suppresion and 99.8% *E. coli* growth suppression for procyanidine-AuNPs	*S. aureus* and *E. coli*	[59]
Quercetin	30 nm (HRTEM)	reduction of $HAuCl_4$ by sonication	MIC 7.6 μg/mL	*E. coli*	[62]
Chrysin	20 nm	chemical reduction of $HAuCl_4$ in the acidic pH and high temperature	IC_{50} 0.8 μg/mL	*L. donovani*	[63]
EGCG	125 ± 13 nm	chemical reduction of $HAuCl_4$ following with carbodiimide-mediated cross-linking of EGCG	50% cell growth inhibition at EGCG concentrations of 2.2 μM	Pancreatic cancer cells	[64]

Table 2. *Cont.*

Flavonoid/ Compound	Size of NPs	Synthesis Method	MIC/ID$_{50}$	Action	Ref.
EGCG, ECG	Not specified	chemical reduction of $HAuCl_4$	Not specified	*E. coli*	[65]
Catechin modified with linker	1.7 nm	chemical reduction of $HAuCl_4$ in the presence of glutathione	Not applicable	Fluorescent nanoprobe for detection of dopamine	[66]
Hesperidin	18–32 nm	chemical reduction method with trisodium citrate	degradation rate > 90% (methyl orange, methylene blue, bromocresol green, 4-nitrophenol); IC$_{50}$ 37.16 µg/mL	Photocatalyst for the treatment of industrial wastewater; antioxidant activity	[67]
Catechin	Ag/Au NPs Not specified	chemical reduction of $HAuCl_4$ and $AgNO_3$ by NaOH	detection limit of 0.4 µM	glucose detection	[68]

3.3. Metal Oxide Nanoparticles

Many studies present connections of flavonoids with metal oxide nanoparticles, mainly zinc oxide, ceria, iron oxide-based, titanium-based, CuO/ZnO, and titanium oxide-gold nanocomposites, which resulted in materials of interesting physicochemical and biological properties (Figure 9). Of great interest are the potential antibacterial, antifungal, antiviral, and anticancer properties of metal oxide nanoparticles as well as their applications as drug delivery platforms for improved osseointegration in orthopedic implants. In this regard, metal oxide nanoparticles were functionalized with quercetin, eupatorin, silibinin, EGCG, baicalin, luteolin, chrysin, morin, icariin hesperetin 7-rutinoside, flavanone-7-*O*-glucoside to study their prospective medical applications.

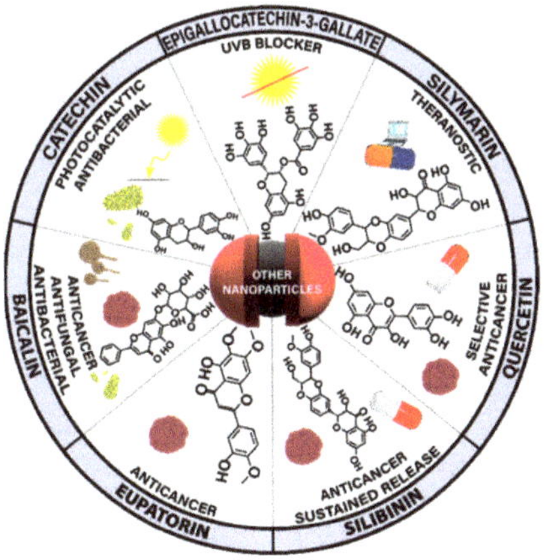

Figure 9. Spectrum of flavonoid-metal oxide nanoparticle conjugates and hybrids with their most striking biological features.

3.3.1. Iron Oxide-Based Nanoparticles

In 2022, Askar et al. reported the preparation of quercetin-conjugated magnetite nanoparticles (QMNPs) that exhibited salient antitumor abilities toward breast cancer during the in vitro and in vivo studies [73]. QMNPs derived from iron(II) sulfate solution that was enzymatically converted into iron(II,III) oxide (Fe_3O_4) cores of about 11 nm in diameter under treatment by extracellular bio extract from *Aspergillus oryzae* fungus. The obtained MNPs were clad with quercetin, resulting in nanospheres having a diameter of

40 nm (Figure 10). The particles were characterized with the use of UV-vis spectroscopy, FTIR spectroscopy, TEM, and XRD analysis, confirming the positive experimental outcome. In vitro studies were conducted in respect of the ability of QMNPs to inhibit the growth of cancer cell lines MCF-7, HePG-2, and A-459 showing effective inhibition after the 24-h incubation period with IC_{50} values in nanomolar ranges. The authors also described the radiosensitization effect of new particles on the example of MCF-7 cells, demoing the cell decline by 91.2%, compared with the survived cell count after exposure to 6 Gy gamma-ray irradiation (the optimal irradiation dose was estimated experimentally) or 11 nM/mL of QMNPs alone. During the in vivo studies, it was concluded that QMNPs significantly enhanced lateral radiotherapy of the *N*-methyl-*N*-nitrosourea-induced breast cancer in white albino rats through upregulation of pro-apoptotic proteins and downregulation of antiapoptotic proteins of the mitochondrial apoptotic pathway, while preserving nontoxic nature towards hematological, hepatic, and renal markers.

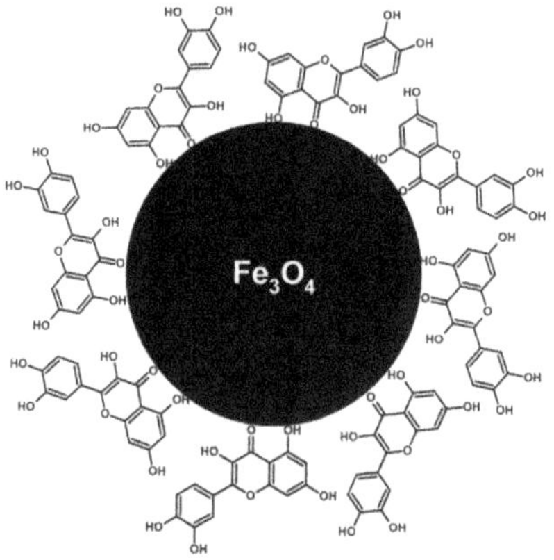

Figure 10. Iron(II,III) oxide nanoparticles clad with quercetin.

In 2021, Ahmadi et al. developed a nanocomposite of Fe_2O_3, CHI, and montmorillonite (MMT) for the encapsulation of QUR as a less harmful alternative to chemical antitumor agents [74]. To obtain the desired Fe_2O_3/CS/MMT@QUR, the solution of $Fe(NO_3)_3 \cdot 9H_2O$ and $NH_3 \cdot H_2O$ was autoclaved for 10 h to obtain pure Fe_2O_3 nanoparticles, which were then mixed with CHI and MMT to be subsequently loaded with QUR. The resulting nanostructure was then crosslinked with glutaraldehyde before being added to a solution of paraffin oil and Span 80. After stabilization with PVA, the nanocarrier was centrifuged from the aqueous solution. According to the XRD study, the Fe_2O_3 component exhibited an inverse spinel structure with an average size of 11.79 nm, and the FTIR study indicated the existence of Fe_2O_3, QUR, MMT, and CHI in the nanocomposite. As defined by FESEM, Fe_2O_3/CS/MMT@QUR had an average size of 148.2 nm, which corresponded to the hydrodynamic radius determined by DLS analysis (161.3 nm). Vibrating sample magnetometry analysis demonstrated the decline in magnetization value after the magnetic nanoparticles were coated with CHI. The NCs exhibited high EE (94%) and pH-sensitive QUR release, with a higher release rate in a slightly acidic environment (pH 5.4). The Weibull model accurately described the drug release data, indicating that the interaction between the carrier and the drug influences the loading and release processes. The NCs also exhibited a regulated release profile, with rapid initial release followed by sustained release. In vitro cytotoxicity tests on MCF-7 cancer cells showed that Fe_2O_3/CS/MMT@QUR had lower toxicity compared to free QUR by about 22% due to controllable drug release. As was determined by flow cytometry, the percentages of viable MCF-7 breast cells treated with Fe_2O_3/CS/MMT were not substantially different from those in the control group, portraying the carrier as biocompatible and non-toxic. Apoptosis studies for cancer cell lines showed lower viability in the free QUR group, indicating a better function of the QUR-loaded Fe_2O_3/CS/MMT nanocomposite, confirming the cytotoxicity experiment, and implying consistency with the release profile. Furthermore, under the presence of Fe_2O_3, total apoptotic cells increased by 31.25% as compared to CS/MMT@QUR. In another study, Garfias evaluated the potential of using polyelectrolyte-coated iron oxide nanoparticles as quercetin drug carriers for targeted chemotherapy in ovarian cancer [75]. The synthesis of iron oxide nanoparticles (Fe_2O_3) was performed through co-precipitation in an aqueous solution. These nanoparticles were then coated with alternating layers of polyelectrolytes through a layer-by-layer technique. This was performed by adding a volume of an anionic polyelectrolyte solution—poly(styrene sulfonate) (PSS) or sodium carboxymethylcellulose (CMC)—to a suspension of the Fe_2O_3 nanoparticles in water, followed by washing and then adding a volume of a cationic polyelectrolyte solution—poly(allylamine hydrochloride) (PAH) or CHI—containing quercetin. This process was repeated to create multiple layers on Fe_2O_3 so that quercetin was included in the even layer numbers along with cationic components. The samples were labeled based on the type of polyelectrolyte systems used (synthetic PSS/PAH or natural CMC/CHI, referred to as P or C), the number of layers, and the presence or absence of quercetin in the layers, with QUR indicating the presence of quercetin. The size, structure, and surface properties of the nanoparticles were analyzed using various techniques including DLS, TEM, XRD, FTIR spectroscopy, and vibrating sample magnetometry. The nanoparticles were found to have a mean size of 8.7 nm and a crystallite size of 10.4 nm. Zeta potential measurements and the pH at which the isoelectric point occurred indicated that the polyelectrolytes and quercetin were effectively coated onto the nanoparticles. The nanoparticles showed no cytotoxicity on their own, but when quercetin was included, there was a statistically significant reduction in the viability of human ovarian carcinoma cells. The authors found that the layer-by-layer technique was effective for encapsulating quercetin, with 64.7% efficiency using synthetic polyelectrolytes and 87.7% using natural polyelectrolytes. The authors also observed that the release of quercetin from Fe_2O_3 was pH-dependent, with faster release at basic pH and slower release at acidic or neutral pH. Based on the above-mentioned findings, the proposed nanoformulations were proposed for use as targeted drug delivery vehicles for cancer chemotherapy. In 2021, Tousi et al. conducted a study aimed to investigate methoxy poly(ethylene glycol) (mPEG)-b-PLGA coated iron oxide nanoparticles as a carrier of eupatorin in the treatment of prostate cancer [76]. The synthesis started with magnetic nanoparticles (Fe_3O_4) fabrication and functionalization with oleic acid. After that, eupatorin-loaded Fe_3O_4@mPEG-b-PLGA nanoparticles were generated utilizing a nano-precipitation approach that involved combining Fe_3O_4-oleic acid nanoparticles with eupatorin and mPEG-b-PLGA in acetone and stirring the mixture for 10 min. The formulation was then dropwise added to deionized water before being freeze-dried to remove the water. The same method was used to make control samples of Fe_3O_4@mPEG-b-PLGA nanoparticles without eupatorin. DLS technique showed that the average particle size was 58.5 nm with a PDI of 0.167. An SAED pattern indicated that the nanoparticles had a consistent spherical form and were thoroughly coated with mPEG-b-PLGA. In the in vitro drug release studies, nanoparticles revealed the rapid initial release of eupatorin (30% over the first 24 h), followed by sustained release over 200 h. The drug content and EE of the nanoparticles were 8.28% and 90.99%, respectively. Eupatorin-loaded Fe_3O_4@mPEG-b-PLGA nanoparticles were produced and examined for their capacity to transport eupatorin in a sustained manner to human prostate cancer cell lines DU-145 and LNCaP. The nanoparticles were shown to be efficient in lowering the growth rate of cancer cells in a dose-dependent manner with IC_{50} values of 100 mM and 75 mM for DU-145 and LNCaP, respectively. These values were lower than the IC_{50} values for free eupatorin, showing that the nanoparticles were able to increase the therapeutic benefits of eupatorin. The nanoparticles also caused apoptosis in the cancer cells. The authors also applied flow cytometry to study the effect of free eupatorin and nano-eupatorin on the distribution of cells in different cell cycle stages in DU-145 and LNCaP human prostate cancer cell lines. Free eupatorin produced cell arrest during the G2-M interphase, but nano-eupatorin increased the percentage of cells in the sub-G1 phase, evidencing DNA destruction and apoptosis. The annexin V-PI test revealed that free eupatorin enhanced the number of necrotic and late apoptotic cells in both cell lines. While nano-eupatorin showed similar effects on DU-145 cells, it greatly decreased the number of necrotic cells. There was no significant difference in necrosis or late apoptosis rates in LNCaP cells with either treatment; however, nano-eupatorin was more successful in increasing the proportion of cells undergoing early apoptosis. The annexin V-PI test revealed that nano-eupatorin could considerably lower the rate of necrosis in DU-145 cells while increasing the rate of early apoptosis in LNCaP cells. Nano-eupatorin was considerably more effective, raising the Bax/Bcl-2 ratio in DU-145 and LNCaP to 13.5 and 20.5, respectively. Furthermore, nano-eupatorin enhanced cleaved caspase-3 levels, although no significant change was seen in free eupatorin-treated cells compared to the control group.

In their study from 2021, Takke and Shende used iron oxide nanoparticles to create biocompatible nanopolymeric carriers of PLGA-encapsulated silibinin (SLB-MPNPs) for sustained release in kidney cancer cells [77]. The first synthetic stage involved the addition of H_2O_2 to $FeCl_2$ solution to form a black precipitate, which was separated, washed, and dried to create iron oxide nanoparticles (MNPs). SLB-MPNPs were then created utilizing a two-fold emulsion process that combined PLGA dissolved in acetone, MNPs, and dissolved SLB in ethanol. The resulting primary emulsion was sonicated with aqueous PVA, and the NPs were purified using centrifugation and drying. The PLGA concentration in the SLB-MPNP formulations ranged from 50 to 200 mg, and blank MPNPs were also created. Blank MNPs were measured for particle size and zeta potential and found to have an average particle size of 206.4 nm and zeta potential of −21.1 mV, indicating a stable formulation. SLB-MPNPs M3 (particles composed of 150 mg of PLGA), with a particle size of 285.9 nm and a zeta potential of −14.71 mV, were picked for further investigation. As the PLGA concentration grew, the % EE of the formulations increased from 72.16% to 88.20%. MNPs revealed a greater saturation magnetization (36.35 emu/g) than SLB-MPNPs (12.78 emu/g). The in vitro release analysis of SLB from SLB-MPNPs in PBS at pH 7.4 revealed roughly 48% release after 24 h, with the overall volume of SLB released reaching 65.21% over 2 days. The nanoparticles with the highest encapsulation (200 mg PLGA) demonstrated a delayed and consistent release of 98.04% over 15 days due to drug diffusion from the PLGA core. On A-498 human kidney cancer cells, SLB-MPNPs demonstrated stronger cytotoxicity than plain SLB at all tested doses, with an IC_{50} of 3 µg/mL compared to an IC_{50} of 5 µg/mL for SLB. During the in vivo acute toxicity research, no abnormalities or behavioral changes were detected in the mice, and no tremors, convulsions, salivation, diarrhea, or lethargy were observed. Body mass, food consumption, and water intake did not change significantly between the experimental and control groups, and hematological, biochemical, and organ weight characteristics did not differ significantly either. Histopathological examinations revealed no evidence of tissue injury. For stability testing, the SLB-MPNP formulations were kept at 4 °C preserved from light. The test held for 0, 1, 2, and 3 months revealed no significant alterations, signifying that the formulations remained stable.

In 2021, Qin et al. managed to successfully synthesize photothermally active iron-polyphenol nanoparticles with tunable size and ion content using different polyphenols as a ligand, including EGCG, epicatechin (EC), and proanthocyanidin (PAC) among tested substances [78]. The nanoparticles were synthesized through a sol-gel process that involved dissolving PVP in a mixture of water, ethanol, and ammonia, with the subsequent addition of a specific polyphenol, formaldehyde, $FeSO_4$, and eventual autoclaving. Finally, the products underwent dialysis and were isolated through freeze-drying. The DLS investigation revealed that the nanoparticles had hydrodynamic diameters of 21 nm for Fe-EGCG, 27 nm for Fe-EC, and 30 nm for Fe-PAC. All of the nanoparticles were highly dispersed in water and demonstrated good photothermal conversion efficiency varying from 35.2% to 43.6% when irradiated at 808 nm for 10 min. The zeta potentials were noticed to negatively depend on the amount of PVP polymer present. The iron content appeared to impact the photothermal performance of nanoparticles, with higher iron content resulting in better performance, reaching the best results at an iron content of 85.7 mg/g, whereas higher amounts prompted agglomeration of the nanoparticles caused by excess iron acting as a cross-linker. Furthermore, as the size of the nanoparticles decreased, the temperature increased, most likely due to the higher absorption and scattering ratios of smaller particles, resulting in a more efficient light-to-heat conversion. The photothermal performance of colloidal solutions was also affected by both concentration and power density, with higher values leading to increased temperature increases after applying laser while retaining high photostability after multiple cycles of irradiation. Iron-polyphenol colloidal nanoparticles exhibited low cytotoxicity but effectively killed cancer Hela cells under MTT assay through photothermal therapy in vitro when irradiated with lasers. These results were confirmed by in vivo animal studies, which showed a significant increase in the temperature at the tumor site with effective inhibition of the cancer growth in mice after nanoparticles were injected intravenously and exposed to laser NIR irradiation. In terms of biodistribution, the iron content was measured in various organs at different time points after injection of colloidal solution, showing effective accumulation in the tumor due to enhanced permeability and retention effects. The in vivo toxicity of colloidal nanoparticles was assessed using hematological and histochemical analyses, which revealed that the nanoparticles were safe for mice at the current experimental dosage. The hematological analysis found that the white blood cell count was decreased in the colloidal solution group but recovered after 3 days. Histochemical study of multiple organs revealed no evidence of inflammation or injury.

The study of Sadegha et al. aimed to investigate the potential use of super-paramagnetic iron oxide nanoparticles (SPIONs), coated with mesoporous silica ($mSiO_2$) and loaded with curcumin (CUR) and silymarin (SIL), as a theranostic asset for breast cancer treatment (Figure 11) [11]. SPIONs were synthesized via a reverse microemulsion method. Next, the resulting nanoparticles were coated with $mSiO_2$ using CTAB and tetraethyl orthosilicate to create mesoporous silica-coated SPIONs ($mSiO_2$@ SPIONs), which were then suspended in solutions of CUR and SIL to incorporate the cargo molecules. The average hydrodynamic diameter of SPIONs was 25.50 nm. Following coating with $mSiO_2$, the size grew to 57.00 nm but did not alter appreciably after additional processing to remove CTAB. The PDI fell considerably from bare SPIONs to coated and CTAB-free SPIONs. The size and PDI of the SPIONs did not alter much with polyphenols addition. The amount of CTAB employed in the coating process influenced SPION size, with higher levels resulting in bigger sizes. The best quantity of CTAB for adequate size and monodispersity was determined to be 25 mg. As validated by DLS measurements, SEM pictures of the produced NPs revealed that they were spherical and moderately monodispersed, with a size of 60–80 nm. CUR had the highest DL at 0.5 and 1.25 mg/mL, whereas SIL had the highest DL at 1.25 mg/mL. At 0.25 mg/mL, both polyphenols revealed the maximum EE (above 90%). CUR/SIL-loaded $mSiO_2$@SPIONs released different amounts of CUR and SIL at different pH values. The MTT experiment on MCF-7 cells revealed that the IC_{50} for the combination of CUR and SIL was lower than that noted for the individual substances. Magnetic resonance imaging (MRI) revealed that the NPs exhibited a high T2-weighted contrast and might be employed in the in vitro diagnosis of early-stage breast cancer.

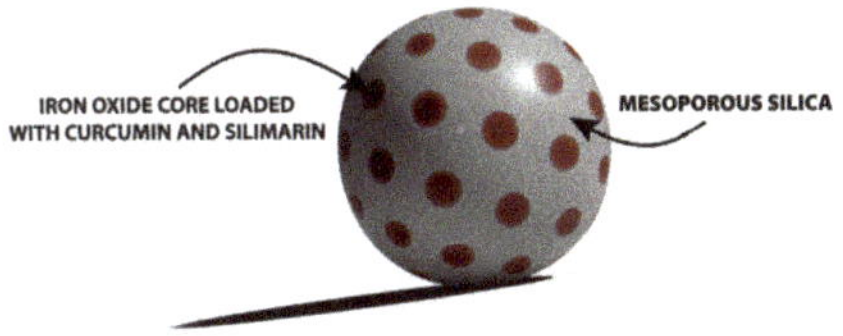

Figure 11. Super-paramagnetic iron oxide nanoparticles coated with mesoporous silica and loaded with the cytotoxic agents curcumin and silymarin.

This kind of potential application of nanoparticles was also presented in another study. Shubhra et al. developed a smart drug delivery system (DDS) marked ICGOx@IO-Dox-EGCG-PPP NPs for multimodal synergistic cancer therapy using magnetic photothermal agents synthesized from iron oxide (IO) nanoparticles with covalently attached indocyanine green (ICG) and glucose oxidase (GOx), coencapsulated with doxorubicin and EGCG inside PLGA nanoparticles, and modified with arginylglycylaspartic acid (RGD) peptides for dual targeting [79]. To make IO NPs, $FeCl_2$ and $FeCl_3$ were coprecipitated with NH_4OH at room temperature under vigorous stirring to yield Fe_3O_4 NPs, which were subsequently oxidized with NaClO under sonication to produce more stable γ-Fe_2O_3 NPs. To stabilize the system even more, citric acid was applied to IO NPs, and massive aggregates were centrifuged from the resultant colloid. Next, to make ICGOx@IO NPs, ICG, and GOx were co-loaded onto the surface of IO NPs. Using a multiple emulsion solvent evaporation process, these NPs were subsequently encased in a PEG–PLGA matrix with Dox and EGCG to form ICGOx@IO-Dox-EGCG-PP NPs. To make ICGOx@IO-Dox-EGCG-PPP NPs, the PLGA NPs in this formulation were tagged with RGD peptide. The hydrodynamic size of the nanoparticles was 209 nm, and their strongly negative zeta potential contributed to their excellent dispersibility and stability. The size distribution was single-point, indicating that the NPs did not aggregate. The nanoparticles were characterized using SEM, TEM (NPs were spherical, with most having a size of 150–180 nm), and FTIR. The saturation magnetization of the ICGOx@IO-Dox-EGCG-PPP NPs was 11.8 emu/g, which was considered sufficient for magnetic targeting. ICGOx@IO-Dox-EGCG-PPP NPs exhibited photothermal properties when irradiated with 808 nm laser light. Irradiation showed an acute increase in temperature to 53.9 °C, while the control samples without ICG showed no significant temperature increase. When external magnets were used for magnetic targeting, the temperature rise reached 55 °C and simultaneously increased the enzymatic activity of GOx. A biphasic release profile was observed for both Dox and EGCG, with an initial rapid release followed by a gradual, sustained release, with a larger rate at pH 5.5 than at pH 7.4, possibly due to enhanced drug solubility at the lower pH and hydrolysis of the PLGA polymer at the acidic pH. At pH 7.4, laser irradiation combined with a magnetic field increased drug release, with up to 73% of Dox and 65% of EGCG released after 48 h. After 5 min of laser irradiation, the PLGA NPs reached the glass transition temperature range of PLGA, promoting drug release as the physical state of PLGA transformed from firm to soft rubbery. Confocal microscopy and flow cytometry were used to investigate the cellular uptake, demonstrating that NPs may successfully transport drugs to B16F10 cells even in the absence of a magnetic field or peptide targeting. When cells were treated with dual-targeted NPs (i.e., with magnet), the highest fluorescence was witnessed, proving that the combination of peptide and magnetic targeting was most successful at delivering the NPs to cells. Single targeting using a peptide or a magnetic field increased cellular uptake over non-targeted ICGOx@IO-Dox-EGCG-PP NPs. Since the magnetic field can only enable NP connection with cells, cell-penetrating peptides directly facilitate NP entrance into cells. The flow cytometry data were supported by a quantitative evaluation of cellular uptake using ICP-MS. DDS was more effective for the reduction in the viability of cancer B16F10 cells than either Dox or EGCG alone, and dual targeting with laser exposure was the most effective for the reduction in the cell viability by 90%. The DDS did not show significant cytotoxicity to non-cancerous HEK293 cells and rendered apoptosis in cancer cells treated with it. ICGOx@IO-Dox-EGCG-PPP NPs were also more effective at declining cell viability in the presence of glucose, possibly due to the enzymatic H_2O_2 formation by GOx, and also elevated ROS production. When subjected to both a magnetic field and laser irradiation in vivo, DDS proved efficient at reducing tumor volume and extending mice longevity up to complete tumor eradication, while specimens given free Dox had a lower survival rate, which might be attributed to the negative effects of free Dox. When compared to the control and free Dox groups, the DDS boosted intratumoral H_2O_2 concentration and the production of apoptosis-related cytokines in tumor tissues. The NPs accumulative potential in tumor tissues could be expressed in the following order: ICGOx@IO-Dox-EGCG-PPP

(+magnet + laser) > ICGOx@IO-Dox-EGCG-PPP (+magnet) > ICGOx@IO-Dox-EGCG-PPP > ICGOx@IO-Dox-EGCG-PP (+magnet) > ICGOx@IO-Dox-EGCG-PP. The researchers discovered that free Dox induced large increases in creatine kinase MB and LDH levels in mouse plasma, signaling cardiac damage, but DDS did not present such an effect. Doxol, a harmful metabolite of Dox, was also found in substantially lower concentrations in the hearts of mice treated with the ICGOx@IO-Dox-EGCG-PPP NPs, which coincided with histological heart tissue damage examinations. Iron concentrations in cardiac tissue were the lowest in mice given the dual-targeted DDS, indicating effective tumor targeting. These findings imply that the introduction of EGCG in the nanoformulations reduced Dox cardiotoxicity by suppressing the activity of carbonyl reductase 1, involved in the synthesis of Doxol. In 2022, Rahmati et al. published a study on the use of QUR-loaded magnetic nano-micelles (QMNMs) as a multifunctional drug delivery platform [80]. QMNMs were prepared using modified oil-in-water emulsion methods. QUR, magnetic nanoparticles (MNPs) of Fe_3O_4, and a methoxy poly(ethylene glycol)-block-poly(ε-caprolactone) (mPEG-PCL) copolymer were dissolved in chloroform and added to an aqueous solution containing PVA. The mixture was stirred and homogenized, then allowed to evaporate. The resulting nanoparticles were collected, washed, and dried. The DL and EE of quercetin in the QMNMs were 17.1% and 95.9%, respectively. The stability of the micelles was confirmed by the low critical micelle concentration of 50 mg/L. The characterization of nanoparticles was performed using UV-vis spectrophotometry, DLS measurements (a monodisperse distribution with a polydispersity index of 0.269 and an average hydrodynamic size of 85 nm), and TEM images (semi-spherical shape and an average size of 10–15 nm). The physical stability of QMNMs was assessed by monitoring their particle size over a 90-day period using DLS analysis, and it was found that there were no significant changes in the average size. The successful incorporation of magnetic nanoparticles into QMNMs was confirmed through energy-dispersive X-ray spectroscopy analysis and XRD patterns. The magnetic properties of both MNPs and QMNMs were compared, with the latter showing a reduced saturation magnetization of 12.2 emu/g due to the coating of the nanoparticles with an mPEG-PCL layer. However, both materials displayed superparamagnetic behavior with no remanent magnetization or coercivity at relatively low fields. The observations of drug release from QMNMs represented their sensitivity to pH changes. In general, a burst QUR release occurred in the first 12 h, followed by a sustained release for the next 5 days. At pH 5.3, 28% of the loaded drug was released after 12 h and 55% after 140 h. At pH 7.4, 15% of the drug was released after 12 h and 28% after 140 h. The levels of reduced GSH in isolated rat mitochondria were not reduced in any of the tested groups, signing that oxidation of thiol groups in mitochondrial permeability transition pores did not occur. This suggests that pure QUR, QMNMs, and MNPs did not cause mitochondrial dysfunction in rats. These results were confirmed by lipid peroxydation (LPO) assay.

In the study from 2022, Hou et al. synthesized a clickable azido derivative of baicalin (BCL-N_3) for the functionalization of alkyne-modified Fe_3O_4@SiO_2 core-shell magnetic nanoparticles (MNPs) to create baicalin affinity nanoparticles (BCL-N_3@MNPs) able to identify proteins that interact with baicalin (Figure 12) [81]. Baicalin was modified by attaching an azide group to its carboxyl via a PEG chain. The resulting BCL-N_3 was then conjugated with MNPs via an azido-alkynyl click reaction. The study examined whether BCL-N_3 preserves the biological activity of BCL in human liver microsomes (HLMs) by comparing its inhibitory activity on human carboxylesterase 1 (hCE1). Although both BCL and BCL-N_3 had similar IC_{50} values, BCL-N3 showed a more potent inhibitory effect at higher concentrations. However, when baicalin was functionalized onto MNPs, its inhibitory activity on hCE1 was even higher at the same concentration as BCL and BCL-N_3. Researchers optimized the capture efficiency of BCL-N_3@MNPs on target proteins of baicalin by comparing the number of proteins seized from protein extracts of human embryonic kidney (HEK293) cells. The results showed that 100 nm BCL-N_3@MNPs had higher capture efficiency and that non-specific protein absorption on BCL-N_3@MNPs could be reduced by washing the probes. Therefore, the two-hour incubation and washes were defined as optimal conditions that led to the capture efficiency of 19.69 μg/mg for BCL-N_3@MNPs (versus 4.71 μg/mg for MNPs). A total of 14 proteins were identified from extracts of HEK293 cells through mass spectrometry as interacting with baicalin, including enzymes, transcription regulators, a transporter, a kinase, a translation regulator, and other proteins. The interactions may be related to baicalin's various pharmacological activities, such as anti-inflammatory and anti-infection effects, and its potential as an anticancer agent. Additionally, baicalin was found to interact with actin-related proteins, which may play a role in its ability to inhibit cancer cell motility, and with peroxiredoxin IV, which may contribute to its antioxidant activity.

To sum up
Pros
• Connections of flavonoids with iron oxide-based nanoparticles resulted in materi-als of interesting physicochemical and anticancer potential, as well as prospective applications as drug delivery platforms. • Iron oxide-based nanoparticles were functionalized with quercetin, eupatorin, silibinin, EGCG, and baicalin to study their prospective medical applications. • Several types of polymers, e.g., PLGA, PEG, PCL, CMC, and PVP, as well as meso-porous silica, were used for surface functionalization of iron oxide-based nano-particles, which increased the stability of flavonoid-nanoparticle conjugates and impacted the controlled release of organic molecules. • The surface functionalization of iron oxide nanoparticles with the use of quercetin reduced their toxicity to healthy cells. • Iron oxide-based nanoparticles, especially super-paramagnetic iron oxide nano-particles (SPIONs) loaded with curcumin and silymarin, demonstrated potential as theranostic agents for anticancer treatment.
Cons
• For most of the nanoparticles directed against cancers, the MIC/ID50 values were not specified. • There are no comparative studies of standard cancer therapy with and without the particular DDS. • There is a lack of information about the selectivity index of the synthesized NPs, which is an important parameter allowing us to assess their effectiveness.

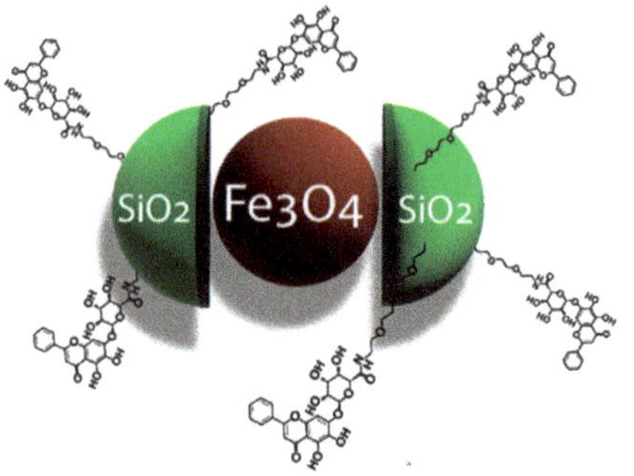

Figure 12. Fe_3O_4@SiO_2 core-shell magnetic nanoparticles with baicalin.

Iron oxide nanoparticles have been produced and formulated with flavonoids for applications in the treatment of ovarian cancer [75] after coating with mesoporous silica [11] or in the form of nanocomposite of Fe_2O_3, CHI, and montmorillonite (MMT) in breast cancer [74]. Methoxy poly(ethylene glycol) (mPEG)-b-PLGA-coated iron oxide nanoparticles seem to be suitable carriers of flavonoid compounds in the treatment of prostate cancer [76], whereas iron oxide NPs can serve as carriers of PLGA-encapsulated silibinin for sustained release in kidney cancer cells [77]. Iron oxide nanoparticles are also components of drug delivery systems applied in multimodal synergistic cancer therapy [79] and formulations used for the identification of selected proteins [81]. The summary of the data related to connections of iron-based nanoparticles with flavonoids, which were presented in the section, was included in Table 3.

Table 3. The summary of the data presented in Section 3.3.1.

Flavonoid Type	Type of NPs	Size of NPs	Synthesis Method	MIC/ID_{50}	Action	Ref.
Curcumin, silymarin	SPIONs	57 nm	reverse microemulsion method	Not specified	MCF-7 cells	[11]
Quercetin	Fe_3O_4	40 nm	enzymatic reduction of $FeSO_4$	Cell viability $\leq$ 9%	MCF-7 cells	[73]
Quercetin	Fe_2O_3	148.2 nm	chemical reduction of $Fe(NO_3)_3 \cdot 9H_2O$ by $NH_3 \cdot H_2O$	Not specified	MCF-7 cancer cells	[74]
Quercetin	Fe_3O_4	8.7 nm	co-precipitation in an aqueous solution	Not specified	Human ovarian carcinoma cells	[75]
Eupatorin	Fe_3O_4	58.5 nm	nano-precipitation approach	IC_{50} 100 mM and 75 mM	Human prostate cancer cell lines DU-145 and LNCaP	[76]
Silibinin	Fe_3O_4	285.9 nm	co-precipitation of $FeCl_2$ in the presence of H_2O_2	IC_{50} 3 µg/mL	A-498 human kidney cancer cells	[77]
EGCG, epicatechin, proanthocyanidin	Fe	21 nm for Fe-EGCG, 27 nm for Fe-EC, and 30 nm for Fe-PAC	sol-gel process	Not specified	HeLa cells	[78]
EGCG	Fe_3O_4	150–180 nm	co-precipitation of $FeCl_2$ and $FeCl_3$ with NH_4OH	Cell viability $\leq$ 10%	B16F10 cells	[79]
Quercetin	Fe_3O_4	10–15 nm	co-precipitation method	non-toxic for isolated rat mitochondria	multifunctional drug delivery platform	[80]
Baicalin	Fe_3O_4	100 nm	Commercially available $Fe_3O_4@SiO_2$ nanoparticles were used	protein capture efficiency of 19.69 µg/mg	Human embryonic kidney cells (HEK293)	[81]

3.3.2. Zinc(II) Oxide-Based Nanoparticles

In a study conducted in 2021, Kollur et al. evaluated the potential anticancer activity of luteolin-functionalized zinc oxide nanoparticles (L-ZnONPs) [82]. L-ZnONPs were made by mixing an aqueous solution of $Zn(OAc)_2$ and luteolin, sifting the resulting white precipitate, washing it with ethanol to remove impurities, and calcining the result. Luteolin's 2′- and 3′-hydroxyls were employed to clad the ZnO nanostructures through the oxidation process with the zinc ions. The obtained nanostructures were characterized using XRD, SEM (nanospheres between 12 and 25 nm in size), and TEM (hexagonal structure with a dimension of roughly 17 nm). L-ZnONPs considerably outperformed individual treatments of luteolin and ZnO in cytotoxicity experiments on MCF-7 cells under hypoxic conditions, lowering the number of viable cells to 15% at a concentration of 40 µM (the IC_{50} value of free L was previously reported as 43 µM for MCF-7 cells). Based on the in silico protein validation, all the proteins chosen for this study (1Q4O, 2FK9, 2LAV, 3PP0, 4RIW, 5YZ0) expressed high levels of favored and allowed residues, allowing them for further molecular interactions. Based on increasing binding affinity, six docked poses for L-ZnONP against a specific protein were achieved, and the results were visualized to investigate interactions between the ligand and protein. The further computations allowed the prediction that L-ZnONPs interact with 1Q4O, 3PP0, and 2LAV proteins via hydrogen bonds with binding affinities of −9.7, −8.3, and −10.1, respectively. The other proteins created less hydrogen bonding with L-ZnONPs. The best docked L-ZnONP poses with the selected proteins featured conserved salt bridges and numerous bonded and non-bonded interactions. L-ZnONPs were assumed to limit MCF-7 cell proliferation via molecular interactions with the human polo-like kinase 1 protein. In another study, Ramalingam et al. investigated ZnO nanoparticles functionalized with quercetin (ZnO@Quercetin) with respect to their anticancer efficacy towards human ovarian cancer cells [83]. ZnO@Quercetin nanoparticles were created by dissolving quercetin in a zinc nitrate solution, refluxing the combination, and then adding KOH. The same approach was used to create a control group of ZnO but without the inclusion of quercetin. QUR was also physically combined with ZnO to generate a different sample in which the two were linked by static contacts, van der Waals forces, or Lewis acid-base interactions. According to SEM and EDS analyses, the produced nanoparticles were evenly dispersed, agglomerated, and devoid of contaminants. Elemental mapping using scanning augur microscopy (SAM) revealed a homogeneous distribution of carbon, zinc, and oxygen in the functionalized ZnO@Quercetin, showing that functionalization was effective. The ZnO nanoparticles produced in this work were discovered to be monocrystalline hexagons with a size range of 12–18 nm and a lattice spacing of 0.21 nm, hinting on the wurtzite type XRD examination revealed the typical peaks of ZnO nanoparticles, while Raman spectroscopy validated the nanoparticles' structural purity. The surface functionalization of QUR with ZnO nanoparticles did not alter the diffraction peaks or lattice planes substantially, showing that the surface functionalization was effective. Functionalization improved the electrochemical performance and surface area of ZnO nanoparticles, as demonstrated by cyclic voltammetry and Brunauer–Emmett–Teller analyses. The functionalized nanoparticles outperformed the non-functionalized ones in terms of redox behavior, conductivity, and surface area. ZnO@Quercetin showed increased concentration-dependent cytotoxicity against human ovarian cancer cells, with a lower concentration of ZnO@Quercetin resulting in increased cytotoxicity (an IC_{50} was about 10 μg/mL), compared to ZnO or QUR alone. This toxicity was expressed in significant morphological alterations induced in cancer cells, including reduced density, detaching, clumping, and floating. The functionalization of quercetin on the surface of ZnO nanoparticles proved to be more effective at causing cancer cells to lose their structural integrity. ZnO nanoparticles were discovered to promote the formation of ROS in cancer cell mitochondria, as seen by weak fluorescence emission in a DHE staining experiment and verified by spectrofluorimetry analysis. QUR boosted ROS creation as well, albeit to a lower amount than ZnO@Quercetin, which had significant fluorescence emission and the highest ROS generation as compared to control cells. The ZnO@Quercetin also demonstrated substantial anticancer efficacy through the permeabilization and ROS production in the mitochondrial membrane, regulating key proteins involved in the intrinsic apoptotic cascade, thus predisposing apoptosis in human ovarian cancer cells. When QUR and ZnO nanoparticles were combined, the number of apoptotic cells increased much more than when each component was used alone. The dual staining experiment discovered that both ZnO and QUR therapy produced early death in the cells yet showed poorer performance in comparison to the ZnO@Quercetin formulation. Mahalanobish et al. described in their study from 2022 the development of a zinc oxide nanoparticle-based drug delivery medium for the targeted delivery of a natural bioactive compound - CHY, to lung cancer cells [84]. To achieve final ZnO-PBA-CHY formulations, zinc oxide nanoparticles (ZnO NPs) and amine-conjugated ZnO NPs (NH_2-ZnO) were synthesized first through a series of chemical reactions and mixing steps. BE was activated and then combined with the NH_2-ZNPs to create the ZnO-PBA nanocarrier. Chrysin was loaded onto ZnO-PBA by adding it to the nanocarrier and stirring the mixture. The resulting drug loading content and EE were measured at

30.56% and 44%, respectively. The stability of the nanoconjugate remained relatively constant in a solution with 10% FBS. The results of UV-vis spectra in a dialysis bag experiment showed that the nanoconjugates partially dissolved, releasing almost 59% of the CHY after 48 h at pH 5.0, while only 14% and 9% were released at pH 6.0 and 7.4, respectively. The synthesized nanohybrids were shown to emit blue 4′,6-diamidino-2-phenylindole (DAPI) and green (FITC) fluorescence when viewed under a fluorescent microscope, implying their potential use as a bio-imaging agent. The intake of ZnO and ZnO-PBA nanoparticles in A549 cells was studied using fluorescence-activated cell sorting analysis, which showed that the ZnO-PBA nanoparticles had greater intake efficacy in the cells compared to the ZnO nanoparticles. This increased intake is thought to be due to the interaction of the PBA-tagged nanoparticles with overexpressed sialic acid receptors on the surface of the cancer cells. The cytotoxicity of ZnO-PBA-CHY was examined on A549 and L132 cell lines using the MTT assay, showing dose-dependent cytotoxicity in the A549 cells at doses ranging from 16.3 to 130.8 μg/mL (an equivalent to 5–40 μg/mL of free CHY). The nanohybrids had a greater cytotoxic impact on A549 cells compared to free CHY or ZnO-PBA and did not show significant cytotoxicity in the normal alveolar epithelial L132 cells. The apoptotic effect of the nanohybrids on A549 cells increased the value of apoptotic cells from 6.54% in the control to 12.44%, 22.52%, and 55.62% after treatment with free CHY, ZnO-PBA, and ZnO-PBA-CHY, respectively. Additionally, the nanohybrid successfully stopped the cell cycle in the G0/G1 phase. These findings imply that the nanohybrid triggers the innate cell death mechanisms, causing cellular apoptosis and cell cycle arrest. Finally, by reducing MMP-2 expression and VE-cadherin expression, the nanoparticles inhibited the migration and invasion of A549 cells.

In another study, a quercetin-functionalized CuO/ZnO nanocomposite (CuO/ZnO@Q) was studied in terms of photocatalytic and biocidal activity [85]. CuO/ZnO@Q was formulated by mixing copper acetate and QUR solution, then adding QUR and zinc acetate. The resultant mixture was washed and centrifuged before being dried (yielding CuO/ZnO@Q) and calcined to provide pristine CuO/ZnO for comparative studies. The pure CuO/ZnO XRD pattern revealed the existence of both ZnO and CuO phases, with peaks matching to polycrystalline hexagonal wurtzite structure of ZnO and the monoclinic secondary phase of CuO. The CuO/ZnO@Q pattern was X-ray amorphous, most likely due to the bounded QUR. The FTIR spectrum of CuO/ZnO@Q nanocomposite indicated the majority of the quercetin peaks but with a modest drop in intensity and change in position, indicating surface functionalization. After 600 °C sintering, the FTIR spectrum revealed no peaks that might be ascribed to QUR, but rather the existence of metal–oxygen bonds typical of CuO and ZnO. The nanocomposite was characterized using TEM, EDX, and UV-vis spectroscopy. The photoluminescence (PL) spectra of CuO/ZnO revealed an emission peak at 434 nm, which was ascribed to excitonic band-to-band radiative emission. PL intensity of CuO/ZnO@Q declined, suggesting efficient suppression of charge carrier recombination and higher separation of electron and hole pairs. When exposed to UV light, the excited electrons from RhB and QUR molecules were transported to the CuO/ZnO conduction band, allowing for more effective charge carrier separation and oxidation of the dye molecules. After 75 min of UV irradiation, CuO/ZnO@Q demonstrated nearly full degradation of RhB (99.91%), whereas pure CuO/ZnO reached only 70.78%. CuO/ZnO@Q caused 99.98% degradation after 90 min, whereas pure CuO/ZnO decomposed the dye to 96.5% over the same period. By improving the capacity of CuO/ZnO@Q to absorb UV light at 256 nm and 365 nm, QUR can boost its photocatalytic activity. The UV absorption spectra showed that the degradation of RhB was enhanced in the presence of QUR. The rate of dye degradation rises when the catalyst concentration (CuO/ZnO@Q or CuO/ZnO) increases from 20 mg/L to 30 mg/L (the optimal catalyst concentration). When the concentration was raised to 40 mg/L, however, the solution became turbid, and the rate of degradation decreased. The rate of dye degradation grows as the dye concentration increases from 20 mg/L to 30 mg/L (the optimal dye concentration). However, the degradation rate decreases when the dye concentration approaches 40 mg/L. The recyclability of the CuO/ZnO@Q was demonstrated through five cycles of 75 min long degradation reaction, decomposing RhB with 99.9%, 97.4%, 95.7%, 95.12%, and 94.6% efficacy in the first to fifth cycles, respectively. This demonstrates just a little decrease in dye degradation % throughout the five cycles, proving the stability and recyclability of the produced NCs. During the biocidal evaluations, CuO/ZnO@Q inhibited bacterial strains (*Escherichia coli*, *Staphylococcus aureus*, *Shigella*, and *Bacillus subtilis*) better than CuO/ZnO. The functionalized nanoformulations also revealed substantial antifungal activity against *Aspergillus niger* and *Candida albicans*, unlike CuO/ZnO. The presence of QUR biomolecules hindered the development of the tested bacteria considerably well.

In 2021, Nisar et al. released a study on quercetin-loaded zinc oxide nanoparticles (quercetin@ZnO NPs) as a promising candidate for use in antiphotoaging therapeutics (Figure 13) [86]. To create quercetin@ZnO NPs, quercetin was combined and homogenized with acetone before being diluted in various ratios with as-prepared ZnO NPs. To remove superfluous water, the mixture was ultrasonicated and centrifuged, yielding pure quercetin@ZnO NPs. SEM imaging revealed that quercetin was successfully loaded onto ZnO NPs in the form of flower clusters. The best quercetin/ZnO ratio was determined to be 1:10, and when quercetin concentration grew in different quercetin/ZnO NP ratios, the ZnO structure changed and became less prominent. The optimal quercetin/ZnO mass ratio was discovered to be 10:1, providing the adsorption rate of 90.61% and the loading capacity of 29.35%, potentially allowing for maximum drug penetration. It was noticed that UVA radiation boosted drug release, with 88.71% of quercetin released after 8 h of exposure to a 150 kJ/m^2 UVA dose. Only 10.29% of quercetin was released from NPs held in the dark over the same time, revealing the stimulatory effect of UVA irradiation on drug release from NPs, most likely due to hydrophobic/hydrophilic transitions. The ability of quercetin to bind iron was demonstrated using cyclic voltammogram studies, and this was confirmed by spectrophotometry through testing different $Fe(NO_3)_3$ concentrations. These findings imply that quercetin can lower ROS generation while also protecting against UV damage. During the test on UVA-exposed HaCaT cells, quercetin@ZnO NPs influenced the reduction of ROS levels and inflammatory factors. These findings suggested that quercetin@ZnO NPs could be employed to minimize the harmful effects of UVA on the skin, such as photoaging. In a cytotoxicity assay, HaCaT cells were exposed to an optimal value of 150 kJ/m^2 UVA, resulting in approximately 80% cell viability after 24 h. When the cells were treated with different concentrations of quercetin@ZnO NPs at 8 and 24 h after UVA exposure, no cytotoxicity was observed, and the cells showed proliferative behavior, indicating excellent biocompatibility.

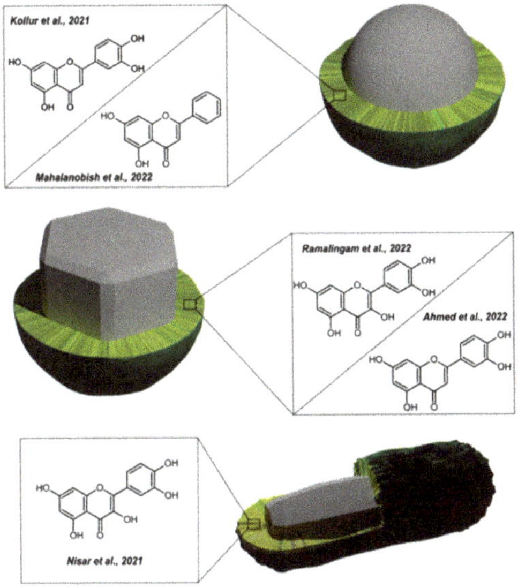

Figure 13. Representative studies on flavonoid-loaded zinc oxide nanoparticles; Mahalanobish et al. 2022 [84], Kollur et al. 2021 [82], Ramalingam et al. 2022 [83], Ahmed et al. 2022 [87], Nisar et al. 2021 [86].

In a related study, Salaheldin et al. investigated the potential use of three innovative (−)-epigallocatechin-3-gallate (EGCG) nanoformulations as natural chemopreventive agents against ultraviolet beam (UVB) radiation-induced DNA damage in keratinocytes [88]. In this study, one of the examined nanosystems was EGCG-CHI/zinc oxide (ZnO)-poly(lactic-co-glycolic acid) (PLGA), which was referred to as Nano 1 (Figure 14). To assemble it, firstly, ZnO nanoparticles were produced via co-precipitation, in which zinc nitrate hexahydrates and sodium hydroxide were combined under stirring. The resulting white precipitate of zinc hydroxide nanoparticles was washed and heated to create a white powder of ZnO nanoparticles. Then, PLGA was dissolved in water with ZnO nanoparticles added to form a ZnO-PLGA suspension. EGCG-CHI was made by mixing ascorbic acid, CHI, and EGCG, to which the ZnO-PLGA suspension was added. The solution was then sonicated and allowed to stir. The resulting EGCG-CHI/ZnO-PLGA was purified through dialysis. Dynamic Light Scattering (DLS) and Electrophoretic Light Scattering (ELS) techniques performed on Nano 1 revealed an average size distribution of 152 nm and Zeta potential of +10 mV. Its EGCG encapsulation efficiency (EE) was 99%, with a loading ratio of 8.4%. Nanoformulation 1 demonstrated slow EGCG release in phosphate-buffered saline (PBS), with a release rate of around 7% even after 24 h. However, in fetal bovine serum (FBS), the release of EGCG from Nano 1 was faster, with a maximum release rate of around 65% in the first hour, followed by a decrease to approximately 40% for the rest of the 24-h period. The stability of nanoformulations was confirmed in this study through the lack of significant change in EGCG concentration during the study period, as well as the stable physical properties, including homogeneity, precipitation, aggregation, and color change. The absorption of UVB radiation by the skin is known to result in the formation of cyclobutane pyrimidine dimers (CPDs) and 6-4 photoproducts (6-4PPs), as well as an increase in the levels of various cytokines and chemokines. To study these effects, the researchers conducted tests on human-immortalized HaCaT keratinocytes. Pretreatment with EGCG and its nanoformulation 1 did not effectively prevent the formation of UVB-induced CPDs or 6-4PPs and did not exhibit significant protective abilities during chemokine/cytokine quantification in the course of in vitro studies. During the in vivo studies, the preventive abilities of nanoformulations were tested on UVB-exposed SKH-1 hairless mice. In terms of protection against UVB-induced DNA damage, nanoformulations showed a 20–45% reduction of UVB-generated CPDs and a 20–48% reduction of 6-4PPs. During the topical application of EGCG and its nanoformulations, the concentration of EGCG in the epidermis was higher for the free EGCG treatment compared to Nano 1. However, the concentration of EGCG in the dermis and hypodermis was higher for Nano 1 treatment compared to the free EGCG treatment. This suggests that the EGCG-CHI/ZnO-PLGA system has higher skin permeability and stability compared to the free EGCG, which may contribute to their protective effect against UVB-induced skin damage.

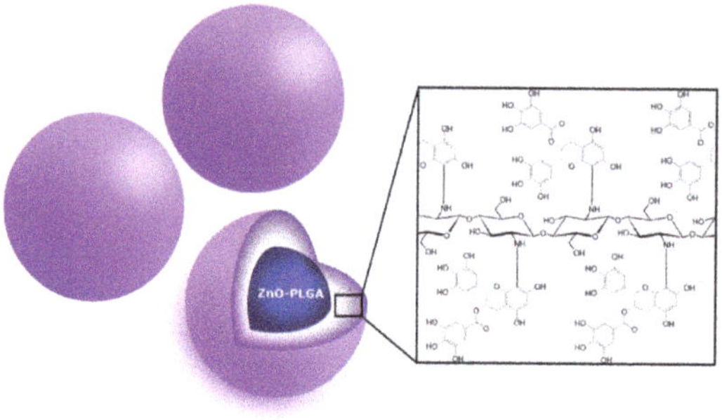

Figure 14. EGCG-CHI/zinc oxide (ZnO)-poly(lactic-co-glycolic acid) (PLGA) nanosystems.

Another application of zinc oxide nanoparticles was proposed by Ahmed et al., who investigated the potential of luteolin/zinc oxide nanoparticles (Lut/ZnO NPs) to improve insulin resistance and treat non-alcoholic fatty liver disease (NAFLD) in a diabetic rat model [87]. To prepare Lut/ZnO NPs, luteolin was dissolved in water and then combined with zinc acetate and sodium hydroxide under heat and stirring. The solution was washed and exposed to a dose of 10 KGy before being sealed. The properties of the resulting Lut/ZnO nanodispersions were characterized, unveiling a hexagonal shape with a mean size of approximately 172.6 nm as per DLS results (17 nm according to TEM) composed of single crystalline phase of hexagonal ZnO with a mean particle size of 174.7 nm confirmed by the XRD pattern. UV-vis spectroscopy also showed the presence of luteolin with absorbance maxima at 340 and 390 nm. The LD50 test on female rats determined that the safe dose for intraperitoneal injection of Lut/ZnO NPs was 12 mg/kg body weight. The nanodispersions were categorized as safe and effective at reducing blood glucose, insulin levels, and insulin resistance in rats with high-fat diet (HFD)-induced obesity and type 2 diabetes mellitus. The Luteolin/ZnO nano-dispersions also increased the expression of proteins involved in the insulin signaling pathway in the liver, including IRS, PI3K, and AKT, as well as reduced the expression of FoxO1 and its downstream target G6Pase. They were also effective at reducing the expression of SREBP1c, a protein involved in the regulation of lipid and cholesterol metabolism. Lut/ZnO NPs were shown to improve hyperlipidemia in NAFLD and type 2 diabetes mellitus (T2DM) through reducing levels of total cholesterol and triglycerides in both serum and liver tissues, as well as decreasing levels of free fatty acids and increasing levels of high-density lipoprotein cholesterol. It was considered that the antihyperlipidemic effect of Lut/ZnO NPs may indirectly improve insulin resistance by reducing the accumulation of fatty acids and triglycerides in the liver. The nanoparticles reduced lipid peroxidation and improved the antioxidant status in the livers of rats fed HFD and those with T2DM. These effects were indicated by decreases in the levels of MDA and oxidized glutathione as well as increases in the levels of reduced GSH and the expression of heme oxygenase-1. These findings imply that Lut/ZnO NPs may have a protective effect against oxidative stress in the livers of these rats. They also significantly improved liver function in rats with HFD and T2DM, as indicated by a reduction in the activities of liver enzymes—alanine transaminase and aspartate transaminases, manifesting a possible protective effect on the liver. As per a histopathological examination, treatment significantly improved liver damage in rats with HFD and T2DM, as indicated by a reduction in the presence of fatty deposits, necrosis, and hyperplasia of Kupffer cells in the liver tissue, along with an improvement in the regular arrangement of hepatic cords and a decrease in intracellular micro-vesicular steatosis and dilated hepatic sinusoids.

3.3.3. Titanium-Based Nanoparticles

In the work of Zhu et al., the authors investigated the use of strontium, and ICA-loaded TiO_2 nanotube coatings to promote the osseointegration and early implant loading of titanium implants in ovariectomized rats (Figure 15) [89]. The preparation included the formation of TiO_2 nanotubes on titanium plates (Ti) by anodizing oxidation, consecutive thermal Sr coating, and the introduction of ICA onto the coating via chemical deposition. For analysis, the plates were separated into four groups. At different magnifications, SEM images of the surface morphology displayed that the Ti group had a smooth surface with some mechanical scrapes, the TiO_2 group consisted of "honeycomb" nanotubes with ruptures on the surface, while TiO_2 + Sr and TiO_2 + Sr + ICA groups were irregular and different in size nanotubes with thickened walls and reduced tube holes. The EDS investigation confirmed the suggested composition for all groups. XRD investigation revealed the formation of pure Ti and anatase phases of TiO_2 in all four groups, as well as the rutile phase of TiO_2 in TiO_2-bearing groups. In Sr-containing groups, the $SrTiO_3$ was also detected. When compared to the Ti group, the TiO_2-bearing groups showed higher surface roughness and hydrophilicity, and Sr^{2+} was released swiftly in the first week, followed by a constant release until day 30, whereas ICA had a burst release in the first 12 h and low thereafter. When compared to the Ti group, the TiO_2-containing groups enhanced MC3T3-E1 cell adhesion, proliferation, alkaline phosphatase activity, mineralization, and osteogenic gene expression, with the TiO_2 + Sr + ICA group having the highest results. In vivo tests back up these findings, revealing that the TiO_2 + Sr + ICA group had the greatest bone-to-implant contact, bone density, and bone strength when compared to the other groups.

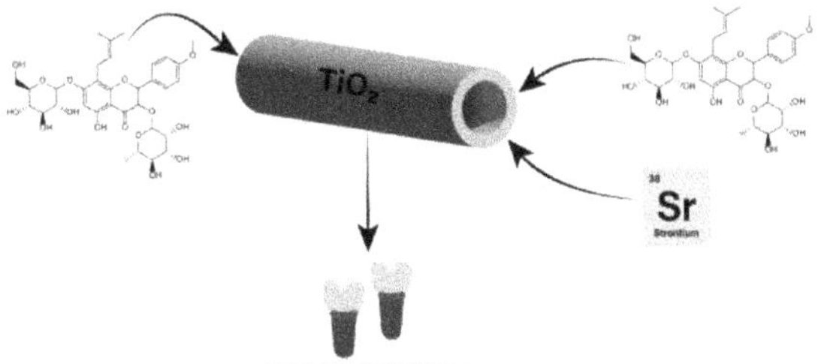

Figure 15. Strontium (Sr) and icariin-loaded TiO_2 nanotube coatings to promote the osseointegration and early implant loading of titanium implants.

In their study from 2022, León-Gutiérrez et al. investigated the use of flavonoids, such as hesperetin 7-rutinoside (H7R) and flavanone-7-*O*-glucoside (F7G) adsorbed onto functionalized titanium dioxide nanoparticles (FTNP) as a potential antiviral against coronaviruses HCoV 229E and SARS-CoV-2 [90]. TiO_2 nanoparticles (NPs) were prepared through adsorption and functionalized by adding various solutions containing functional groups to the NP suspension. The resulting functionalized NPs were characterized using XRD and TEM, which showed that they had an anatase structure with an average grain size of 2 nm. Organic extracts rich in H7R, F7G, and residual terpenes were then adsorbed onto the surface of the functionalized NPs through an impregnation process by stirring a suspension of the NPs in an aqueous solution of the extracts. Molecular docking studies were conducted to investigate how flavonoids or terpenes may interfere with the coupling between the SARS-CoV-2 spike protein and the human angiotensin-converting enzyme

2 (ACE-2) receptor. The study indicated that H7R and F7G had a high binding affinity to different sites on the spike protein with low-affinity energy values. Flavonoids were found to bind to specific regions of the spike protein, including the bottom and top of the S1 domain, near the binding site with ACE-2. The researchers hypothesized that these compounds might disrupt the interactions between the spike protein and ACE-2 or inhibit the necessary movement of the top region of the spike protein, potentially hindering the correct exposition of the binding site with ACE-2 and preventing COVID-19 infection. Molecular dynamics simulations were conducted under three conditions: free protein, protein/H7R, and protein/F7G, and lasted for 100 ns. An analysis of root mean square deviation (RMSD) and root mean square fluctuation (RMSF) showed that certain regions of the spike protein, including the amino and carboxyl-terminal regions and a loop in the middle section, had more movement than others. Cluster analysis was used to identify representative structures for each simulation, and the middle structures of these clusters were analyzed for protein–ligand interactions. The results showed that certain amino acids found in the docking analysis remained in close molecular interaction with the ligands throughout the simulations, constituting the binding site for H7R and F7G. These interactions included hydrogen bonds, with H7R forming more hydrogen bonds than F7G. The effects of FTNPs on the infectivity of CHoV-229E and SARS-CoV-2 were investigated in vitro. The incubation of CHoV-229E-infected MRC-5 cells and SARS-CoV-2-infected VERO.E6 cells with FTNP significantly reduced infection and viral replication. When FTNP was pre-incubated with SARS-CoV-2 before being added to VERO.E6 cells, there was a clear dose-dependent reduction in viral infectivity, while the pre-incubation with CHoV-229E could essentially eliminate any sign of infection in MRC-5 cells already under 1:2 dilution. FTNP also significantly increased the metabolic activity of SARS-CoV-2-infected cells, implying enhanced cell viability. To confirm the mechanism behind the protective effects of FTNPs against SARS-CoV-2 infection, the authors conducted a luciferase assay in co-transfected CHO-K1 cells and found that FTNPs significantly inhibited cell fusion. An increase in affinity between FTNPs and SARS-CoV-2 spikes was observed in dose-response studies, suggesting that FTNPs interfere with ACE-2 receptor-SARS-CoV-2 interactions.

Unnikrishnan et al. developed a method for synthesizing photocatalytic titanium oxide–gold nanocomposites (TiO_x–Au NCs) with polycatechin shell [91]. To achieve these structures, $HAuCl_4$, CT, and $TiCl_3$ were combined at room temperature, framing target TiO_x–Au nanocomposites (NCs). To make various NCs, the molar ratios of Ti^{3+} to Au^{3+} in the reaction mixture differed: 1:1, 5:1, 50:1, and 100:1. The resultant NCs were centrifuged and rinsed with ultrapure water. A similar procedure was used to create polycatechin-coated Au nanoparticles (Au@PC NPs) without the use of $TiCl_3$ (the 0:1 reference). The generated nanoparticles displayed a variety of morphologies, ranging from a deformed sphere (TiO_x–Au $NCs_{(1:1)}$) to a sphere with tiny projections (TiO_x–Au $NCs_{(5:1)}$) or a star-like structure (TiO_x–Au $NCs_{(50:1)}$) and eventually to an urchin-like shape (TiO_x–Au $NCs_{(100:1)}$) when the Ti^{3+}/Au^{3+} ratio increased. The light absorption characteristics of these TiO_x–Au NCs altered considerably as the Ti^{3+} concentration increased. Au@PC NPs exhibited absorption at ~285 nm and ~410 nm, conditioned by the oxidized polycatechin shell, as well as the strong surface plasmon resonance (SPR) absorption at ~580 nm, which faded in favor of increased absorption at a longer wavelength with an increase in the Ti^{3+} presence. The TiO_x–Au NCs series exhibited high light absorption in the NIR region due to a rise in the number of spikes with varying diameters and lengths as the Ti^{3+}/Au^{3+} ratio increased. The photocatalytic capabilities of TiO_x–Au NCs were investigated by evaluating the degradation of dyes methylene blue, rhodamine B (RhB), and malachite green under the illumination of a Xe Arc lamp. The dye degradation was examined between commercial TiO_2 nanoparticles P25, Au@PC NPs, and all TiOx–Au NCs. TiOx–Au NCs had stronger photocatalytic activity than P25, most likely due to their localized surface plasmon resonance (LSPR) characteristics and polymer supramolecular "host–guest" chemistry (CT). The photocatalytic effectiveness rose for the series of NCs from TiOx–Au NCs (1:1) to TiOx–Au $NCs_{(50:1)}$, and subsequently dropped for all three dyes at the greater Ti^{3+}

concentrations. The broad range absorption of TiO_x–Au NCs and localized surface plasmon resonance resulted in efficient hot electron production, which might provide more stable photogenerated electrons and holes for the breakdown of organic dye molecules. The inclusion of Au in the nanocomposite improved the conductance of electrons in the TiO_2 conduction band to the surface, which facilitated electron transfer with dissolved O_2 and dye molecules in the solution, advancing the degradation efficiency even more. TiO_x–Au $NCs_{(50:1)}$ were investigated further and shown to be effective for photocatalytic disinfection of both *E. coli* and MRSA bacteria in the presence of light. After 60 min of light exposure, the viability of these bacteria was decreased to less than 5% utilizing TiO_x–Au $NCs_{(50:1)}$, whereas P25 and the control revealed a minimal change in viability. The evidence of membrane shrinking and disruption of bacteria treated with TiO_x–Au $NCs_{(50:1)}$ was noted on TEM pictures after 30 min of light exposure. Intracellular ROS generation was also detected in bacteria treated with TiO_x–Au $NCs_{(50:1)}$ and subjected to light, but P25 and the control produced little ROS. The high levels of ROS produced during the photocatalytic process most likely caused oxidative stress in the bacteria, resulting in cell death. In the absence of light, the TiO_x–Au $NCs_{(50:1)}$ demonstrated antibacterial activity against MRSA, perhaps due to physical contact between the nanoparticles and the bacterial cell wall. The thin coating of polycatechin atop the TiO_x–Au $NCs_{(50:1)}$ may further contribute to antibacterial activity against Gram-positive bacteria via bacterial wall breakdown. Overall, the photocatalytic disinfection effectiveness of TiOx-Au $NCs_{(50:1)}$ was good for both Gram-negative and Gram-positive bacteria. The PrestoBlue cell viability assay revealed that the cell viability of all four tested cell lines (human liver cancer HepG2, human embryonic kidney HEK-293 T, adenocarcinomic human alveolar basal epithelial A549, and mouse embryonic fibroblast NIH-3 T3) remained above 80%, even at 0.5 mg/mL of TiO_x–Au $NCs_{(50:1)}$, while sustaining the insignificant hemolytic effect. According to these findings, the nanomaterial is biocompatible with mammalian cells. TiO_x–Au $NCs_{(50:1)}$ were also reported to exhibit strong antibacterial activity against MRSA in a wound infection rat model. When compared to the control group and the group treated with P25 and light irradiation, the TiO_x–Au $NCs_{(50:1)}$-treated group with light irradiation demonstrated considerably quicker wound healing. Histological examination of infected tissues revealed that the TiO_x–Au $NCs_{(50:1)}$-treated group had much less inflammation and better-ordered tissue structure than the control group. These findings imply that TiO_x–Au $NCs_{(50:1)}$ have the potential to be a useful and effective antibacterial agent for wound healing. In the study published in 2022, Negrescu et al. explored the use of titanium dioxide nanotubes (TNT) loaded with icariin (ICA) through a layer of polydopamine (PDA) as a drug delivery platform for improved osseointegration in orthopedic implants [92]. As synthetic prerequisites for the fabrication of TiO_2 nanotubes, titanium plates were prepared through a process of mechanical polishing and cleaning. The TNTs were then produced via anodic oxidation in an electrolyte solution and rinsed before being annealed. An adhesive intermediate layer of DP was applied to the nanotubes, followed by the immobilization of ICA through physical adsorption. The resulting samples were washed and used in further studies alongside flat titanium and unsheathed nanotubes. The characterization with SEM revealed that the resulting TNTs were found to be uniform and well-organized, with an inner diameter of around 65–70 nm and a wall thickness of 6–7 nm. When coated with DP, the tube diameter was reduced to 50–55 nm, and the thickness increased to 14–15 nm. AFM images of the uncoated and coated TNTs showed that the surface roughness increased from 54 nm for pure Ti to 95 nm for TiO_2 nanotubes. After functionalization with a layer of DP and loading with ICA, the surface roughness increased to 159 nm, indicating that surface functionalization may alter the surface topography by elevating roughness and potentially influencing cell behavior. The hydrophilic nature and surface energy of unmodified and modified TiO_2 nanotubes were evaluated using contact angle measurements. It was found that the surface energy increased from approximately 40 mJ/m^2 for plain Ti to roughly 70 mJ/m^2 for TNTs. The addition of a DP-ICA coating did not significantly affect the surface energy, indicating that the surface nanostructure has a greater impact on the final value than the coating. Electrochemical impedance spectroscopy (EIS) was used to assess the electrochemical behavior of the samples, revealing two different electrical circuits that were ascribed to the unmodified Ti sample and to the TNT and TNT-DP-Ica supports. The native TiO_2 layer contributed highly to corrosion resistance, with values in the hundreds of kΩ for all three samples. Adding TiO_2 in the form of nanotubes introduced a lower supplementary resistance of around 3 kΩ, due to the open nanotubular oxide structure. However, the addition of DP increased these supplementary values to 12 kΩ, possibly due to DP's non-conductive nature. ICA was rapidly released from the TNT and TNT-DP samples in the first hour, with a rate of 10 μg/h and 6.5 μg/h, respectively. The release slowed significantly over the next 5 h, reaching rates of 5 μg/h for both samples. Over the next 168 h, the TNT support showed nearly no further ICA release, while the TNT-DP sample still presented a slow release rate of 0.04 μg/h. The impact of the tested surfaces on the survival rate of RAW 264.7 macrophages was examined using the Cell Counting Kit 8 assay and Live & Dead Cell Viability/Cytotoxicity Assay. The results showed that the coated TNT-DP-ICA sample had a lower number of viable metabolic active cells compared to the flat Ti and bare TNT substrates, but statistically, no significant differences were found. However, the TNT-DP-ICA sample had a lower cellular density and inhibited cell proliferation, while the bare TNT surface had no effect on cell survival or density. The effects of the tested surfaces on RAW 264.7 macrophages were investigated using Alex Flour 546-conjugated phalloidin in the actin cytoskeleton labeling. In the presence of *Escherichia coli* LPS, macrophages on flat Ti surfaces demonstrated increased spreading and an activated, migratory phenotype. However, TiO_2 nanotubes protected against LPS-induced morphological changes, and the addition of ICA significantly reduced these changes. Quantitative analysis also showed that coated TNT surfaces inhibited the activation of macrophages. The secretion of pro-inflammatory mediators such as IL-6, TNF-α, and MCP-1 by RAW 264.7 cells was measured after 24 h of culture using an ELISA. The results showed that the levels of these cytokines increased significantly in the presence of LPS, with the exception of IL-6, which was only detectable after LPS stimulation. The levels of IL-6 and MCP-1 were lower in cells grown on the TNT-DP-ICA surface compared to the flat Ti and bare TNT surfaces, while the levels of TNF-α were similar across all samples. The ability of the samples to stimulate the release of NO by macrophages in the presence of LPS was also analyzed by measuring the levels of nitrite in the culture media. In standard culture conditions, the levels of NO were low and comparable across all samples. However, in the presence of LPS, the cells grown on the flat Ti surface released the highest levels of NO, while the TNT-DP-ICA surface elicited the lowest levels of NO compared to the flat Ti and bare TNT surfaces. In an in vivo study using rats, the effects of various titanium implants on bone formation were assessed. The rats received implants in their posterior legs; some received unmodified Ti implants, others received Ti implants with a layer of TiO_2 nanotubes, and others received Ti implants with a layer of TiO_2 nanotubes and ICA functionalization through an intermediate DP layer. X-ray and histological analyses of the harvested bone tissue were performed at 1- and 90-days post-implantation. The results showed that the Ti implants with TiO_2 nanotubes and ICA functionalization revealed the most pronounced bone formation, while the unmodified Ti implants had minimal bone formation and dense fibrous tissue. Inflammation was not observed in any of the implants.

3.3.4. Ceria-Based Nanoparticles

Thakur et al. studied the usage of amine-functionalized ceria nanoparticles (CeO_2-NH_2NPs) as a medication delivery vehicle for a natural flavonoid morin against both Gram-negative and Gram-positive bacteria [93]. CeO_2NPs were made by combining ammonium cerium nitrate, urea, and sodium hydroxide in water and refluxing it for 5 h. After that, the NPs were functionalized with 3-aminopropyl triethoxysilane (APTES) by combining them with APTES and refluxing the mixture for 12 h. Morin-CeO_2-NH_2NPs were made by combining morin with a dispersed solution of amine-functionalized NPs in ethanol and swirling the mixture continuously for 12 h. CeO_2NPs with a size of

3–4 nm were characterized using TEM, XPS, and XRD. The nanoparticles were found to have a cubic fluorite structure and were surface functionalized with APTES, as confirmed by FTIR. The hydrodynamic size of the nanoparticles was determined to be 86.11 nm using DLS, and their zeta potential was found to be +23.2 mV. The UV-visible spectra of the nanoparticles displayed strong absorption in the 300–325 nm range. Morin was also incorporated into the nanoparticles, resulting in an absorption band in the 300–400 nm range as determined by UV-vis spectroscopy. Morin drug loading content on the nanoceria surface was around 10%. Morin was released from the nanohybrid in a time-dependent manner, with 30.65%, 43.79%, 45.98%, and 54.74% released after 6, 12, 24, and 48 h, respectively. The nanohybrid demonstrated concentration-dependent radical scavenging activity in DPPH, hydrogen peroxide, and superoxide radical scavenging experiments, with maximal scavenging of 50%, 87%, and 70%, respectively. When normal kidney epithelial (NKE) cells were treated with the nanohybrid before being exposed to tertiary butyl hydroperoxide, an ROS inducer, the nanohybrid was observed to lower intracellular ROS levels when compared to free morin or CeO_2-NH_2NPs. The nanohybrid triggered dose-dependent cytotoxicity in NKE cells, with cell viability reaching 80.67% at 50 µg/mL and surpassing 65% at 100 µg/mL. Morin-CeO_2-NH_2NPs inhibited growth in both Gram-positive (*S. aureus*) and Gram-negative (*E. coli*) bacteria and exhibited the highest growth inhibitory efficacy among the formulations evaluated. Their MIC values against the studied microorganisms were 120.75 µg/mL for *S. aureus* and 165.28 µg/mL for *E. coli*, respectively. Among the investigated formulations, the nanohybrid produced the highest DNA degradation in both S. aureus and *E. coli* bacteria, and the degree of degradation was greater in *S. aureus*. It raised ROS levels in *S. aureus* and *E. coli* bacteria by 4- and 5.5-fold, respectively, compared to control groups, and increased membrane permeability, depolarization, and cell shrinkage, while elevating LDH release in both types of bacteria. Overall, the results demonstrate that morin-CeO_2-NH_2NPs have strong antioxidant and antibacterial capabilities, with low toxicity toward NKE cells at 50 µg/mL concentration.

To sum up
Pros
• Connections of flavonoids with metal oxide nanoparticles, mainly zinc oxide, ce-ria, titanium-based, CuO/ZnO, and titanium oxide–gold nanocomposites, resulted in materials of interesting physicochemical and biological properties, including antibacterial, antifungal, antiviral, anticancer, and prospective applications as drug delivery platforms for improved osseointegration in orthopedic implants. • Metal oxide nanoparticles, such as zinc oxide, ceria, titanium-based, CuO/ZnO, and titanium oxide–gold, were functionalized with quercetin, EGCG, baicalin, lu-teolin, chrysin, morin, icariin hesperetin 7-rutinoside, flavanone-7-O-glucoside to study their prospective medical applications. • The ability of the UV light absorption of zinc oxide nanoparticles indicates their perspective application in antiphotoaging therapy.
Cons
• The effects of ZnO-NPs that could be safely applied in the biomedical field have been intensively studied to understand the interplay between their physicochem-ical properties and toxic effects. The safety and toxicity issues of ZnO-NPs have become a field of increased public attention [94,95]. • Titanium dioxide is a commonly used food, cosmetic, and drug additive or ingre-dient. Everyday use of nanosize TiO2 raises concerns about safety. There are vari-ous data demonstrating the toxic effects of titania in animal models. An agree-ment on the safety of titania has not yet been reached among researchers [96].

CuO/ZnO NCs functionalized with quercetin presented biocidal effectiveness [85]. ZnO nanosystems combined with EGCG, CHI, and PLGA were found to prevent UVB-induced skin damage [88], and quercetin-loaded ZnO NPs show promising potential for use in antiphotoaging therapeutics [86]. Although ZnO NPs are often applied as nanocarriers in anticancer procedures such as the ZnO-PBA (phenylboronic acid) to lung cancer cells [84] or for action potentiation of active flavonoid such as ZnO@Quercetin towards human ovarian cancer [83], they can also be combined with luteolin to improve insulin resistance and reduce the activity of liver enzymes [87]. Apart from iron or zinc oxide NPs, nanoforms of titanium dioxide are often investigated. A drug delivery platform consisting of titanium dioxide nanotubes loaded with ICA through a layer of PDA contributed to improved osseointegration in orthopedic implants [92], whereas flavonoid compounds adsorbed on FTNP revealed antiviral potential against coronaviruses HCoV 229E and SARS-CoV-2 [90]. The summary of the data related to connections of selected metal oxide nanoparticles with flavonoids, which were presented in the section, was included in Table 4.

Table 4. The summary of the data presented in Sections 3.3.2–3.3.4.

Flavonoid Type	Type of NPs	Size of NPs	Synthesis Method	MIC/ID_{50}	Action	Ref.
Luteolin	ZnO	17 nm	chemical reduction of zinc acetate	Cell viability = 15% at 40 μM	MCF-7 cells	[82]
Quercetin	ZnO	12–18 nm	chemical reduction of zinc nitrate by KOH	IC_{50} = 10 μg/mL	human ovarian cancer cells	[83]
Chrysin	ZnO	25–30 nm	chemical reduction of zinc acetate	cytotoxicity observed in the A549 cells at 16.3 μg/mL	A549 and L132 lung cancer cell lines	[84]
Quercetin	CuO/ZnO	Not specified	co-precipitation of copper acetate, zinc acetate and quercetin; calcination	Zone inhibition method (in mm) for conc. 500 μg/mL	*E. coli*, *S. aureus*, *Shigella*, and *Bacillus subtilis*; *Aspergillus niger* and *Candida albicans*	[85]
Quercetin	ZnO	Not specified	chemical reduction of zinc acetate	Cell viability = 80% after 24 h	antiphotoaging therapy HaCaT cells	[86]
EGCG	ZnO	152 nm	co-precipitation method	20–45% reduction of UVB-generated CPDs and 20–48% reduction of 6-4PPs	UVB-exposed SKH-1 hairless mice (chemopreventive agent)	[88]
Luteolin	ZnO	17 nm (TEM)	chemical reduction of zinc acetate by sodium hydroxide	Not applicable	decrease in the insulin resistance	[87]
Icariin	TiO_2	50 nm diameter	anodizing oxidation	Not applicable	drug delivery platform for improved osseointegration	[89]
Hesperetin 7-rutinoside (H7R) and flavanone-7-O-glucoside (F7G)	TiO_2	2 nm	adsorption process	dose-dependent reduction in viral infectivity	antiviral agents against coronaviruses HCoV 229E and SARS-CoV-2	[90]
Catechin	TiOx–Au NCs	144.3–338.9 nm (hydrodynamic sizes depending on core composition)	Reaction of $HAuCl_4$, catechin and $TiCl_3$ at rt	Cell viability $\leq$ 5%	*E. coli* and MRSA	[91]
Icariin	TiO_2	50–55 nm × 14–15 nm	mechanical polishing and cleaning process	Not applicable	drug delivery platform for improved osseointegration	[92]
Morin	CeO_2	86.11 nm	chemical reduction of cerium nitrate by NaOH in the presence of urea	MIC = 120.75 μg/mL for *S. aureus* and 165.28 μg/mL for *E. coli*	*S. aureus* *E. coli*	[93]

4. Summary and Conclusions

Broad interest in hybrid materials based on metal nanoparticles and flavonoids has increased due to their unique physical, chemical, and biological properties. The widespread applications of nanomaterials based on flavonoids and metal nanoparticles constitute a promising perspective for both interesting basic study and practical applications.

Flavonoid-modified metal nanoparticles demonstrate their utilities from the initial preparation step to advanced medical and pharmaceutical studies. The review combines knowledge of multiple possibilities of synthesis of flavonoid-metal nanoparticle conjugates and hybrids via their characterization, biological properties, and medical applications. Many research results have led to the conclusion that the combination of flavonoids with metal nanoparticles results in the improvement of their individual properties and applications. The combinations of the flavonoid quercetin with various nanoparticles discussed in this review can be used to indicate the effects caused by the nanoparticle on the flavonoid molecule, and vice versa. The obtained nanoparticles functionalized with flavonoids have not only been endowed with modified physicochemical properties, but their biological activity has also been extended, and thus the prospects for their applications in medical sciences (Figure 16).

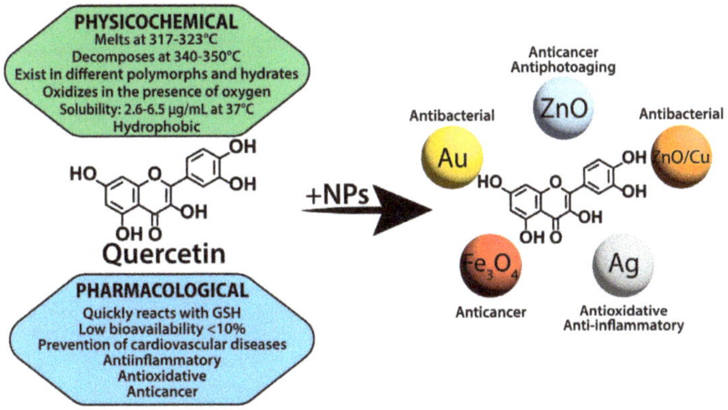

Figure 16. Quercetin and its physicochemical and pharmacological [97–99] properties as well as prospective biological and medical activities in connections with nanoparticles.

Silver nanoparticles conjugated with flavonoids exerted satisfying antibacterial, anti-inflammatory, antioxidant, anticancer, and antifungal activities, as well as diagnostic applications. Moreover, silver nanoparticles modified with zinc or gold revealed antiviral and antiparasitic activity potential. Functionalizing gold nanoparticles with flavonoids improved their properties and broadened the spectrum of applications. Gold nanorods with metal-phenolic networks used with NIR-irradiation significantly accelerated the healing of infected wounds and proved their bactericidal effectiveness. Selected flavonoid-conjugated nano-gold particles exerted antibacterial, antiparasitic, antiangiogenic, anticancer, and antioxidant activity, as well as being studied for diagnostic applications. Iron oxide nanoparticles were produced and formulated with flavonoids for applications in the treatment of various cancers. They were also applied as components of drug delivery systems for multimodal synergistic cancer therapy. Zinc oxide nanosystems combined with selected flavonoids were found to prevent UVB-induced skin damage and demonstrated promising potential for use in antiphotoaging therapeutics. They were also proposed as nanocarriers in anticancer procedures and metabolic studies. Titanium dioxide connections with flavonoids were studied against viruses and proposed as a drug delivery platform of the greatest bone-to-implant contact allowing improved osseointegration in orthopedic implants.

To sum up, flavonoid-metal nanoparticles have potential as agents for the treatment of various diseases due to anti-inflammatory, bactericidal, antifungal, antiviral, or antiparasitic properties, as well as for the preparation of modern DDSs and diagnostic agents for computed tomography and magnetic resonance imaging. Although novel nanoparticles offer many advantages to health therapies, some of their properties, such as biodegradability and elimination from the human organism, raise objections and still require more research.

Author Contributions: Conceptualization: S.S., R.L. and T.G.; Resources: S.S., B.C.-G. and T.G.; Writing—original draft preparation: S.S. and B.C.-G.; Writing—review and editing: S.S., T.K., P.S., W.S., D.T.M., B.C.-G., R.L. and T.G.; Visualization: S.S., D.T.M. and P.S.; Supervision: R.L. and T.G.; Project administration: T.G.; Funding acquisition: T.G. All authors have read and agreed to the published version of the manuscript.

Funding: The research was supported by the National Science Center, Poland—grant number 2019/35/B/NZ7/01165.

Data Availability Statement: Not applicable.

Acknowledgments: Stepan Sysak is a participant of the STER Internationalisation of Doctoral Schools Programme from NAWA Polish National Agency for Academic Exchange No. PPI/STE/2020/1/00014/DEC/02. Parts of the figures were drawn by using pictures from Servier Medical Art. Servier Medical Art by Servier is licensed under a Creative Commons Attribution 3.0 Unported License (https://creativecommons.org/licenses/by/3.0/) (accessed on 25 April 2023). Parts of the paper were checked in terms of grammar and language with the assistance of ChatGPT (GPT-3), an artificial intelligence language model developed by OpenAI (https://chat.openai.com/) (accessed on 26 April 2023).

Conflicts of Interest: The authors declare no conflict of interest.

Abbreviations

47DHF, 4′,7-dihydroxyflavone; 4NP, 4-nitrophenol; Ag@QCNPs, silver@quercetin nanoparticles; Ag-Hes NP, hesperidin-capped silver nanoparticle; AgNPs-Iso, isoorientin-loaded silver nanoparticles; AIEgen, aggregation-induced emission luminogen; AP, apigenin; BE, phenylboronic acid; C-Au NC, catechin-functionalized gold nanocluster; CC_{50}, 50% cytotoxic concentration; CeO_2-NH_2NP, amine-functionalized ceria nanoparticle; CHI, chitosan; CHY, chrysin; CHY-AuNPs, chrysin-gold nanoparticles; CMC, carboxymethylcellulose; CT, catechin; CTAB, cetyltrimethylammonium bromide; DA, dopamine; DMY, dihydromyricetin; DPPH, 2,2-diphenyl-1-picrylhydrazyl; EC, epicatechin; ECG, (−)-epicatechin gallate; EDS, energy dispersive spectroscopy; EDX, energy-dispersive X-ray; EGCG, epigallocatechin gallate; F7G, flavanone-7-*O*-glucoside; FBS, fetal bovine serum; FESEM, emission scanning electron microscopy; GA, gallic acid; GNP, gold nanoparticle; GNR, gold nanorods; GOx, glucose oxidase; GSH, glutathione; GTP, green tea polyphenols; H7R, hesperetin 7-*O*-rutinoside; Hes, hesperidin; HP-AgNPs, silver nanoparticles coated with hesperidin and pectin; ICA, icariin; ICG, indocyanine green; IO, iron oxide; Iso, isoorientin; KH-AgNPs, silver nanoparticles conjugated with kaempferol and hydrocortisone; LDH, lactate dehydrogenase; LPS, lipopolysaccharide; LSPR, localized surface plasmon resonance; Lut, luteolin; MDA, malondialdehyde; MIC, minimum inhibitory concentration; MNPs, iron oxide nanoparticles; mPEG, methoxy poly(ethylene glycol); MY, mycerin; OPC, procyanidin; PAC, proanthocyanidin; PC, phenolic compound; PCL, poly(ε-caprolactone); PDA, polydopamine; PI, propidium iodine; PVA-Alg, poly(vinyl alcohol)-sodium alginate; QMNPs, quercetin-conjugated magnetite nanoparticles; QUR, quercetin; RSC, radical scavenging capacity; RSV, resveratrol; SAED, selected area electron diffraction; SC-SNPs, sodium citrate-based silver nanoparticles; SI, selectivity index; SIL, silymarin; SLB, silibinin; SLB-MPNP, PLGA-encapsulated silibinin; SPION, super-paramagnetic iron oxide nanoparticle; SPR, surface plasmon resonance; TA, tannic acid; TEOS, tetraethyl orthosilicate; TiOx–Au NC, titanium oxide-gold nanocomposite; TNT, titanium dioxide nanotubes.

References

1. Kopustinskiene, D.M.; Jakstas, V.; Savickas, A.; Bernatoniene, J. Flavonoids as Anticancer Agents. *Nutrients* **2020**, *12*, 457. [CrossRef] [PubMed]
2. Niedzwiecki, A.; Roomi, M.W.; Kalinovsky, T.; Rath, M. Anticancer Efficacy of Polyphenols and Their Combinations. *Nutrients* **2016**, *8*, 552. [CrossRef] [PubMed]
3. Fresco, P.; Borges, F.; Diniz, C.; Marques, M.P.M. New Insights on the Anticancer Properties of Dietary Polyphenols. *Med. Res. Rev.* **2006**, *26*, 747–766. [CrossRef]
4. Talib, W.H.; Abuawad, A.; Thiab, S.; Alshweiat, A.; Mahmod, A.I. Flavonoid-Based Nanomedicines to Target Tumor Microenvironment. *OpenNano* **2022**, *8*, 100081. [CrossRef]
5. Dobrzynska, M.; Napierala, M.; Florek, E. Flavonoid Nanoparticles: A Promising Approach for Cancer Therapy. *Biomolecules* **2020**, *10*, 1268. [CrossRef] [PubMed]
6. Sathishkumar, P.; Gu, F.L.; Zhan, Q.; Palvannan, T.; Mohd Yusoff, A.R. Flavonoids Mediated 'Green' Nanomaterials: A Novel Nanomedicine System to Treat Various Diseases—Current Trends and Future Perspective. *Mater. Lett.* **2018**, *210*, 26–30. [CrossRef]
7. Dikshit, P.K.; Kumar, J.; Das, A.K.; Sadhu, S.; Sharma, S.; Singh, S.; Gupta, P.K.; Kim, B.S. Green Synthesis of Metallic Nanoparticles: Applications and Limitations. *Catalysts* **2021**, *11*, 902. [CrossRef]
8. Klębowski, B.; Depciuch, J.; Parlińska-Wojtan, M.; Baran, J. Applications of Noble Metal-Based Nanoparticles in Medicine. *Int. J. Mol. Sci.* **2018**, *19*, 4031. [CrossRef]
9. Bordiwala, R.V. Green Synthesis and Applications of Metal Nanoparticles—A Review Article. *Results Chem.* **2023**, *5*, 100832. [CrossRef]

10. Wijesinghe, W.P.S.L.; Mantilaka, M.M.M.G.P.G.; Ruparathna, K.A.A.; Rajapakshe, R.B.S.D.; Sameera, S.A.L.; Thilakarathna, M.G.G.S.N. Filler matrix interfaces of inorganic/biopolymer composites and their applications. In *Interfaces in Particle and Fibre Reinforced Composites*; Goh, K.L., Aswathi, M.K., De Silva, R.T., Thomas, Eds.; Woodhead Publishing: Cambridge, UK, 2020; ISBN 978-0-08-102665-6. [CrossRef]
11. Sadegha, S.; Varshochian, R.; Dadras, P.; Hosseinzadeh, H.; Sakhtianchi, R.; Mirzaie, Z.H.; Shafiee, A.; Atyabi, F.; Dinarvand, R. Mesoporous Silica Coated SPIONs Containing Curcumin and Silymarin Intended for Breast Cancer Therapy. *DARU J. Pharm. Sci.* **2022**, *30*, 331–341. [CrossRef]
12. Cherukuri, P.; Glazer, E.S.; Curley, S.A. Targeted Hyperthermia Using Metal Nanoparticles. *Adv. Drug Deliv. Rev.* **2010**, *62*, 339–345. [CrossRef] [PubMed]
13. Xu, J.-J.; Zhang, W.-C.; Guo, Y.-W.; Chen, X.-Y.; Zhang, Y.-N. Metal Nanoparticles as a Promising Technology in Targeted Cancer Treatment. *Drug Deliv.* **2022**, *29*, 664–678. [CrossRef] [PubMed]
14. Chandrakala, V.; Aruna, V.; Angajala, G. Review on Metal Nanoparticles as Nanocarriers: Current Challenges and Perspectives in Drug Delivery Systems. *Emergent Mater.* **2022**, *5*, 1593–1615. [CrossRef]
15. Singh Tuli, H. (Ed.) *Current Aspects of Flavonoids: Their Role in Cancer Treatment*; Springer: Singapore, 2019; ISBN 9789811358739.
16. Zhang, H.; Wang, M.; Chen, L.; Liu, Y.; Liu, H.; Huo, H.; Sun, L.; Ren, X.; Deng, Y.; Qi, A. Structure-Solubility Relationships and Thermodynamic Aspects of Solubility of Some Flavonoids in the Solvents Modeling Biological Media. *J. Mol. Liq.* **2017**, *225*, 439–445. [CrossRef]
17. Kumar, S.; Pandey, A.K. Chemistry and Biological Activities of Flavonoids: An Overview. *Sci. World J.* **2013**, *2013*, 162750. [CrossRef]
18. Biesaga, M. Influence of Extraction Methods on Stability of Flavonoids. *J. Chromatogr. A* **2011**, *1218*, 2505–2512. [CrossRef]
19. Plaza, M.; Pozzo, T.; Liu, J.; Gulshan Ara, K.Z.; Turner, C.; Nordberg Karlsson, E. Substituent Effects on in Vitro Antioxidizing Properties, Stability, and Solubility in Flavonoids. *J. Agric. Food Chem.* **2014**, *62*, 3321–3333. [CrossRef] [PubMed]
20. Ioannou, I.; Chekir, L.; Ghoul, M. Effect of Heat Treatment and Light Exposure on the Antioxidant Activity of Flavonoids. *Processes* **2020**, *8*, 1078. [CrossRef]
21. Stevenson, D.E.; Scheepens, A.; Hurst, R.D. Bioavailability And Metabolism Of Dietary Flavonoids – Much Known – Much More To Discover. In *Flavonoids: Biosynthesis, Biological Effects and Dietary Sources*; Keller, R.B., Ed.; Nutrition and Diet Research Progress Series; Nova Science Publishers: New York, NY, USA, 2009; ISBN 978-1-60741-622-7.
22. Pérez-Cano, F.; Massot-Cladera, M.; Rodríguez-Lagunas, M.; Castell, M. Flavonoids Affect Host-Microbiota Crosstalk through TLR Modulation. *Antioxidants* **2014**, *3*, 649–670. [CrossRef]
23. Kanakis, C.D.; Tarantilis, P.A.; Polissiou, M.G.; Diamantoglou, S.; Tajmir-Riahi, H.A. DNA Interaction with Naturally Occurring Antioxidant Flavonoids Quercetin, Kaempferol, and Delphinidin. *J. Biomol. Struct. Dyn.* **2005**, *22*, 719–724. [CrossRef]
24. Nafisi, S.; Shadaloi, A.; Feizbakhsh, A.; Tajmir-Riahi, H.A. RNA Binding to Antioxidant Flavonoids. *J. Photochem. Photobiol. B Biol.* **2009**, *94*, 1–7. [CrossRef]
25. Hanrahan, J.R.; Chebib, M.; Johnston, G.A.R. Flavonoid Modulation of $GABA_A$ Receptors. *Br. J. Pharmacol.* **2011**, *163*, 234–245. [CrossRef] [PubMed]
26. Donadio, G.; Mensitieri, F.; Santoro, V.; Parisi, V.; Bellone, M.L.; De Tommasi, N.; Izzo, V.; Dal Piaz, F. Interactions with Microbial Proteins Driving the Antibacterial Activity of Flavonoids. *Pharmaceutics* **2021**, *13*, 660. [CrossRef] [PubMed]
27. Speisky, H.; Shahidi, F.; Costa De Camargo, A.; Fuentes, J. Revisiting the Oxidation of Flavonoids: Loss, Conservation or Enhancement of Their Antioxidant Properties. *Antioxidants* **2022**, *11*, 133. [CrossRef] [PubMed]
28. Augustyniak, A.; Bartosz, G.; Čipak, A.; Duburs, G.; Horáková, L.; Łuczaj, W.; Majekova, M.; Odysseos, A.D.; Rackova, L.; Skrzydlewska, E.; et al. Natural and Synthetic Antioxidants: An Updated Overview. *Free Radic. Res.* **2010**, *44*, 1216–1262. [CrossRef] [PubMed]
29. Brunetti, C.; Di Ferdinando, M.; Fini, A.; Pollastri, S.; Tattini, M. Flavonoids as Antioxidants and Developmental Regulators: Relative Significance in Plants and Humans. *Int. J. Mol. Sci.* **2013**, *14*, 3540–3555. [CrossRef] [PubMed]
30. Ginwala, R.; Bhavsar, R.; Chigbu, D.I.; Jain, P.; Khan, Z.K. Potential Role of Flavonoids in Treating Chronic Inflammatory Diseases with a Special Focus on the Anti-Inflammatory Activity of Apigenin. *Antioxidants* **2019**, *8*, 35. [CrossRef]
31. Badshah, S.L.; Faisal, S.; Muhammad, A.; Poulson, B.G.; Emwas, A.H.; Jaremko, M. Antiviral Activities of Flavonoids. *Biomed. Pharmacother.* **2021**, *140*, 111596. [CrossRef]
32. Tuli, H.S.; Garg, V.K.; Bhushan, S.; Uttam, V.; Sharma, U.; Jain, A.; Sak, K.; Yadav, V.; Lorenzo, J.M.; Dhama, K.; et al. Natural Flavonoids Exhibit Potent Anticancer Activity by Targeting MicroRNAs in Cancer: A Signature Step Hinting towards Clinical Perfection. *Transl. Oncol.* **2023**, *27*, 101596. [CrossRef]
33. Ayaz, M.; Sadiq, A.; Junaid, M.; Ullah, F.; Ovais, M.; Ullah, I.; Ahmed, J.; Shahid, M. Flavonoids as Prospective Neuroprotectants and Their Therapeutic Propensity in Aging Associated Neurological Disorders. *Front. Aging Neurosci.* **2019**, *11*, 155. [CrossRef]
34. Li, X.; Huang, W.; Tan, R.; Xu, C.; Chen, X.; Li, S.; Liu, Y.; Qiu, H.; Cao, H.; Cheng, Q. The Benefits of Hesperidin in Central Nervous System Disorders, Based on the Neuroprotective Effect. *Biomed. Pharmacother.* **2023**, *159*, 114222. [CrossRef] [PubMed]
35. Zulkefli, N.; Che Zahari, C.N.M.; Sayuti, N.H.; Kamarudin, A.A.; Saad, N.; Hamezah, H.S.; Bunawan, H.; Baharum, S.N.; Mediani, A.; Ahmed, Q.U.; et al. Flavonoids as Potential Wound-Healing Molecules: Emphasis on Pathways Perspective. *Int. J. Mol. Sci.* **2023**, *24*, 4607. [CrossRef] [PubMed]

36. Li, W.; Kandhare, A.D.; Mukherjee, A.A.; Bodhankar, S.L. Hesperidin, a Plant Flavonoid Accelerated the Cutaneous Wound Healing in Streptozotocin-Induced Diabetic Rats: Role of TGF-β/Smads and Ang-1/Tie2 Signaling Pathways. *EXCLI J.* **2018**, *17*, 399–419. [CrossRef] [PubMed]
37. Su, L.; Li, X.; Wu, X.; Hui, B.; Han, S.; Gao, J.; Li, Y.; Shi, J.; Zhu, H.; Zhao, B.; et al. Simultaneous Deactivation of FAK and Src Improves the Pathology of Hypertrophic Scar. *Sci. Rep.* **2016**, *6*, 26023. [CrossRef] [PubMed]
38. Hwang, S.-L.; Yen, G.-C. Modulation of Akt, JNK, and P38 Activation Is Involved in Citrus Flavonoid-Mediated Cytoprotection of PC12 Cells Challenged by Hydrogen Peroxide. *J. Agric. Food Chem.* **2009**, *57*, 2576–2582. [CrossRef] [PubMed]
39. Hou, B.; Cai, W.; Chen, T.; Zhang, Z.; Gong, H.; Yang, W.; Qiu, L. Vaccarin Hastens Wound Healing by Promoting Angiogenesis via Activation of MAPK/ERK and PI3K/AKT Signaling Pathways in Vivo. *Acta Cir. Bras.* **2019**, *34*, e201901202. [CrossRef]
40. Fan, J.; Liu, H.; Wang, J.; Zeng, J.; Tan, Y.; Wang, Y.; Yu, X.; Li, W.; Wang, P.; Yang, Z.; et al. Procyanidin B2 Improves Endothelial Progenitor Cell Function and Promotes Wound Healing in Diabetic Mice via Activating Nrf2. *J. Cell. Mol. Med.* **2021**, *25*, 652–665. [CrossRef]
41. Beken, B.; Serttas, R.; Yazicioglu, M.; Turkekul, K.; Erdogan, S. Quercetin Improves Inflammation, Oxidative Stress, and Impaired Wound Healing in Atopic Dermatitis Model of Human Keratinocytes. *Pediatr. Allergy Immunol. Pulmonol.* **2020**, *33*, 69–79. [CrossRef]
42. Fuentes, R.G.; Arai, M.A.; Ishibashi, M. Natural Compounds with Wnt Signal Modulating Activity. *Nat. Prod. Rep.* **2015**, *32*, 1622–1628. [CrossRef]
43. Zhao, Z.-Y.; Li, P.-J.; Xie, R.-S.; Cao, X.-Y.; Su, D.-L.; Shan, Y. Biosynthesis of Silver Nanoparticle Composites Based on Hesperidin and Pectin and Their Synergistic Antibacterial Mechanism. *Int. J. Biol. Macromol.* **2022**, *214*, 220–229. [CrossRef]
44. Ren, X.; Hu, Y.; Chang, L.; Xu, S.; Mei, X.; Chen, Z. Electrospinning of Antibacterial and Anti-Inflammatory Ag@hesperidin Core-Shell Nanoparticles into Nanofibers Used for Promoting Infected Wound Healing. *Regen. Biomater.* **2022**, *9*, rbac012. [CrossRef] [PubMed]
45. Kannanoor, M.; Lakshmi, B.A.; Kim, S. Synthesis of Silver Nanoparticles Conjugated with Kaempferol and Hydrocortisone and an Evaluation of Their Antibacterial Effects. *3 Biotech* **2021**, *11*, 317. [CrossRef]
46. Scroccarello, A.; Molina-Hernández, B.; Della Pelle, F.; Ciancetta, J.; Ferraro, G.; Fratini, E.; Valbonetti, L.; Chaves Copez, C.; Compagnone, D. Effect of Phenolic Compounds-Capped AgNPs on Growth Inhibition of Aspergillus Niger. *Colloids Surf. B Biointerfaces* **2021**, *199*, 111533. [CrossRef] [PubMed]
47. Saadh, M.J.; Aldalaen, S.M. Inhibitory Effects of Epigallocatechin Gallate (EGCG) Combined with Zinc Sulfate and Silver Nanoparticles on Avian Influenza A Virus Subtype H5N1. *Eur. Rev. Med. Pharmacol. Sci.* **2021**, *25*, 2630–2636. [CrossRef]
48. Sasidharan, S.; Saudagar, P. Gold and Silver Nanoparticles Functionalized with 4′,7-Dihydroxyflavone Exhibit Activity against Leishmania Donovani. *Acta Trop.* **2022**, *231*, 106448. [CrossRef] [PubMed]
49. Zarei, M.; Karimi, E.; Oskoueian, E.; Es-Haghi, A.; Yazdi, M.E.T. Comparative Study on the Biological Effects of Sodium Citrate-Based and Apigenin-Based Synthesized Silver Nanoparticles. *Nutr. Cancer* **2021**, *73*, 1511–1519. [CrossRef] [PubMed]
50. Anwer, S.T.; Mobashir, M.; Fantoukh, O.I.; Khan, B.; Imtiyaz, K.; Naqvi, I.H.; Rizvi, M.M.A. Synthesis of Silver Nano Particles Using Myricetin and the In-Vitro Assessment of Anti-Colorectal Cancer Activity: In-Silico Integration. *Int. J. Mol. Sci.* **2022**, *23*, 11024. [CrossRef] [PubMed]
51. Li, Z.; Ali, I.; Qiu, J.; Zhao, H.; Ma, W.; Bai, A.; Wang, D.; Li, J. Eco-Friendly and Facile Synthesis of Antioxidant, Antibacterial and Anticancer Dihydromyricetinmediated Silver Nanoparticles. *Int. J. Nanomed.* **2021**, *16*, 481–492. [CrossRef]
52. Li, Y.; Xiao, D.; Li, S.; Chen, Z.; Liu, S.; Li, J. Silver@quercetin Nanoparticles with Aggregation-Induced Emission for Bioimaging In Vitro and In Vivo. *Int. J. Mol. Sci.* **2022**, *23*, 7413. [CrossRef] [PubMed]
53. Kumawat, M.; Madhyastha, H.; Singh, M.; Revaprasadu, N.; Srinivas, S.P.; Daima, H.K. Double Functionalized Haemocompatible Silver Nanoparticles Control Cell Inflammatory Homeostasis. *PLoS ONE* **2022**, *17*, e0276296. [CrossRef]
54. Wang, X.; Yuan, L.; Deng, H.; Zhang, Z. Structural Characterization and Stability Study of Green Synthesized Starch Stabilized Silver Nanoparticles Loaded with Isoorientin. *Food Chem.* **2021**, *338*, 127807. [CrossRef] [PubMed]
55. Shao, X.-R.; Wei, X.-Q.; Song, X.; Hao, L.-Y.; Cai, X.-X.; Zhang, Z.-R.; Peng, Q.; Lin, Y.-F. Independent Effect of Polymeric Nanoparticle Zeta Potential/Surface Charge, on Their Cytotoxicity and Affinity to Cells. *Cell Prolif.* **2015**, *48*, 465–474. [CrossRef] [PubMed]
56. Park, M.V.D.Z.; Neigh, A.M.; Vermeulen, J.P.; de la Fonteyne, L.J.J.; Verharen, H.W.; Briedé, J.J.; van Loveren, H.; de Jong, W.H. The Effect of Particle Size on the Cytotoxicity, Inflammation, Developmental Toxicity and Genotoxicity of Silver Nanoparticles. *Biomaterials* **2011**, *32*, 9810–9817. [CrossRef] [PubMed]
57. Yeh, Y.-C.; Creran, B.; Rotello, V.M. Gold Nanoparticles: Preparation, Properties, and Applications in Bionanotechnology. *Nanoscale* **2012**, *4*, 1871–1880. [CrossRef]
58. Alhadrami, H.A.; Orfali, R.; Hamed, A.A.; Ghoneim, M.M.; Hassan, H.M.; Hassane, A.S.I.; Rateb, M.E.; Sayed, A.M.; Gamaleldin, N.M. Flavonoid-Coated Gold Nanoparticles as Efficient Antibiotics against Gram-Negative Bacteria-Evidence from in Silico-Supported in Vitro Studies. *Antibiotics* **2021**, *10*, 968. [CrossRef]
59. Zhang, C.; Huang, L.; Sun, D.-W.; Pu, H. Interfacing Metal-Polyphenolic Networks upon Photothermal Gold Nanorods for Triplex-Evolved Biocompatible Bactericidal Activity. *J. Hazard. Mater.* **2022**, *426*, 127824. [CrossRef]
60. Nikoobakht, B.; El-Sayed, M.A. Preparation and Growth Mechanism of Gold Nanorods (NRs) Using Seed-Mediated Growth Method. *Chem. Mater.* **2003**, *15*, 1957–1962. [CrossRef]

61. Fu, G.; Sun, D.-W.; Pu, H.; Wei, Q. Fabrication of Gold Nanorods for SERS Detection of Thiabendazole in Apple. *Talanta* **2019**, *195*, 841–849. [CrossRef]
62. Das, S.; Pramanik, T.; Jethwa, M.; Roy, P. Flavonoid-Decorated Nano-Gold for Antimicrobial Therapy Against Gram-Negative Bacteria Escherichia Coli. *Appl. Biochem. Biotechnol.* **2021**, *193*, 1727–1743. [CrossRef]
63. Raj, S.; Sasidharan, S.; Tripathi, T.; Saudagar, P. Biofunctionalized Chrysin-Conjugated Gold Nanoparticles Neutralize Leishmania Parasites with High Efficacy. *Int. J. Biol. Macromol.* **2022**, *205*, 211–219. [CrossRef]
64. Cunha, L.; Coelho, S.C.; Do Carmo Pereira, M.; Coelho, M.A.N. Nanocarriers Based on Gold Nanoparticles for Epigallocatechin Gallate Delivery in Cancer Cells. *Pharmaceutics* **2022**, *14*, 491. [CrossRef] [PubMed]
65. Panda, A.; Karhadkar, S.; Acharya, B.; Banerjee, A.; De, S.; Dasgupta, S. Enhancement of Angiogenin Inhibition by Polyphenol-Capped Gold Nanoparticles. *Biopolymers* **2021**, *112*, e23429. [CrossRef] [PubMed]
66. Liu, Y.; Liu, Y.; Zhang, J.; Zheng, J.; Yuan, Z.; Lu, C. Catechin-Inspired Gold Nanocluster Nanoprobe for Selective and Ratiometric Dopamine Detection via Forming Azamonardine. *Spectrochim. Acta A Mol. Biomol. Spectrosc.* **2022**, *274*, 121142. [CrossRef] [PubMed]
67. Pradhan, S.P.; Swain, S.; Sa, N.; Pilla, S.N.; Behera, A.; Sahu, P.K.; Chandra Si, S. Photocatalysis of Environmental Organic Pollutants and Antioxidant Activity of Flavonoid Conjugated Gold Nanoparticles. *Spectrochim. Acta A Mol. Biomol. Spectrosc.* **2022**, *282*, 121699. [CrossRef]
68. Scroccarello, A.; Della Pelle, F.; Ferraro, G.; Fratini, E.; Tempera, F.; Dainese, E.; Compagnone, D. Plasmonic Active Film Integrating Gold/Silver Nanostructures for H_2O_2 Readout. *Talanta* **2021**, *222*, 121682. [CrossRef]
69. Sani, A.; Cao, C.; Cui, D. Toxicity of Gold Nanoparticles (AuNPs): A Review. *Biochem. Biophys. Rep.* **2021**, *26*, 100991. [CrossRef]
70. Pleus, S.; Jendrike, N.; Baumstark, A.; Mende, J.; Haug, C.; Freckmann, G. Evaluation of Analytical Performance of Three Blood Glucose Monitoring Systems: System Accuracy, Measurement Repeatability, and Intermediate Measurement Precision. *J. Diabetes Sci. Technol.* **2019**, *13*, 111–117. [CrossRef]
71. Lage, A.C.P.; Ladeira, L.O.; do Camo, P.H.F.; Amorim, J.M.; Monte-Neto, R.L.; Santos, D.A.; Tunes, L.G.; Castilho, R.O.; Moreira, P.O.; Ferreira, D.C.; et al. Changes in Antiparasitical Activity of Gold Nanorods According to the Chosen Synthesis. *Exp. Parasitol.* **2022**, *242*, 108367. [CrossRef]
72. Lee, D.G.; Lee, M.; Go, E.B.; Chung, N. Resveratrol-Loaded Gold Nanoparticles Enhance Caspase-Mediated Apoptosis in PANC-1 Pancreatic Cells via Mitochondrial Intrinsic Apoptotic Pathway. *Cancer Nanotechnol.* **2022**, *13*, 34. [CrossRef]
73. Askar, M.A.; El-Nashar, H.A.S.; Al-Azzawi, M.A.; Rahman, S.S.A.; Elshawi, O.E. Synergistic Effect of Quercetin Magnetite Nanoparticles and Targeted Radiotherapy in Treatment of Breast Cancer. *Breast Cancer Basic Clin. Res.* **2022**, *16*, 11782234221086728. [CrossRef]
74. Ahmadi, M.; Pourmadadi, M.; Ghorbanian, S.A.; Yazdian, F.; Rashedi, H. Ultra PH-Sensitive Nanocarrier Based on Fe2O3/Chitosan/Montmorillonite for Quercetin Delivery. *Int. J. Biol. Macromol.* **2021**, *191*, 738–745. [CrossRef] [PubMed]
75. Garfias, A.F.P.; Jardim, K.V.; Ruiz-Ortega, L.I.; Garcia, B.Y.; Báo, S.N.; Parize, A.L.; Sousa, M.H.; Beltrán, C.M. Synthesis, Characterization, and Cytotoxicity Assay of γ-Fe2O3 Nanoparticles Coated with Quercetin-Loaded Polyelectrolyte Multilayers. *Colloid Polym. Sci.* **2022**, *300*, 1327–1341. [CrossRef]
76. Tousi, M.S.; Sepehri, H.; Khoee, S.; Farimani, M.M.; Delphi, L.; Mansourizadeh, F. Evaluation of Apoptotic Effects of MPEG-b-PLGA Coated Iron Oxide Nanoparticles as a Eupatorin Carrier on DU-145 and LNCaP Human Prostate Cancer Cell Lines. *J. Pharm. Anal.* **2021**, *11*, 108–121. [CrossRef] [PubMed]
77. Takke, A.; Shende, P. Magnetic-Core-Based Silibinin Nanopolymeric Carriers for the Treatment of Renal Cell Cancer. *Life Sci.* **2021**, *275*, 119377. [CrossRef] [PubMed]
78. Qin, J.; Liang, G.; Cheng, D.; Liu, Y.; Cheng, X.; Yang, P.; Wu, N.; Zhao, Y.; Wei, J. Controllable Synthesis of Iron-Polyphenol Colloidal Nanoparticles with Composition-Dependent Photothermal Performance. *J. Colloid Interface Sci.* **2021**, *593*, 172–181. [CrossRef] [PubMed]
79. Shubhra, Q.T.H.; Guo, K.; Liu, Y.; Razzak, M.; Serajum Manir, M.; Moshiul Alam, A.K.M. Dual Targeting Smart Drug Delivery System for Multimodal Synergistic Combination Cancer Therapy with Reduced Cardiotoxicity. *Acta Biomater.* **2021**, *131*, 493–507. [CrossRef]
80. Rahmati, M.-A.; Rashidzadeh, H.; Hosseini, M.-J.; Sadighian, S.; Kermanian, M. Self-Assembled Magnetic Polymeric Micelles for Delivery of Quercetin: Toxicity Evaluation on Isolated Rat Liver Mitochondria. *J. Biomater. Sci. Polym. Ed.* **2022**, *33*, 279–298. [CrossRef]
81. Hou, Y.; Liang, Z.; Qi, L.; Tang, C.; Liu, X.; Tang, J.; Zhao, Y.; Zhang, Y.; Fang, T.; Luo, Q.; et al. Baicalin Targets HSP70/90 to Regulate PKR/PI3K/AKT/ENOS Signaling Pathways. *Molecules* **2022**, *27*, 1432. [CrossRef]
82. Kollur, S.P.; Prasad, S.K.; Pradeep, S.; Veerapur, R.; Patil, S.S.; Amachawadi, R.G.; Rajendra Prasad, S.; Lamraoui, G.; Al-Kheraif, A.A.; Elgorban, A.M.; et al. Wluteolin-Fabricated Zno Nanostructures Showed Plk-1 Mediated Anti-Breast Cancer Activity. *Biomolecules* **2021**, *11*, 385. [CrossRef]
83. Ramalingam, V.; Muthukumar Sathya, P.; Srivalli, T.; Mohan, H. Synthesis of Quercetin Functionalized Wurtzite Type Zinc Oxide Nanoparticles and Their Potential to Regulate Intrinsic Apoptosis Signaling Pathway in Human Metastatic Ovarian Cancer. *Life Sci.* **2022**, *309*, 121022. [CrossRef]
84. Mahalanobish, S.; Kundu, M.; Ghosh, S.; Das, J.; Sil, P.C. Fabrication of Phenyl Boronic Acid Modified PH-Responsive Zinc Oxide Nanoparticles as Targeted Delivery of Chrysin on Human A549 Cells. *Toxicol. Rep.* **2022**, *9*, 961–969. [CrossRef]

85. Sandhya, J.; Kalaiselvam, S. UV Responsive Quercetin Derived and Functionalized CuO/ZnO Nanocomposite in Ameliorating Photocatalytic Degradation of Rhodamine B Dye and Enhanced Biocidal Activity against Selected Pathogenic Strains. *J. Environ. Sci. Health Part A Toxic Hazard. Subst. Environ. Eng.* **2021**, *56*, 835–848. [CrossRef]
86. Nisar, M.F.; Yousaf, M.; Saleem, M.; Khalid, H.; Niaz, K.; Yaqub, M.; Waqas, M.Y.; Ahmed, A.; Abaid-Ullah, M.; Chen, J.; et al. Development of Iron Sequester Antioxidant Quercetin@ZnO Nanoparticles with Photoprotective Effects on UVA-Irradiated HaCaT Cells. *Oxidative Med. Cell. Longev.* **2021**, *2021*, 6072631. [CrossRef] [PubMed]
87. Ahmed, E.S.A.; Mohamed, H.E.; Farrag, M.A. Luteolin Loaded on Zinc Oxide Nanoparticles Ameliorates Non-Alcoholic Fatty Liver Disease Associated with Insulin Resistance in Diabetic Rats via Regulation of PI3K/AKT/FoxO1 Pathway. *Int. J. Immunopathol. Pharmacol.* **2022**, *36*, 03946320221137435. [CrossRef]
88. Salaheldin, T.A.; Adhami, V.M.; Fujioka, K.; Mukhtar, H.; Mousa, S.A. Photochemoprevention of Ultraviolet Beam Radiation-Induced DNA Damage in Keratinocytes by Topical Delivery of Nanoformulated Epigallocatechin-3-Gallate. *Nanomed. Nanotechnol. Biol. Med.* **2022**, *44*, 102580. [CrossRef] [PubMed]
89. Zhu, Y.; Zheng, T.; Wen, L.-M.; Li, R.; Zhang, Y.-B.; Bi, W.-J.; Feng, X.-J.; Qi, M.-C. Osteogenic Capability of Strontium and Icariin-Loaded TiO_2 Nanotube Coatings in Vitro and in Osteoporotic Rats. *J. Biomater. Appl.* **2021**, *35*, 1119–1131. [CrossRef] [PubMed]
90. León-Gutiérrez, G.; Elste, J.E.; Cabello-Gutiérrez, C.; Millán-Pacheco, C.; Martínez-Gómez, M.H.; Mejía-Alvarez, R.; Tiwari, V.; Mejía, A. A Potent Virucidal Activity of Functionalized TiO_2 Nanoparticles Adsorbed with Flavonoids against SARS-CoV-2. *Appl. Microbiol. Biotechnol.* **2022**, *106*, 5987–6002. [CrossRef]
91. Unnikrishnan, B.; Gultom, I.S.; Tseng, Y.-T.; Chang, H.-T.; Huang, C.-C. Controlling Morphology Evolution of Titanium Oxide–Gold Nanourchin for Photocatalytic Degradation of Dyes and Photoinactivation of Bacteria in the Infected Wound. *J. Colloid Interface Sci.* **2021**, *598*, 260–273. [CrossRef]
92. Negrescu, A.-M.; Mitran, V.; Draghicescu, W.; Popescu, S.; Pirvu, C.; Ionascu, I.; Soare, T.; Uzun, S.; Croitoru, S.M.; Cimpean, A. TiO2 Nanotubes Functionalized with Icariin for an Attenuated In Vitro Immune Response and Improved In Vivo Osseointegration. *J. Funct. Biomater.* **2022**, *13*, 43. [CrossRef]
93. Thakur, N.; Kundu, M.; Chatterjee, S.; Singh, T.A.; Das, J.; Sil, P.C. Morin-Loaded Nanoceria as an Efficient Nanoformulation for Increased Antioxidant and Antibacterial Efficacy. *J. Nanopart. Res.* **2022**, *24*, 176. [CrossRef]
94. Chong, C.L.; Fang, C.M.; Pung, S.Y.; Ong, C.E.; Pung, Y.F.; Kong, C.; Pan, Y. Current Updates On the In Vivo Assessment of Zinc Oxide Nanoparticles Toxicity Using Animal Models. *BioNanoScience* **2021**, *11*, 590–620. [CrossRef]
95. Jin, M.; Li, N.; Sheng, W.; Ji, X.; Liang, X.; Kong, B.; Yin, P.; Li, Y.; Zhang, X.; Liu, K. Toxicity of Different Zinc Oxide Nanomaterials and Dose-Dependent Onset and Development of Parkinson's Disease-like Symptoms Induced by Zinc Oxide Nanorods. *Environ. Int.* **2021**, *146*, 106179. [CrossRef] [PubMed]
96. Musial, J.; Krakowiak, R.; Mlynarczyk, D.T.; Goslinski, T.; Stanisz, B.J. Titanium Dioxide Nanoparticles in Food and Personal Care Products—What Do We Know about Their Safety? *Nanomaterials* **2020**, *10*, 1110. [CrossRef]
97. Borghetti, G.S.; Carini, J.P.; Honorato, S.B.; Ayala, A.P.; Moreira, J.C.F.; Bassani, V.L. Physicochemical Properties and Thermal Stability of Quercetin Hydrates in the Solid State. *Thermochim. Acta* **2012**, *539*, 109–114. [CrossRef]
98. Wang, W.; Sun, C.; Mao, L.; Ma, P.; Liu, F.; Yang, J.; Gao, Y. The Biological Activities, Chemical Stability, Metabolism and Delivery Systems of Quercetin: A Review. *Trends Food Sci. Technol.* **2016**, *56*, 21–38. [CrossRef]
99. Kandemir, K.; Tomas, M.; McClements, D.J.; Capanoglu, E. Recent Advances on the Improvement of Quercetin Bioavailability. *Trends Food Sci. Technol.* **2022**, *119*, 192–200. [CrossRef]

nanomaterials

MDPI

Review

Current Applications of Liposomes for the Delivery of Vitamins: A Systematic Review

Matheus A. Chaves [1,2,†], Letícia S. Ferreira [1,†], Lucia Baldino [3,*], Samantha C. Pinho [1] and Ernesto Reverchon [3]

1 Laboratory of Encapsulation and Functional Foods (LEnAlis), Department of Food Engineering, School of Animal Science and Food Engineering, University of São Paulo, Av. Duque de Caxias Norte, 225, Pirassununga 13635900, SP, Brazil; matheus.chaves@usp.br (M.A.C.); leticia2.ferreira@usp.br (L.S.F.); samantha@usp.br (S.C.P.)

2 Laboratory of Molecular Morphophysiology and Development (LMMD), Department of Veterinary Medicine, School of Animal Science and Food Engineering, University of São Paulo, Av. Duque de Caxias Norte, 225, Pirassununga 13635900, SP, Brazil

3 Department of Industrial Engineering, University of Salerno, Via Giovanni Paolo II, 132, 84084 Fisciano, Italy; ereverchon@unisa.it

* Correspondence: lbaldino@unisa.it; Tel.: +39-089964356

† These authors contributed equally to this work.

Abstract: Liposomes have been used for several decades for the encapsulation of drugs and bioactives in cosmetics and cosmeceuticals. On the other hand, the use of these phospholipid vesicles in food applications is more recent and is increasing significantly in the last ten years. Although in different stages of technological maturity—in the case of cosmetics, many products are on the market—processes to obtain liposomes suitable for the encapsulation and delivery of bioactives are highly expensive, especially those aiming at scaling up. Among the bioactives proposed for cosmetics and food applications, vitamins are the most frequently used. Despite the differences between the administration routes (oral for food and mainly dermal for cosmetics), some challenges are very similar (e.g., stability, bioactive load, average size, increase in drug bioaccessibility and bioavailability). In the present work, a systematic review of the technological advancements in the nanoencapsulation of vitamins using liposomes and related processes was performed; challenges and future perspectives were also discussed in order to underline the advantages of these drug-loaded biocompatible nanocarriers for cosmetics and food applications.

Keywords: vitamins; nanoencapsulation; nanodispersions; phospholipid vesicles; liposomes; cosmetics; food application

Citation: Chaves, M.A.; Ferreira, L.S.; Baldino, L.; Pinho, S.C.; Reverchon, E. Current Applications of Liposomes for the Delivery of Vitamins: A Systematic Review. *Nanomaterials* **2023**, *13*, 1557. https://doi.org/10.3390/nano13091557

Academic Editors: Andrei Honciuc and Mirela Teodorescu

Received: 7 April 2023
Revised: 27 April 2023
Accepted: 2 May 2023
Published: 5 May 2023

1. Why Encapsulating Vitamins for Food and Cosmetics Application?

Vitamins are organic compounds required for metabolic processes, and not produced by the human body in sufficient amounts. Each vitamin has specific functions in the body and cannot be replaced by any other substance [1]. Having different chemical structures, vitamins can be classified in two main groups: fat-soluble and water-soluble compounds. Thirteen main vitamins have been identified; they are characterized by different ways of action and different beneficial roles in the body. In addition, vitamins play a key role in transforming energy and regulating the metabolism pathways of the human body [2].

Vitamins A, D, E, and K are compounds that are soluble in fat, meaning that they need a certain amount of fat in the diet to be properly absorbed by the body. In contrast, group B vitamins and vitamin C are soluble in water and cannot be stored in the body, requiring daily intake. Figure 1 provides a schematic representation of the main differences between fat-soluble (hydrophobic) and water-soluble (hydrophilic) vitamins. Vitamins are typically added to food products to fortify them, to replace vitamin losses during processing, or to serve as antioxidants and natural colorants. Table 1 summarizes the chemical structure, sources, and ways of obtainment of the main vitamins. Vitamins are also used in cosmetics for skin care, hair care, and oral health [2]. According to the European Union legislation, vitamins A, E, K, B, and C can all be used in cosmetics as ingredients, except for vitamin D_3 (cholecalciferol), which is restricted. However, in the USA and Japan, there are no such restrictions [2].

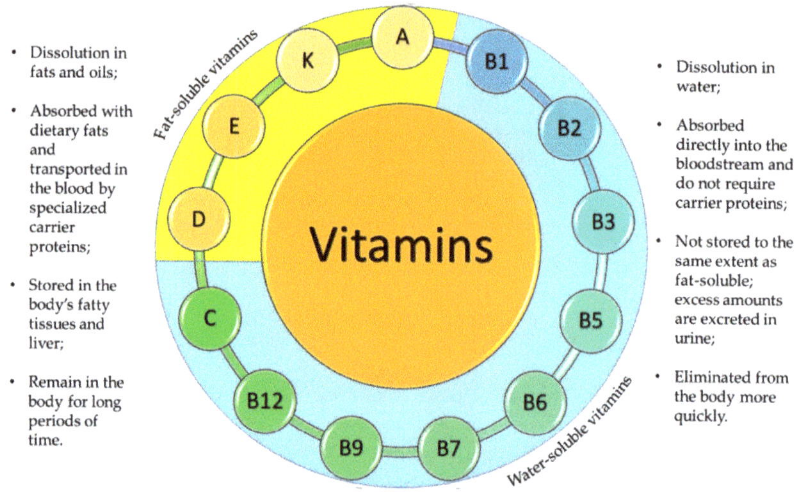

Figure 1. Schematic representation of the main differences between fat-soluble and water-soluble vitamins. Fat-soluble vitamins (A, D, E, and K) are stored in the body's fatty tissues and liver, whereas water-soluble vitamins (B-complex and C) are not. The differences in solubility, absorption, storage, and excretion can affect the way these vitamins act in the body.

In order to improve their effectiveness, stability, and palatability, vitamins can be encapsulated into delivery carrier systems. This can be especially beneficial for individuals who struggle to obtain adequate amounts of vitamins through their diet or who have specific nutrient deficiencies that require a targeted delivery of these essential nutrients. Encapsulation offers a way to protect vitamins from degradation, to enhance their bioavailability, and to mask any unpleasant taste or odor. This technique refers to the process of enclosing one or more active ingredients, such as vitamins, in a protective carrier, such as liposomes, micelles, or nanoparticles. The encapsulated active ingredients are typically surrounded by a protective layer or membrane that can help shield them from degradation, improve their solubility, target them to specific tissues or cells, and enhance their bioavailability and efficacy. Overall, encapsulation offers several advantages over other delivery methods, such as improved stability, protection from external factors, controlled release, and targeted delivery, making it a promising approach for the development of more effective and efficient delivery systems for vitamins and other bioactive compounds.

Table 1. Chemical structure, ways of obtainment, and main sources of vitamins in foods.

Vitamin	Chemical Structure	Way of Obtainment	Food Sources
A (retinol)	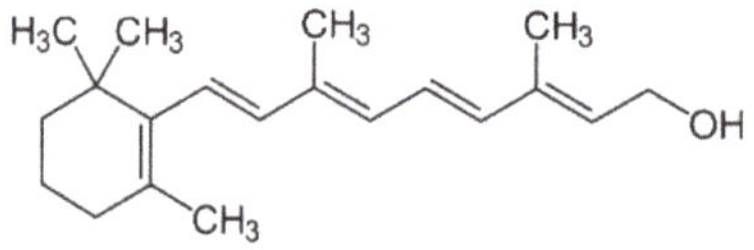	Intake of dietary sources or vitamin supplements	Fish, meat, eggs, whole milk, carrots, spinach, and mango

Table 1. *Cont.*

Vitamin	Chemical Structure	Way of Obtainment	Food Sources
B_1 (thiamine)	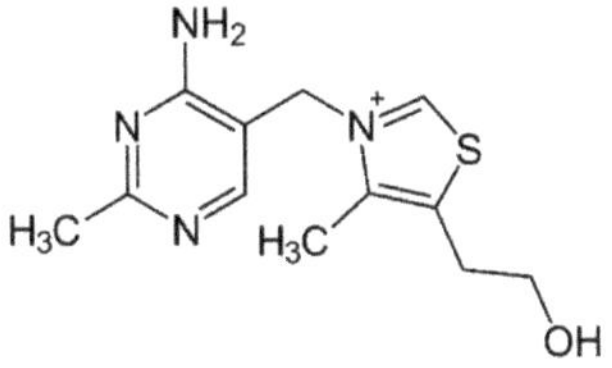	Intake of dietary sources or vitamin supplements	Brewer's dried yeast, pork, lamb, beef, poultry, whole grains, nuts, vegetables, and legumes
B_2 (riboflavin)	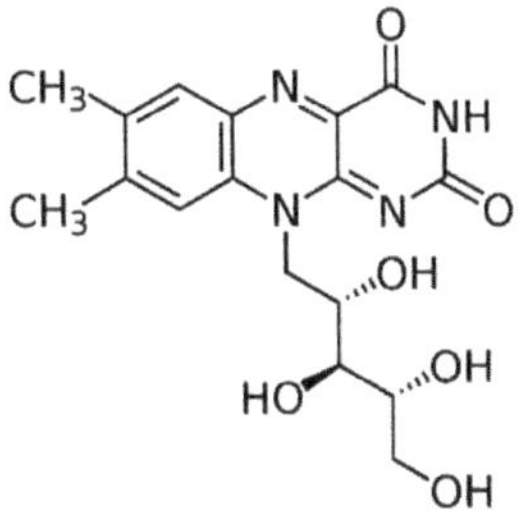	Plants and animal cells	Milk and milk products, meat, eggs, and leafy green vegetables
B_3 (niacin)	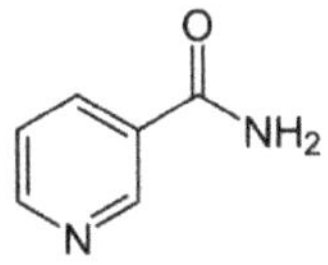	Intake of dietary sources or vitamin supplements	Yeast, liver, poultry, lean meats, nuts, and legumes
B_5 (pantothenic acid)	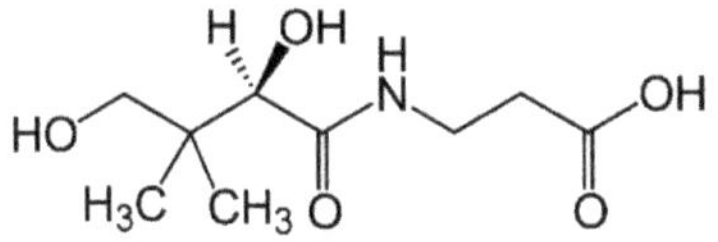	Intake of dietary sources or cosmetic products	Organ of animals, eggs, milk, vegetables, legumes, and whole grain cereals
B_6 (pyridoxine)	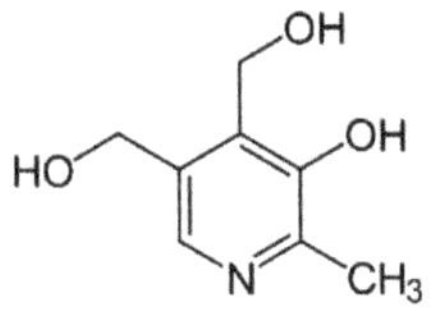	Intake of dietary sources or vitamin supplements	Fruits, vegetables, and grains
B_7 (biotin)	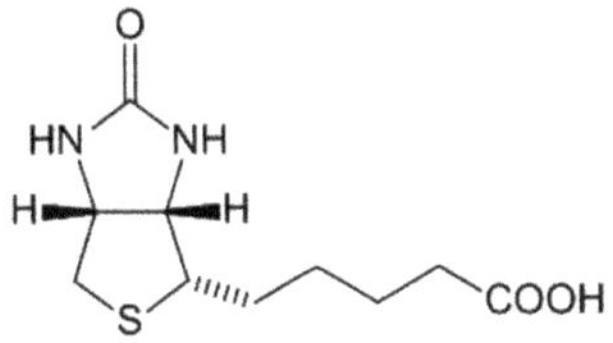	Intake of dietary sources or vitamin supplements	Chicken liver, beef liver, egg yolk, peanuts, sunflower seeds, almonds, salmon, and pork chop
B_9 (folacin)	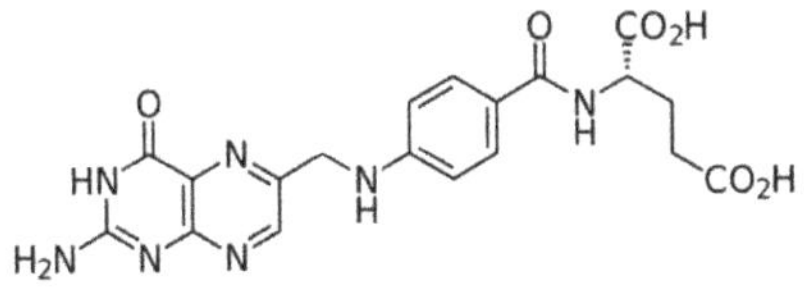	Intake of dietary sources, fortified foods or vitamin supplements	Peanuts, sunflower seeds, lentils, chickpeas, asparagus, spinach, chicken liver, calf liver, cheese, hazelnuts, and avocados

Table 1. *Cont.*

Vitamin	Chemical Structure	Way of Obtainment	Food Sources
B_{12} (cobalamin)	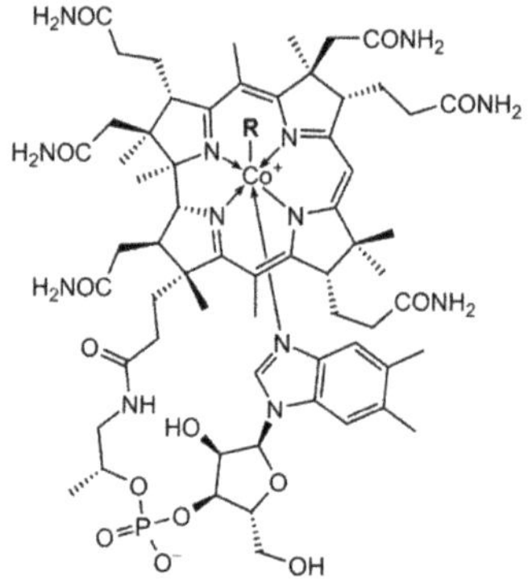 **R** = 5'-deoxyadenosyl, CH_3, OH, CN	Intake of dietary sources or vitamin supplements	Milk and dairy products, eggs, and salmon
C (ascorbic acid)	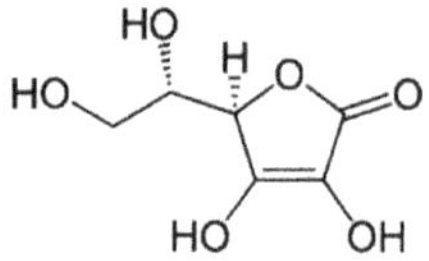	Intake of dietary sources or vitamin supplements	Citric fruits, currants, peppers, parsley, cauliflower, potatoes, sweet potatoes, broccoli, brussels sprouts, strawberries, guava, and mango
D_2 (ergocalciferol)	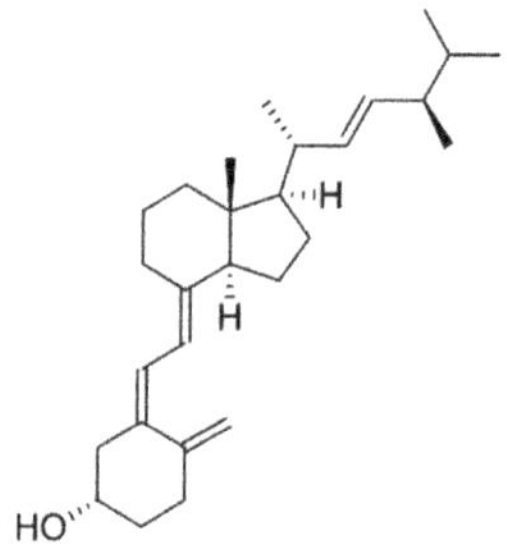	Intake as vitamin D supplement	Mushrooms (portobello, crimini, shitake)
D_3 (cholecalciferol)	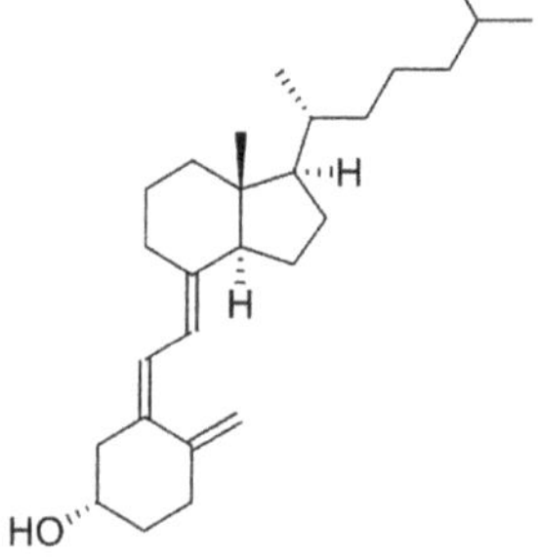	Synthesized by the human epidermis or consumed in the form of supplements or fortified foods	Fish, as salmon and sardines, butter, and eggs
E (α-tocopherol)	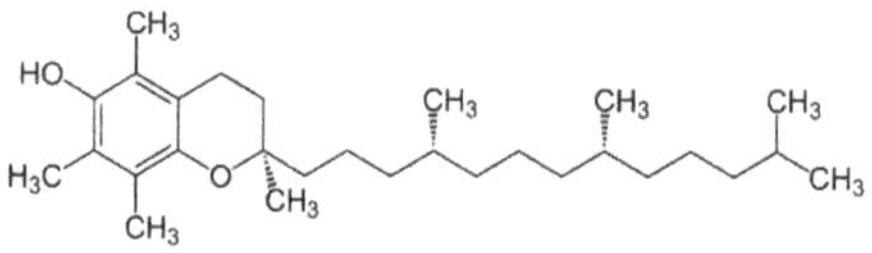	Intake of dietary sources or vitamin supplements	Vegetable oils of peanut, soya, palm, corn, safflower, sunflower, and wheat germ

Table 1. *Cont.*

Vitamin	Chemical Structure	Way of Obtainment	Food Sources
K_1 (phylloquinone)	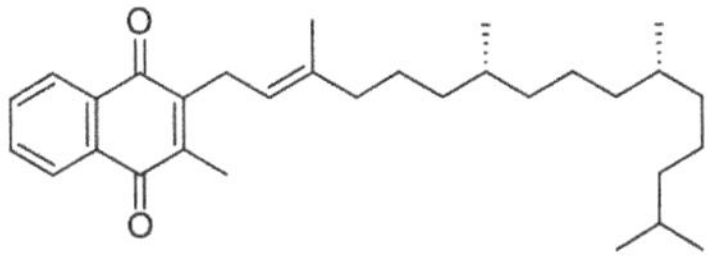	Plants	Green leafy vegetables, liver, lean meat, cow's milk, egg yolks, and whole wheat products
K_2 (menaquinone)	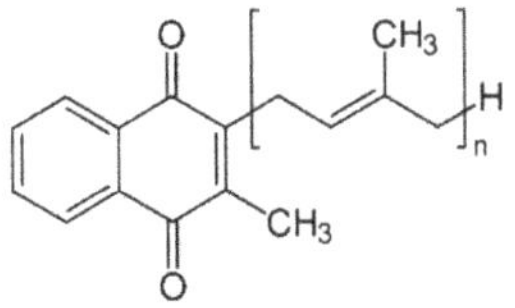	Synthesized by bacteria in the human and animal intestines	-
K_3 (menadione)	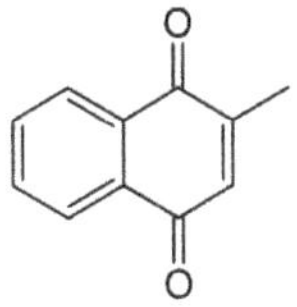	Converted to K_2 in the intestinal tract	-

1.1. Fat-Soluble Vitamins

Vitamin A can be found as retinoids (retinol, retinal, and retinoic acid) and provitamin A carotenoids (mainly β-carotene) [3]. It is important to emphasize that vitamin A is not produced endogenously. The retinol is the first precursor of two vital active metabolites: retinal, which plays an important role in the development of vision, and retinoic acid which acts as an intracellular signal that alters the transcription of a range of genes. Vitamin A is not found directly in plants; however, plants can contain carotenoids, such as β-carotene, which is transformed in vitamin A in the intestine and other body tissues. Therefore, the vitamin A supply is necessarily obtained through ingestion (dietary sources or vitamin supplements) [4]. Foods such as fish, meat (mainly liver), eggs, and whole milk are animal sources of retinol [2,3]. Fruits and vegetables such as carrots, spinach, and mango are sources of carotenoids. Vitamin A is crucial for the growth and development of children and for maintaining the immune system's function [5]. In adults, a deficiency in vitamin A can negatively impact the immune system, reproductive function, and eyesight, resulting in conditions such as night blindness [3,4,6]. However, overexposure to vitamin A can lead to harmful health effects, including teratogenicity [7]. In the cosmetic industry, vitamin A is widely used due to its ability to delay photoaging effects. Being the main bioactive for skin treatment, it promotes the regeneration of the skin aged by UV radiation, reduces wrinkles, and improves skin elasticity [2]. Vitamin A is not degraded by heat, but it is easily oxidized, and care is required during its processing. To reduce this undesired effect, antioxidants are added to vitamin A products such as, for example, vitamin E [2].

Group D vitamins are composed of ergocalciferol (vitamin D_2) and cholecalciferol (vitamin D_3). Vitamin D_3, or cholecalciferol, is a fat-soluble compound synthesized by the human epidermis by the irradiation of UV light on 7-dehydrocholesterol [8]. The precursor molecule of vitamin D is ergosterol (or 7-dehydrocholesterol), a rigid structure that is inserted by the body when absorbed by the lipid layer of the plasmatic membrane. The production of provitamin D occurs after solar incidence on the aromatic ring of ergosterol. The structure, then, becomes less rigid, promoting an increase in its permeability and, thus, allowing the incorporation of numerous ions into its interior, including calcium. It can be synthesized by the human epidermis or consumed in the form of supplements or fortified foods. A few foods naturally contain vitamin D in their composition: fish, such as salmon and sardines, butter, and eggs being the main sources. Consuming vitamin D_3

can provide several benefits, including improving calcium absorption in the intestine and maintaining proper levels of calcium in the bloodstream, which helps prevent conditions such as osteoporosis [9]. Unfortunately, a significant portion of the urban population today is deficient in vitamin D_3, primarily due to a lack of sun exposure and the loss of 7-hydrocholesterol reserves in the epidermis as a result of aging [10]. Vitamin D deficiency can lead to bone diseases and other health problems, such as cancer, asthma, arthritis, hypertension, osteoporosis, and cardiovascular issues. Symptoms of a deficiency may include bone pain and muscle weakness [11]. On the other hand, possible adverse effects regarding the excessive amounts of vitamin D can cause calcium buildup in the blood, leading to nausea, vomiting, and muscle weakness. Long-term overdoses can result in kidney damage [8]. Vitamin D is used in the cosmetic industry as it prevents photo-damage, wrinkles, and other morphological skin changes [2]. Some studies indicated its topical application for the treatment of psoriasis [12]. However, it has an adverse effect on calcium metabolism and limits its use for topical applications [13].

Vitamin E is a term used to describe several compounds, including tocopherols and tocotrienols, which are differentiated by the prefixes α, β, γ, and δ. Among them, α-tocopherol is the most common and has a higher bioavailability than other forms of vitamin E [14]. Vegetable oils, such as peanut, soy, palm, corn, safflower, sunflower, and wheat germ are the most important dietary sources of vitamin E. This vitamin plays an essential role in several physiological functions, including acting as an antioxidant, regulating immunity, and providing anti-inflammatory and neuroprotective benefits [15]. Moreover, vitamin E helps protect body tissues from oxidation caused by metabolic processes and external agents while also assisting in the synthesis of vitamin A [2]. High doses of vitamin E, however, can lead to bleeding, particularly in individuals taking blood-thinning medications [14]. However, its incorporation in foods can be challenging, as it is extremely sensitive to high temperatures, light, oxygen, and alkaline conditions, and has low solubility in water [15]. Encapsulation techniques can help to overcome these obstacles allowing the application of vitamin E in foods, cosmetics, and nutraceuticals. In the cosmetic industry, vitamin E is generally used as an antioxidant for the skin, aiding in softening and promoting hydration [2].

Vitamin K can exist in three forms: vitamin K_1 (phylloquinone, phytonadione, phytomenadione), vitamin K_2 (menaquinone), or vitamin K_3 (menadione). Vitamin K_1 is commonly found in plants, K_2 is synthesized by bacteria in the human and animal intestines, and K_3 is a synthetic compound that is converted to K_2 in the intestinal tract [2,16,17]. Green leafy vegetables, such as spinach, kale, broccoli, and cauliflower, are excellent sources of this vitamin, which is also found in smaller amounts in liver, lean meat, cow's milk, egg yolks, and whole wheat products [2]. Although structurally different, both vitamins K_1 and K_2 can act as cofactors for the enzyme gamma-glutamylcarboxylase, with hepatic and extrahepatic activity. Additionally, vitamin K_2 plays a vital role in regulating osteoporosis, atherosclerosis, cancer, and inflammatory diseases, with no risk of negative side effects or overdosing [16,17]. Additionally, vitamin K is effective in treating dark circles and bruises on the face, and its application to reduce the effects of bruising after certain dermatological procedures has also been studied [18]. Careful attention must be paid to excessive amounts of vitamin K, as it can interfere with the effectiveness of blood-thinning medications and increase the risk of blood clots [16].

1.2. Water-Soluble Vitamins

Group B vitamins are thiamine (B_1), riboflavin (B_2), niacin (B_3), pantothenic acid (B_5), pyridoxine (B_6), biotin (B_8), folacin (B_9), and cobalamin (B_{12}). Vitamin B_1 can be found in small amounts in brewer's dried yeast, pork, lamb, beef, poultry, whole grains, nuts, vegetables, and legumes. It is important for carbohydrate breakdown, nerve and muscle function, and healthy skin [2]. Vitamin B_2 can be sourced from various food items such as milk, dairy products, meat, eggs, and leafy green vegetables. It plays a crucial role in releasing energy from food and promoting the development of healthy skin, vision quality, and growth. Vitamin B_3 is found in yeast, liver, poultry, lean meats, nuts, and legumes, and is used for the treatment of lipid disorders and cardiovascular diseases, which are essential for growth and hormone synthesis [19]. Pantothenic acid is necessary for the release of energy from food for the production of antibodies and healthy growth, and is present in almost every type of food, and particularly abundant in yeast and animal organs. Vitamin B_{12} is essential for DNA synthesis and red blood cell production, and its deficiency can result in anemia, cognitive impairment, and neurological abnormalities. Vitamins B_5 and B_{12} are commonly used in skin and hair care products due to their moisturizing, anti-inflammatory, and wound healing properties [20]. Main potential risks regarding the overconsumption of vitamin B may lead to skin flushing, itching (vitamin B_3), numbness, and tingling sensations (vitamin B_6), and nerve damage and anemia (vitamin B_9) [19].

Vitamin C, also known as ascorbic acid, is a commonly used ingredient in both cosmetic and pharmaceutical products due to its powerful antioxidant properties. However, incorporating vitamin C into products poses a significant challenge as its stability must be maintained and delivery to the desired site improved [21]. Vitamin C is naturally occurring in a variety of fruits and vegetables, such as citrus fruits, currants, peppers, parsley, cauliflower, potatoes, sweet potatoes, broccoli, Brussels sprouts, strawberries, guava, and mangoes. In addition to its use as an antioxidant to prevent food and beverage spoilage, vitamin C is essential for the production of collagen, connective tissue, and protein fibers, which provide strength to teeth, gums, muscles, blood vessels, and skin. It also plays an important role in the immune system by aiding white blood cells in fighting infections and facilitating iron absorption within the body [2,21]. Overconsumption of vitamin C can lead to diarrhea, nausea, and abdominal cramps. Long-term excess consumption can lead to kidney stones [21].

2. Why Choose Liposomes for the Encapsulation of Vitamins?

In 1965, it was reported, for the first time, that phospholipid molecules were able to instantaneously form closed bilayer vesicles in aqueous media due to the amphiphilic nature of phospholipids [22]. Liposomes are vesicular structures formed by one or more phospholipid bilayers that encapsulate part of the aqueous medium in which they are dispersed [23]. Their average diameters range from 20 nm to several microns. Phospholipids are the main constituents of vesicles, being an amphiphilic molecule in which the hydrophilic polar head groups are oriented towards the aqueous phase and the hydrophobic non-polar hydrocarbon tails are oriented towards each other in an ordered bilayer structure. A wide variety of phospholipids can be used for the production of liposomes, such as eggs, soy, and milk, which are natural and safe sources. The most used phospholipid for the production of liposomes in the food and cosmetic industry is phosphatidylcholine (PC). In addition to PC, phospholipids such as lysophosphatidylcholine (LPC), phosphatidylinositol (PI), phosphatidylethanolamine (PE), and phosphatidylglycerol (PG) can also be used.

When introduced into water, phospholipids tend to group together to form lipid bilayers due to their insolubility in water. By providing the system with sufficient energy through external methods, such as sonication, heating, or homogenization, the negative interaction between the fatty acid molecules and water can be eliminated, allowing the bilayers to organize themselves in a favorable manner. As a result of this process, liposomes can effectively encapsulate hydrophilic compounds in their aqueous core, as well as hydrophobic compounds in the internal regions of the lipid bilayer [24].

While being formed, liposomes acquire various sizes and structural characteristics, such as the number of bilayers [25–27]. Size and number of bilayers determine the classification of liposomes in two main types: unilamellar vesicles and multilamellar vesicles, as shown in Table 2. Small unilamellar vesicles (SUVs) have a size ranging from 20 to 200 nm and only contain one bilayer membrane. Large unilamellar vesicles (LUVs) are larger than 200 nm and have a single bilayer membrane. Giant unilamellar vesicles (GUVs) have a size exceeding 1 µm. Multilamellar vesicles (MLVs) contain several concentrically arranged vesicles with a size between 0.5 to 5 µm, whereas multivesicular vesicles (MVVs) have smaller vesicles inside a larger vesicle [12,24]. The size characteristics of liposomes are determined by the production method and the type(s) of phospholipid(s) utilized. SUVs have a higher surface area to volume ratio compared to MLVs. This can result in faster release kinetics and better cellular uptake due to their smaller size and increased surface area. However, they are generally less stable than MLVs, particularly under harsh environmental conditions, which can result in aggregation and fusion with other liposomes. On the other hand, MLVs have a higher loading capacity due to their multiple lamellar layers. In terms of in vitro performance, the choice between SUVs and MLVs will depend on the specific application and the desired characteristics of the liposomes. For example, if rapid drug release and cellular uptake is important, SUVs may be preferred. If long-term stability and high loading capacity are required, MLVs may be preferred [23].

Table 2. Classification of liposomes based on size and lamellarity.

Type of Liposome	Abbreviation	Characteristic	Diameter	Schematic Representation
Small unilamellar vesicles	SUV	Small unilamellar vesicles	20 to 200 nm	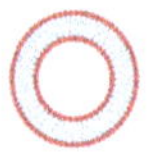
Large unilamellar vesicles	LUV	Unilamellar liposomes with average diameters higher than SUV	Above 200 nm	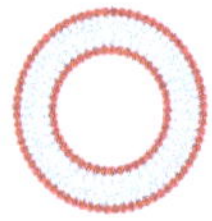
Giant unilamellar vesicles	GUV	Unilamellar liposomes with average diameters higher than LUV	Higher than 1 μm	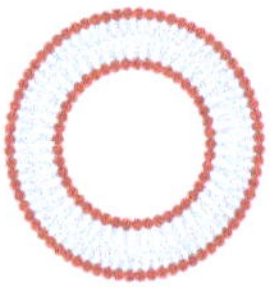
Multilamellar vesicles	MLV	Several concentrically arranged vesicles	0.5 to 5 μm	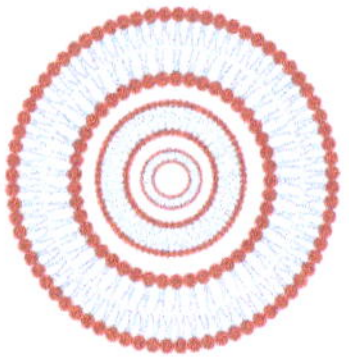
Multivesicular vesicles	MVV	Smaller vesicles within a vesicle of large size	Higher than 1 μm	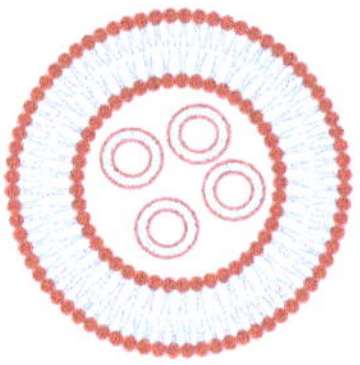

In general, liposomes are one of the most commonly commercialized lipid carriers used for nutraceutical purposes. They are also increasingly studied for their potential to be incorporated into foods for functional applications [28,29]. In the vitamin field, other lipid systems have also been used for encapsulation purposes, such as emulsions, micro/nanoemulsions and solid lipid particles [24]. In this context, some advantages can be pointed out regarding the use of liposomes over these other systems: (i) biocompatibility: the phospholipids used for liposomes production are similar to the phospholipids found in cell membranes and are, thus, less likely to cause adverse reactions; (ii) versatility: liposomes can encapsulate all types of vitamins, hydrophilic, lipophilic, or even both in the same structure due to the amphiphilicity of their structure; (iii) targeted delivery: liposomes can be modified with targeting ligands, such as antibodies or peptides, to enhance their specificity for a particular cell type or tissue; (iv) controlled release: these vesicles can be engineered to release their contents at a specific time or location, improving the therapeutic effect of the vitamins and reducing the need for frequent dosing; and (v) ease of preparation: liposomes can be easily prepared using simple techniques, which makes them a cost-effective drug delivery system compared to others previously mentioned [24,28].

However, despite their versatility, liposomes are physicochemically unstable. This is due to the fact that the lipids present in their structure can undergo natural degradation through oxidation or hydrolysis, or even because the particles can form agglomerates. Liposomes originally show repulsive forces between their particles that provide a certain physical stability, but external factors, such as high temperatures or pH changes, can affect their structure and change the permeability of the bilayer, causing the release of the encapsulated compound or the formation of agglomerates. In order to reduce these undesired effects, a possible solution is the coating of liposomes with biopolymers that can increase their physical stability through steric and electrostatic factors, thus creating a hybrid system. Among the biopolymers used for the coating of liposomes, starches, gelatin, proteins, cellulose, pectin, and chitosan can be mentioned. Other possibilities to increase their overall stability include the use of lyophilization, the incorporation of hydrogenated phospholipids in the lipid bilayers, and the use of cross-linking agents [24,28].

3. Methods for the Production of Vitamin-Loaded Liposomes

Generally, a successful encapsulation of bioactive molecules in liposomes, with desired and specific size and structure, depends on several factors, such as the correct choice of the main lipid, the affinity between the liposome and the bioactive of interest, and, most of all, the production method. The latter is considered to be extremely important when encapsulating vitamins for several reasons that may include: (i) the efficiency of encapsulation, as some methods may result in higher encapsulation efficiencies than others; (ii) the stability of the vitamin, as certain methods may be less harmful to them, which is particularly important for sensitive vitamins that can be easily degraded by heat, light, or exposure to oxygen; (iii) the control of particle size and distribution, which are directly related to the release rate of the vitamin from the encapsulating material and also to its bioaccessibility and bioavailability; and (iv) the process scalability, to ensure both consistent quality and efficient production.

Conventional methods of liposome production for vitamin encapsulation are still in the spotlight for many researchers due to their simplicity, recognition, and wide range of applications. On the other hand, novel methods have arisen due to their advantages over the conventional ones, such as higher encapsulation efficiency, reduced toxicity, versatility, and scalability. Overall, the production method used for encapsulating vitamins plays a critical role in ensuring the stability, efficacy, and quality of the final product. Figure 2 presents all the methods that will be discussed in this review, whereas Table 3 summarizes the advantages and disadvantages of each one.

Table 3. Advantages and disadvantages related to the methods for liposome production.

Production Methods		Advantages	Disadvantages
Conventional techniques	Thin-film hydration method	✓ Easy to scale up ✓ Straightforward approach ✓ Compatibility with different lipids ✓ Simple method with low cost	✓ Multiple steps ✓ Post-processing is required for size reduction ✓ Water-soluble bioactives may exhibit low EE% ✓ Removal of organic solvents is needed ✓ Sterilization is required

Table 3. *Cont.*

Production Methods		Advantages	Disadvantages
	Reverse-phase evaporation method	✓ Simple and rapid process ✓ High encapsulation efficiency	✓ Time-consuming ✓ Large amounts of organic solvents are needed ✓ Post-processing is required for size reduction ✓ Sterilization is required
	Injection methods	✓ Easy to scale up ✓ Ethanol injection is simple, rapid and reproducible ✓ Ether injection forms a concentrated liposome with high EE%	✓ Removal of solvents is needed ✓ Post-processing is required for size reduction ✓ Macromolecules may inactivate in presence of ethanol
	Detergent removal method	✓ Simple process ✓ Suitable particle size distribution	✓ Low concentration of liposomes ✓ Low EE% for hydrophobic molecules ✓ Time-consuming
	Double emulsion method	✓ Biodegradability ✓ Versatility ✓ Suitable particle size distribution	✓ Multiple steps ✓ Possible leakage of hydrophilic bioactives during production ✓ Post-processing is required for size reduction
	Hydration of proliposomes	✓ Versatility ✓ Easy to scale up ✓ Higher stability when in dried form	✓ Heterogeneity of vesicles ✓ Production variability ✓ Limited loading capacity
Novel techniques	Heating method	✓ Simple and rapid process ✓ No organic solvents ✓ Sterilization is not needed	✓ Multiple steps ✓ Possible degradation of thermosensitive molecules ✓ Post-processing is required for size reduction
	Membrane contactor method	✓ Simple and fast process ✓ Suitable particle size distribution ✓ No organic solvents	✓ Possibility of clogging the pores ✓ High temperature ✓ Membrane/filter fragility ✓ Sterilization is required
	Electroformation	✓ Rapid process ✓ High purity ✓ Adaptable conditions ✓ Formation of GUVs with high EE%	✓ Unknown underlying mechanism ✓ High cost of electrodes ✓ Only suitable for very low ionic strength
	Nanoprecipitation	✓ Simple method ✓ Use of biocompatible solvent	✓ Post-processing is required for size reduction ✓ Sterilization/aseptic processing is required
	Microfluidic method	✓ Simple method ✓ Suitable particle size distribution	✓ Difficulty to remove the organic solvents ✓ Unsuitable for bulk production ✓ High cost of microfluidic channels
	Dual asymmetric centrifugation	✓ Easy to operate ✓ Reproducibility ✓ Suitable particle size distribution ✓ High EE% for water-soluble bioactives ✓ No organic solvents	✓ High amount of phospholipids is required ✓ Batch scale production

Table 3. *Cont.*

Production Methods		Advantages	Disadvantages
	Crossflow filtration detergent depletion	✓ Rapid process ✓ Suitable particle size distribution ✓ Sterilization is not required ✓ Water filtrate can be recycled	✓ Use of detergents ✓ Difficulty to scale up
	Inkjet method	✓ Suitable particle size distribution ✓ High EE%	✓ Specific equipment is needed ✓ Removal of ethanol is required
	Supercritical technologies	✓ Versatility ✓ Small number of unit operations ✓ Low solvent residue ✓ Used for bioactives with low solubility ✓ High efficiency ✓ Environmentally friendly ✓ Suitable particle size distribution ✓ Reproducibility	✓ High pressure ✓ Possible agglomeration of particles at the end of the process ✓ May involve nozzle blockage ✓ May involve high cost for implementation

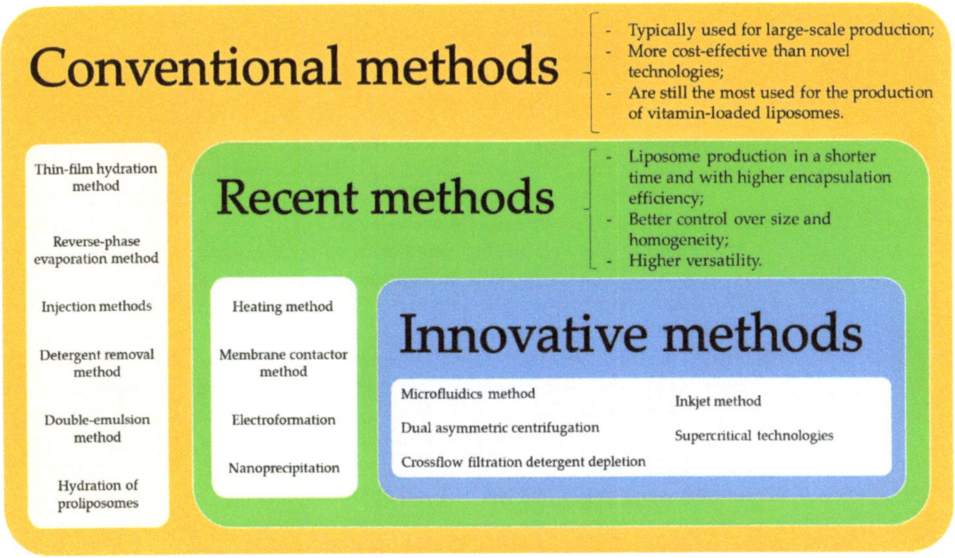

Figure 2. Overview of the conventional and novel methods for liposome production covered in this review. Advantages of each group (conventional vs. recent/innovative) are summarized within the braces.

3.1. Conventional Methods

In the conventional methods, five main steps are basically followed: (i) lipids are dissolved in an organic solvent (ethanol, ether, chloroform, dichloromethane); (ii) the organic solvent is removed using evaporation, rotary evaporation, or distillation; (iii) the resultant lipid layer is hydrated using an aqueous medium (distilled water, buffer solutions, serum-containing media, or physiological saline solutions) and agitated; (iv) vesicles are analyzed and eventually treated by downsizing steps (see Supplementary Material for a detailed description of these methods), depending on the liposome final use; and (v) post-formation processing (purification or sterilization) is carried out to increase the stability of the liposomes [30]. Vitamins must be dissolved in the liquid medium in which they are easily solubilized, i.e., in the organic solvent if lipophilic (A, D, E, and K) or in the hydration medium if hydrophilic (complex B vitamins and C) vitamins are selected.

3.1.1. Thin-Film Hydration (TFH) Method (Bangham Method)

The thin-film hydration technique, also referred as the Bangham method due to the British biochemist that first described it in 1965, is the simplest method used for the preparation of MLVs. It is mostly used for the encapsulation of lipophilic molecules and is based on the production of a thin lipid film by the evaporation at 45–60 °C of an organic solvent from a lipid solvent solution during flask rotation under vacuum and its further hydration using an aqueous media. When small volumes of liposomes are desired, the organic solvent can be dried by using argon steam or dry nitrogen in a fume hood [31]. Hydration processes should be performed above the phase transition temperature of the lipids (e.g., 60–70 °C) for a duration of at least 1–2 h [30]. In addition to the ability to encapsulate both hydrophilic and hydrophobic molecules, the TFH method presents some limitations, such as low encapsulation efficiencies and batch-to-batch variability. The production of liposomes using TFH method is usually followed by a downsizing process to obtain SUVs.

Bi et al. [32] used egg PC, cholesterol (Chol), and vitamin D_3 (3:1:1 *w*/*w*) to produce vitamin D_3-loaded MLVs by the TFH method followed by high-pressure homogenization. Obtained liposomes presented a particle size of 169 nm and an encapsulation efficiency (EE%) of 62%. Vitamin B_{12}-loaded liposomes were produced by Andrade et al. [33] by the TFH technique followed by ultrasonication, using 1,2-distearoyl-sn-glycero-3-phosphocholine (DSPC), Chol, polyethylene glycol 2000 (PEG200), and amine 1,2-distearoyl-sn-glycero-3-phosphoethanolamine-N-[amino(polyethylene glycol)-2000] (DSPE-PEG2000) in a 52:45:3:0.06 molar ratio. LUVs showed sizes of 116 nm and an EE% of 14%. The authors stated that the low EE% was due to the high hydrophilicity of vitamin B_{12}. Campani et al. [34] produced a liposome-based formulation loaded with vitamin K_1 by the TFH technique followed by extrusion through polycarbonate membranes with decreasing porosity (400, 200, and 100 nm). Lipid films were prepared using soy phosphatidylcholine (SPC) and were rehydrated by PBS at pH 7.4 or aqueous solution at 0.01% *w*/*v* of benzalkonium chloride. Samples produced in these conditions were all in the nanometric range (131–147 nm) and presented vitamin K_1 encapsulation ratios of 3.4–154.0 μg VK1/mg SPC. Vitamin E-loaded nanoliposomes were efficiently prepared by Qu et al. [35] by using egg lecithin, cholesterol, sodium deoxycholate, and vitamin E, in a 5.8:1:1.1:1.8 mass ratio, by the TFH method followed by sonication and extrusion. The obtained vesicles presented an average hydrodynamic diameter of 231 nm, EE% of 97%, a narrow size distribution (polydispersity index, PDI = 0.217) and high zeta potential (−52.4 mV).

3.1.2. Reverse-Phase Evaporation (RPE) Method

This technique was first described by Szoka and Papahadjopoulos in 1978 [36] and is generally used to encapsulate large amounts of hydrophilic bioactives with high encapsulation efficiencies [37]. Similar to the TFH method, lipids are first dissolved in an organic solvent, such as diethyl ether, chloroform, isopropyl ether, or a mixture of two solvents, in order to form inverted micelles [38]. After the addition of an amount of aqueous phase, water-in-oil (W/O) microemulsions are formed due to the rearrangement of lipids at the interface between water and oil. During this step, a large amount of the aqueous phase is encapsulated within the microemulsion, as well as the hydrophilic molecules. The organic solvent can be slowly removed using a rotary evaporator under vacuum until the conversion of the micelles to a semi-solid viscous gel-like structure is obtained. The gradual removal of the solvent favors the disruption of the gel and promotes the formation of LUVs [30]. As discussed for the TFH method, a downstream step, such as sonication or extrusion, is required to reduce the average size of liposomes and to obtain a narrow size distribution [39]. The main drawbacks related to this technique include: (i) the use of a large amount of organic solvents and the presence of residual solvent at the end of the process; (ii) the high complexity and difficulty to industrial scaling up; (iii) the unsuitability to be employed for the encapsulation of sensitive molecules due to the long-lasting contact with the organic solvent; (iv) the time-consuming process; and (v) the sterile boundary is quite hard to establish [40].

There are a few studies exploring this method for the encapsulation of vitamins, mainly due to their high degradability rates. Favarin et al. [41] employed the RPE method to encapsulate vitamin C into liposomes. In their study, the aqueous phase was composed of polysorbate 80, at pH 3.65, and vitamin C, whereas the organic phase was composed of phospholipid Lipoid® S100, cholesterol, and ethanol. The inverted micelles were formed after the addition of the aqueous phase to the organic phase under ultrasonic conditions. The mixture was, then, submitted to a slow evaporation in a rotary evaporator operating at 80 rpm and 35 °C for the removal of the organic solvent, and the formation of the gel-like structure. Liposomes were produced after the addition of the aqueous phase under agitation. To homogenize the vesicles size, samples were extruded using membranes with pores of 0.45 µm and 0.22 µm. Vitamin C-loaded liposomes showed a mean size of 160 nm and PDI of 0.23, besides a slight negative zeta potential of −7.3 mV and an EE% equal to 19%.

3.1.3. Injection Methods

The injection methods are based on the dissolution of lipids into an organic solvent (ethanol or diethyl ether) and the further injection of the resulting solution into an aqueous phase. Among all the liposome production techniques, the injection methods are suitable to operate continuously [42].

- Ethanol injection (EI) method: this method was first described by Batzri and Korn [43] and is based on the dissolution of phospholipids in ethanol and, then, the injection of the solution into a rapidly stirred aqueous phase [43]. Part of the ethanol evaporates upon contact with the aqueous phase, producing a lipid film that forms liposomes upon hydration. Vesicles are formed due to the immediate diffusion of the ethanol in the aqueous medium, leading the lipid molecules to precipitate and to form bilayered planar structures that tend to liposomes [44]. A change in the solubility of lipids leads to the spontaneous formation of vesicles that encapsulate a small volume of the aqueous phase. This method is relatively simple and produces liposomes characterized by a high entrapment efficiency of hydrophobic molecules. SUVs can be formed if proper process parameters, such as low lipid concentrations and fast rate of injection, are used. A disadvantage of EI is that the encapsulation efficiency of hydrophilic compounds is low, and the control of both size and size distribution of the resulting liposomes can be challenging. On the other hand, some advantages of this method include: (i) it is a straightforward method, which makes it a popular choice for liposome production; (ii) its high reproducibility; and (iii) its easy scale up, making it a practical choice for commercial production [44]. However, this method is limited by the need for subsequent processing to solvent evaporation and the residual content [45]. The use of the EI method to produce liposomes for food or cosmetics is usually hindered by the low encapsulation efficiency of hydrophilic bioactive molecules such as vitamin C. This occurs because hydrophilic bioactives tend to be preferentially retained in the external aqueous phase of vesicles instead of in their small aqueous core [46]. Charcosset et al. [47] developed a continuous process for the ethanol injection method coupled with membrane extrusion, in which vitamin E-loaded MLVs were produced using Lipoid® S100 as the source of phospholipids. Liposomes were stored in a vacuumed double jacketed reactor in which the ethanol residue was constantly removed. Using this method, volumes between 60 mL to 3 L of liposomes with sizes ranging from 89 to 118 nm were produced;
- Ether injection method: this method requires the dissolution of lipids in diethyl ether and, then, a slow injection of the solution into an aqueous phase under high pressure. The ether rapidly evaporates upon contact with the aqueous phase under warming, resulting in a lipid film that forms liposomes upon hydration [48]. This method results in a concentrated liposomal (LUVs) product with a narrow size distribution and high entrapment

efficiency [45]. Some differences can be pointed out between this method and EI: (i) ethanol is a polar solvent, unlike ether that is non-polar; this can affect the lipid solubility and the self-assembly of lipids in liposomes, resulting in differences in size, size distribution, and other properties of liposomes; (ii) ether injection method may be more complex and time-consuming than EI method; and (iii) different effects on the stability of lipids and liposomes as the use of ether may cause a larger lipid degradation or oxidation than the use of ethanol due to the formation of peroxides [30]. However, an advantage of using the ether injection method consists of the more efficient removal of the organic solvent from the final product [30]. For food purposes, it is generally preferable to use ethanol instead of ether. Ether is highly flammable and can pose a fire hazard, which may be a concern in a food production setting. Ethanol, on the other hand, is a safer solvent that is commonly used in food processing and is generally recognized as safe (GRAS) by the US Food and Drug Administration (FDA). Ethanol is also a polar solvent, thus, more compatible with the polar environment of aqueous foods. The ether injection method is not commonly used for the encapsulation of vitamins in liposomes, basically because of its high toxicity [49].

3.1.4. Detergent Removal (Depletion) Method

The detergent depletion method is a mild process capable of producing highly homogeneous liposomes. This method is based on the formation of mixed micelles of detergents and lipids and the further removal of the detergent to form LUVs. The size of vesicles is based on the rate at which the detergent is removed from the formulation and the initial detergent to phospholipid ratio [50]. Detergents, such as sodium cholate, sodium deoxycholate, and octyl glycoside, are often used in the initial stages of liposome preparation to solubilize the lipids and to form a homogeneous solution. However, detergents can also destabilize the liposome membrane and lead to aggregation and fusion of the liposomes. To remove the detergent and to obtain stable liposomes, various methods are used. One of the most commonly used methods is dialysis [51]. Using this method, the detergent containing the liposome dispersion is submitted to a dialysis step. This process is typically carried out over several hours to several days, depending on the amount of detergent and the desired degree of detergent removal [52]. Another method for detergent removal is the use of adsorbents such as Bio-Beads [53]. The obtained liposomes are, then, separated from the beads containing adsorbed detergent molecules by filtration or centrifugation, resulting in detergent-free liposomes [53].

Vitamin-loaded liposomes can also be produced by the detergent depletion method. In this case, vitamins should be solubilized with the lipids in a detergent-containing solution. Hydrophilic compounds are generally encapsulated more efficiently in liposomes produced by detergent removal than hydrophobic bioactives since the latter may be more prone to being removed along with the detergent during the depletion process. However, in the case of food applications, the choice of detergents used in the liposome production should be carefully considered; some safety concerns or regulatory restrictions may apply. Secondly, this process can be time-consuming and labor-intensive, and may not be easily scalable to commercial production levels. Some detergents, such as Tween 20, Tween 80, and Brij-35, are considered safe for food applications, as they are approved by FDA and European Food Safety Authority (EFSA). Triton X-100 cannot longer be used for food applications, but its use in cosmetics is still feasible due to the regulation approved by several agencies. It is important to notice that the choice of the detergent can affect the properties of the resulting liposomes, including their size, stability, and encapsulation efficiency [45]. Therefore, it is necessary to carefully optimize the detergent concentration and preparation method to achieve the desired properties of the liposomes for the intended final application. Additionally, it is important to ensure that any residual detergent is removed from the liposomes during the detergent depletion process to avoid potential health risks [54]. When producing liposomes for cosmetic applications, it is important to choose a detergent that is safe for use on the skin and does not cause irritation or other adverse effects on humans.

3.1.5. Double Emulsion Method

The double emulsion method, also known as the water-in-oil-in-water (W/O/W) emulsion method, is a technique for encapsulating both hydrophilic and hydrophobic molecules in liposomes [55]. Hydrophilic bioactives, such as proteins, are more suitable to be encapsulated using this technique due to the multiple aqueous phases involved in the emulsification process. In this context, the method involves two emulsions, one inside the other one [56]. The first step is to prepare a W/O emulsion. This is performed by mixing a small amount of an aqueous solution containing the hydrophilic compound of interest with an immiscible organic solvent, such as chloroform or toluene, and a lipid solution. The mixture is vigorously shaken to create small droplets of the aqueous solution surrounded by a lipid bilayer. The second step is to create a second emulsion by adding the W/O emulsion to a large amount of the aqueous phase containing a stabilizer, such as a surfactant or a polymer. This is typically carried out by sonication or homogenization. The resulting W/O/W emulsion contains small droplets of the aqueous phase plus the hydrophilic compound, surrounded by a lipid bilayer, which in turn is surrounded by a second aqueous phase. Liposomes are formed after the removal of the organic solvent by stripping gas or vacuum pressure, which leads to the direct contact between the external and internal oil–water phases and the formation of the lipid bilayer [57]. The double emulsion method presents some advantages over the other liposome production methods, such as high encapsulation efficiency of hydrophilic compounds, drug-controlled release, and versatility. However, this process can be time-consuming since it requires several steps to form liposomes, besides the low yield, and the need for specialized equipment, such as sonicators or homogenizers [39].

A modification of this method, named "freeze-drying of double emulsions", has been efficiently used to produce liposomes characterized by improved stability and prolonged shelf life. This technique involves three main steps: (i) the formation of a double emulsion by the emulsification of an aqueous phase with a lipid phase, followed by a new emulsification with an external aqueous phase; (ii) the freezing of the double emulsion to produce a frozen matrix of liposomes, which helps to stabilize the liposomes and to prevent aggregation or coalescing phenomena; and (iii) the freeze-drying process, in which the frozen matrix is submitted to freeze drying for the removal of water through sublimation under vacuum. This step transforms the frozen matrix of liposomes into a dry powder that can be easily reconstituted with water or other appropriate solvents. Some advantages related to this method include: (i) improved stability, making them suitable for long-term storage and transportation; (ii) preservation of structural integrity, thus, maintaining their biological activity; and (iii) flexibility, as liposomes can be produced at different sizes and compositions depending on the final destination. However, this technique shows some drawbacks, such as the requirement for specialized equipment, the potential for loss of encapsulated compounds during the freeze-drying process, and some damage to the freeze-dryer apparatus due to residual solvents [58].

Li et al. [59] produced complex liposomes containing medium-chain fatty acids (MCFAs) and vitamin C using the double emulsion method followed by dynamic high pressure microfluidization. A mass ratio of 100:25:4 soybean phospholipids, cholesterol, and vitamin E, altogether with an ethanolic solution containing MCFAs, was used throughout the production. The procedure was followed by the injection of a small volume of twice-distilled water under vigorous stirring at 50 °C. The primary emulsion was formed after the evaporation of part of the solvent under reduced pressure. The aqueous phase composed of twice-distilled water, Tween 80, and vitamin C was, then, incorporated into the organic phase. After agitation, the residual solvent was removed through rotary evaporation under reduced pressure. These authors obtained liposomes with a mean diameter of approximately 93 nm, encapsulation efficiencies of MCFAs and vitamin C of 49% and 64%, respectively, and a good stability over 90 days at 4 °C. In another study, Yang et al. [60] used the same protocol as Li et al. [59] to produce complex liposomes encapsulating MCFAs and vitamin C, but these authors added a freeze-drying step at the end of the process to increase vesicle stability. The reconstituted liposomes presented a mean diameter of 110 nm and an encapsulation efficiency of MCFAs and vitamin C of 44% and 62%, respectively. Vesicles remained stable for 60 days at 4 °C. In a more recent study, Pattnaik and Mishra [61] produced a multivitamin (A, D, B_9, and B_{12}) liposome using the double emulsion technique and a mix of soy lecithin and vegetable oil blend, in addition to a polymer solution containing milk protein isolate and trehalose at different concentrations. Liposomes were stable, with size varying from 143 to 396 nm and zeta potential from −20 to −33.5 mV. Interestingly, these authors observed that hydrophilic vitamins showed lower entrapment efficiency than the hydrophobic vitamins (vitamin B_9, EE% = 78–97.6% and vitamin B_{12}, EE% = 96–99.9% vs. vitamin A, EE% > 99.9% and vitamin D, EE% > 98%) and stated that this behavior was probably due to the interaction between the lipophilic vitamins with the hydrophobic tails of phospholipids during freeze-drying, which might have caused a protective effect over the vitamins.

3.1.6. Hydration of Proliposomes

This method of production uses a powder mixture of dry phospholipids and bioactives, called proliposomes. Several techniques can be used to obtain these powders, including fluidized bed, spray drying, freeze-drying, coating of micronized sugars, milling, and supercritical techniques. Proliposomes are defined as dry, free-flowing powders that contain the bioactives to be encapsulated [62]. Upon hydration under appropriate conditions, MLVs are formed. Their solid form confers stability and offers advantages, such as improved transport convenience, storage, distribution, and dosage. Proliposomes can be manufactured using industrialized procedures as tablets or capsules, which eliminates the stability problems of liquid liposomes and may increase the oral bioavailability of bioactives [63].

Spray drying is a commonly used technique in the pharmaceutical and food industries for the production of powdered materials. This method is widely used for the encapsulation of oils, flavors, and fragrances. Its large application is mostly due to its ability to evaporate moisture rapidly from a sample, maintaining a low temperature in the particles [64]. Wall materials, such as polysaccharides and proteins, are commonly used throughout the process [65]. In the liposome field, spray drying acts as a post-processing step to convert the liquid form of dispersions to a high stable solid form [66]. First, the liposome dispersion must be produced, selecting the desired bioactive molecule by any method of interest. This solution is, then, atomized in small droplets using a spray nozzle and dried using a hot gas stream. The resulting powder is collected and can be used for further processing or packaging. Spray drying has several advantages for the production of bioactive-loaded liposomes. It is a fast and efficient method for producing large quantities of powdered material, and it can be easily scaled up for commercial production [67]. The resulting powder is also stable and can be stored for long periods of time without degradation. However, there are some challenges associated with spray drying regarding the encapsulation of vitamins. This process can cause damage to the liposome structure, leading to a loss of the encapsulated material. Additionally, the high temperatures used during spray drying can degrade sensitive vitamins [68]. However, some recent studies on thermosensitive vitamins, such as vitamin B_1, B_9, and vitamin B_{12}, showed that they may not experience significant degradation during the spray drying process, depending on the wall material and process parameters [69,70]. The microencapsulation of lipophilic vitamins such as vitamin D using spray drying is rare due to the mandatory water-dispersed form that is needed for the processing. In addition, the porosity of the resultant material can act as an advertisement about the risk of degradation of more sensitive materials due to oxygen exposure [68]. The use of lower temperatures or shorter drying times during the spray drying process may prevent early thermal degradation.

Spray drying was used to produce β-carotene-loaded proliposomes by Toniazzo et al. [71]. Phospholipids, sucrose, and β-carotene were solubilized in anhydrous ethanol and treated by spray drying using an inlet temperature of 90 °C and outlet temperature of 85 °C, at a flow rate of 10 mL/min in a 1 mm diameter spray nozzle with co-current airflow. Proliposomes were, then, rehydrated with deionized water and thickeners xanthan and guar gums were used as stabilizers. The MLVs produced by this method were useful to protect β-carotene over 95 days of storage with an EE% up to 96% when a 0.10% *w*/*w* mixture of xanthan and guar guns was incorporated during the hydration step.

3.2. Recent and Innovative Methods

Novel methods for liposome production have been developed and improved over the years, with the aim of overcoming the drawbacks related to the traditional methods. Differently from the liposomes produced using the conventional methods, vesicles produced by these new methods are mainly unilamellar and show a more homogeneous size distribution (lower values of PDI) [72]. Therefore, post-processing techniques, such as sonication or extrusion, are rarely required. Moreover, conventional methods are based on the use of organic solvents and detergents, which can limit their use in foods and cosmetics, and are often not environmentally friendly. Currently, research efforts around liposomes focused on the optimization of green technologies that can be scaled up to industrial levels.

3.2.1. Heating Method

This method was first described by Mozafari et al. [73] and involves the hydration of phospholipids with a 3% *v*/*v* glycerol solution at increasing temperature up to 60 °C or 120 °C. The processing temperature depends on the absence or presence of Chol. The use of glycerol is due to its physicochemical characteristics, such as water solubility and non-toxicity, besides its ability to avoid sedimentation or coagulation of vesicles. Sterilization of the resulting liposomes is not required due to the high temperature already used during processing (120 °C), leading to cost reduction. The heating method has been improved over the years resulting in a new method called the Mozafari method, in which large-scale production of liposomes can be achieved without the need for the prehydration step of raw materials and without the use of toxic solvents or detergents [74]. Although vitamins are mostly thermosensitive, equally sensible molecules such as DNA are being incubated at room temperature with preformed liposomes produced by the heating method [73].

3.2.2. Membrane Contactor Method

In this technique, a porous membrane with a defined pore size is used to separate two fluid compartments [75,76]. The organic phase (generally ethanol + lipids) is flowed across one side of the membrane, while an aqueous solution is flowed across the other side. The two fluids come in contact at the pores of the membrane, and liposomes are formed thanks to the presence of water. Liposomes then diffuse tangentially through the pores of the membrane and are collected on the surface of the other size. The organic solvent is removed by evaporation under reduced pressure. This method offers some advantages over traditional liposome production methods. Firstly, it is a continuous process, able to be scaled up, thus allowing for the production of large quantities of liposomes. Secondly, it is a gentle method, with minimal shear stress, which reduces the potential damage to liposomes or to the encapsulated molecules. Thirdly, it offers better control over the size and size distribution of the liposomes, as the pore size of the membrane can be precisely controlled [77]. Additionally, membranes can be regenerated by washing using a water/ethanol mixture.

This method was already used to produce vitamin E-loaded liposomes [78]. Vesicles were produced using 20–50 mg/mL of 1-palmitoyl-2-oleoyl-sn-glycero-3-phosphocholine (POPC) or Lipoid® E80, 5–12.5 mg/mL of a stabilizer (Chol, stearic acid or cocoa butter), and up to 5 mg/mL of vitamin E. Shear stress on the membrane surface ranged from 0.80 to 16 Pa. Mean particle size under optimal conditions was 84 and 59 nm for Lipoid® E80 and POPC liposomes, respectively. EE% up to 99% was obtained in MLVs produced using Lipoid® E80.

3.2.3. Electroformation

This method was developed by Angelova and Dimitrov [79] and consists of the formation of GUVs under electric fields by the hydration of a lipid film deposited on electrodes. This method relies on the ability of lipids to self-assemble in bilayer structures in the presence of an electric field. The electric field creates a potential difference across the lipid solution, causing the lipid molecules to move towards the electrode, where they self-assemble in bilayer structures. The electroformation process typically involves the following steps: (i) preparation of the lipid solution, typically by mixing lipids in a suitable organic solvent; (ii) deposition of the lipid solution on an electrode or a conductive substrate, typically made of glass, quartz, or indium tin oxide; (iii) application of an electric field to the lipid solution by using two electrodes, typically made of platinum or gold; and (iv) collection of the vesicles from the electrode surface and further purification using size exclusion chromatography, ultracentrifugation, or dialysis [80]. The main disadvantages of this method include its low production yield and limited scalability.

Vitamin E-loaded GUVs were prepared by Di Pasquale et al. [81] in which the influence of vitamin E on the membrane organization was corroborated by SANS and fluorescence microscopy. Samples were produced using 1,2-dipalmitoyl-sn-glycero-3-phosphocholine (DPPC), 1,2-dioleoyl-sn-glycero-3-phosphocholine (DOPC), and Chol at a 37.5:37.5:25 mole ratio and vitamin E at a mole fraction of 0.10.

3.2.4. Nanoprecipitation

This method for liposome production can be easily scaled up and allows for the creation of liposomes in a single step without the need to homogenize the vesicles to achieve a uniform size. The procedure involves the mixing of a lipid solution with an aqueous solution, and the lipid molecules come together spontaneously to create liposomes since they are less soluble in the aqueous medium. By altering the composition and processing parameters, the size of the liposomes can be controlled. Due to the self-assembly mechanism, the resulting liposomes are characterized by a narrow size range and can be manufactured with a high degree of consistency [82].

Jash and Rizvi [83] utilized the nanoprecipitation method to produce coated liposomes loaded with both vitamin C and vitamin E. They employed polyethylene glycol and acetic acid sodium acetate buffer with a pH of 4.5 as solvent and non-solvent, respectively. The surface of the liposomes was coated with a polyanionic block copolymer Eudragit® S100, and the resulting solution was slowly added dropwise to the acetic acid-sodium acetate buffer. The coated liposomes were then concentrated using centrifugation. The study's primary findings revealed that Eudragit® S100 safeguarded the encapsulated cargo from the harsh gastric environment and enabled targeted pH-triggered release in simulated intestinal conditions.

3.2.5. Microfluidics Method

Microfluidics refers to a collection of techniques that use narrow channels with micrometric cross-sectional dimensions (ranging from 5 to 500 μm) to manipulate fluid flows. It offers several advantages, such as axial mixing regulated by diffusion and continuous operation at low volumes. Microfluidic techniques have been proven to produce uniformly dispersed liposomes and allow for precise control over liposome size by adjusting either the volumetric flow rate or the total flow rate [84]. The process operates at low Reynolds numbers (laminar flow) and requires diffusive mass transfer. A detailed description of liposome formation via microfluidics can be found in Carugo et al. [85]. In brief, the process involves pushing a stream of lipids dissolved in alcohol through the central channel of a microfluidic device, where it is sheathed and crossed by two lateral streams of a water phase. By adjusting the volumetric flow rate ratio and the total flow rate, the size of the focused stream can be adjusted. Liposome formation occurs when the lipids solubilized in alcohol diffuse into the water, and the water diffuses into the alcohol until the alcohol concentration decreases below the lipid solubility limit, resulting in self-assembly of lipids to form liposomes [85]. However, there are some limitations to the microfluidic approach, including the use of organic solvents, the requirement for delicate mechanical agitation, and the challenges associated with scaling-up production.

Dalmoro et al. [86] prepared vitamin D_3 and vitamin K_2-loaded uncoated and chitosan-coated nanoliposomes by a novel simil-microfluidic device. PC, Chol, and each vitamin were diluted with alcohol, whereas deionized water was used as the hydration solution. Higher encapsulation efficiencies were obtained for vitamin D_3-loaded liposomes (EE% = 88–98%) and vitamin K_2-loaded liposomes (EE% = 95–98%) at chitosan concentrations of 0.01% and 0.005% w/v, respectively.

3.2.6. Dual Asymmetric Centrifugation (DAC)

The DAC method employs a distinctive kind of centrifugation where a vial holding the mixture of liposomes is spun around its own center and the main axis. While the primary rotation forces the sample material outward, the rotation of the vial around its own center moves the sample material inward as it adheres to the vial. This inward movement is effective when the material is viscous enough and can adhere to the vial material. The resulting vesicular phospholipid gel can then be diluted to produce liposomes. These gels are particularly useful for obtaining a high content of liposomes suitable for use in products such as creams, lotions, or hydrogels [87]. The DAC method has also been shown to produce a highly uniform population of liposomes with improved drug release properties compared to other methods [88]. Furthermore, the equipment is of a small size, easy to operate, and offers good reproducibility. Liposomes with high EE% for water-soluble bioactives can be produced by this method, which does not require the use of organic solvents. Drawbacks of this method include the high shear force and the need for formulations that contain a high amount of phospholipids to increase the viscosity and to form the vesicular gel [89]. To the best of our knowledge, there is no literature on the encapsulation of vitamins by the DAC method; however, it appears as a suitable method to encapsulate mainly water-soluble vitamins.

3.2.7. Cross-Flow Filtration Detergent Depletion

Peschka et al. [90] developed a method that combines the conventional detergent depletion technique with a cross-flow filtration system to provide a fast solution for detergent removal. This system comprises a starting reservoir, a pump, a filtration device with a membrane system, tubing with an integrated rotary slide valve, and a manometer to monitor the retentate pressure. Increasing the pressure on the membrane leads to the rapid removal of the detergent. The mixed micelle solution in the starting reservoir undergoes tangential filtration through a single membrane or membrane cassettes with a selected molecular weight cutoff. This cross-flow filtration process enables the production of liposomes with uniform size, homogeneity, and high stability [89]. Compared with other detergent removal methods, large quantities of liposomes can be produced in a significantly shorter time [91]. This method also allows for the production of sterile products by using sterile filtered mixed micelles and autoclaved devices. Additionally, the wasted filtrate can be recycled to minimize production costs [91].

3.2.8. Inkjet Method

The ethanol injection method has been modernized by the development of the inkjet method, which allows for precise control over liposome size and has the potential for scalable production [92]. Amphiphilic compounds are dissolved in ethanol and printed into an aqueous solution using an inkjet device, resulting in uniform liposome droplets in the range of 20–100 nm. Additionally, the inkjet method has been utilized to produce unilamellar lipid vesicles by transforming a bioactive solution into a jet of uniform droplets, which then collide with a solution containing a lipid bilayer membrane at the liquid/air interface. This sequence of events leads to the formation of each lipid vesicle, as the membrane undergoes deformation, collapse, and separation [93]. Although it is not currently used for the encapsulation of vitamins in liposomes, inkjets have been used to produce edible films for oral delivery of B-complex vitamins [94].

3.2.9. Supercritical Technologies

Supercritical fluid technologies have emerged as attractive methods for producing liposomes [56]. CO_2 is the most commonly used gas in supercritical fluid technology for several reasons, including low cost, non-toxicity, safety, and large availability. It is therefore considered a safe gas for use in food, as well as pharmaceutical and cosmetic applications [56]. Most processes of liposome production that use supercritical CO_2 (sc-CO_2) involve dissolving a lipid mixture in a solution of CO_2 at high pressure and temperature. The CO_2 solution then acts as a solvent for the lipids, leading them to self-assemble and form vesicles. The pressure and temperature of the CO_2 solution can be easily controlled to modify the size and shape of the samples. One of the key advantages of sc-CO_2-based processes is the absence of organic solvents, which can be toxic and difficult to remove from the final product. In addition to being non-toxic, CO_2 readily evaporates, leaving no residue in the final product. Over the years, many supercritical technologies for liposomes production have been explored, and some of them are discussed below.

- Depressurization of expanded liquid organic solution into aqueous solution (DELOS-susp): this is a compressed fluid-based method that enables the reproducible and scalable production of nanovesicular systems with exceptional physicochemical properties, including uniformity, morphology, and particle size [95]. To prepare the samples, the reagents are initially dissolved in an organic solvent and subsequently treated using pressurized CO_2 until saturation. Then, the sample is rapidly depressurized from the bottom, experiencing a significant pressure drop from 10 MPa to ambient pressure. CO_2 molecules are released from phospholipid bilayers to temporarily disrupt them into highly dispersed phospholipids that undergo a rapid reorganization due to hydrophobic and van der Waals interactions, and then packing themselves into liposomes [96,97]. This process produces small and uniform liposomes due to the high rate of depressurization and can be used to encapsulate thermo-sensitive materials since it works under slight conditions [98]. Adaptations of this method have been proposed to remove the need for organic solvents or surfactants during production, some of them still resulting in SUVs at high storage stability [99];

- Depressurization of expanded solution into aqueous media (DESAM): this is an alternative dense gas technology, in which pressure requirements for liposome production are reduced to 4–5.5 MPa [100]. This method involves dissolving lipids in a solvent and pressurizing the solution with a dense gas to create an expanded lipid solution. The expanded solution is, then, released in a controlled manner into heated aqueous media, with pressure maintained through the addition of more dense gas. Care is taken to keep the pressurization and expansion below a certain threshold to avoid solute precipitation. The dense gas and solvent can be separated and reused. The resulting fine droplets disperse the lipid in the aqueous phase and improve component interaction, resulting in uniform liposome formation. The high temperature in the vesicle formation chamber can also aid in removing organic solvent from the product [100]. Liposomes produced by this method are mainly unilamellar, with sizes ranging from 50 to 200 nm, PDI not exceeding 0.29, and is highly stable for periods of 8 months [100]. Recently, a continuous process named nano-carrier by a continuous dense gas (NADEG) technique appeared as an evolution of the DESAM method [101,102];

- Rapid expansion of supercritical solution (RESS): this is a technology currently used for micronization, co-precipitation, and encapsulation. Lipids are dissolved in a mixture of sc-CO_2 plus 5–10% v/v of ethanol within an extractor. This primary dissolution in a co-solvent (ethanol) is strictly necessary because the natural phospholipids are poorly soluble in sc-CO_2 [103]. This solution is then released through a heated small nozzle in a low-pressure chamber and mixed with an aqueous solution. A rapid depressurization follows, and the pressure drop results in the lipids desolvation, which favors the formation of layers around the droplets due to solute supersaturation. Small particles are obtained from the gas stream [104,105]. This process produces small particles with a uniform size. However, this method shows problems such as the difficult separation between vesicles and co-solvents during depressurization, which increases production costs [106]. Nevertheless, the RESS method is one of the most studied supercritical technologies for vitamin-loaded liposome production [107–112]. Han et al. [112] produced vitamin E acetate-loaded liposomes using an optimized RESS process without any organic solvent. Operating conditions were controlled using the single-factor analysis and the response surface methodology combined with Box–Behnken design. Samples were produced using polyvinyl acetate grafted phospholipids and vitamin E acetate in a 6.35:1 mass ratio and resulted in vesicles characterized by EE% = 93%, size of 247 nm, PDI of 0.295, and zeta potential of −42.5 mV. Jiao et al. [108] also used the Box–Behnken design to optimize the process parameters for the production of vitamin C-loaded liposomes using PC as wall material. Vesicles presented size of 270 nm, PDI of 0.254, zeta potential of −41.7 mV, and EE% of 75%. Sharifi et al. [111] produced ironized multivitamin-loaded liposomes containing lecithin, Chol, iron sulfate, and hydrophilic and hydrophobic vitamins (C and E, respectively) by a new venturi-based method called Vent-RESS, in which RESS was combined with Bernoulli principles. Liposomes with unimodal size distribution were obtained and EE% of bioactive molecules were improved when operating pressure increased from 12 to 18 MPa. The Vent-RESS process was also used by Jash, Ubeyitogullari, and Rizvi [110] to produce vitamin C–vitamin E co-loaded liposomes in milk fat globule membrane phospholipids (MFGM) or sunflower phosphatidylcholine (SFPC). These authors verified that MFGM-based ULVs were smaller in size than SFPC-based ones (533 nm vs. 761 nm, respectively), with higher zeta potential (−57 mV vs. −37 mV, respectively);

- Supercritical reverse phase evaporation process (scRPE): this is a batch method developed by Otake et al. [113] that enables the efficient formation of liposomes using a one-step process. It acts similarly to the RPE method but, in this case, sc-CO_2 substitutes the organic solvent. It involves the mixing of sc-CO_2, lipids, and ethanol and then, the introduction of small amounts of water to generate a liposome dispersion through an emulsion formation. The procedure is carried out in a stirred volume cell at a temperature above the lipid phase transition temperature. As the aqueous solution is gradually added to the reactor, sc-CO_2 is released, resulting in the formation of liposomes upon depressurization [97,114]. Some years later, Otake et al. [115] optimized the scRPE method in a way that organic solvents were no longer needed. Zhao and Temelli [116] developed a similar process in which liposomes are formed by simple pressurization and depressurization of a sc-CO_2 lipid-based aqueous solution. Liposomes size can vary from 100 nm to 1.2 μm by using this method, being SUVs or MLVs [117];

- Supercritical antisolvent (SAS): it involves a continuous spraying of an organic solution of phospholipids into sc-CO_2, which serves as an antisolvent for phospholipid precipitation. As soon as sc-CO_2 contacts the liquid phospholipid phase, it quickly diffuses and divides the liquid phase in tiny droplets. At the same time, the organic solvent evaporates from the droplets as a consequence of the dissolution in sc-CO_2. This mass transfer creates a supersaturation of phospholipids within the droplets, leading to the formation of small phospholipid particles through nucleation and aggregation. A pure CO_2 washing step can be performed to remove any trace of the organic solvent. Finally, spherical micro- or nanoliposomes are formed upon hydration with an aqueous buffer. Xia et al. [118] produced vitamin D_3-loaded proliposomes using a SAS based technology. Hydrogenated PC was used as a lipid source. Conditions such as T = 45 °C, P = 8 MPa, and 15% *w*/*w* lipid to vitamin D_3 ratio resulted in hydrated liposomes with an EE% = 100% and an effective loading of 12.9%;

- Supercritical assisted liposome formation (SuperSomes): this is a continuous sc-CO_2-based process proposed by Reverchon and co-workers [119] in which, differently from the other methods, water particles are first formed by atomization and then, covered by lipids dissolved in an expanded liquid mixture. The expanded liquid mixture is composed of phospholipids, ethanol, and sc-CO_2. The main idea is that lipids reorganize themselves around the water droplets forming inverted micelles, which tend to form liposomes as soon as they come in contact with a water pool located at the bottom of the vessel [120,121]. Process parameters, as water flow rate, injector diameter, phospholipid concentration, pressure, and gas to liquid ratio, have been constantly optimized during SuperSomes studies [122]. This apparatus has been efficiently used to encapsulate both hydrophobic and hydrophilic bioactive molecules [123–126]. The process is reproducible, therefore, allowing for a good control of vesicle size distribution, and nanometric vesicles at high EE%. Recent studies showed the feasibility of using SuperSomes apparatus to encapsulate vitamin D_3 into nanoliposomes [127,128]. In both studies, liposomes were produced using different ratios of hydrogenated soy and nonhydrogenated egg yolk phospholipids. In Chaves et al. [127], samples produced using only egg yolk phosphatidylcholine presented sizes of 132 nm. Furthermore, a 10 mL/min water flow rate also led to highly homogeneous vesicles produced using a maximum of 20% of hydrogenated soy phospholipids with a size of 218 nm, PDI of 0.253, and an EE% of 89%. In Chaves et al. [128], the effect of the incorporation of vitamin D_3 in curcumin-loaded liposomes was investigated. The addition of vitamin D_3 reduced the overall size of liposomes from approximately 220 nm to 130 nm, but also promoted a decrease in EE% of curcumin. The authors stated that this behavior was probably due to the competition between the two hydrophobic bioactives for the inner region of lipid bilayers;

- Aerosol solvent extraction system (ASES): the ASES method, initially intended for the creation of a sterile product composed of a biodegradable carrier and a molecule embedded within it, involves the spraying of organic liquids through a nozzle into a bulk of sc-CO_2 [129]. This facilitates the rapid precipitation of solutes from the solution, which can be easily dried with circulating sc-CO_2, allowing for the simple removal of residual solvent from the precipitates. This technique has been applied to liposome preparation, resulting in the production of dry and reconstitutable pharmaceutical liposomes that are suitable for large-scale manufacturing [130,131];

- Particles from gas saturated solution (PGSS): this is a cutting-edge method that uses supercritical fluids to produce particles of a precise size. This process operates at mild temperatures, generally between 40–60 °C, in an inert environment, and uses CO_2 and water as solvents [132]. The PGSS-drying method provides an alternative to conventional techniques such as spray drying and freeze drying. One of its significant benefits is the efficient atomization achieved through rapid gas release and expansion during depressurization from supercritical to ambient conditions. Additionally, this method allows drying at lower temperatures in the spray tower, minimizing the exposure of the bioactive material to high temperatures that may cause damage [133]. Another advantage of this technique is its ability to process carrier materials with low melting temperatures, which conventional spray drying cannot handle.

In summary, the use of traditional methods for creating and reducing the size of liposomes is still popular due to their ease of implementation and lack of requirement for advanced equipment. Although some of these methods can be effective for processing certain vitamins, they may also cause structural changes that could impact their functionality. Additionally, these methods are not always easily scalable for an industrial production. In recent years, there is a growing need for innovative manufacturing techniques in nanotechnology that can encapsulate both hydrophilic and hydrophobic molecules without the use of organic solvents or complex equipment, which is of interest considering the need for multivitamin products. Additionally, the challenge of maintaining system stability for liposomes loaded with hydrophilic materials requires new strategies to achieve optimal loading while targeting delivery to the intended site. The lack of scalability in some conventional liposome formation processes and low encapsulation efficiencies are still major obstacles to mass production. To address these challenges, it is essential to explore novel formation methods and to integrate new technologies to produce advanced liposome formulations that are not only suitable for industrial-scale production, but also highly effective for clinical applications. Therefore, much work still needs to be performed in terms of scaling, designing, controlling, and optimizing liposome formation processes [30].

4. How to Measure the Effectiveness of Liposomes in Encapsulating Vitamins?

In 2018, the Center for Drug Evaluation and Research of Food and Drug Administration (CDER-FDA, Silver Spring, MD, USA) developed a guidance summarizing the main characterization techniques that producers should consider when developing liposomes at an industrial scale. In this context, the liposome formulation should be able to contain and retain as much as possible the molecule in the appropriate structure. Vesicles should be characterized in terms of morphology, surface characteristics, encapsulation efficiency, bioactive loading, particle size, phase transition temperature, in vitro release of the molecule, leakage rate from liposomes throughout shelf life, and integrity changes in response to changes in factors, such as salt concentration, pH, temperature, or the addition of other excipients.

Liposome composition, production method, and bilayer membrane rigidity can all impact EE%. EE% is calculated as the percentage of bioactives inside the liposomes compared to the total amount of bioactive used. Generally, an encapsulation efficiency of over 50% is considered good. Separation of the free bioactive is necessary to quantify the amount within liposomes. Several techniques are used for separation, including chromatography, gravitation or centrifugation, dialysis membrane, and ultracentrifugation. Indirect and direct methods are used to determine EE%. Conventional techniques for measuring drug concentration include UV–Vis and fluorescence spectroscopy, enzyme or protein-based assays, HPLC, UPLC, LC-MS, and GC-MS. ESR and ^{1}H NMR can also be used for bioactive quantification. Table 4 summarizes some studies in which EE% tests were performed on vitamin-loaded liposomes. Solvents and other chemicals commonly used to disrupt the liposome membranes, in addition to the techniques carried out to quantify the amount of hydrophobic and/or hydrophilic vitamins released from vesicles, are also presented.

Table 4. Chemicals and methods currently used for the extraction and quantification of vitamins in liposomes.

Vitamin	Liposome Type/Size	Chemical Used to Demulsify/ Precipitate the Liposomes	Method for Vitamin Quantification	EE%	Reference
Vitamin A	SUV	Methanol	HPLC	50.6–56.2%	[134]
Vitamin A	MLV	Chloroform/Methanol 2:1 *v*/*v*	UV–Vis spectrophotometry	99%	[135]
Vitamin A	MLV/SUV	Chloroform	HPLC	15.8%	[136]
Vitamins A, D, E, and K	n.s.	Tween 80	HPLC	20–100%	[137]
Vitamin B_1	MLV	n-octyl β-D-glucopyranoside	Optical density	31.2%	[138]
Vitamin B_1	n.s.	PBS	HPLC	97%	[139]
Vitamin B_2	n.s.	Triton X-100	Indirect method (photolysis)	42.3%	[140]
Vitamin B_5	n.s.	Chloroform/Methanol	Direct method (mass weight)	75%	[141]
Vitamin B_9	n.s.	Ethanol	Ultrafiltration in centrifuge tube followed by HPLC	87.4%	[142]
Vitamin B_{12}	MLV	Methanol/Water 1:10 *v*/*v*	HPLC	70%	[143]
Vitamin B_{12}	LUV	-	Centrifugation followed by UV–Vis spectroscopy	27%	[33]
Vitamins B_{12}, D_2, and E	SUV	Ethanol	UV–Vis spectrophotometry	56–76%	[144]

Table 4. *Cont.*

Vitamin	Liposome Type/Size	Chemical Used to Demulsify/ Precipitate the Liposomes	Method for Vitamin Quantification	EE%	Reference
Vitamin C	n.s.	Ethanol	Dialysis followed by UV–Vis spectrophotometry	77.9%	[145]
Vitamin C	n.s.	Methanol, chloroform, and ammonium buffer	HPLC	99.2%	[146]
Vitamin C	n.s.	Ethanol	HPLC	75.4%	[108]
Vitamins C and E	n.s.	Triton X-100/DMSO	UV–Vis spectrophotometry	93–95%	[83]
Vitamins C and E	n.s.	Assay kit BC1235	Indirect method	98.5%	[147]
Vitamin C	SUV/LUV	Meta-phosphoric acid	Titration with indophenol solution	94.2%	[148]
Vitamin C	SUV/MLV	-	Gel permeation chromatography	5–16%	[149]
Vitamins C and E	MLV/MVV	Protamine sulfate solution followed by Triton X-100	Protamine aggregation method followed by UV–Vis spectrophotometry	12–88%	[111]
Vitamins C and E	MLV	Acetic acid and protamine solution	UV–Vis spectrophotometry	38%	[150]
Vitamin C	n.s.	-	Dialysis followed by UV–Vis spectrophotometry	100%	[151]
Vitamin D	MLV	Triton X-100	Ultracentrifugation followed by HPLC	61.5%	[152]
Vitamin D_3	n.s.	Methanol	Ultrafiltration in centrifuge tube followed by HPLC	74%	[127]
Vitamin D_3	SUV/LUV	Methanol	HPLC	90.2%	[72]
Vitamin D_3	SUV/MLV	Water	Ultrafiltration in centrifuge tube followed by RP-HPLC	100%	[153]
Vitamins D_3 and K_2	SUV/LUV	Ethanol	UV–Vis spectrophotometry	98%	[86]
Vitamin E	n.s.	Methanol	Dialysis followed by HPLC	92.5%	[154]
Vitamin E	n.s.	Sephadex®G25 M solution (10%, *w*/*v*) in double distilled water	Minicolumn centrifugation	98-101%	[155]
Vitamin E	n.s.	Chloroform/Methanol 2:1 *v*/*v*	UV–Vis spectrophotometry	78%	[156]
Vitamin E	n.s.	Water	Centrifugation followed by HPLC	94%	[157]
Vitamin E	n.s.	-	Centrifugation followed by UV–Vis spectrophotometry	83.8%	[158]
Vitamin E	n.s.	Chloride acid, Tween 80:ethanol and n-pentane	UV–Vis spectrophotometry	97%	[159]
Vitamin K_1	n.s.	-	Ultracentrifugation followed by HPLC	79.2%	[160]
Vitamin K_1	n.s.	Methanol	HPLC	3.4–154 μg/mg	[34]

n.s.: non specified.

Table 4 shows that different authors have obtained different encapsulation efficiencies when using liposomes to encapsulate the same vitamin. Although the methods for vitamin quantification are normally well-established, sometimes the method used to produce the liposomes is not the most suitable for a particular vitamin. In this case, the polarity of the molecule plays a crucial role in the choice. Fan et al. [137] attributed the low encapsulation of vitamin A in liposomes to its reactive polyolefin structure, but found that increasing the loading content of vitamin A improved EE%. Conversely, they found that an increase in loading rate decreased the EE% of vitamin E due to a resultant higher viscosity of the vesicles. Pezeshky et al. [136] attributed the low EE% of vitamin A palmitate in nanoliposomes produced by the TFH method to the low hydrophilicity and the size of its structure. The authors also noted that vitamin A palmitate is sensitive to light, oxygen, and organic solvents, making degradation during the process more likely. Cansell, Moussaoui, and Lefrançois [138] observed low entrapment of vitamin B_1 in liposomes using the TFH method, attributing this behavior to the kind of vesicles formed during the process (MLVs), which usually present low EE% at low lipid concentrations. Xanthan gum was found to help retain vitamin B_1 by coating the outer surface of the liposomes and thus prisoning a higher content of the vitamin through hydrogen bounding and/or electrostatic interactions. Farhang et al. [149] attributed low values of EE% for liposomes containing water-soluble ascorbic acid to the formation of MLVs when using milk phospholipids and a dehydration/rehydration method. Andrade et al. [33] found that the ethanol injection method was more efficient in producing vitamin B_{12}-loaded liposomes with higher encapsulation efficiency than the TFH method, also due to the hydrophilicity of this vitamin. Sharifi et al. [111] found that increasing pressure resulted in increased EE% of vitamin C and vitamin E in loaded liposomes using a supercritical method, which was attributed to a higher solubility of coating materials in sc-CO_2 at higher pressures. Additionally, higher EE% of vitamin C was observed when using higher amounts of aqueous cargo throughout the process.

5. Applications of Vitamin-Loaded Liposomes

5.1. In Foods

Liposomes have become increasingly popular in the food industry for functional purposes due to their ability to increase bioactive dissolution rate and bioavailability, protect sensitive ingredients, improve stability during processing, storage and digestion, and to confine undesirable flavors. They have also been used to achieve controlled release to specific targets. The encapsulation of food bioactives using liposomes has been investigated, and there is a wide range of opportunities for research in real applications of liposomes in different food formulations. A proper application of bioactive-loaded liposomes in the food industry should involve excipient materials and bioactives that are generally recognized as safe, fully incorporated within the liposomal structure, and not reactive with the core ingredients. Liposomes can be used to encapsulate an aqueous phase in order to decrease the vapor pressure of a matrix, allowing for the lowering of the water activity without decreasing the moisture content, and thus, preventing the growth of microorganisms in foods that contain nutrients such as proteins or sugars [161]. Studies on the evaluation of shelf-life and sensorial acceptance, as well as oral processing, digestibility, and bioaccessibility of encapsulated compounds in liposomes designed for food applications, should receive more attention in the literature.

Vitamins are not easily incorporated into foods. Liposoluble vitamins are hardly introduced into aqueous-based food formulations and are easily oxidized in the air. Most vitamins are also thermolabile and can be readily degraded after thermal treatments such as pasteurization. On the other hand, the addition of vitamins into liposomes can protect their activity, in addition to increasing their absorption and bioavailability [86]. Some studies have been carried out in order to test the effects of the incorporation of vitamin-loaded liposomes in food matrices. Marsanasco et al. [150] produced SPC-based liposomes encapsulating vitamins E and C for orange juice fortification. The hydration of a thin-film lipid method was used to prepare these vesicles. To substitute cholesterol, which is related to health issues as atherosclerosis and high blood pressure, the authors tested other molecules to increase the rigidity of membranes such as stearic acid (SA) and calcium stearate (CaS). The lipid films were constituted of mixtures of SPC and SA or CaS in a 1:0.25 molar ratio and vitamin E from a stock solution, and were further hydrated with acetic acid 3% *v*/*v* containing vitamin C to form liposome dispersions. These dispersions were, then, added to orange juice samples, which were pasteurized at 65 °C for 30 min and stored at 4 °C. In another study, Marsanasco et al. [162] incorporated SPC:SA- and SPC:CaS-based vitamin E-folic acid co-loaded liposomes to enrich chocolate milk. Vesicles were prepared using the thin lipid film hydration at a 1:0.25 molar ratio of SPC and SA or CaS. Vitamin E was incorporated in the lipids before the hydration step, whereas folic acid was incorporated in the hydration buffer. Samples were pasteurized and then, added in a 1/100 ratio in chocolate milk kept at 4 °C. These authors observed a protective action of vitamin E over folic acid as oxidative stability of folic acid remained unchanged in the presence of vitamin E after pasteurization.

Banville, Vuillemard, and Lacroix [152] produced vitamin D-loaded liposomes aiming to fortify cheddar cheese. Vesicles were prepared by hydrating a proliposome mixture (Pro-Lipo-Duo^{TM}, Lucas Meyer, Chelles, France) using a vitamin D solution. The resulting MLVs were then incorporated into raw milk, which was employed for the cheese manufacturing. The results showed that vitamin D-loaded liposomes promoted a higher final concentration of the vitamin in the processed cheeses than a commercial water-soluble vitamin D emulsion (Vitex-D). Wechtersbach, Ulrih, and Cigić [163] incorporated vitamin C-loaded liposomes based in DPPC and Chol in both apple juice and fermented milk that underwent pasteurization at 72 °C; they observed higher retentions of vitamins after encapsulation than when added in free form. These authors also verified a higher retention of the molecule in vesicles containing DPPC and Chol than those containing DPPC alone. Mohammadi, Ghanbarzadeh, and Hamishehkar [164] produced vitamin D_3-loaded liposomes aiming at beverage fortification. Samples were produced using SPC and Chol at various concentrations by the thin-film hydration method. The resulting MLVs were subjected to homogenization at 60 °C for 15 min and to probe sonication at 20 kHz and 70% of strength in order to decrease their size and turn them undetectable by the human eye. Size of vesicles were up to approximately 87 nm after 30 days of refrigerated storage for samples produced at a 50:10 PC:Chol ratio with an EE% up to 95%. Vitamin D_3-loaded liposomes were also incorporated into dark and white chocolate by Didar [165,166]. The ethanol injection method was used to prepare these liposomes based on phospholipids and Chol. A concentration of 5 µg/10 g of vitamin D_3 in chocolate was fixed for free form-enriched and encapsulated-enriched samples. The author verified a reduction of vitamin D_3 after 60 days of storage in samples fortified with the free form of the vitamin instead of encapsulation in liposomes. Moreover, no impacts on rheological, colorimetric or sensory characteristics of chocolate were observed in the samples enriched with liposomes. Another study regarding the enrichment of foods using vitamin D_3 was conducted by Chaves et al. [167], in which pineapple yogurt was fortified with nanoliposomes co-encapsulating vitamin D_3 and curcumin. The nanovesicles were produced using different ratios of purified and unpurified phospholipids by the hydration of proliposomes, which in turn were produced by coating of micronize sucrose. Dispersions were produced upon hydration of proliposomes, which were stabilized by xanthan and guar gums at a 0.02% *w*/*v* concentration. Liposomes were incorporated into formulated pineapple yogurts at 5% *v*/*v* and maintained under refrigeration at 4 °C. These authors verified that an increased concentration of purified phospholipids was beneficial to increase the EE% of both bioactives in dispersions. Liu et al. [168] produced vitamin C-loaded nanoliposomes by self-assembly of alginate and chitosan to enrich mandarin juice. Samples were prepared by the thin lipid film hydration method combined with microfluidization at 120 MPa for 2 cycles. The main ingredients included SPC, Chol, Tween 80, and vitamin E at a 6:1:1.8:0.12 ratio. Samples were rehydrated using PBS 0.05 M at pH 7.4 containing 5 mg/mL of vitamin

C. Vesicles were, then, coated with alginate and chitosan using electrostatic deposition. Finally, liposomes were incorporated into mandarin juice samples and pasteurized at 90 °C for 10 s and stored at 4 °C. Liposome coverage by alginate and chitosan was responsible for an increase in size and a reduction in zeta potential of the samples, in addition to promoting aggregation of vesicles at day 90. However, lipid peroxidation and vitamin C release were lower than non-coated samples, suggesting a protective effect. It is worth mentioning that several other bioactive molecules that are not vitamins have been incorporated into liposomes and tested in food matrices, such as *Clove* oil in tofu [169], nisin in Minas frescal cheese [170], rutin for chocolate coating [171], quercetin in cornstarch [172], betanin in gummy candy [173], baicalin in mushrooms [174], basil essential oil in pork [175], ginger extract in wheat bread [176], and fish oil in yogurts [177].

It is important to note that the risk of allergenicity due to liposomes depends on the source and purity of the phospholipids used in the liposome preparation. To minimize this undesired effect, liposome preparations can undergo rigorous purification steps to remove potential allergenic components. For example, liposomes can be subjected to multiple purification steps to remove any residual allergenic proteins or other components [26].

5.2. In Cosmetics

Nanotechnology-based approaches in cosmetics are growing exponentially with the aim of developing novel formulations that can confer aesthetic and therapeutic benefits to people [178,179]. In particular, such cosmetic formulations are referred to as "cosmeceuticals" when they have both cosmetic and medicinal functions [180]. Liposome-based nanoformulations are gaining particular interest since they can represent a promising strategy to prepare antiperspirants, creams, lipsticks, deodorants, moisturizers, hair care products, etc., and can also be successfully used to deliver vitamins, such as vitamin A, B_{12}, E, K, antioxidants (such as coenzyme Q10, lycopene, carotenoids, etc.), and other bioactive molecules [46,143,181].

In the literature, in vitro and in vivo studies demonstrated the improved cosmeceutical effect of bioactives when incorporated into vesicles, e.g., in reducing pigmentation disorders, skin aging, and solar exposure problems. These positive effects are due to the incorporation of the bioactives into vesicles that favor the penetration through the stratum corneum of the skin and promotes their activity at the damaged site [182,183]. For instance, the loading of vitamin C into liposomes, by facilitating its percutaneous transport, can increase its concentration in the dermis more than five times compared with the application of the pure vitamin [184]. The encapsulation of vitamin B_{12} in vesicles and its slow release enhances the bioactive absorption and bioavailability, in addition to the protection of the vitamin from the degradation induced by heat, light, air, and improper storage [143]. Jiao et al. [142], to improve vitamin C (VC) and folic acid (FA) stability, co-loaded these antioxidant molecules in liposomes (VCFA-Lip) and chitosan-coated liposomes (CS-VCFA-Lip). The mean vesicles size of VCFA-Lip and CS-VCFA-Lip was 138 nm and 249 nm, respectively, whereas the encapsulation efficiency of both drugs in CS-VCFA-Lip was much higher than that measured for VCFA-Lip. Moreover, the stability study revealed that the chitosan coating can efficiently improve the physical stability of VCFA-Lip. According to Dreier et al. [185], the primary mechanism of action for liposomes is their rupture and fusion with the lipids in the stratum corneum. However, the efficacy of liposomes in delivering bioactives to the skin is impacted by various factors, including their size, surface charge, number of layers (uni- or multi-lamellar vesicles), and the flexibility of the bilayer. A research study conducted by Touti et al. [186] demonstrated that systems possessing higher lipid bilayer flexibility can pass through narrow skin constrictions more efficiently. In contrast, multi-layered liposomes with high flexibility and larger dimensions can penetrate deeper into the skin than smaller vesicles due to abrasion and the loss of outer layers during skin diffusion. Additionally, ultra-deformable lipid carriers that are less than 150 nm in size can penetrate the skin and release active molecules in the deeper layers without adsorption on keratinocytes. The positive charge of phospholipid carriers can also enhance transdermal transport through the skin by promoting stronger electrostatic interactions with the skin surface, which facilitates liposome accumulation on or within hair follicles, enabling them to diffuse more easily. Choi et al. [187] showed that flexible cationic liposomes can promote a larger penetration as compared to their less charged counterparts.

Currently, the two most popular techniques for synthesizing liposomes for cosmeceuticals are the same used for foods, namely, the TFH method and the ethanol injection method. Both are low-cost, easy to use, and versatile methods that create liposomes with different properties. However, as aforementioned throughout this review, the first method produces liposomes that are heterogeneous in size and shape, thus, requiring further steps such as sonication or extrusion to create small, uniform vesicles, making it difficult to use on a large scale. The second offers advantages such as simplicity, scalability, and the ability to produce small-sized liposomes, but it is limited in its ability to encapsulate hydrophilic antioxidants, such as vitamin C [46]. To overcome these limitations, advanced techniques, such as supercritical and microfluidic technologies, that can control size, encapsulation efficiency, and the structural heterogeneity of liposomes through automatic and programmable systems, have shown promise, being considered particularly effective alternatives for liposome assembly to both food and cosmetical applications [46].

6. Discussion and Concluding Remarks

Based on the data and methods presented in this review, it appears that exploring the field of vitamin encapsulation in liposomes is a relevant avenue for both the food and cosmetics industries. Several methods of liposome production, from the most conventional to the highly innovative ones, have been described and the application of liposomes in vitamin incorporation was found and contextualized. However, the scale up and the use of liposomes in food and cosmetic products are in distinct technological stages. It is worth mentioning that while there are already liposome-based cosmetic products available on the market, the same cannot be said for food products.

It is quite important to highlight that the number of publications on this subject has been constantly growing over the last 20 years, but particularly in the last ten years, according to Figure 3.

In the case of foods, the oral route is the main challenge to be faced, mainly due to the known sensitivity of liposomes to the gastric environment. Most vitamins should be released at the end of the gastric phase or only in the intestine to be absorbed, which often requires the liposomes to be gastro resistant. This is one of the main challenges in vitamin-loaded liposomes that needs to be overcome, and some approaches have been tried, such as the coating using polysaccharides (e.g., chitosan and pectin) [188,189]. However, studies about the digestion (both in vitro and in vivo) to determine the bioaccessibility and bioavailability of vitamin-loaded liposomes in food products are still scarce. Other challenges that can be cited for the liposomes to be used in foods are: (i) the high price of phospholipids when considering a large scale; and (ii) restrictions about the use of organic solvents, which is a limitation for the use of some of the most common methods to produce liposomes already scaled up.

On the other hand, as the main route for the application of vitamin-loaded liposomes in cosmetics is through the skin, the elasticity and flexibility of the vesicles are pointed out among the main challenges, together with their average size. Some important rewards come from the topical administration in comparison to the oral route, such as the possibility of lower fluctuations in plasma bioactive levels, site-specific delivery, avoidance of first-pass metabolism, and better patient compliance [190]. Furthermore, as the skin is the first defensive barrier against external factors and prevents several bioactives from penetrating the underlying layers or going into systemic circulation, there are some limitations to deliver the active ingredient to the target site [191]. Ultradeformable liposomes, as transfersomes, ethosomes, niosomes, and transethosomes are the new generation of elastic liposomes suitable for a more effective transdermal delivery of therapeutics, including vitamins [190].

Other challenges related to the use of liposomes in cosmetic products are the low drug loading, physical and chemical instability, and scarce reproducibility of the results [192].

Therefore, although both types of utilization (foods and cosmetics) of vitamin-loaded liposomes are in different stages of technological maturity, the development of new technologies to face the challenges presented in this work can be mutually beneficial.

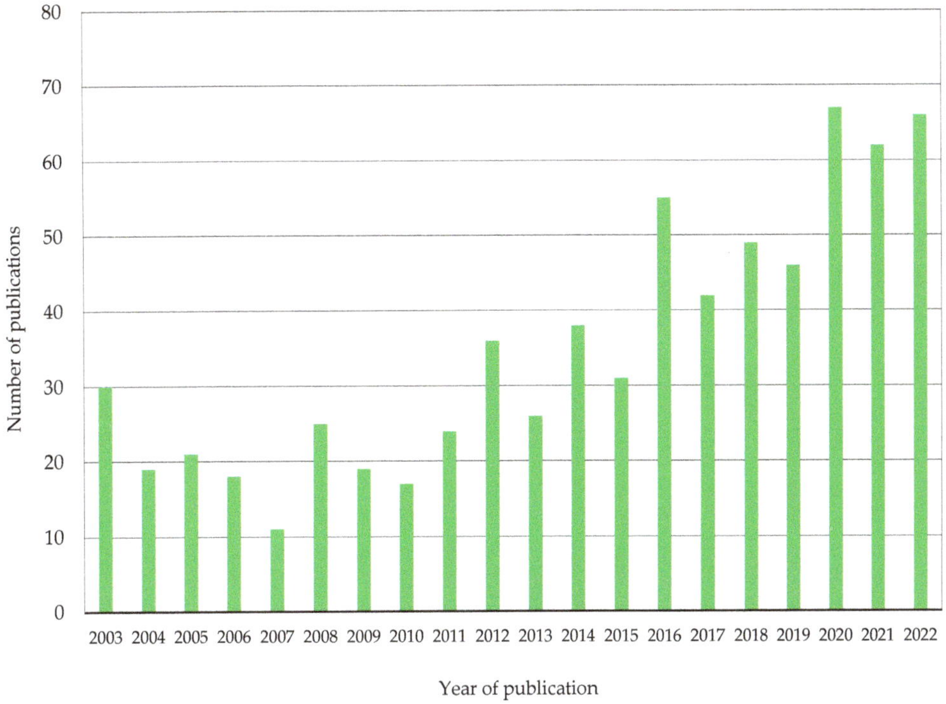

Figure 3. Evolution of the publications about the encapsulation of vitamins in liposomes over the last 20 years.

Supplementary Materials: The following supporting information can be downloaded at: https://www.mdpi.com/article/10.3390/nano13091557/s1, 1. Post-processing methods that can be used during the production of vitamin-loaded liposomes (1.1 sonication, 1.2. extrusion, 1.3. high-pressure homogenization, 1.4 microfluidization, 1.5 dialysis, 1.6 freeze-thaw, and 1.7 lyophilization). References [193–214] are cited in the main text.

Author Contributions: M.A.C.: conceptualization, formal analysis, writing—original draft preparation, writing—review and editing; L.S.F.: conceptualization, formal analysis, writing—original draft preparation; L.B.: conceptualization, formal analysis, writing—original draft preparation, writing—review and editing; S.C.P.: conceptualization, formal analysis, writing—original draft preparation, writing—review and editing, supervision, funding acquisition; E.R.: conceptualization, formal analysis, writing—review and editing, supervision. All authors have read and agreed to the published version of the manuscript.

Funding: This research is supported by the São Paulo Research Foundation (FAPESP, grant numbers 2017/10954-2; 2019/08345-3, and 2022/01236-7 awarded to M.A.C.), the Coordination of Superior Level Staff Improvement (CAPES, finance code 001 awarded to L.S.F.) and the National Council for Scientific and Technological Development (CNPq, grant number: 306816/2019-9 awarded to S.C.P.).

Institutional Review Board Statement: Not applicable.

Data Availability Statement: The data presented in this study are available on request from the corresponding author.

Conflicts of Interest: The authors declare no conflict of interest.

References

1. Tolonen, M. Vitamins. In *Vitamins and Minerals in Health and Nutrition*; Woodhead Publishing: Cambridge, UK, 1990; pp. 99–147. [CrossRef]
2. Casas, C. Vitamins. In *Analysis of Cosmetic Products*, 1st ed.; Salvador, A., Chisvert, A., Eds.; Elsevier: Amsterdam, The Netherlands, 2007; pp. 364–379. [CrossRef]
3. Cilla, A.; Zanirato, V.; Rodriguez-Estrada, M.T.; Garcia-Llatas, G. Nutriential Hazards: Micronutrients: Vitamins and Minerals. In *Encyclopedia of Food Safety*; Motarjemi, Y., Ed.; Academic Press: Waltham, MA, USA, 2014; pp. 86–94. [CrossRef]
4. Hon, S.L.; Vitamin, A. *Encyclopedia of Toxicology*, 3rd ed.; Wexler, P., Ed.; Academic Press: Cambridge, MA, USA, 2014; pp. 960–961. [CrossRef]
5. Al Tanoury, Z.; Piskunov, A.; Rochette-Egly, C. Vitamin A and retinoid signaling: Genomic and nongenomic effects. *J. Lipid Res.* **2013**, *54*, 1761–1775. [CrossRef] [PubMed]
6. Ross, A.C.; Stephensen, C.B. Vitamin A and retinoids in antiviral responses. *FASEB J.* **1996**, *10*, 979–985. [CrossRef] [PubMed]
7. Rothman, K.J.; Moore, L.L.; Singer, M.R.; Nguyen, U.S.; Mannino, S.; Milunsky, A. Teratogenicity of high vitamin A intake. *N. Engl. J. Med.* **1995**, *333*, 1369–1373. [CrossRef]
8. Holick, M.F. The vitamin D deficiency pandemic: Approaches for diagnosis, treatment and prevention. *Rev. Endocr. Metab. Disord.* **2017**, *18*, 153–165. [CrossRef] [PubMed]
9. Caroli, A.; Poli, A.; Ricotta, D.; Banfi, G.; Cocchi, D. Invited review: Dairy intake and bone health: A viewpoint from the state of the art. *J. Dairy Sci.* **2011**, *94*, 5249–5262. [CrossRef]
10. Chaves, M.A.; Oseliero-Filho, P.L.; Jange, C.G.; Sinigaglia-Coimbra, R.; Oliveira, C.L.P.; Pinho, S.C. Structural characterization of multilamellar liposomes coencapsulating curcumin and vitamin D_3. *Colloids Surf. A Physicochem. Eng. Asp.* **2018**, *549*, 112–121. [CrossRef]
11. Barkoukis, H. Nutrition recommendations in elderly and aging. *Med. Clin. N. Am.* **2016**, *100*, 1237–1250. [CrossRef]
12. Kira, M.; Kobayashi, T.; Yoshikawa, K. Vitamin D and the skin. *J. Dermatol.* **2003**, *30*, 429–437. [CrossRef] [PubMed]
13. Mitani, H.; Naru, E.; Yamashita, M.; Arakane, K.; Suzuki, T.; Imanari, T. Ergocalciferol promotes in vivo differentiation of keratinocytes and reduces photodamage caused by ultraviolet irradiation in hairless mice. *Photodermatol. Photoimmunol. Photomed.* **2004**, *20*, 215–223. [CrossRef]
14. Jiang, Q. Metabolism of natural forms of vitamin E and biological actions of vitamin E metabolites. *Free Radic. Biol. Med.* **2022**, *179*, 375–387. [CrossRef] [PubMed]
15. Ribeiro, A.M.; Estevinho, B.N.; Rocha, F. The progress and application of vitamin E encapsulation—A review. *Food Hydrocoll.* **2021**, *121*, 106998. [CrossRef]
16. Halder, M.; Petsophonsakul, P.; Akbulut, A.C.; Pavlic, A.; Bohan, F.; Anderson, E.; Maresz, K.; Kramann, R.; Schurgers, L. Vitamin K: Double bonds beyond coagulation insights into differences between vitamin K_1 and K_2 in health and disease. *Int. J. Mol. Sci.* **2019**, *20*, 896. [CrossRef] [PubMed]
17. Weish, J.; Bak, M.J.; Narvaez, C.J. New insights into vitamin K biology with relevance to cancer. *Trends Mol. Med.* **2022**, *28*, 864–881. [CrossRef]
18. Shah, N.S.; Lazarus, M.C.; Bugdodel, R.; Hsia, S.L.; He, J.; Duncan, R.; Baumann, L. The effects of topical vitamin K on bruising after laser treatment. *J. Am. Acad. Dermatol.* **2002**, *47*, 241–244. [CrossRef]
19. Kamanna, V.S.; Kashyap, M.L. Mechanism of action of niacin. *Am. J. Cardiol.* **2008**, *101*, S20–S26. [CrossRef] [PubMed]
20. Rémond, D.; Shakar, D.R.; Gille, D.; Pinto, P.; Kachal, J.; Peyron, M.-A.; dos Santos, C.N.; Walther, B.; Bordoni, A.; Dupont, D.; et al. Understanding the gastrointestinal tract of the elderly to develop dietary solutions that prevent malnutrition. *Oncotarget.* **2015**, *6*, 13858–13898. [CrossRef]
21. Caritá, A.C.; Fonseca-Santos, B.; Schultz, J.D.; Michniak-Kohn, B.; Chorilli, M.; Leonardi, G.R. Vitamin C: One compound, several uses. Advances for delivery, efficiency and stability. *Nanomed. Nanotechnol. Biol. Med.* **2020**, *24*, 102117. [CrossRef]
22. Bangham, A.D.; Standish, M.M.; Watkins, J.C. Diffusion of univalent ions across the lamellae of swollen phospholipids. *J. Mol. Biol.* **1965**, *13*, 238–252. [CrossRef]
23. Lasic, D.D. Novel applications of liposomes. *Trends Biotechnol.* **1998**, *16*, 307–321. [CrossRef]
24. Emami, S.; Azadmard-Damirchi, S.; Peighambardoust, S.H.; Valizadeh, H.; Hesari, J. Liposomes as carrier vehicles for functional compounds in food sector. *J. Exp. Nanosci.* **2016**, *11*, 737–759. [CrossRef]
25. Thompson, A.K.; Mozafari, M.R.; Singh, H. The properties of liposomes produced from milk fat globule membrane material using different techniques. *Lait* **2007**, *87*, 349–360. [CrossRef]
26. Van Hoogevest, P.; Wendel, A. The use of natural and synthetic phospholipids as pharmaceutical excipients. *Eur. J. Lipid Sci. Technol.* **2014**, *116*, 1088–1107. [CrossRef] [PubMed]
27. Priya, S.; Desai, V.M.; Singhvi, G. Surface modification of lipid-based nanocarriers: A potential approach to enhance targeted drug delivery. *ACS Omega* **2023**, *8*, 74–86. [CrossRef] [PubMed]
28. Shukla, S.; Haldorai, Y.; Hwang, S.K.; Bajpai, V.K.; Huh, Y.S.; Han, Y.-K. Current demands for food-approved liposome nanoparticles in food and safety sector. *Front. Microbiol.* **2017**, *8*, 2398. [CrossRef]
29. Khorasani, S.; Danaei, M.; Mozafari, M.R. Nanoliposome technology for the food and nutraceutical industries. *Trends Food Sci.* **2018**, *79*, 106–115. [CrossRef]

30. Lombardo, D.; Kiselev, M.A. Methods of liposomes preparation: Formation and control factors of versatile nanocarriers for biomedical and nanomedicine application. *Pharmaceutics* **2022**, *14*, 543. [CrossRef] [PubMed]
31. Liu, P.; Chen, G.; Zhang, J. A review of liposomes as a drug delivery system: Current status of approved products, regulatory environments, and future perspectives. *Molecules* **2022**, *27*, 1372. [CrossRef] [PubMed]
32. Bi, H.; Xue, J.; Jiang, H.; Gao, S.; Yang, D.; Fang, Y.; Shi, K. Current developments in drug delivery with thermosensitive liposomes. *Asian J. Pharm.* **2019**, *14*, 365–379. [CrossRef]
33. Andrade, S.; Ramalho, M.J.; Loureiro, J.A.; Pereira, M.C. Transferrin-functionalized liposomes loaded with vitamin VB12 for Alzheimer's disease therapy. *Int. J. Pharm.* **2022**, *626*, 122167. [CrossRef]
34. Campani, V.; Marchese, D.; Pitaro, M.T.; Pitaro, M.; Grieco, P.; de Rosa, G. Development of a liposome-based formulation for vitamin K_1 nebulization on the skin. *Int. J. Nanomed.* **2014**, *9*, 1823–1832. [CrossRef]
35. Qu, Y.; Tang, J.; Liu, L.; Song, L.; Chen, S.; Gao, Y. α-Tocopherol liposome loaded chitosan hydrogel to suppress oxidative stress injury in cardiomyocytes. *Int. J. Biol. Macromol.* **2019**, *125*, 1192–1202. [CrossRef]
36. Szoka Jr., F.; Papahadjopoulos, D. Procedure for preparation of liposomes with large internal aqueous space and high capture by reverse-phase evaporation. *Proc. Natl. Acad. Sci. USA* **1978**, *75*, 4194–4198. [CrossRef] [PubMed]
37. Shah, S.; Dhawan, V.; Holm, R.; Nagarsenker, M.S.; Perrie, Y. Liposomes: Advancements and innovation in the manufacturing process. *Adv. Drug Deliv. Rev.* **2020**, *154–155*, 102–122. [CrossRef]
38. Guimarães, D.; Cavaco-Paulo, A.; Nogueira, E. Design of liposomes as drug delivery system for therapeutic applications. *Int. J. Pharm.* **2021**, *601*, 120571. [CrossRef]
39. Bigazzi, W.; Penoy, N.; Evrard, B.; Piel, G. Supercritical fluid methods: An alternative to conventional methods to prepare liposomes. *Chem. Eng. J.* **2020**, *383*, 123106. [CrossRef]
40. Andra, V.V.S.N.L.; Pammi, S.V.N.; Bhatraju, L.V.K.P.; Ruddaraju, L.K. A comprehensive review on novel liposomal methodologies, commercial formulations, clinical trials and patents. *BioNanoScience* **2022**, *12*, 274–291. [CrossRef] [PubMed]
41. Favarin, F.R.; Gündel, S.S.; Ledur, C.M.; Roggia, I.; Fagan, S.B.; Gündel, A.; Fogaça, A.O.; Ourique, A.F. Vitamin C as a shelf-life extender in liposomes. *Braz. J. Pharm. Sci.* **2022**, *58*, e20492. [CrossRef]
42. Laouini, A.; Jaafar-Maalej, C.; Limayem-Blouza, I.; Charcosset, C.; Fessi, H. Preparation, characterization and applications of liposomes: State of the art. *J. Colloid Sci. Biotechnol.* **2012**, *1*, 147–168. [CrossRef]
43. Batzri, S.; Korn, E.D. Single bilayer liposomes prepared without sonication. *Biochim. Biophys. Acta* **1973**, *16*, 1015–1019. [CrossRef]
44. Tsai, W.-C.; Rizvi, S.S.H. Liposomal microencapsulation using the conventional methods and novel supercritical fluid processes. *Trends Food Sci. Technol.* **2016**, *55*, 61–71. [CrossRef]
45. Woodle, M.C.; Papahadjopoulos, D. Liposome preparation and size characterization. *Meth. Enzymol.* **1989**, *171*, 193–217. [CrossRef]
46. Van Tran, V.; Moon, J.-Y.; Lee, Y.-C. Liposomes for delivery of antioxidants in cosmeceuticals: Challenges and development strategies. *J. Control. Release* **2019**, *300*, 114–140. [CrossRef] [PubMed]
47. Charcosset, C.; Juban, A.; Valour, J.-P.; Urbaniak, S.; Fessi, H. Preparation of liposomes at large scale using the ethanol injection method: Effect of scale-up and injection devices. *Chem. Eng. Res. Des.* **2015**, *94*, 508–515. [CrossRef]
48. Deamer, D.W.; Bangham, A.D. Large volume liposomes by an ether vaporization method. *Biochim. Biophys. Acta* **1973**, *443*, 629. [CrossRef]
49. Farag, M.A.; Elimam, D.M.; Afifi, S.M. Outgoing and potential trends of the omega-3 rich linseed oil quality characteristics and rancidity management: A comprehensive review for maximizing its food and nutraceutical Applications. *Trends Food Sci. Technol.* **2021**, *114*, 292–309. [CrossRef]
50. Maherani, B.; Arab-Tehrany, E.; Mozafari, M.R.; Gaiani, C.; Linder, M. Liposomes: A review of manufacturing techniques and targeting strategies. *Curr. Nanosci.* **2011**, *7*, 136–452. [CrossRef]
51. Cardoza, J.D.; Kleinfeld, A.M.; Stallcup, K.C.; Merscher, M.F. Hairpin configuration of H-2Kk in liposomes formed by detergent dialysis. *Biochemistry* **1984**, *23*, 4401–4409. [CrossRef]
52. Zumbuehl, O.; Weder, H.G. Liposomes of controllable size in the range of 40 to 180 nm by defined dialysis of lipid/detergent mixed micelles. *Biochim. Biophys. Acta Biomembr.* **1981**, *640*, 252–262. [CrossRef]
53. Lasch, J.; Weissig, V.; Brandl, M. Preparation of liposomes. In *Liposomes—A practical approach*, 2nd ed.; Torchilin, V.P., Weissig, V., Eds.; Oxford University Press: Oxford, UK, 2003; pp. 3–30.
54. Mastrogiacomo, D.; Lenucci, M.S.; Bonfrate, V.; di Carolo, M.; Piro, G.; Valli, L.; Rescio, L.; Milano, F.; Comparelli, R.; de Leo, V.; et al. Lipid/detergent mixed micelles as a tool for transferring antioxidant power from hydrophobic natural extracts into bio-deliverable liposome carriers: The case of lycopene rich oleoresins. *RSC Adv.* **2015**, *5*, 3081–3093. [CrossRef]
55. Shum, H.C.; Lee, D.; Yoon, I.; Kodger, T.; Weitz, D.A. Double emulsion templated monodisperse phospholipid vesicles. *Langmuir* **2008**, *24*, 7651–7653. [CrossRef]
56. Meure, L.A.; Foster, N.R.; Dehghani, F. Conventional and dense gas techniques for the production of liposomes: A review. *AAPS PharmSciTech* **2008**, *9*, 798–809. [CrossRef] [PubMed]
57. Filipczak, N.; Pan, J.; Yalamarty, S.S.K.; Torchilin, V.P. Recent advancements in liposome technology. *Adv. Drug Deliv. Rev.* **2020**, *156*, 4–22. [CrossRef] [PubMed]

58. Kuroiwa, T.; Horikoshi, K.; Suzuki, A.; Neves, M.A.; Kobayashi, I.; Uemura, K.; Nakajima, M.; Kanazawa, A.; Ichikawa, S. Efficient encapsulation of a water-soluble molecule into lipid vesicles using W/O/W multiple emulsions via solvent evaporation. *J. Am. Oil Chem. Soc.* **2016**, *93*, 421–430. [CrossRef]
59. Li, T.; Yang, S.; Liu, W.; Liu, C.; Liu, W.; Zheng, H.; Zhou, W.; Tong, G. Preparation and characterization of nanoscale complex liposomes containing medium-chain fatty acids and vitamin C. *Int. J. Food Prop.* **2015**, *18*, 113–124. [CrossRef]
60. Yang, S.; Liu, C.; Liu, W.; Yu, H.; Zheng, H.; Zhou, W.; Hu, Y. Preparation and characterization of nanoliposomes entrapping medium-chain fatty acids and vitamin C by lyophilization. *Int. J. Mol. Sci.* **2013**, *14*, 19763–19773. [CrossRef]
61. Pattnaik, M.; Mishra, H.N. Effect of ultrasonication and wall materials on the stability, rheology, and encapsulation efficiency of vitamins in a lipid-based double emulsion template. *J. Food Process Eng.* **2022**, e14201. [CrossRef]
62. Payne, N.I.; Timmins, P.; Ambrose, C.V.; Ward, M.D.; Ridgway, F. Proliposomes: A novel solution to an old problem. *J. Pharm. Sci.* **1986**, *75*, 325–329. [CrossRef]
63. Xu, H.; He, L.; Nie, S.; Guan, J.; Zhang, X.; Yang, X.; Pan, W. Optimized preparation of vinpocetine proliposomes by a novel method and in vivo evaluation of its pharmacokinetics in New Zealand rabbits. *J. Control. Release* **2009**, *140*, 61–68. [CrossRef] [PubMed]
64. Dhakal, S.P.; He, J. Microencapsulation of vitamins in food applications to prevent losses in processing and storage: A review. *Food. Res. Int.* **2020**, *137*, 109326. [CrossRef] [PubMed]
65. Lee, S.J.; Wong, M. Nano- and Microencapsulation of Phytochemicals. In *Nano- and Microencapsulation for Foods*; Kwak, H.-S., Ed.; John Wiley & Sons, Ltd.: Hoboken, NJ, USA, 2014; pp. 117–165. [CrossRef]
66. Yu, J.Y.; Chuesiang, P.; Shin, G.H.; Park, H.J. Post-processing techniques for the improvement of liposome stability. *Pharmaceutics* **2021**, *13*, 1023. [CrossRef]
67. Has, C.; Sunthar, P. A comprehensive review on recent preparation techniques of liposomes. *J. Liposome Res.* **2020**, *30*, 336–365. [CrossRef] [PubMed]
68. Maurya, V.K.; Bashir, K.; Aggarwal, M. Vitamin D microencapsulation and fortification: Trends and technologies. *J. Steroid Biochem. Mol. Biol.* **2020**, *196*, 105489. [CrossRef] [PubMed]
69. Carlan, I.C.; Estevinho, B.N.; Rocha, F. Production of vitamin B_1 microparticles by a spray drying process using different biopolymers as wall materials. *Can. J. Chem. Eng.* **2020**, *98*, 1682–1695. [CrossRef]
70. Estevinho, B.N.; Mota, R.; Leite, J.P.; Tamagnini, P.; Gales, L.; Rocha, F. Application of a cyanobacterial extracellular polymeric substance in the microencapsulation of vitamin B_{12}. *Powder Technol.* **2019**, *343*, 644–651. [CrossRef]
71. Toniazzo, T.; Berbel, I.F.; Cho, S.; Fávaro-Trindade, C.S.; Moraes, I.C.F.; Pinho, S.C. β-carotene-loaded liposome dispersions stabilized with xanthan and guar gums: Physico-chemical stability and feasibility of application in yogurt. *LWT* **2014**, *59*, 1265–1273. [CrossRef]
72. Chaves, M.A.; Pinho, S.C. Unpurified soybean lecithins impact on the chemistry of proliposomes and liposome dispersions encapsulating vitamin D_3. *Food Biosci.* **2020**, *37*, 100700. [CrossRef]
73. Mozafari, M.R.; Reed, C.J.; Rostron, C.; Kocum, C.; Piskin, E. Constructing of stable anionic liposome-plasmid particles using the heating method: A preliminary investigation. *Cell. Mol. Biol. Lett.* **2002**, *7*, 923–928.
74. Mozafari, M.R. Liposomes: Na overview of manufacturing techniques. *Cell. Mol. Biol. Lett.* **2005**, *10*, 711–719.
75. Charcosset, C.; Limayem, I.; Fessi, H. The membrane emulsification process—A review. *J. Chem. Technol. Biotechnol.* **2004**, *79*, 209–218. [CrossRef]
76. Jaafar-Maalej, C.; Charcosset, C.; Fessi, H. A new method for liposome preparation using a membrane contactor. *J. Liposome Res.* **2011**, *21*, 213–220. [CrossRef]
77. Laouini, A.; Jaafar-Maalej, C.; Sfar, S.; Charcosset, C.; Fessi, H. Liposome preparation using a hollow fiber membrane contactor—Application to spironolactone encapsulation. *Int. J. Pharm.* **2011**, *415*, 53–61. [CrossRef] [PubMed]
78. Laouini, A.; Charcosset, C.; Fessi, H.; Holdich, R.G.; Vladisavljević, G.T. Preparation of liposomes: A novel application of microengineered membranes—Investigation of the process parameters and application to the encapsulation of vitamin E. *RSC Adv.* **2013**, *3*, 4985–4994. [CrossRef]
79. Angelova, M.I.; Dimitrov, D.S. Liposome electroformation. *Faraday Discuss.* **1986**, *81*, 303–311. [CrossRef]
80. Angelova, M.I.; Soléau, S.; Méléard, P.; Faucon, F.; Bothorel, P. Preparation of giant vesicles by external AC electric fields. Kinetics and applications. In *Trends in Colloid and Interface Science VI*; Lösche, M., Möhwald, H., Helm, C., Eds.; Progress in Colloid & Polymer Science; Steinkopff: Darmstadt, Germany, 1992; Volume 89. [CrossRef]
81. DiPasquale, M.; Nguyen, M.H.L.; Rickeard, B.W.; Cesca, N.; Tannous, C.; Castillo, S.R.; Katsaras, J.; Kelley, E.G.; Heberle, F.A.; Masquardt, D. The antioxidant vitamin E as a membrane raft modulator: Tocopherols do not abolish lipid domains. *Biochim. Biophys. Acta Biomembr.* **2020**, *1862*, 183189. [CrossRef] [PubMed]
82. Cheung, C.C.L.; Monaco, I.; Kostevšek, N.; Franchini, M.C.; Al-Jamal, W.T. Nanoprecipitation preparation of low temperature-sensitive magnetoliposomes. *Colloids Surf. B* **2021**, *198*, 111453. [CrossRef]
83. Jash, A.; Rizvi, S.S.H. Heat-stable liposomes from milk fat globule membrane phospholipids for pH-triggered delivery of hydrophilic and lipophilic bioactives. *Innov. Food Sci. Emerg. Technol.* **2022**, *79*, 103030. [CrossRef]
84. Ottino, J.M.; Wiggins, S. Introduction: Mixing in microfluidics. *Phil. Trans. R. Soc. Lond. A* **2004**, *362*, 923–935. [CrossRef]
85. Carugo, D.; Bottaro, E.; Owen, J.; Stride, E.; Nastruzzi, C. Liposome production by microfluidics: Potential and limiting factors. *Sci. Rep.* **2016**, *6*, 25876. [CrossRef]

86. Dalmoro, A.; Bochicchio, S.; Lamberti, G.; Bertoncin, P.; Janssens, B.; Barba, A.A. Micronutrients encapsulation in enhanced nanoliposomal carriers by a novel preparative technology. *RSC Adv.* **2019**, *9*, 19800–19812. [CrossRef]
87. Ingebrigtsen, S.G.; Škalko-Basnet, N.; Holsæter, A.M. Development and optimization of a new processing approach for manufacturing topical liposomes-in-hydrogel drug formulations by dual asymmetric centrifugation. *Drug Dev. Ind. Pharm.* **2016**, *42*, 1375–1383. [CrossRef]
88. Massing, U.; Cicko, S.; Ziroli, V. Dual asymmetric centrifugation (DAC)—A new technique for liposome preparation. *J. Control. Release* **2008**, *125*, 16–24. [CrossRef]
89. Huang, Z.; Li, X.; Zhang, T.; Song, Y.; She, Z.; Li, J.; Deng, Y. Progress involving new techniques for liposome preparation. *Asian J. Pharm.* **2014**, *9*, 176–182. [CrossRef]
90. Peschka, R.; Dennehy, C.; Szoka Jr., F.C. A simple in vitro model to study the release kinetics of liposome encapsulated material. *J. Control. Release* **1998**, *56*, 41–51. [CrossRef]
91. Kapoor, B.; Gupta, R.; Gulati, M.; Singh, S.K.; Khursheed, R.; Gupta, M. The Why, Where, Who, How, and What of the vesicular delivery systems. *Adv. Colloid Interface Sci.* **2019**, *271*, 101985. [CrossRef] [PubMed]
92. Hauschild, S.; Lipprandt, U.; Rumplecker, A.; Borchert, U.; Rank, A.; Schubert, R.; Förster, S. Direct preparation and loading of lipid and polymer vesicles using inkjets. *Small* **2005**, *1*, 1177–1180. [CrossRef] [PubMed]
93. Bnyan, R.; Cesarini, L.; Khan, I.; Roberts, M.; Ehtezazi, T. The effect of ethanol evaporation on the properties of inkjet produced liposomes. *DARU J. Pharm.* **2020**, *28*, 271–280. [CrossRef] [PubMed]
94. Eleftheriadis, G.; Monou, P.K.; Andriotis, E.; Mitsouli, E.; Moutafidou, N.; Markopoulou, C.; Bouropoulos, N.; Fatouros, D. Development and characterization of inkjet printed edible films for buccal delivery of b-complex vitamins. *Pharmaceutics* **2020**, *13*, 203. [CrossRef]
95. Merlo-Mas, J.; Tomsen-Melero, J.; Corchero, J.-L.; González-Mira, E.; Font, A.; Pedersen, J.N.; García-Aranda, N.; Cristóbal-Lecina, E.; Alcaina-Hernando, M.; Mendoza, R.; et al. Application of Quality by Design to the robust preparation of a liposomal GLA formulation by DELOS-susp method. *J. Supercrit. Fluids* **2021**, *173*, 105204. [CrossRef]
96. Cano-Sarabia, M.; Ventosa, N.; Sala, S.; Patiño, C.; Arranz, R.; Veciana, J. Preparation of uniform rich cholesterol unilamellar nanovesicles using CO_2-expanded solvents. *Langmuir* **2008**, *24*, 2433–2437. [CrossRef]
97. Leitgeb, M.; Knez, Ž.; Primožič, M. Sustainable technologies for liposome preparation. *J, Supercrit. Fluids* **2020**, *165*, 104984. [CrossRef]
98. Zhao, L.; Temelli, F. Preparation of liposomes using a modified supercritical process via depressurization of liquid phase. *J. Supercrit. Fluids* **2015**, *100*, 110–120. [CrossRef]
99. Elizondo, E.; Moreno, E.; Cabrera, I.; Córdoba, A.; Sala, S.; Veciana, J.; Ventosa, N. Liposomes and other vesicular systems: Structural characteristics, methods of preparation, and use in nanomedicine. *Prog. Mol. Biol. Transl. Sci.* **2011**, *104*, 1–52. [CrossRef] [PubMed]
100. Meure, L.A.; Knott, R.; Foster, N.R.; Dehgani, F. The depressurization of an expanded solution into aqueous media for the bulk production of liposomes. *Langmuir* **2009**, *25*, 326–337. [CrossRef]
101. Beh, C.C.; Mammucari, R.; Foster, N.R. Process intensification: Nano-carrier formation by a continuous dense gas process. *Chem. Eng. J.* **2015**, *266*, 320–328. [CrossRef]
102. Beh, C.C.; Wong, M.G.; Olet, V.; Foster, N. Development of a novel continuous dense gas process for the production of residual solvent-free self-assembled nano-carriers. *Chem. Eng. Process.* **2019**, *143*, 107589. [CrossRef]
103. Karn, P.R.; Cho, W.; Hwang, S.-J. Liposomal drug products and recent advances in the synthesis of supercritical fluid-mediated liposomes. *Nanomedicine* **2013**, *8*, 1529–1548. [CrossRef]
104. Frederiksen, L.; Anton, K.; van Hoogevest, P.; Keller, H.R.; Leuenberger, H. Preparation of liposomes encapsulating water-soluble compounds using supercritical carbon dioxide. *J. Pharm. Sci.* **2000**, *86*, 921–928. [CrossRef]
105. Wen, Z.; Liu, B.; Zheng, Z.K.; You, X.K.; Pu, Y.T.; Li, Q. Preparation of liposome particle of atractylone by supercritical carbon dioxide process. *Adv. Mater. Res.* **2010**, *92*, 177–182. [CrossRef]
106. Trucillo, P.; Campardelli, R.; Reverchon, E. Liposomes: From Bangham to supercritical fluids. *Processes* **2020**, *8*, 1022. [CrossRef]
107. Tsai, W.-C.; Rizvi, S.S.H. Simultaneous microencapsulation of hydrophilic and lipophilic bioactives in liposomes produced by an ecofriendly supercritical fluid process. *Food Res. Int.* **2017**, *99*, 256–262. [CrossRef]
108. Jiao, Z.; Wang, X.; Han, S.; Zha, X.; Xia, J. Preparation of vitamin C liposomes by rapid expansion of supercritical solution process: Experiments and optimization. *J. Drug Deliv. Sci. Technol.* **2019**, *51*, 1–6. [CrossRef]
109. Jiao, Z.; Han, S.; Wang, W.; Song, J.; Cheng, J. Preparation and optimization of Vitamin E acetate liposomes using a modified RESS process combined with response surface methodology. *Part. Sci. Technol.* **2020**, *38*, 863–875. [CrossRef]
110. Jash, A.; Ubeyitogullari, A.; Rizvi, S.S.H. Liposomes for oral delivery of protein and peptide-based therapeutics: Challenges, formulation strategies, and advances. *J. Mater. Chem. B* **2021**, *9*, 4773–4792. [CrossRef]
111. Sharifi, F.; Jash, A.; Abbaspourrad, A.; Rizvi, S.S.H. Generation of ironized and multivitamin-loaded liposomes using venturi-based rapid expansion of a supercritical solution (Vent-RESS). *Green Chem.* **2020**, *22*, 1618–1629. [CrossRef]
112. Han, Y.; Cheng, J.; Ruan, N.; Jiao, Z. Preparation of liposomes composed of supercritical carbon dioxide-philic phospholipids using the rapid expansion of supercritical solution process. *J. Drug Deliv. Sci. Technol.* **2021**, *64*, 102568. [CrossRef]
113. Otake, K.; Imura, T.; Sakai, H.; Abe, M. Development of a new preparation method of liposomes using supercritical carbon dioxide. *Langmuir* **2001**, *17*, 3898–3901. [CrossRef]

114. Imura, T.; Otake, K.; Hashimoto, S.; Gotoh, T.; Yuasa, M.; Yokoyama, S.; Sakai, H.; Rathman, J.F.; Abe, M. Preparation and physicochemical properties of various soybean lecithin liposomes using supercritical reverse phase evaporation method. *Colloids Surf. B* **2003**, *27*, 133–140. [CrossRef]
115. Otake, K.; Shimomura, T.; Goto, T.; Imura, T.; Furuya, T.; Yoda, S.; Takebayashi, Y.; Sakai, H.; Abe, M. Preparation of liposomes using an improved supercritical reverse phase evaporation method. *Langmuir* **2006**, *22*, 2543–2550. [CrossRef]
116. Zhao, L.; Temelli, F. Preparation of anthocyanin-loaded liposomes using an improved supercritical carbon dioxide method. *Innov. Food Sci. Emerg. Technol.* **2017**, *39*, 119–128. [CrossRef]
117. Lesoin, L.; Boutin, O.; Crampon, C.; Badens, E. CO_2/water/surfactant ternary systems and liposome formation using supercritical CO_2: A review. *Colloids Surf. A Physicochem. Eng. Asp.* **2011**, *377*, 1–14. [CrossRef]
118. Xia, F.; Jin, H.; Zhao, Y.; Guo, X. Supercritical antisolvent-based technology for preparation of vitamin D_3 proliposome and its characteristics. *Chin. J. Chem. Eng.* **2011**, *19*, 1039–1046. [CrossRef]
119. Espírito-Santo, I.; Campardelli, R.; Albuquerque, E.C.; de Melo, S.V.; della Porta, G.; Reverchon, E. Liposomes preparation using a supercritical fluid assisted continuous process. *Chem. Eng. J.* **2014**, *249*, 153–159. [CrossRef]
120. Campardelli, R.; Espírito-Santo, I.; Albuquerque, E.C.; de Melo, S.V.; della Porta, G.; Reverchon, E. Efficient encapsulation of proteins in submicro liposomes using a supercritical fluid assisted continuous process. *J. Supercrit. Fluids* **2016**, *107*, 163–169. [CrossRef]
121. Espírito-Santo, I.; Campardelli, R.; Albuquerque, E.C.; de Melo, S.A.B.V.; Reverchon, E.; della Porta, G. Liposomes size engineering by combination of ethanol injection and supercritical processing. *J. Pharm. Sci.* **2015**, *104*, 3842–3850. [CrossRef] [PubMed]
122. Trucillo, P.; Campardelli, R.; Scognamiglio, M.; Reverchon, E. Control of liposomes diameter at micrometric and nanometric level using a supercritical assisted technique. *J. CO2 Util.* **2019**, *32*, 119–127. [CrossRef]
123. Campardelli, R.; Trucillo, P.; Reverchon, E. A supercritical fluid-based process for the production of fluorescein-loaded liposomes. *Ind. Eng. Chem. Res.* **2016**, *55*, 5359–5365. [CrossRef]
124. Trucillo, P.; Campardelli, R.; Reverchon, E. Supercritical CO_2 assisted liposomes formation: Optimization of the lipidic layer for an efficient hydrophilic drug loading. *J. CO2 Util.* **2017**, *18*, 181–188. [CrossRef]
125. Trucillo, P.; Campardelli, R.; Reverchon, E. Production of liposomes loaded with antioxidants using a supercritical CO_2 assisted process. *Powder Technol.* **2018**, *323*, 155–162. [CrossRef]
126. Trucillo, P.; Campardelli, R.; Reverchon, E. A versatile supercritical assisted process for the one-shot production of liposomes. *J. Supercrit. Fluids* **2019**, *146*, 136–143. [CrossRef]
127. Chaves, M.A.; Baldino, L.; Pinho, S.C.; Reverchon, E. Supercritical CO_2 assisted process for the production of mixed phospholipid nanoliposomes: Unloaded and vitamin D_3-loaded vesicles. *J. Food. Eng.* **2022**, *316*, 110851. [CrossRef]
128. Chaves, M.A.; Baldino, L.; Pinho, S.C.; Reverchon, E. Co-encapsulation of curcumin and vitamin D_3 in mixed phospholipid nanoliposomes using a continuous supercritical CO_2 assisted process. *J. Taiwan Inst. Chem. Eng.* **2022**, *132*, 104120. [CrossRef]
129. Bleich, J.; Müller, B.W. Production of drug loaded microparticles by the use of supercritical gases with the Aerosol Solvent Extraction System (ASES) process. *J. Microencapsul.* **1996**, *13*, 131–139. [CrossRef] [PubMed]
130. Kunastitchai, S.; Pichert, L.; Sarisuta, N.; Müller, B.W. Application of aerosol solvent extraction system (ASES) process for preparation of liposomes in a dry and reconstitutable form. *Int. J. Pharm.* **2006**, *316*, 93–101. [CrossRef] [PubMed]
131. Kunastitchai, S.; Sarisuta, N.; Panyarachun, B.; Müller, B.W. Physical and chemical stability of miconazole liposomes prepared by Supercritical Aerosol Solvent Extraction System (ASES) process. *Pharm. Dev.* **2007**, *12*, 361–370. [CrossRef] [PubMed]
132. De Paz, E.; Martín, Á.; Cocero, M.J. Formulation of β-carotene with soybean lecithin by PGSS (Particles from Gas Saturated Solutions)-drying. *J. Supercrit. Fluids* **2012**, *72*, 125–133. [CrossRef]
133. Varona, S.; Martín, Á.; Cocero, M.J. Liposomal incorporation of lavandin essential oil by a thin-film hydration method and by particles from gas-saturated solutions. *Ind. Eng. Chem. Res.* **2011**, *50*, 2088–2097. [CrossRef]
134. Rovoli, M.; Pappas, I.; Lalas, S.; Gortzi, O.; Kontopidis, G. In vitro and in vivo assessment of vitamin A encapsulation in a liposome–protein delivery system. *J. Liposome Res.* **2019**, *29*, 142–152. [CrossRef]
135. Lee, D.-U.; Park, H.-W.; Lee, S.-C. Comparing the stability of retinol in liposomes with cholesterol, β-sitosterol, and stigmasterol. *Food Sci. Biotechnol.* **2021**, *30*, 389–394. [CrossRef]
136. Pezeshky, A.; Ghanbarzadeh, B.; Hamishehkar, H.; Moghadam, M.; Babazadeh, A. Vitamin A palmitate-bearing nanoliposomes: Preparation and characterization. *Food Biosci.* **2016**, *13*, 49–55. [CrossRef]
137. Fan, C.; Feng, T.; Wang, X.; Xia, S.; Swing, C.J. Liposomes for encapsulation of liposoluble vitamins (A, D, E and K): Comparation of loading ability, storage stability and bilayer dynamics. *Food Res. Int.* **2023**, *163*, 112264. [CrossRef]
138. Cansell, M.; Moussaoui, N.; Lefrançois, C. Stability of marine lipid based-liposomes under acid conditions. Influence of xanthan gum. *J. Liposome Res.* **2001**, *11*, 229–242. [CrossRef]
139. Fathima, S.J.; Fathima, I.; Abhishek, V.; Khanum, F. Phosphatidylcholine, an edible carrier for nanoencapsulation of unstable thiamine. *Food Chem.* **2016**, *197*, 562–570. [CrossRef] [PubMed]
140. Ahmad, I.; Arsalan, A.; Ali, S.A.; Sheraz, M.A.; Ahmed, S.; Anwar, Z.; Munir, I.; Shah, M.R. Formulation and stabilization of riboflavin in liposomal preparations. *J. Photochem. Photobiol. B Biol.* **2015**, *153*, 358–366. [CrossRef]
141. Ota, A.; Istenič, K.; Skrt, M.; Šegatin, N.; Žnidaršič, N.; Kogej, K.; Ulrih, N.P. Encapsulation of pantothenic acid into liposomes and into alginate or alginate–pectin microparticles loaded with liposomes. *J. Food Eng.* **2018**, *229*, 21–31. [CrossRef]

142. Jiao, Z.; Wang, X.; Yin, Y.; Xia, J.; Mei, Y. Preparation and evaluation of a chitosan-coated antioxidant liposome containing vitamin C and folic acid. *J. Microencapsul.* **2018**, *35*, 272–280. [CrossRef] [PubMed]
143. Marchianò, V.; Matos, M.; Serrano, E.; Álvarez, J.R.; Marcet, I.; Blanco-López, M.C.; Gutiérrez, G. Lyophilised nanovesicles loaded with vitamin B_{12}. *J. Mol. Liq.* **2022**, *365*, 120129. [CrossRef]
144. Bochicchio, S.; Barba, A.A.; Grassi, G.; Lamberti, G. Vitamin delivery: Carriers based on nanoliposomes produced via ultrasonic irradiation. *LWT* **2016**, *69*, 9–16. [CrossRef]
145. Liu, X.; Wang, P.; Zou, Y.-X.; Luo, Z.-G.; Tamer, T.M. Co-encapsulation of vitamin C and β-Carotene in liposomes: Storage stability, antioxidant activity, and in vitro gastrointestinal digestion. *Food Res. Int.* **2020**, *136*, 109587. [CrossRef] [PubMed]
146. Łukawski, M.; Dałek, P.; Borowik, T.; Foryś, A.; Langner, M.; Witkiewicz, W.; Przybyło, M. New oral liposomal vitamin C formulation: Properties and bioavailability. *J. Liposome Res.* **2020**, *30*, 227–234. [CrossRef]
147. Huang, Z.; Brennan, C.S.; Zhao, H.; Liu, J.; Guan, W.; Mohan, M.S.; Stipkovits, L.; Zheng, H.; Kulasiri, D. Fabrication and assessment of milk phospholipid-complexed antioxidant phytosomes with vitamin C and E: A comparison with liposomes. *Food Chem.* **2020**, *324*, 126837. [CrossRef]
148. Kunthia, A.; Kumar, R.; Premjit, Y.; Mitra, J. Release behavior of vitamin C nanoliposomes from starch–vitamin C active packaging films. *J. Food Process Eng.* **2022**, *45*, e14075. [CrossRef]
149. Farhang, B.; Kakuda, Y.; Corredig, M. Encapsulation of ascorbic acid in liposomes prepared with milk fat globule membrane-derived phospholipids. *Dairy Sci. Technol.* **2012**, *92*, 353–366. [CrossRef]
150. Marsanasco, M.; Márquez, A.L.; Wagner, J.R.; Alonso, S.V.; Chiaramoni, N.S. Liposomes as vehicles for vitamins E and C: An alternative to fortify orange juice and offer vitamin C protection after heat treatment. *Food Res. Int.* **2011**, *44*, 3039–3046. [CrossRef]
151. Liu, H.; Meng, X.; Li, L.; Hu, X.; Fang, Y.; Xia, Y. Synergistic effect on antioxidant activity of vitamin C provided with acidic vesiculation of hybrid fatty acids. *J. Funct. Foods* **2021**, *85*, 104647. [CrossRef]
152. Banville, C.; Vuillemard, J.C.; Lacroix, C. Comparison of different methods for fortifying Cheddar cheese with vitamin D. *Int. Dairy J.* **2000**, *10*, 375–382. [CrossRef]
153. Dałek, P.; Drabik, D.; Wołczańska, H.; Foryś, A.; Jagas, M.; Jędruchniewicz, N.; Przybyło, M.; Witkiewicz, W.; Langner, M. Bioavailability by design—Vitamin D_3 liposomal delivery vehicles. *Nanomed. Nanotechnol. Biol. Med.* **2022**, *43*, 102552. [CrossRef]
154. Xu, T.; Zhang, J.; Jin, R.; Cheng, R.; Wang, X.; Yuan, C.; Gan, C. Physicochemical properties, antioxidant activities and in vitro sustained release behaviour of co-encapsulated liposomes as vehicle for vitamin E and β-carotene. *J. Sci. Food Agric.* **2022**, *102*, 5759–5767. [CrossRef]
155. Padamwar, M.N.; Pokharkar, V.B. Development of vitamin loaded topical liposomal formulation using factorial design approach: Drug deposition and stability. *Int. J. Pharm.* **2006**, *320*, 37–44. [CrossRef] [PubMed]
156. Amiri, S.; Ghanbarzadeh, B.; Hamishehkar, H.; Hosein, M.; Babazadeh, A.; Adun, P. Vitamin E loaded nanoliposomes: Effects of gammaoryzanol, polyethylene glycol and lauric acid on physicochemical properties. *Colloids Interface Sci. Commun.* **2018**, *26*, 1–6. [CrossRef]
157. Ma, Q.H.; Kuang, Y.Z.; Hao, X.Z.; Gu, N. Preparation and characterization of tea polyphenols and vitamin E loaded nanoscale complex liposome. *J. Nanosci. Nanotechnol.* **2009**, *9*, 1379–1383. [CrossRef] [PubMed]
158. Souri, J.; Almasi, H.; Hamishehkar, H.; Amjadi, S. Sodium caseinate-coated and β-cyclodextrin/vitamin E inclusion complex-loaded nanoliposomes: A novel stabilized nanocarrier. *LWT* **2021**, *151*, 112174. [CrossRef]
159. Xia, S.; Tan, C.; Xue, J.; Lou, X.; Zhang, X.; Feng, B. Chitosan/tripolyphosphate-nanoliposomes core-shell nanocomplexes as vitamin E carriers: Shelf-life and thermal properties. *Int. J. Food Sci.* **2013**, *49*, 1367–1374. [CrossRef]
160. Samadi, N.; Azar, P.A.; Husain, S.W.; Maibach, H.I.; Nafisi, S. Experimental design in formulation optimization of vitamin K_1 oxide-loaded nanoliposomes for skin delivery. *Int. J. Pharm.* **2020**, *579*, 119136. [CrossRef]
161. Kim, H.-H.Y.; Baianu, I.C. Novel liposome microencapsulation techniques for food applications. *Trends Food Sci. Technol.* **1991**, *2*, 55–61. [CrossRef]
162. Marsanasco, M.; Márquez, A.L.; Wagner, J.R.; Chiaramoni, N.S.; Alonso, S.V. Bioactive compounds as functional food ingredients: Characterization in model system and sensory evaluation in chocolate milk. *J. Food Eng.* **2015**, *166*, 55–63. [CrossRef]
163. Wechtersbach, L.; Ulrih, N.P.; Cigić, B. Liposomal stabilization of ascorbic acid in model systems and in food matrices. *LWT* **2012**, *45*, 43–49. [CrossRef]
164. Mohammadi, M.; Ghanbarzadeh, B.; Hamishehkar, H. Formulation of nanoliposomal vitamin D_3 for potential application in beverage fortification. *Adv. Pharm. Bull.* **2014**, *4*, 569–575. [CrossRef] [PubMed]
165. Didar, Z. Enrichment of dark chocolate with vitamin D_3 (free or liposome) and assessment quality parameters. *J. Food Sci. Technol.* **2021**, *58*, 3065–3072. [CrossRef]
166. Didar, Z. Inclusion of vitamin D_3 (free or liposome) into white chocolate and an investigation of its stability during storage. *J. Food Process. Preserv.* **2021**, *45*, e15231. [CrossRef]
167. Chaves, M.A.; Franckin, V.; Sinigaglia-Coimbra, R.; Pinho, S.C. Nanoliposomes coencapsulating curcumin and vitamin D_3 produced by hydration of proliposomes: Effects of the phospholipid composition in the physicochemical characteristics of vesicles and after incorporation in yoghurts. *Int. J. Dairy Technol.* **2021**, *74*, 107–117. [CrossRef]
168. Liu, W.; Tian, M.; Kong, Y.; Lu, J.; Li, N.; Han, J. Multilayered vitamin C nanoliposomes by self-assembly of alginate and chitosan: Long-term stability and feasibility application in mandarin juice. *LWT* **2017**, *75*, 608–615. [CrossRef]

169. Cui, H.; Zhao, C.; Lin, L. The specific antibacterial activity of liposome-encapsulated Clove oil and its application in tofu. *Food Control* **2015**, *56*, 128–134. [CrossRef]
170. Malheiros, P.S.; Sant'Anna, V.; Barbosa, M.S.; Brandelli, A.; Franco, B.D.G.M. Effect of liposome-encapsulated nisin and bacteriocin-like substance P34 on *Listeria monocytogenes* growth in Minas frescal cheese. *Int. J. Food Microbiol.* **2012**, *156*, 272–277. [CrossRef]
171. Lopez-Polo, J.; Silva-Weiss, A.; Zamorano, M.; Osorio, F.A. Humectability and physical properties of hydroxypropyl methylcellulose coatings with liposome-cellulose nanofibers: Food application. *Carbohydr. Polym.* **2020**, *231*, 115702. [CrossRef]
172. Toniazzo, T.; Galeskas, H.; Dacanal, G.C.; Pinho, S.C. Production of cornstarch granules enriched with quercetin liposomes by aggregation of particulate binary mixtures using high shear process. *J. Food Sci.* **2017**, *82*, 2626–2633. [CrossRef]
173. Amjadi, S.; Ghorbani, M.; Hamishehkar, H.; Roufegarinejad, L. Improvement in the stability of betanin by liposomal nanocarriers: Its application in gummy candy as a food model. *Food Chem.* **2018**, *256*, 156–162. [CrossRef]
174. Lu, S.; Tao, J.; Liu, X.; Wen, Z. Baicalin-liposomes loaded polyvinyl alcohol-chitosan electrospinning nanofibrous films: Characterization, antibacterial properties and preservation effects on mushrooms. *Food Chem.* **2022**, *371*, 131372. [CrossRef] [PubMed]
175. Li, C.; Bai, M.; Chen, X.; Hu, W.; Cui, H.; Lin, L. Controlled release and antibacterial activity of nanofibers loaded with basil essential oil-encapsulated cationic liposomes against *Listeria monocytogenes*. *Food Biosci.* **2022**, *46*, 101578. [CrossRef]
176. Pinilla, C.M.B.; Thys, R.C.S.; Brandelli, A. Antifungal properties of phosphatidylcholine-oleic acid liposomes encapsulating garlic against environmental fungal in wheat bread. *Int. J. Food Microbiol.* **2019**, *293*, 72–78. [CrossRef] [PubMed]
177. Ghorbanzade, T.; Jafari, S.M.; Akhavan, S.; Hadavi, R. Nano-encapsulation of fish oil in nano-liposomes and its application in fortification of yogurt. *Food Chem.* **2017**, *216*, 146–152. [CrossRef]
178. Hua, S. Lipid-based nano-delivery systems for skin delivery of drugs and bioactives. *Front. Pharmacol.* **2015**, *6*, 219. [CrossRef]
179. Bozzuto, G.; Molinari, A. Liposomes as nanomedical devices. *Int. J. Nanomed.* **2015**, *10*, 975–999. [CrossRef] [PubMed]
180. US Food and Drug Administration (FDA), Cosmeceutical. 2018. Available online: https://www.fda.gov/cosmetics/labeling/claims/ucm127064.htm (accessed on 23 February 2023).
181. Dubey, S.K.; Dey, A.; Singhvi, G.; Pandey, M.M.; Singh, V.; Kesharwani, P. Emerging trends of nanotechnology in advanced cosmetics. *Colloids Surf. B* **2022**, *214*, 112440. [CrossRef] [PubMed]
182. Figueroa-Robles, A.; Antunes-Ricardo, M.; Guajardo-Flores, D. Encapsulation of phenolic compounds with liposomal improvement in the cosmetic industry. *Int. J. Pharm.* **2021**, *593*, 120125. [CrossRef] [PubMed]
183. Dymek, M.; Sikora, E. Liposomes as biocompatible and smart delivery systems—The current state. *Adv. Colloid Interface Sci.* **2022**, *309*, 102757. [CrossRef] [PubMed]
184. Maione-Silva, L.; de Castro, E.G.; Nascimento, T.L.; Cintra, E.R.; Moreira, L.C.; Cintra, B.A.S.; Valadares, M.C.; Lima, E.M. Ascorbic acid encapsulated into negatively charged liposomes exhibits increased skin permeation, retention and enhances collagen synthesis by fibroblasts. *Sci. Rep.* **2019**, *9*, 1–14. [CrossRef]
185. Dreier, J.; Sørensen, J.A.; Brewer, J.R. Superresolution and fluorescence dynamics evidence reveal that intact liposomes do not cross the human skin barrier. *PLoS ONE* **2016**, *11*, e0146514. [CrossRef]
186. Touti, R.; Noun, M.; Guimberteau, F.; Lecomte, S.; Faure, C. What is the fate of multilamellar liposomes of controlled size, charge and elasticity in artificial and animal skin? *Eur. J. Pharm. Biopharm.* **2020**, *151*, 18–31. [CrossRef]
187. Choi, J.U.; Lee, S.W.; Pangeni, R.; Byun, Y.; Yoon, I.S.; Park, J.W. Preparation and in vivo evaluation of cationic elastic liposomes comprising highly skin-permeable growth factors combined with hyaluronic acid for enhanced diabetic wound-healing therapy. *Acta Biomater.* **2017**, *57*, 197–215. [CrossRef] [PubMed]
188. Caddeo, C.; Manconi, M.; Fadda, A.M.; Lai, F.; Lampis, S.; Diez0Sales, O.; Sinico, C. Nanocarriers for antioxidant resveratrol: Formulation approach, vesicle self-assembly and stability evaluation. *Colloids Surf. B* **2013**, *111*, 327–332. [CrossRef]
189. Nguyen, S.; Alund, S.J.; Hiorth, M.; Kjøniksen, A.-L.; Smistad, G. Studies on pectin coating of liposomes for drug delivery. *Colloids Surf. B* **2011**, *88*, 664–673. [CrossRef] [PubMed]
190. Nadaf, S.J.; Killedar, S.G. Ultradeformable liposomal nanostructures: Role in transdermal delivery of therapeutics. In *Nanoscale Processing*; Thomas, S., Balakrishnan, P., Eds.; Elsevier: Amsterdam, The Netherlands, 2021; pp. 492–516. [CrossRef]
191. Ahmadi Ashtiani, H.R.; Bishe, P.; Lashgari, N.; Nilforoushzadeh, M.A.; Zare, S. Liposomes in Cosmetics. *J. Skin Stem Cell* **2016**, *3*, e65815. [CrossRef]
192. Shaw, T.K.; Paul, P.; Chatterjee, B. Research-based findings on scope of liposome-based cosmeceuticals: An updated review. *Futur. J. Pharm. Sci.* **2022**, *8*, 46. [CrossRef]
193. Taylor, T.M.; Weiss, J.; Davidson, P.M.; Bruce, B.D. Liposomal nanocapsules in food science and agriculture. *Crit. Rev. Food Sci. Nutr.* **2005**, *45*, 587–605. [CrossRef]
194. Barenholz, Y.; Gibbes, D.; Litman, B.J.; Goll, J.; Thompson, T.E.; Carlson, F.D. A simple method for the preparation of homogeneous phospholipid vesicles. *Biochemistry* **1977**, *16*, 2806–2810. [CrossRef]
195. Amiri, S.; Rezazadeh-Bari, M.; Alizadeh-Khaledabad, M.; Amiri, S. New formulation of vitamin C encapsulation by nano-liposomes: Production and evaluation of particle size, stability and control release. *Food Sci. Biotechnol.* **2019**, *28*, 423–432. [CrossRef]
196. Hope, M.J.; Bally, M.B.; Webb, G.; Cullis, P.R. Production of large unilamellar vesicles by a rapid extrusion procedure: Characterization of size distribution, trapped volume and ability to maintain a membrane potential. *Biochim. Biophys. Acta.* **1985**, *812*, 55–65. [CrossRef]

197. Barnadas-Rodríguez, R.; Sabés, M. Factors involved in the production of liposomes with a high-pressure homogenizer. *Int. J. Pharm.* **2001**, *213*, 175–186. [CrossRef]
198. He, W.; Guo, X.; Feng, M.; Mao, N. In vitro and in vivo studies on ocular vitamin A palmitate cationic liposomal in situ gels. *Int. J. Pharm.* **2013**, *458*, 305–314. [CrossRef]
199. Barnadas-Rodríguez, R.; Sabés Xamaní, M. Liposomes prepared by high-pressure homogenizers. *Methods Enzymol.* **2003**, *367*, 28–46. [CrossRef]
200. Yang, S.; Liu, W.; Liu, C.; Liu, W.; Tong, G.; Zheng, H.; Zhou, W. Characterization and bioavailability of vitamin C nanoliposomes prepared by film evaporation-dynamic high pressure microfluidization. *J. Dispers. Sci. Technol.* **2012**, *33*, 1608–1614. [CrossRef]
201. Zhou, W.; Liu, W.; Zou, L.; Liu, W.; Liu, C.; Liang, R.; Chen, J. Storage stability and skin permeation of vitamin C liposomes improved by pectin coating. *Colloids Surf. B.* **2014**, *117*, 330–337. [CrossRef]
202. Bosworth, M.E.; Hunt, C.A.; Pratt, D. Liposome dialysis for improved size distributions. *J. Pharm. Sci.* **1982**, *71*, 806–812. [CrossRef] [PubMed]
203. Zhu, T.F.; Szostak, J.W. Preparation of large monodisperse vesicles. *PLoS ONE* **2009**, *4*, e5009. [CrossRef] [PubMed]
204. Adamala, K.; Engelhart, A.E.; Kamat, N.P.; Jin, L.; Szostak, J.W. Construction of a liposome dialyzer for the preparation of high-value, small-volume liposome formulations. *Nat. Protoc.* **2015**, *10*, 927–938. [CrossRef]
205. Castile, J.D.; Taylor, K.M.G. Factors affecting the size distribution of liposomes produced by freeze–thaw extrusion. *Int. J. Pharm.* **1999**, *188*, 87–95. [CrossRef]
206. Kirby, C.; Gregoriadis, G. Dehydration-rehydration vesicles: A simple method for high yield drug entrapment in liposomes. *Nat. Biotechnol.* **1984**, *2*, 979–984. [CrossRef]
207. Crowe, J.H.; Crowe, L.M.; Carpenter, J.F.; Wistrom, C.A. Stabilization of dry phospholipid bilayers and proteins by sugars. *Biochem. J.* **1987**, *242*, 1–10. [CrossRef]
208. Hudiyanti, D.; Hamidi, N.I.; Anugrah, D.S.B.; Salimah, S.N.M.; Siahaan, P. Encapsulation of vitamin C in sesame liposomes: Computational and experimental studies. *Open Chem.* **2019**, *17*, 537–543. [CrossRef]
209. Unden, G.; Mörschel, E.; Bokranz, M.; Kröger, A. Structural properties of the proteoliposomes catalyzing electron transport from formate to fumarate. *Biochim. Biophys. Acta Bioenerg.* **1983**, *725*, 41–48. [CrossRef]
210. Battistelli, M.; Salucci, S.; Falcieri, E. Morphological evaluation of liposomal iron carriers. *Microsc. Res. Tech.* **2018**, *81*, 1295–1300. [CrossRef] [PubMed]
211. Li, C.; Deng, Y. A novel method for the preparation of liposomes: Freeze drying of monophase solutions. *J. Pharm. Sci.* **2004**, *93*, 1403–1414. [CrossRef] [PubMed]
212. Guimarães, D.; Noro, J.; Silva, C.; Cavaco-Paulo, A.; Nogueira, E. Protective effect of saccharides on freeze-dried liposomes encapsulating drugs. *Front. Bioeng. Biotechnol.* **2019**, *7*, 424. [CrossRef] [PubMed]
213. Koudelka, Š.; Mašek, J.; Neuzil, J.; Turánek, J. Lyophilised liposome-based formulations of α-tocopheryl succinate: Preparation and physico-chemical characterisation. *J. Pharm. Sci.* **2010**, *99*, 2434–2443. [CrossRef] [PubMed]
214. Hua, Z.-Z.; Li, B.-G.; Liu, Z.-J.; Sun, D.-W. Freeze-drying of liposomes with cryoprotectants and its effect on retention rate of encapsulated ftorafur and vitamin A. *Dry. Technol.* **2003**, *21*, 1491–1505. [CrossRef]

nanomaterials

MDPI

Article

PEG-Coated MnZn Ferrite Nanoparticles with Hierarchical Structure as MRI Contrast Agent

Sedigheh Cheraghali [1], Ghasem Dini [2,*], Isabella Caligiuri [3], Michele Back [1,*] and Flavio Rizzolio [1,3]

1 Department of Molecular Sciences and Nanosystems, Ca' Foscari University of Venice, 30172 Venice, Italy
2 Department of Nanotechnology, Faculty of Chemistry, University of Isfahan, Isfahan 81746-73441, Iran
3 Pathology Unit, Centro di Riferimento Oncologico di Aviano (CRO) IRCCS, 33081 Aviano, Italy
* Correspondence: g.dini@sci.ui.ac.ir (G.D.); michele.back@unive.it (M.B.); Tel.: +98-31-3793-4914 (G.D.); Fax: +98-379-32700 (G.D.)

Abstract: In this work, MnZn ferrite nanoparticles with hierarchical morphology were synthesized hydrothermally, and their surface characteristics were improved by the PEGylation process. In vitro MRI studies were also conducted to evaluate the ability of the synthesized nanoparticles as a contrast agent. All results were compared with those obtained for MnZn ferrite nanoparticles with normal structure. Microstructural evaluations showed that in ferrite with hierarchical morphology, the spherical particles with an average size of ~20 nm made a distinctive structure consisting of rows of nanoparticles which is a relatively big assembly like a dandelion. The smaller particle size and dandelion-like morphology led to an increase in specific surface area for the hierarchical structure (~69 m^2/g) in comparison to the normal one (~30 m^2/g) with an average particle size of ~40 nm. In vitro MRI, cytotoxicity and hemocompatibility assays confirmed the PEG-coated MnZn ferrite nanoparticles with hierarchical structure synthesized in the current study can be considered as an MRI contrast agent.

Keywords: MnZn ferrite; nanoparticles; hierarchical nanostructure; MRI contrast agent; PEGylation process

check for updates

Citation: Cheraghali, S.; Dini, G.; Caligiuri, I.; Back, M.; Rizzolio, F. PEG-Coated MnZn Ferrite Nanoparticles with Hierarchical Structure as MRI Contrast Agent. *Nanomaterials* **2023**, *13*, 452. https://doi.org/10.3390/nano13030452

Academic Editor: Jiyan Dai

Received: 8 December 2022
Revised: 19 January 2023
Accepted: 20 January 2023
Published: 22 January 2023

1. Introduction

Magnetic resonance imaging (MRI), a common non-invasive imaging method in molecular medicine for diagnosing diseases, is commonly used to scan soft tissues (mainly malignancies) and ensure the availability for the precise examination of the discrepancies between healthy and cancerous cells. High resolution, non-ionizing radiation, and precise anatomical area delineation seem to be just a few advantages of this method. However, some efforts have been performed to ameliorate its inherent limitations, specifically the low signal sensitivity as a significant constraint [1–3].

High-resolution magnetic resonance images can be produced using contrast agents. Over 300 million doses of gadolinium-based contrast agents (GBCAs) as T_1 positive-contrast agents have been used worldwide for approximately 30 years. Very few gadolinium-related toxicities have been documented and these agents have an outstanding safety profile overall; however, nephrogenic systemic fibrosis (NSF), a rare but serious disease found in patients with severe renal failure has been reported for GBCAs [4,5].

Superparamagnetic nanoparticle synthesis has grown significantly, in part due to its scientific utility, which includes magnetic storage media, biosensors, and medical applications such as drug carriers, T_2 negative-contrast agents for MRI, etc. Because the characteristics of nanocrystals greatly depend on the size of nanoparticles, controlling the monodisperse size is crucial. Compared to conventional paramagnetic MRI contrast agents based on gadolinium compounds, ferrite (Fe_3O_4) nanoparticles exhibit excellent magnetization and longer circulation times [6,7]. The most attractive ferrites are multi-substituted ones because of their potential for medical diagnosis and treatment [8,9]. The magnetic properties in these ferrite nanoparticles depend on the type and the percentages of each component along with the particle size and colloidal stability. Comparing different ferrite nanoparticles has indicated that manganese-zinc (MnZn) ferrite nanoparticles with a chemical composition of $Mn_{0.5}Zn_{0.5}Fe_2O_4$ are a great choice as an MRI contrast agent because of their high saturation magnetization (Ms) and low coercivity compared to ferrite nanoparticles and other oxides [8–11]. In addition, the improvement of the magnetic properties of ferrite nanoparticles via the synthesis method by controlling the processing parameters may result in a reduction in the proportion of nanoparticles required for MRI applications [11]. For example, hydrothermal synthesis is an interesting technique, particularly for biomedical fields, because the features of produced powder can be easily controlled by processing parameters such as pH, time, and the temperature of the reaction [12–15].

On the other hand, ferrite nanoparticles frequently require to be coated with biocompatible materials for successful MRI applications to stabilize their dispersions in a liquid under physiological circumstances. In order to achieve colloidal stability, a variety of natural and synthetic compounds, including polyethylene glycol (PEG), dextran, chitosan, and oleic acid have been utilized as a coating on the surface of nanoparticles [16–18]. The most widely used polymer among them is PEG. The low toxicity and immunogenicity of PEG, together with its hydrophilic property, make the PEG-coated nanoparticles undetectable to the defense mechanism. The PEG-coated nanoparticles are appealing for biological applications due to these characteristics [18,19].

As far as is known, although there are several studies on the hydrothermal synthesis of individually dispersed MnZn ferrite nanoparticles as MRI contrast agents [10–12], none of these studies investigated the efficacy of these nanoparticles as MRI contrast agents with a hierarchical structure. Hierarchical structures are assemblies of nanomaterials glued together, forming various morphologies, and leading to the formation of multifunctional materials with unique properties for different applications. In other words, the production of three-dimensional hierarchical nanostructures results from the right organization of various nanomaterials as building blocks with two or more levels ranging from the nanoscale to the macroscopic size. These structures have new uses in biology, environmental protection, material science, and other fields. The concept of hierarchical nanostructures has gained popularity and will be actively researched in the upcoming years [20,21].

Therefore, the purpose of this study was to explore the possibility of employing MnZn ferrite nanoparticles with hierarchical morphology produced hydrothermally as MRI contrast agents. A hierarchically MnZn ferrite structure was synthesized and then physically coated with PEG polymer. In vitro MRI studies were conducted to evaluate the ability of the synthesized nanoparticles as an MRI contrast agent. The toxicity and hemocompatibility of the products were also examined. All the results were compared with the results obtained for MnZn ferrite nanoparticles with a normal structure (i.e., individually dispersed nanoparticles).

2. Materials and Methods

2.1. Materials

Materials such as ferric chloride ($FeCl_3{\cdot}6H_2O$), zinc chloride ($ZnCl_2$), manganese chloride ($MnCl_2{\cdot}4H_2O$), ascorbic acid ($C_6H_8O_6$), urea ($CO(NH_2)_2$), ammonia solution (NH_4OH, 25%), phosphate-buffered saline (PBS), and the PEG (MW = 6 kDa) were purchased from Merck Co. Double distilled water (DDW) was also used for all procedures.

2.2. Synthesis of MnZn Ferrites

Hierarchically MnZn ferrite powder with the chemical composition of $Mn_{0.5}Zn_{0.5}Fe_2O_4$ was synthesized according to the procedure proposed by Budhiraja [22] via hydrothermal technique. First, 3.24 g $FeCl_3{\cdot}6H_2O$, 0.65 g $ZnCl_2$, and 0.97 g $MnCl_2{\cdot}4H_2O$ were dissolved in 50 mL DDW, and then 3.17 g $C_6H_8O_6$, and 3.3 g $CO(NH_2)_2$ were added. The prepared solution was stirred at room temperature for 30 min and then poured into a 200 mL-Teflon-lined stainless-steel autoclave and the reactor was kept in an electrical oven for 6 h at 160 °C.

The resulting precipitates were centrifuged at 10,000 rpm and then washed several times until the pH of the suspension reached the pH of DDW (~6.5). Finally, the product was dried for 10 h at 70 °C followed by annealing for 2 h at 500 °C in an electrical furnace.

MnZn ferrite with identical composition and normal morphology was also synthesized according to Ref. [23]. For this reason, 3.24 g $FeCl_3{\cdot}6H_2O$, 0.65 g $ZnCl_2$, and 0.97 g $MnCl_2{\cdot}4H_2O$ were dissolved in 30 mL DDW. The pH of the solution was adjusted to 10 by dropwise adding NH_4OH. The homogeneous solution was then transferred to the autoclave, and it was kept at 180 °C in the electric oven for 3 h. After cooling to room temperature, the product was collected and rinsed several times with DDW until the pH of the suspension reached the pH of DDW (~6.5).

2.3. *Coating of PEG Polymer on MnZn Ferrites*

To coat MnZn ferrite powders with PEG by physical bonding method, 0.5 g of both products (i.e., normal and hierarchical morphologies) was first dispersed in 50 mL of DDW at room temperature for 30 min via an ultrasonic bath. Also, 2 g of PEG was poured into 50 mL of DDW and stirred for 1 h. Then, this solution was added to the previous suspensions and stirred for 24 h. After separation by centrifugation, the coated products were dried at room temperature for 1 day.

2.4. *Characterization*

X-ray diffraction (XRD) patterns were recorded using an Asenware AW-DX 300 diffractometer (Dandong, China, λ = 1.54184 Å). MAUD program [24] was used to determine the phase content of each powder via the Rietveld refinement. A scanning electron microscope (SEM, ZEISS SIGMA 500 VP, Germany) and a transmission electron microscope (TEM, ZEISS EM10C, Germany) were used to investigate the morphology and particle size of the synthesized powders. Fourier-transform infrared spectroscopy (FTIR, JASCO-6300, Japan) and a carbon, hydrogen, nitrogen, sulfur (CHNS) elemental analyzer (Leco 932) were used to confirm the presence of PEG polymer around the particles and measured its content, respectively. Additionally, an inductively coupled plasma optical-emission spectrometer (ICP-OES, Perkin-Elmer Optima 7300 DV, USA) and an energy dispersive X-ray spectrometer (EDS, Oxford Instruments, UK) available in the SEM system were used to evaluate the Fe, Mn, and Zn concentrations of the produced particles. A Horiba SZ-100 series analyzer based on dynamic light scattering (DLS) was used to evaluate the stability of dispersions (~1 mg/mL) of the synthesized powders in PBS (pH = ~7.4). A vibrating sample magnetometer (VSM, Meghnatis Daghigh Kavir Co., Kashan, Iran) was used to evaluate the magnetic features of the PEG-coated and uncoated MnZn ferrites at room temperature. The specific surface area of the powders was determined via the Brunauer–Emmett–Teller (BET) method by adsorption/desorption of N_2 gas at liquid nitrogen temperature (~77 K) using a Series BEL SORP mini II. A 1.5 T clinical MRI system (Philips, Ingenia) with repetition time = 3 s, 32 echoes with 110 ms, acquisition time of 5 min, field of view = 200 × 200 mm^2, matrix size = 128 × 128 mm^2, slice thickness = 3 mm was used for in vitro MRI tests to investigate the efficacy of the synthesized MnZn ferrites as an MRI contrast agent. In order to achieve this, colloids containing products at concentrations of 1 to 4 mM of metal (i.e., Mn + Zn + Fe) were prepared and poured into a 24-well plate placed in the iso-center of the magnet. Then, T_2-weighted MR images of colloids were recorded. T_2 relaxation time was measured by the Carr-Purcell-Meiboom-Gill (CPMG) sequence at room temperature and results were inverted to obtain the R_2 relaxation rates in s^{-1}. The effectiveness of each contrast agent was expressed in terms of T_2 relaxivity, which is denoted as r_2 [$mM^{-1}s^{-1}$]. Then, according to the slopes of relaxation rate curves and the concentration of all metal components, r_2 was calculated (i.e., $\Delta R_2 = R_2 - R_{2,0} = 1/T_2 - 1/T_{2,0}$, and $T_{2,0}$ = 367 ms was the relaxation time of pure PBS solution).

2.5. Complete Blood Count (CBC) Test

Blood samples from a healthy volunteer were mixed with different powders to create colloids at concentrations of 0.1, 0.2, and 0.3 mg/mL. Then, using a Mindray BC-6800 automated hematology analyzer, the red blood cell (RBC) count of the colloids was determined and compared with the blood sample without nanoparticles as the control sample.

2.6. In Vitro Blood Coagulation Test

First, a platelet-poor plasma (PPP) was obtained by centrifuging the blood sample at 3000 rpm. PPP was then mixed with the powders to create colloids at concentrations of 0.1, 0.2, and 0.3 mg/mL. Prothrombin time (PT) and activated partial thromboplastin time (*a*PTT), two blood coagulation factors of the colloids, were tested using an automation blood coagulation analyzer (STA Compact Stago, USA) and results were compared with PPP without nanoparticles as the control sample.

2.7. Cytotoxicity Assay

The cytotoxicity test was carried out for 96 h on MRC5 (normal fibroblast), OVCAR3, Kuramochi (ovarian cancer), and DLD-1 (colon cancer) cell lines. Cells were plated on 96 multiwells and treated with a serial dilution of nanoparticles starting from 1×10^{-2} to 1×10^{3} µg/mL. The viability was evaluated with CellTiter-Glo® Luminescent Cell Viability Assay in a Tecan M1000 instrument. Data were analyzed with a nonlinear regression (curve fit) model with Prism software.

2.8. Statistical Analysis

SPSS software (version 22) was used for statistical analysis. Significant differences were identified by one-way analysis of variance (ANOVA) and the multiple-comparison (Tukey's) test with n = 3. The results were reported as mean ± standard deviation (SD) with a significant value of $p < 0.05$.

3. Results and Discussion

The XRD patterns of the synthesized powders are shown in Figure 1a. In both products, all the major peaks are corresponding to the cubic structure of (Mn, Zn) Fe_2O_4 (ICDD PDF no. 01-074-2401). Although some minor peaks can be assigned to Fe_2O_3 (ICDD PDF no. 01-089-0597), according to the quantitative phase analysis via the Rietveld refinement (Figure 1a,b), the content of this structure in both synthesized powders is less than 10 wt.% (i.e., 5 wt.% in MnZn ferrite with normal morphology, and 7 wt.% in MnZn ferrite with hierarchical morphology). The same results have been presented elsewhere and it has been suggested that the purity of the synthesized ferrite could be improved by optimization of the hydrothermal parameters such as time and temperature [25,26].

Figures 2 and 3 show the SEM micrographs and corresponding EDS results of ferrites with normal and hierarchical morphology. In both powders, the particles are almost spherical and have a relatively uniform distribution. The SEM images also show that the average size of the particles is on the nanoscale. However, it clearly can be seen that the arrangement of the spherical nanoparticles is completely different in both powders. Although ferrite with normal morphology consists of individual nanoparticles, in ferrite with hierarchical morphology, the particles make a distinctive structure consisting of rows of nanoparticles which is a relatively big assembly like a dandelion on the micron scale. It has been suggested that the presence of ascorbic acid and urea leads to a special arrangement of nanoparticles in various directions during the hydrothermal synthesis of this powder [22]. Moreover, the EDS results (Figures 2c and 3d) show that in both powders, Fe, Mn, Zn, and O elements are present. The ICP data (Table 1) also shows that both synthesized MnZn ferrites have approximately designed stoichiometric ratios for the Mn, Zn, and Fe elements.

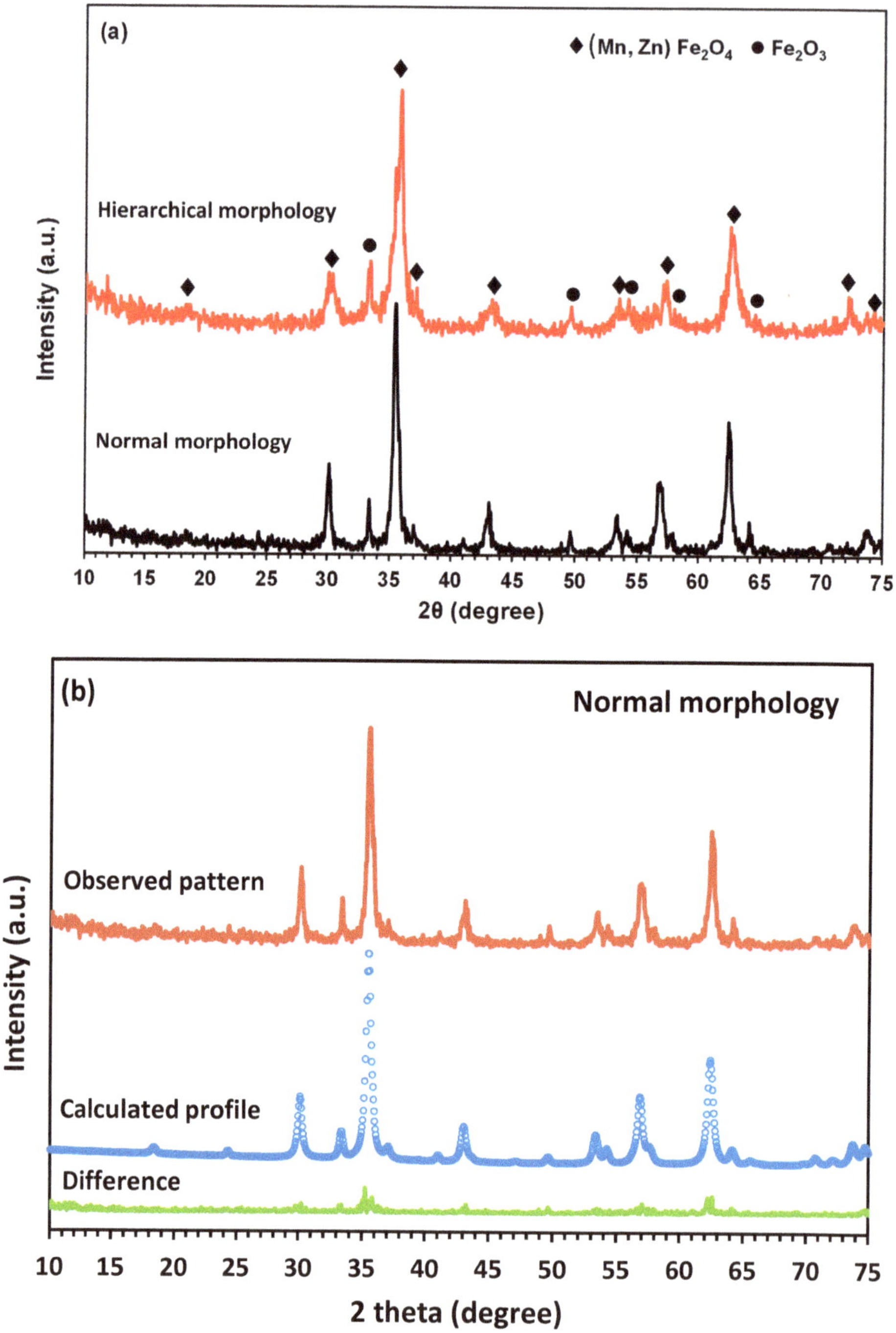

Figure 1. *Cont.*

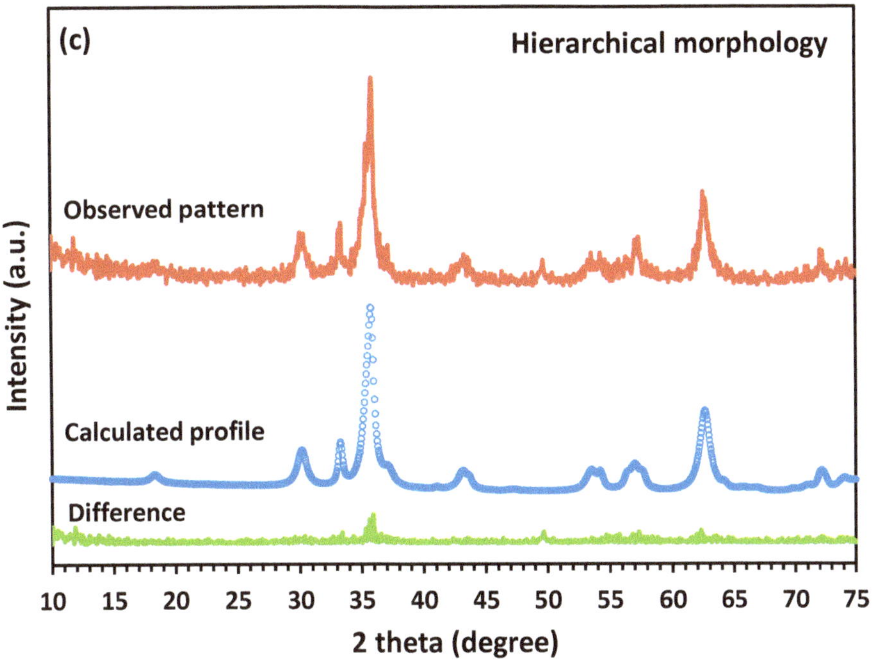

Figure 1. (**a**) XRD patterns of synthesized MnZn ferrites with different morphologies, and (**b**), and (**c**) corresponding Rietveld refined XRD patterns.

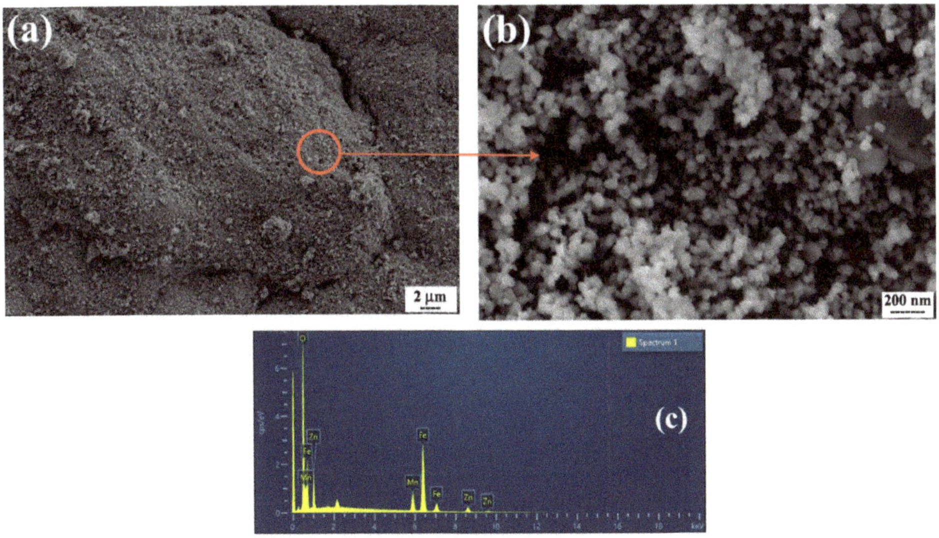

Figure 2. (**a**,**b**) SEM micrographs of synthesized MnZn ferrite with normal morphology at different magnifications, and (**c**) corresponding EDS results.

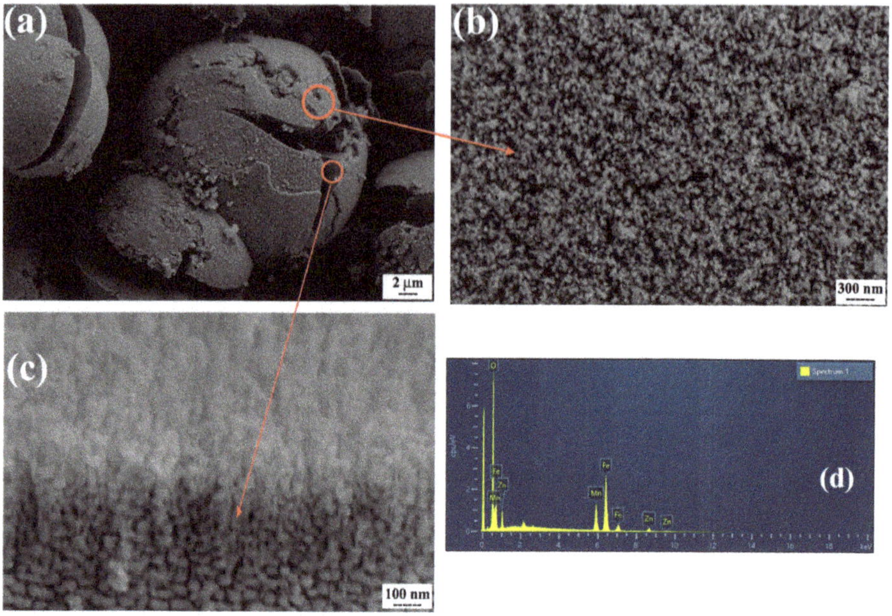

Figure 3. (**a–c**) SEM micrographs of synthesized MnZn ferrite with hierarchical morphology at different magnifications, and (**d**) corresponding EDS results.

Table 1. ICP date for both synthesized ferrite powders.

Powder	Fe (wt.%)	Mn (wt.%)	Zn (wt.%)	Molar Ratio (Fe/Mn + Zn)
Ferrite with normal morphology	65.1	18.3	16.6	1.99
Ferrite with hierarchical morphology	63.9	17.6	18.5	1.90

One of the key characteristics that demonstrates a material's capacity to interact with its environment is surface area. A bigger surface area allows the material to absorb more molecules, which improves performance significantly in applications such as drug release activated by alternating magnetic fields [27]. A decrease in particle size leads to increasing the surface area, and as a direct consequence, different arrangements of nanoparticles in a hierarchical 3D nanostructure could significantly affect the textural properties of the material [21,27]. The N_2 adsorption-desorption isotherms (Figure 4) confirmed that the MnZn ferrite with hierarchical morphology has a relatively high BET surface area (~69 m^2/g) in comparison to its counterpart (~30 m^2/g) with normal morphology. This enhancement provides some useful characteristics for MnZn ferrite with hierarchical morphology such as higher drug loading capacity for drug delivery applications.

To better study the shape and size of the synthesized nanoparticles and also to confirm the presence of PEG coating, TEM analysis was performed and the results are presented in Figure 5. As can be seen from Figure 5a,c, the average sizes of semi-spherical nanoparticles of MnZn ferrite with normal and hierarchical morphologies are about 40 and 20 nm, respectively. PEG coating has also appeared as a light gray substance around the clusters of nanoparticles in both powders (Figure 5b,d). Since PEG-coated nanoparticles must be dried for TEM analysis, so the particles were slightly agglomerated. Additionally, FTIR analysis was used to identify PEG in coated-ferrite powders. The FTIR spectra of bare nanoparticles, PEG polymer as well as coated nanoparticles are illustrated in Figure 6a,b. In the FTIR spectra of PEG, the -C-O-C- bending vibration peak may be found near 1100 and 1350 cm^{-1}, respectively. Additionally, the absorption peaks at 950 cm^{-1} are due to the out-of-plane bending vibration of -CH, whereas the peaks around 1280 and 1470 cm^{-1} are due to the bending vibrations of -CH2 [15,18,19]. Furthermore, the main absorbance of the ether stretch and the vibrational peaks that are seen in the FTIR spectrum of PEG is present in the FTIR spectrum of the PEG-coated MnZn ferrites with normal and hierarchical. In addition, the organic content in the bare and PEG-coated MnZn ferrite nanoparticles with different morphologies was determined by the CHNS analysis. The results showed that the organic content in the PEG-coated MnZn ferrite nanoparticles with normal and hierarchical morphologies was 18.5 and 20.1 wt.%, respectively, and only consisted of C and H elements. Moreover, there was no organic compound in the bare MnZn ferrite nanoparticles. Therefore, grey matter around both coated nanoparticles in corresponding TEM micrographs (i.e., Figure 5b,d) could be ascribed to the PEG polymer.

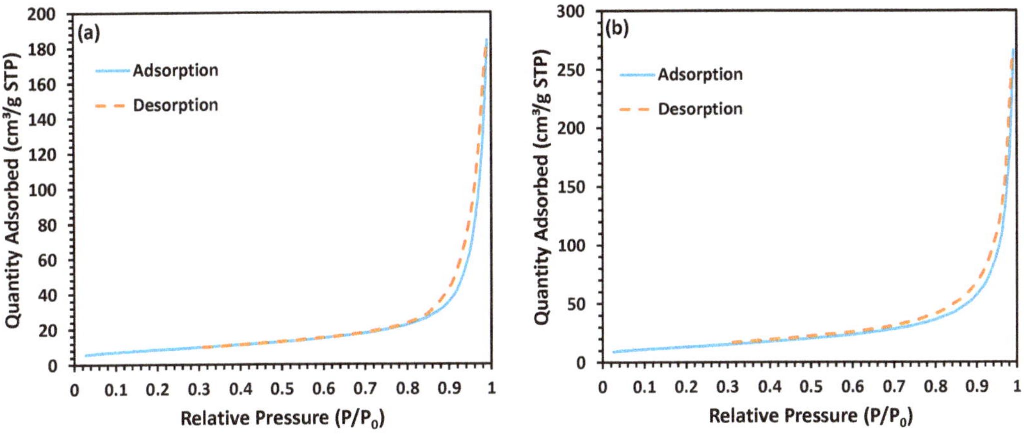

Figure 4. N_2 adsorption/desorption isotherms at 77 K for MnZn ferrite nanopowders with (**a**) normal morphology, and (**b**) hierarchical morphology.

Due to the high specific surface and also magnetic force between particles, ferrite powders could easily be aggregated. Therefore, coating nanoparticles with non-toxic, non-immunogenic polymers such as PEG (i.e., PEGylation process) is an effective way to improve the colloidal stability of dispersions containing ferrite nanoparticles in biological media and a magnetic field. The graphs of DLS and zeta potential analyses of colloids containing both synthesized MnZn ferrites with and without PEG coating are illustrated in Figure 7. Clearly, PEG coating not only leads to a decrease in the hydrodynamic diameter of nanoparticles as a sign of deagglomeration (i.e., 171.5 $\pm$ 10.3 to 133.9 $\pm$ 17.6 nm in ferrite with normal morphology, and 130.3 $\pm$ 15.2 to 81.8 $\pm$ 13.1 nm in ferrite with hierarchical morphology), but also the surface charge of coated nanoparticles shifted to lower values (i.e., -42.3 ± 2.3 to -47.4 ± 3.7 mV in ferrite with normal morphology, and -40.9 ± 2.1 to -55 ± 4.1 mV in ferrite with hierarchical morphology), improving colloidal stability [16,18,28,29]. To investigate the colloidal stability of the prepared suspensions, a new set of DLS experiments was conducted in which the hydrodynamic diameter of nanoparticles in each prepared suspension was measured for one week. The obtained results (Figure 7e) confirmed the PEG-coated MnZn ferrite nanoparticles with different morphologies and also the bare MnZn ferrite nanoparticles with hierarchical morphology remained stable in PBS medium over a week, while the bare MnZn ferrite nanoparticles with normal morphology started to sediment after three days.

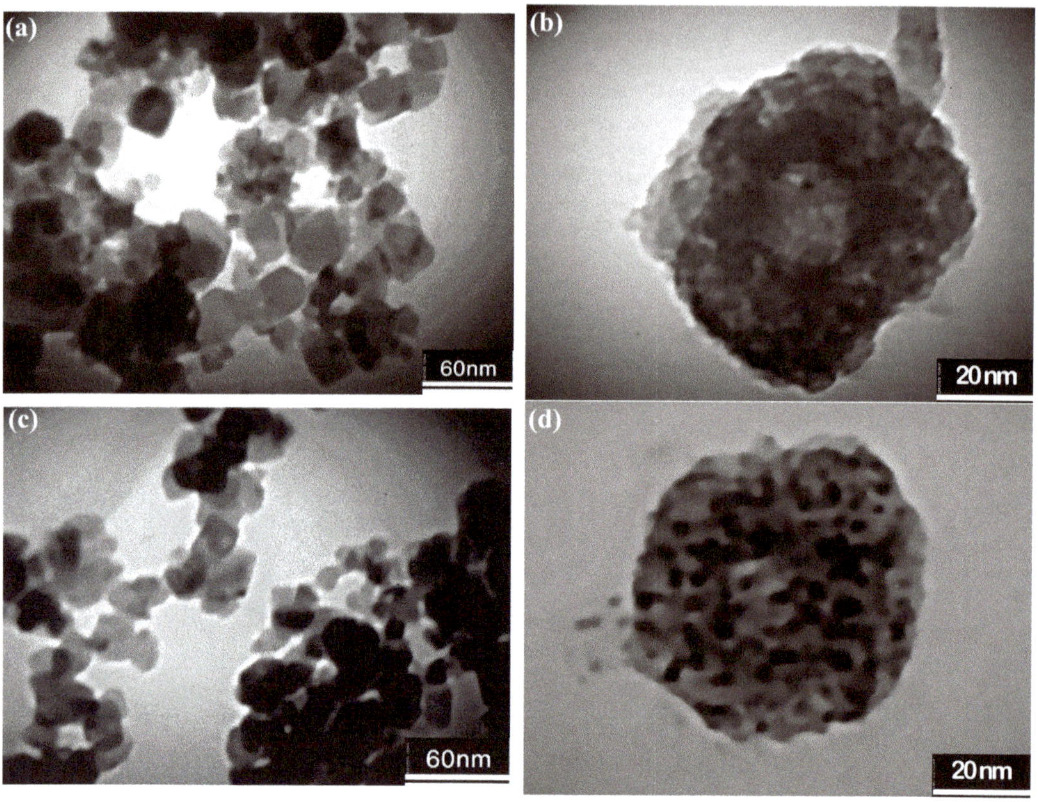

Figure 5. TEM micrographs of bare and PEG-coated MnZn ferrite nanoparticles, (**a**,**b**) with normal morphology, and, (**c**,**d**) with hierarchical morphology.

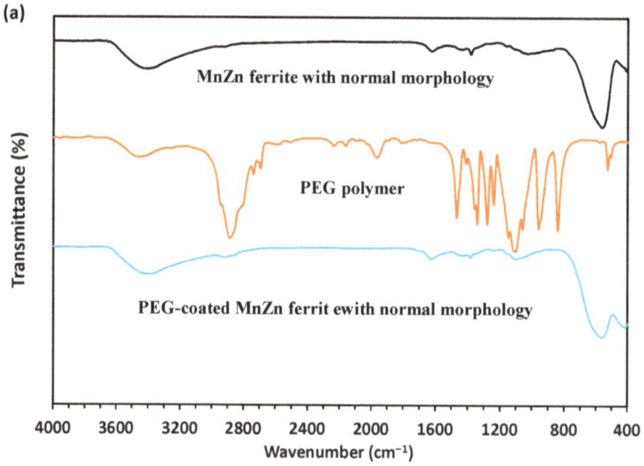

Figure 6. *Cont.*

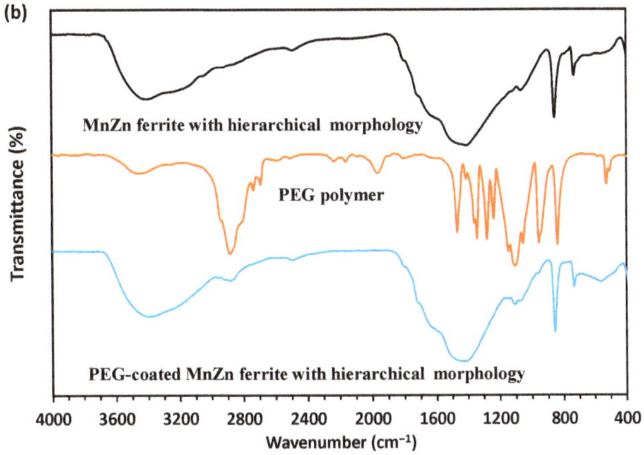

Figure 6. (**a**,**b**) FTIR spectra of bare, PEG-coated MnZn ferrite nanoparticles with different morphologies.

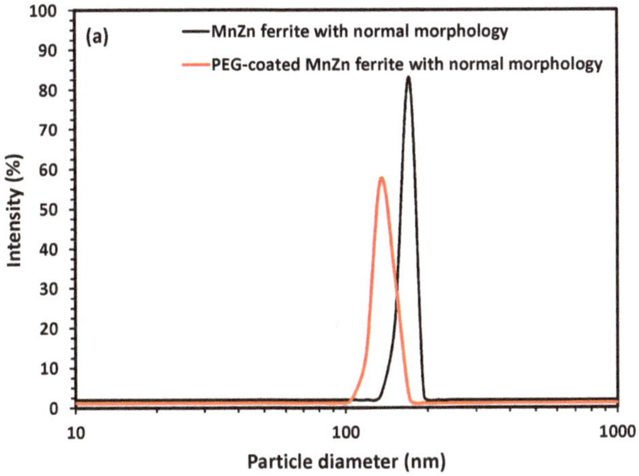

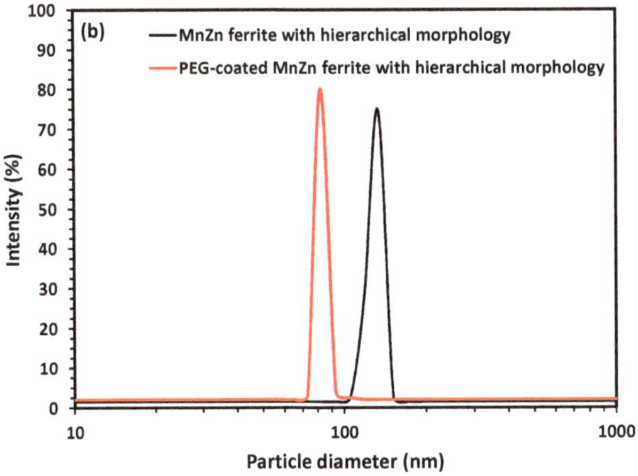

Figure 7. *Cont.*

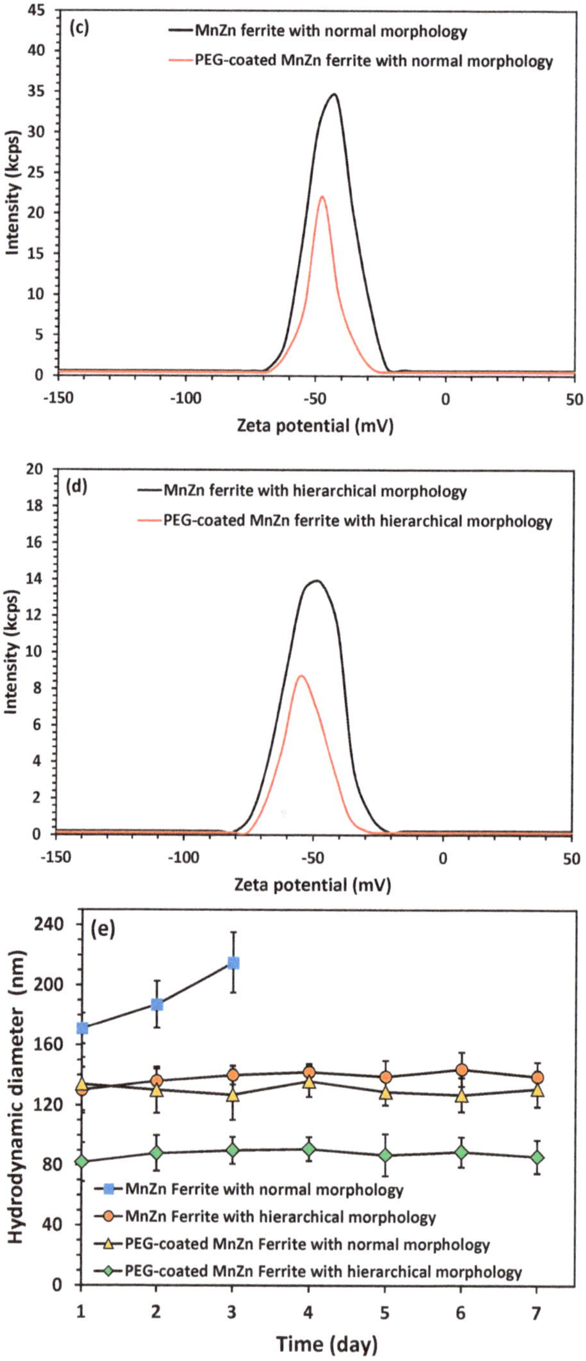

Figure 7. (**a**,**b**) Graphs of DLS, (**c**,**d**) graphs of zeta potential analyses, and (**e**) results of colloidal stability test for suspensions containing PEG-coated and uncoated MnZn ferrite nanoparticles with different morphologies. Error bars are standard deviation (n = 3).

In comparison to the TEM results, it should be noted that the observed difference in the average size obtained via the DLS and TEM methods is due to the hydrodynamic diameter of particles in the DLS measurements. This may result from the "hair layer" theory which is a reaction to the formation of a hairy layer due to molecular chains of PEG on the surface of nanoparticles [18].

Magnetic hysteresis loops of bare and coated MnZn ferrite nanoparticles with normal and hierarchical structures obtained by VSM are illustrated in Figure 8. Both materials exhibit superparamagnetic behavior (i.e., zero remanence and coercivity). In comparison to Ref. [11], all the synthesized powders in the current study exhibit relatively high Ms values (i.e., 66, 59, 54, and 48 emu/g for bare MnZn ferrite powders, and PEG-coated MnZn ferrite powders with normal and hierarchical morphologies, respectively) that are acceptable in terms of magnetic properties. However, Jang et al. [9] synthesized a series of $(Zn_xMn_{1-x})\ Fe_2O_4$ nanoparticles and they reported a maximum Ms of 175 emu/g for $Zn_{0.4}Mn_{0.6}Fe_2O_3$ nanoparticles. Although the hierarchical structure led to a decrease in Ms in bare MnZn ferrite nanoparticles (66 to 59 emu/g), its value is still significant. The decrease in Ms value for MnZn ferrite with hierarchical morphology could be attributed to its smaller particle size. The magnetic property and subsequent response to magnetic fields can be significantly influenced by particle size. For instance, the Ms value of ferrite particles drops as the particle size decreases, but the particles might still display supermagnetism behavior. As quantitative XRD analysis also confirmed, MnZn ferrite with hierarchical morphology contains a slightly higher content of Fe_2O_3 which may alter the corresponding Ms value. On the other hand, PEG as a polymer coating led to a decrease in Ms values of both coated ferrites. However, it is worth noting that coated ferrites still have a high MS. As mentioned before, the presence of PEG coating improves the stability of the colloid. In addition, according to the proposed method by Tenório-Neto et al. [30], the organic content in the PEG-coated nanoparticles was measured based on the Ms value of the bare nanoparticles and that of PEG-coated nanoparticles. This content was determined at about 19 wt.% for both PEG-coated MnZn ferrites with different morphologies and is almost consistent with those obtained by CHNS analyses, as previously mentioned.

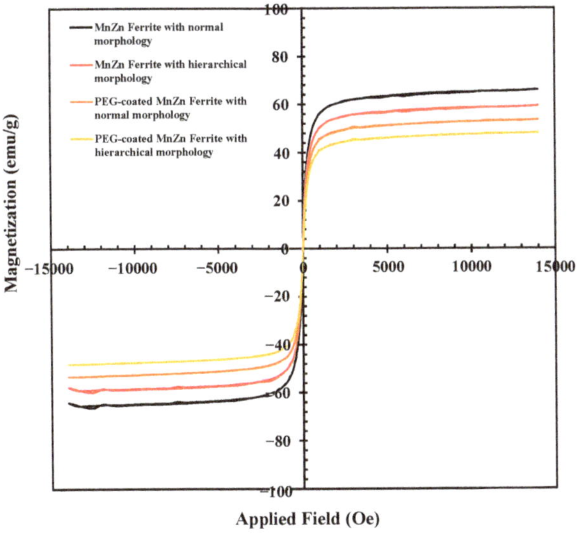

Figure 8. Magnetic hysteresis loops of the PEG-coated and uncoated MnZn ferrite nanoparticles with different morphologies.

In the MRI technique, the contrast improvement directly corresponds to the Ms value of the contrast agent. Superparamagnetic nanoparticles such as MnZn ferrite as contrast agents could produce negative contrast images by reducing the signal intensity to produce dark images. In vitro T_2-weighted MR images (Figure 9a) were taken from PBS media containing different concentrations of both the nanopowders with and without PEG coating and compared with PBS as the control. Figure 9a demonstrates that adding PEG-coated and uncoated MnZn ferrite nanoparticles to PBS as the control considerably improved the contrast quality of MR images. As can be predicted, this improvement got better as the nanoparticle concentration increased in both instances. However, the PEG-coated nanoparticle dispersions as well as bare nanoparticles with hierarchical structure exhibit slightly less improvement in the contrast of MR images than uncoated nanoparticles with normal structure. This result agrees well with the magnetic characteristics determined using VSM (Figure 8). The presence of the non-magnetic PEG coating on the surface of nanoparticles is often related to this, as was previously indicated. The contrasting impact of the produced MnZn ferrite nanoparticles is decreased as a result of this layer's protection against protons [15]. Bare MnZn ferrite nanoparticles with normal morphology also showed the highest relaxivity of 380.5 $mM^{-1}s^{-1}$. In addition, the interesting result is that all the samples synthesized in this work exhibit higher relaxivity (Figure 9b) than reported values for some commercially approved MRI contrast agents of superparamagnetic iron oxide such as Feridex, Resovist, and Combidex at 1.5 T [31]. However, it must be noted that not only some parameters of ferrite nanoparticles such as the crystallinity, composition, size, morphology, and magnetic properties are responsible for T_2 contrast but also the characteristics of nonmagnetic polymers as coatings must be considered. Cho et al. [32] have recently reported that higher T_2 relaxivity could be achieved by manipulating the features of surface coating and ferrite nanoparticles together.

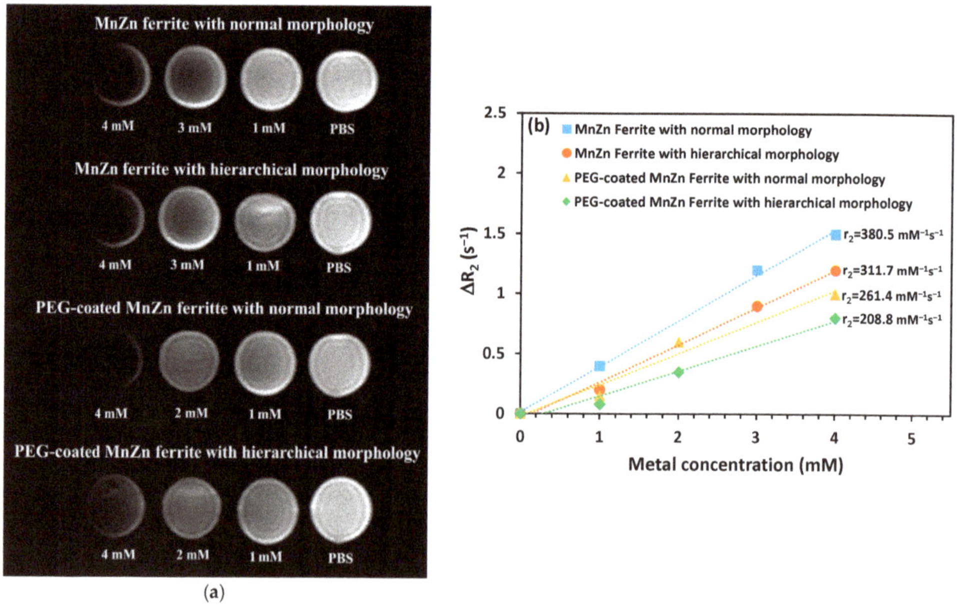

Figure 9. Effect of nanoparticle concentration on (**a**) in vitro T_2-weighted MR images of PBS media containing PEG-coated and uncoated nanopowders with different morphologies, and (**b**) T_2 relaxation rate against metal (i.e., Mn + Zn + Fe) concentration for the PEG-coated and bare MnZn ferrite NPs dispersions with different morphologies.

The results of RBC counts obtained from CBC tests, and also *a*PPT and PT values as coagulation factors for the blood samples treated with PEG-coated and uncoated MnZn ferrite nanoparticles with normal and hierarchical structures at concentrations of 0.1, 0.2, and 0.3 mg/mL are given in Table 2. The statistical analysis showed no significant difference ($p < 0.05$) in all cases from the corresponding control sample. So, it can be concluded that neither the PEG-coated nor the uncoated synthesized MnZn nanoparticles show any procoagulant activity at the tested doses, preventing their usage in intravenous applications. The biocompatibility of the synthesized products was even tested on normal and cancer cell lines but the IC_{50} values were very high (>0.2 mg/mL), indicating low cytotoxicity (Appendix A, Figure A1).

Table 2. Effect of concentration and morphology of MnZn ferrite nanoparticles on values of RBC, *a*PPT, and PT of blood samples.

Powder	Concentration (mg/mL)	RBC * ($10^6/\mu L$)	*a*PPT * (s)	PT * (s)
MnZn ferrite with normal morphology	0.1	5.37 ± 0.02	30.1 ± 0.1	12.9 ± 0.1
	0.2	5.48 ± 0.02	30 ± 0.2	13.0 ± 0.2
	0.3	5.40 ± 0.03	29.5 ± 0.1	12.9 ± 0.2
PEG-coated MnZn ferrite with normal morphology	0.1	5.32 ± 0.05	29.8 ± 0.3	13.1 ± 0.3
	0.2	5.51 ± 0.02	30.4 ± 0.2	12.9 ± 0.3
	0.3	5.48 ± 0.04	29.9 ± 0.2	12.9 ± 0.2
MnZn ferrite with hierarchical morphology	0.1	5.39 ± 0.03	31.4 ± 0.1	13.1 ± 0.2
	0.2	5.42 ± 0.02	30 ± 0.3	12.9 ± 0.2
	0.3	5.52 ± 0.02	30.2 ± 0.2	12.9 ± 0.3
PEG-coated MnZn ferrite with hierarchical morphology	0.1	5.38 ± 0.04	29.5 ± 0.3	12.8 ± 0.1
	0.2	5.47 ± 0.04	29.6 ± 0.3	12.8 ± 0.2
	0.3	5.48 ± 0.03	29.9 ± 0.2	12.9 ± 0.2
Control	-	5.62 ± 0.05	31.4 ± 0.2	12.8 ± 0.3

* Mean values are not significantly different ($p < 0.05$). Errors are standard deviation (n = 3).

4. Conclusions

In this work, the hydrothermal method was used to synthesize MnZn ferrite nanoparticles with hierarchical assembly. The effect of the PEGylation process on the magnetic properties and biocompatibility of the synthesized nanoparticles was also studied. All the results were compared with those obtained for MnZn ferrite nanoparticles with normal structure (i.e., individual particles). The following results were obtained in this study:

1. The average particle size of the synthesized MnZn ferrites with normal and hierarchical structures was about 40, and 20 nm, respectively.
2. PEG coating improved the colloidal stability and biocompatibility of nanoparticles with a slight decrease in Ms values.
3. Adding PEG-coated and uncoated MnZn ferrite nanoparticles to PBS considerably improved the contrast quality of MR images.
4. RBC, blood coagulation and cell cytotoxic studies under laboratory conditions showed that both synthesized MnZn ferrite nanoparticles have no negative effects on blood factors and cell viability, respectively.
5. The PEG-coated MnZn ferrite nanoparticles with hierarchical structure synthesized in the current study can be considered as an MRI contrast agent at concentrations between 0.1 and 0.3 mg/mL.

Author Contributions: Conceptualization, G.D. and F.R.; methodology, G.D.; validation, G.D. and F.R.; formal analysis, S.C. and I.C.; investigation, S.C. and I.C.; resources, G.D. and F.R.; writing—original draft preparation, S.C.; writing—review and editing, G.D., M.B. and F.R.; supervision, F.R.; project administration, G.D. and M.B; funding acquisition, F.R. All authors have read and agreed to the published version of the manuscript.

Funding: This research received no external funding.

Data Availability Statement: The data presented in this study are available on request from the corresponding author.

Acknowledgments: The authors would like to thank the Ca' Foscari University of Venice and the University of Isfahan for making all the necessary resources available for this work.

Conflicts of Interest: The authors declare that there is no conflict of interest.

Appendix A

(a)

Sample	IC_{50} value OVCAR3	Kuramochi	DLD1	MRC5
	Av.	Av.	Av.	Av.
F	NA	NA	NA	NA
HF	NA	NA	NA	NA
FP	NA	>200 μg/mL	NA	NA
HFP	NA	NA	NA	NA

F: MnZn Ferrite with normal morphology, **HF:** MnZn Ferrite with hierarchical morphology
FP: PEG-coated MnZn Ferrite with normal morphology, **HFP:** PEG-coated MnZn Ferrite with hierarchical morphology

(b)

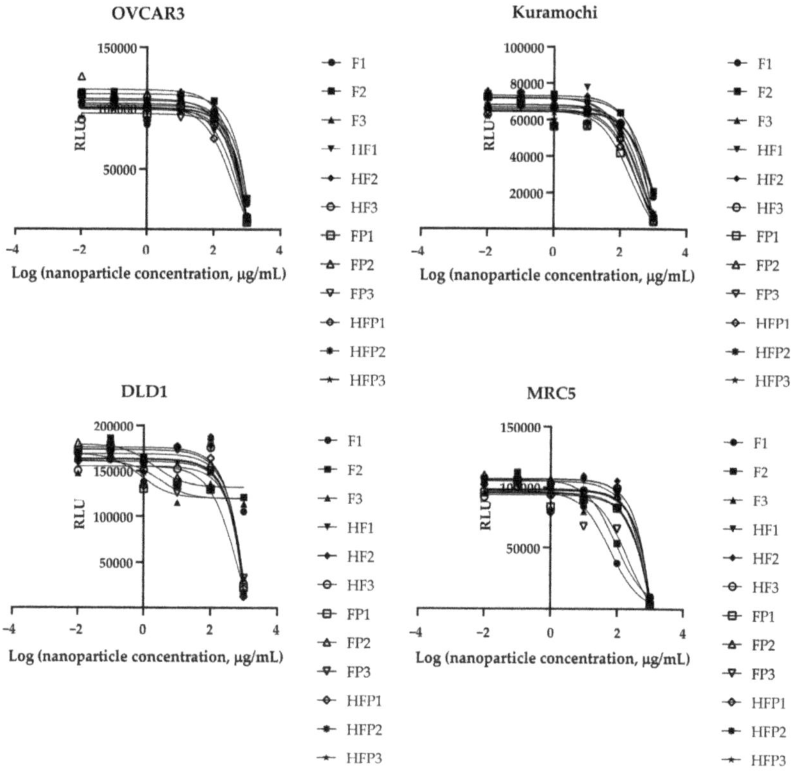

Figure A1. (**a**) Summary of the data presented in b. Values are expressed as μg/mL. Av: average, (**b**) Nonlinear regression analysis of different cells treated with nanoparticles in triplicates. On the Y axis, the relative luminescent values (RLU) are reported. On the X axis, the concentration of nanoparticles on a logarithmic scale is reported.

References

1. Narayan, R. *Encyclopedia of Biomedical Engineering*; Elsevier: Amsterdam, The Netherlands, 2018.
2. Yim, H.; Seo, S.; Na, K. MRI Contrast Agent-Based Multifunctional Materials: Diagnosis and Therapy. *J. Nanomater.* **2011**, *2011*, 19. [CrossRef]
3. Lee, J.-H.; Huh, Y.-M.; Jun, Y.-W.; Seo, J.-W.; Jang, J.-T.; Song, H.-T.; Kim, S.; Cho, E.-J.; Yoon, H.-G.Y.; Suh, J.-S.; et al. Artificially engineered magnetic nanoparticles for ultra-sensitive molecular imaging. *Nat. Med.* **2006**, *13*, 95–99. [CrossRef] [PubMed]
4. Ramalho, J.; Ramalho, M.; Jay, M.; Burke, L.M.; Semelka, R.C. Gadolinium toxicity and treatment. *Magn. Reson. Imaging* **2016**, *34*, 1394–1398. [CrossRef]
5. Gayathri, T.; Sundaram, N.M.; Kumar, R.A. Gadolinium Oxide Nanoparticles for Magnetic Resonance Imaging and Cancer Theranostics. *J. Bionanoscience* **2015**, *9*, 409–423. [CrossRef]
6. Amiri, M.; Salavati-Niasari, M.; Akbari, A. Magnetic nanocarriers: Evolution of spinel ferrites for medical applications. *Adv. Colloid Interface Sci.* **2019**, *265*, 29–44. [CrossRef] [PubMed]
7. Kudr, J.; Haddad, Y.; Richtera, L.; Heger, Z.; Cernak, M.; Adam, V.; Zitka, O. Magnetic Nanoparticles: From Design and Synthesis to Real World Applications. *Nanomaterials* **2017**, *7*, 243. [CrossRef] [PubMed]
8. Cabrera, L.I.; Somoza, Á.; Marco, J.F.; Serna, C.J.; Morales, M.P. Synthesis and surface modification of uniform MFe_2O_4 (M = Fe, Mn, and Co) nanoparticles with tunable sizes and functionalities. *J. Nanopart. Res.* **2012**, *14*, 873. [CrossRef]
9. Jang, J.-T.; Nah, H.; Lee, J.-H.; Moon, S.H.; Kim, M.G.; Cheon, J. Critical Enhancements of MRI Contrast and Hyperthermic Effects by Dopant-Controlled Magnetic Nanoparticles. *Angew. Chem. Int. Ed.* **2009**, *48*, 1234–1238. [CrossRef]
10. Kefeni, K.K.; Msagati, T.A.; Nkambule, T.T.; Mamba, B.B. Spinel ferrite nanoparticles and nanocomposites for biomedical applications and their toxicity. *Mater. Sci. Eng. C* **2019**, *107*, 110314. [CrossRef]
11. Thakur, P.; Chahar, D.; Taneja, S.; Bhalla, N.; Thakur, A. A review on MnZn ferrites: Synthesis, characterization and applications. *Ceram. Int.* **2020**, *46*, 15740–15763. [CrossRef]
12. Xuan, Y.; Li, Q.; Yang, G. Synthesis and magnetic properties of Mn–Zn ferrite nanoparticles. *J. Magn. Magn. Mater.* **2007**, *312*, 464–469. [CrossRef]
13. Xiao, L.; Zhou, T.; Meng, J. Hydrothermal synthesis of Mn–Zn ferrites from spent alkaline Zn–Mn batteries. *Particuology* **2009**, *7*, 491–495. [CrossRef]
14. Vlazan, P.; Vasile, M. Synthesis and characterization $CoFe_2O_4$ nanoparticles prepared by the hydrothermal method. *Optoelectron. Adv. Mater.-Rapid Commun.* **2010**, *4*, 1307–1309.
15. Faraji, S.; Dini, G.; Zahraei, M. Polyethylene glycol-coated manganese-ferrite nanoparticles as contrast agents for magnetic resonance imaging. *J. Magn. Magn. Mater.* **2018**, *475*, 137–145. [CrossRef]
16. Zahraei, M.; Marciello, M.; Lazaro-Carrillo, A.; Villanueva, A.; Herranz, F.; Talelli, M.; Costo, R.; Monshi, A.; Shahbazi-Gahrouei, D.; Amirnasr, M.; et al. Versatile theranostics agents designed by coating ferrite nanoparticles with biocompatible polymers. *Nanotechnology* **2016**, *27*, 255702–255714. [CrossRef]
17. Ali, L.M.; Marzola, P.; Nicolato, E.; Fiorini, S.; Guillamón, M.D.L.H.; Piñol, R.; Gabilondo, L.; Millán, A.; Palacio, F. Polymer-coated superparamagnetic iron oxide nanoparticles as T_2 contrast agent for MRI and their uptake in liver. *Futur. Sci. OA* **2019**, *5*, FSO235. [CrossRef]
18. Masoudi, A.; Hosseini, H.R.M.; Shokrgozar, M.A.; Ahmadi, R.; Oghabian, M.A. The effect of poly(ethylene glycol) coating on colloidal stability of superparamagnetic iron oxide nanoparticles as potential MRI contrast agent. *Int. J. Pharm.* **2012**, *433*, 129–141. [CrossRef]
19. Ruiz, A.; Salas, G.; Calero, M.; Hernández, Y.; Villanueva, A.; Herranz, F.; Veintemillas-Verdaguer, S.; Martínez, E.; Barber, D.; Morales, M. Short-chain PEG molecules strongly bound to magnetic nanoparticle for MRI long circulating agents. *Acta Biomater.* **2013**, *9*, 6421–6430. [CrossRef]
20. Su, B.-L.; Sanchez, C.; Yang, X.-Y. *Insights into Hierarchically Structured Porous Materials: From Nanoscience to Catalysis, Separation, Optics, Energy, and Life Science*, 1st ed.; Wiley VCH: Weinheim, Germany, 2011; pp. 1–27. [CrossRef]
21. Zhang, Q.; Wei, F. (Eds.) *Advanced Hierarchical Nanostructured Materials*; John Wiley & Sons: Hoboken, NJ, USA, 2014.
22. Sapna; Budhiraja, N.; Kumar, V.; Singh, S. Shape-controlled synthesis of superparamagnetic $ZnFe_2O_4$ hierarchical structures and their comparative structural, optical and magnetic properties. *Ceram. Int.* **2018**, *45*, 1067–1076. [CrossRef]
23. Phong, P.T.; Nam, P.; Manh, D.H.; Tung, D.K.; Lee, I.-J.; Phuc, N. Studies of the Magnetic Properties and Specific Absorption of $Mn_{0.3}Zn_{0.7}Fe_2O_4$ Nanoparticles. *J. Electron. Mater.* **2014**, *44*, 287–294. [CrossRef]
24. Lutterotti, L.; Chateigner, D.; Ferrari, S.; Ricote, J. Texture, residual stress and structural analysis of thin films using a combined X-ray analysis. *Thin Solid Films* **2004**, *450*, 34–41. [CrossRef]
25. Faraji, S.; Dini, G.; Zahraei, M. The Effect of pH and Duration of Hydrothermal Process on the Synthesis of Manganese Ferrite Nanoparticles. *J. Adv. Mater. Eng. Esteghlal* **2018**, *37*, 105–111.
26. Karaagac, O.; Köçkar, H. The effects of temperature and reaction time on the formation of manganese ferrite nanoparticles synthesized by hydrothermal method. *J. Mater. Sci. Mater. Electron.* **2020**, *31*, 2567–2574. [CrossRef]
27. Cai, B.; Zhao, M.; Ma, Y.; Ye, Z.; Huang, J. Bioinspired Formation of 3D Hierarchical $CoFe_2O_4$ Porous Microspheres for Magnetic-Controlled Drug Release. *ACS Appl. Mater. Interfaces* **2015**, *7*, 1327–1333. [CrossRef]
28. Suk, J.S.; Xu, Q.; Kim, N.; Hanes, J.; Ensign, L.M. PEGylation as a strategy for improving nanoparticle-based drug and gene delivery. *Adv. Drug Deliv. Rev.* **2016**, *99*, 28–51. [CrossRef]

29. Mannu, R.; Karthikeyan, V.; Velu, N.; Arumugam, C.; Roy, V.A.L.; Gopalan, A.-I.; Saianand, G.; Sonar, P.; Lee, K.-P.; Kim, W.-J.; et al. Polyethylene Glycol Coated Magnetic Nanoparticles: Hybrid Nanofluid Formulation, Properties and Drug Delivery Prospects. *Nanomaterials* **2021**, *11*, 440. [CrossRef]
30. Tenório-Neto, E.; Jamshaid, T.; Eissa, M.; Kunita, M.H.; Zine, N.; Agusti, G.; Fessi, H.; El-Salhi, A.E.; Elaissari, A. TGA and magnetization measurements for determination of composition and polymer conversion of magnetic hybrid particles. *Polym. Adv. Technol.* **2015**, *26*, 1199–1208. [CrossRef]
31. Estelrich, J.; Sanchez-Martin, M.J.; Busquets, M.A. Nanoparticles in magnetic resonance imaging: From simple to dual contrast agents. *Int. J. Nanomed.* **2015**, *10*, 1727–1741. [CrossRef]
32. Cho, M.; Villanova, J.; Ines, D.M.; Chen, J.; Lee, S.S.; Xiao, Z.; Guo, X.; Dunn, J.A.; Stueber, D.D.; Decuzzi, P.; et al. Sensitive T_2 MRI Contrast Agents from the Rational Design of Iron Oxide Nanoparticle Surface Coatings. *J. Phys. Chem. C* **2023**, *127*, 1057–1070. [CrossRef]

nanomaterials

Article

Investigation of the Influence of Wound-Treatment-Relevant Buffer Systems on the Colloidal and Optical Properties of Gold Nanoparticles

Atiđa Selmani [1], Ramona Jeitler [1,2], Michael Auinger [3], Carolin Tetyczka [2], Peter Banzer [4], Brian Kantor [4], Gerd Leitinger [5] and Eva Roblegg [1,2,*]

1 Pharmaceutical Technology & Biopharmacy, Institute of Pharmaceutical Sciences, University of Graz, Universitätsplatz 1, 8010 Graz, Austria; atida.selmani@uni-graz.at (A.S.); ramona.jeitler@uni-graz.at (R.J.)
2 Research Center Pharmaceutical Engineering GmbH, Inffeldgasse 13, 8010 Graz, Austria; carolin.tetyczka@rcpe.at
3 Institute of Chemical Technologies and Analytics, TU Wien, Getreidemarkt 9/164, 1060 Vienna, Austria; michael.auinger@tuwien.ac.at
4 Institute of Physics, NAWI Graz, University of Graz, Universitätsplatz 5, 8010 Graz, Austria; peter.banzer@uni-graz.at (P.B.); brian.kantor@uni-graz.at (B.K.)
5 Division of Cell Biology, Histology and Embryology, Gottfried Schatz Research Center for Cell Signaling, Metabolism and Aging, Medical University of Graz, 8010 Graz, Austria; gerd.leitinger@medunigraz.at
* Correspondence: eva.roblegg@uni-graz.at; Tel.: +43-316-380-8888

Abstract: Biocompatible gold nanoparticles (AuNPs) are used in wound healing due to their radical scavenging activity. They shorten wound healing time by, for example, improving re-epithelialization and promoting the formation of new connective tissue. Another approach that promotes wound healing through cell proliferation while inhibiting bacterial growth is an acidic microenvironment, which can be achieved with acid-forming buffers. Accordingly, a combination of these two approaches appears promising and is the focus of the present study. Here, 18 nm and 56 nm gold NP (Au) were prepared with Turkevich reduction synthesis using design-of-experiments methodology, and the influence of pH and ionic strength on their behaviour was investigated. The citrate buffer had a pronounced effect on the stability of AuNPs due to the more complex intermolecular interactions, which was also confirmed by the changes in optical properties. In contrast, AuNPs dispersed in lactate and phosphate buffer were stable at therapeutically relevant ionic strength, regardless of their size. Simulation of the local pH distribution near the particle surface also showed a steep pH gradient for particles smaller than 100 nm. This suggests that the healing potential is further enhanced by a more acidic environment at the particle surface, making this strategy a promising approach.

Keywords: gold nanoparticles; pH; buffer; ionic strength; size; zeta potential; wound healing

Citation: Selmani, A.; Jeitler, R.; Auinger, M.; Tetyczka, C.; Banzer, P.; Kantor, B.; Leitinger, G.; Roblegg, E. Investigation of the Influence of Wound-Treatment-Relevant Buffer Systems on the Colloidal and Optical Properties of Gold Nanoparticles. *Nanomaterials* **2023**, *13*, 1878. https://doi.org/10.3390/nano13121878

Academic Editors: Andrei Honciuc and Mirela Teodorescu

Received: 25 May 2023
Revised: 14 June 2023
Accepted: 15 June 2023
Published: 17 June 2023

1. Introduction

The microenvironment in the human body is highly complex in a healthy state and can change drastically in a diseased state. This is particularly the case with the skin, considering the progression from injury to wound healing. Wound healing can be classified into four integrated and partly overlapping phases, namely haemostasis, inflammation, proliferation and connective tissue remodelling to regenerate the functional epidermal barrier [1–3].

Thereby, many local wound factors and mediators are involved. One very important factor is the pH. If the skin, which shows a pH of 4–6 under homeostasis, is injured, there is initially increased lactic acid production and hypoxia before the pH at the wound site rises. Acute wounds have a neutral pH ranging from 6.5 to 8.5. This should theoretically reduce buffer capacity, which is also the case in inflammatory skin diseases, although limited data are available [4]. When wound healing is delayed, conditions become alkaline, bacterial growth is promoted, and healing is prevented [5]. At this stage, the wound is known as chronic and has an alkaline pH of 7.2–8.9, which favours bacterial growth. During healing, physiological mechanisms begin to restore the acidic environment by adjusting the pH. This, in turn, triggers cellular processes to restore the epithelial barrier, influences enzyme activity, and adjusts oxygen tension in the wound, which in turn promotes fibroblast growth and collagen synthesis. Cell growth is more active in acidic microenvironments, which are associated with faster migration and proliferation behaviour, eventually leading to wound regeneration. Because of these highly complex conditions, healing wounds remains a major challenge in modern medicine. It is assumed that the enhanced cell proliferation under acidic conditions is probably closely related to the polarity and epithelial potential between injured and uninjured tissue. This is consistent with several studies reporting that acidifying wounds, e.g., with topical creams, gels, dressings or solutions including lactate, acetic acid or citrate, is an effective treatment and also helps to bypass the alkaline microenvironment that promotes bacterial growth [6]. Besides buffered therapeutic systems, nanoparticles (NPs) have become increasingly interesting for therapeutic and diagnostic purposes in wound healing. The NPs used include inorganic/metallic NPs, and lipid- and polymer-based NPs [7]. While the latter are mainly used for drug delivery of enzymatically prone drugs that should be released in a controlled manner, metallic/inorganic NPs may exhibit intrinsic properties in addition to drug delivery, such as antimicrobial activity, anti-inflammatory activity, proangiogenic and antioxidant activity and optical activity. Gold (Au) NPs, for example, are versatile nanomaterials with unique physicochemical and optical properties that are biocompatible and can be fabricated in variable sizes and shapes [8,9]. Due to their excellent physicochemical and optical properties and nontoxic nature, AuNPs exhibit tremendous potential in the biomedical field, such as bio-imaging, diagnostics, photo-induced cancer therapy, tissue engineering and immunology [10]. The optical properties offer enormous opportunities for biosensor applications [11,12], as they exhibit a strong surface plasmon resonance (SPR) absorption band and a high molar absorptivity [13]. Their radical scavenging activity makes them attractive for wound therapy; likewise, they have been found to shorten wound healing time by, for example, improving re-epithelialization and promoting new connective tissue and microscopic blood vessel formation and extracellular matrix deposition. Unlike silver NPs, AuNPs do not possess antimicrobial activity themselves, but they allow efficient coupling with antimicrobial biomolecules and drugs, enhancing the effectiveness against microbes [14–16]. Moreover, they are modifiable and thus suitable for the targeted drug delivery of biomolecules such as DNA, RNA and proteins to, e.g., specific cells. This can be achieved with adsorption or through ionic or covalent binding with the help of linkers. Thereby, drug release can be controlled by various internal (e.g., pH value and enzymes) and external (e.g., light) stimuli, which, again, highlights their enormous potential [17–19].

To take advantage of wound acidification and the intrinsic properties of AuNPs, both strategies should be combined, i.e., AuNPs should be incorporated into different buffer systems relevant to wound treatment. In this respect, the influence of pH and ionic strength (I_c) must be carefully considered, as it is known that pH-induced changes in surface chemistry or fluctuations in electrolyte levels lead to reduced NP stability and, consequently, the altered binding of cellular components or biomolecules [20]. This, in turn, affects cellular interactions, transport mechanisms, accumulation and/or excretion of the therapeutic potential of NPs [21,22]. In addition, local pH shifts caused by chemical reactions on the particle surface must be considered, as these might have a strong impact on the local environment [23] due to stronger pH gradients around smaller NPs [24]. However, steep pH gradients between the dispersant and the particle surface would further acidify the NP environment, which could be exploited in wound healing. The complex interplay of all these factors must be fundamentally understood in order to establish relationships through experiments at the buffer–nano–bio interface. Therefore, as a first step, it is necessary to systematically prepare and investigate biocompatible AuNPs with tuneable properties with respect to influencing parameters such as pH and I_c and to establish meaningful simulation models to study the influence of these conditions at the NP surface.

For this purpose, a robust manufacturing strategy is required, yielding AuNPs with defined initial properties. In general, AuNPs are produced in a batch process using different synthesis methods, such as chemical reduction, plant-assisted synthesis (green synthesis) or with the continuous synthesis in flow reactors [25–27]. Chemical synthesis methods include the Turkevich and Brust method, the seed-mediated growth and digestive ripening method [28,29]. Thereby, the Turkevich method is the most commonly used one-step preparation method, which allows the production of a broad range of spherical AuNP by simply adjusting the reducing agent ratio. The most studied reducing agents for this purpose are ascorbic acid, citrate, UV light and amino acids [30–33]. The Brust method, on the other hand, is also a simple strategy but limits the biological application of NPs produced, and the seed-mediated growth method allows the manufacturing of rod-shaped structures. The digestion ripening method is also a simple fabrication method of monodisperse NPs; however, it is difficult to control particle shape because very high temperatures are required [34]. To overcome environmental limitations, green synthesis has gained more and more attention in recent years. This synthesis strategy is also reported to be a simple, straightforward and cost-effective process in which shape and size can be regulated [35]. Moreover, enhanced bioactivity can be achieved, leading to an apoptotic effect in cancer cells [36]. The difficulty here, however, is that the number of organic components is very high, making it difficult to accurately identify the relevant reactive components.

Based on the advantages and disadvantages of the different production methods, the Turkevich method was chosen in this study to produce spherical AuNPs of different sizes.

To ensure reproducible particle production, the design-of-experiments (DoE) approach was applied. After a risk assessment, the most influential process and formulation parameters were identified, methodically reviewed and correlated. The AuNPs used in this study were considered "bare" since they were coated only with the reductant in order to understand their behaviour without introducing another coating agent that might alter the fundamental surface properties. In a further step, the prepared purified AuNPs were carefully characterized under defined conditions, i.e., dispersed in an aqueous solution of hydrochloric acid/sodium chloride (HCl/NaCl) at an initial pH $\approx$ 3 and different I_c. In the next step, particle behaviour studies were performed with buffer systems adapted for wound treatment, mimicking both the pH and relevant I_c (i.e., therapeutically relevant or high). For this purpose, lactate (LA/NaL) and citrate (H_3Cit/Na_3Cit) buffers, as well as Sørensen's phosphate buffers (PBS), were used as potential dispersants. Characterization methods included size, zeta potential, shape, and (single NP) SPR using dynamic light scattering (DLS), transmission electron microscopy (TEM), atomic force microscopy (AFM), UV-VIS spectrophotometry and extended spectroscopy. The data obtained were then used for simulation studies to predict the behaviour of the AuNP surface in the studied buffers as a function of pH (and I_c).

2. Materials and Methods

2.1. Materials

Tetrachloroauric acid ($HAuCl_4 \cdot 3H_2O$), trisodium citrate dihydrate ($Na_3Cit \cdot 2H_2O$), sodium chloride (NaCl), sodium hydrogen carbonate ($NaHCO_3$) and hydrochloric acid (HCl, w = 36%) were purchased from Carl Roth Gmbh & Co (Karlsruhe, Germany). Ascorbic (AA) and citric acid (H_3Cit) were obtained from Herba Chemosan Apotheker-AG (Graz, Austria). Disodium hydrogen phosphate (Na_2HPO_4), sodium dihydrogen phosphate (NaH_2PO_4) and lactic acid (LA, w = 90%) were purchased from VWR Chemicals (Leuven, Belgium). Sodium lactate (NaL, w = 50%) was purchased from Caesar & Loretz GmbH (Hilden, Germany). Sodium hydroxide (NaOH) and nitric acid (HNO_3, w = 68%) were obtained from Merck KgaA (Darmstadt, Germany). All the chemicals were used as received without additional purification. Ultrapure water (MQ water, resistivity of 18 MΩ·cm, Millipore S.A.S., Molsheim, France) was used for the solution preparation.

2.2. Methods

2.2.1. Synthesis of AuNPs Using Na_3Cit and AA: Design-of-Experiments (DoE) Studies

AuNPs were prepared using two common reductants, Na_3Cit and AA, in various molar ratios. Citrate-capped AuNPs were fabricated with the chemical reduction synthesis of $HAuCl_4$ with Na_3Cit according to the modified protocol of Dong et al. [37]. Alternatively, AA was used as the reducing agent, and particles were prepared following the protocol by Malassis et al. [38]. Briefly, for both procedures, the glassware was cleaned with aqua regia (HCl:HNO_3 = 4:1). For the preparation of the citrate-capped AuNPs, 5 mL of 0.25 mM $HAuCl_4$ was heated to 100 °C under constant stirring on a magnetic stirrer, and different amounts of 34 mM Na_3Cit were added. The synthesis was considered complete when the colour of the suspension stopped changing; the suspension was cooled to room temperature (RT). For the synthesis with AA, 5 mL of 0.5 mM $HAuCl_4$ were mixed with different amounts of 0.1 M AA, and agitated on a magnetic stirrer at RT.

To optimize the syntheses and develop a predictive model, a DoE study was performed. MODDE® software (Version 13.0, Sartorius AG, Göttingen, Germany) was used, and the central composite face-centred (CCF) quadratic experimental design was selected. For the screening of the reduction synthesis with Na_3Cit, the molar ratio of Na_3Cit to $HAuCl_4$ (1.5:1 to 3.7:1), the process temperature (70–100 °C), the pH of the reaction media (3–6), the stirring speed (150–300 rpm) and the reaction time (5–90 min) were selected as input parameters. The resulting particle sizes, polydispersity indices (PdIs) and zeta potential values were used as responses. With the experimental design chosen, MODDE® proposed a total of 29 experiments, 3 of which were used to investigate reproducibility. For the synthesis process involving AA, input parameters such as the molar ratio of AA to $HAuCl_4$ (0.8:1 to 2.5:1), the pH of the reaction media (3–8), the stirring speed (150–300 rpm) and the reaction time (2–60 min) were tested with regard to their influence on the resulting particle size, yield and zeta potential. For this, 27 experiments in total (with 3 centre point runs for reproducibility studies) were required, and the mean values of responses were used for statistical analysis using multiple linear regression (MLR) (MODDE®). Coefficients plots, the summary of fit including R^2 (i.e., percent of the variation of the response explained by the model) and Q^2 (i.e., percent of the variation of the response predicted by the model according to cross-validation) values, the residuals' normal probability and plots of observed vs. predicted values were used to investigate and evaluate the obtained model.

Subsequently, optimized suspensions were prepared according to the results obtained via DoE and were dialyzed for 24 h in MQ water using a dialysis membrane (Carl Roth GmbH & Co, Karlsruhe, Germany) with a molecular weight cutoff from 12 to 14 kDa, and the media was changed twice.

2.2.2. Physicochemical and Optical Characterization of AuNPs

The particles were characterized regarding hydrodynamic particle size (d_h) and zeta potential (ζ) via DLS and electrophoretic light scattering (ELS) after dialysis. For DLS and ELS measurements, the Zetasizer Nano ZS (Malvern, UK) was used with a He–Ne red-light-emitting laser (wavelength, λ = 633 nm). The AuNP suspensions were diluted 1:10 with prefiltered MQ water (Whatman filter, pore size 0.02 μm) to avoid multiple scattering. All measurements were performed in triplicate at RT with the samples remaining in equilibrium for 2 min, considering the refractive indices of AuNPs (0.2) and the dispersant (1.33). During the DLS studies, scattered light was detected at an angle of 173° (i.e., backscatter mode), and data processing was performed with the Zetasizer software 6.32 (Malvern Instruments). The obtained data are presented as volume-based particle-size distribution to avoid overestimation of larger particles due to higher scattered light intensities. The zeta potential of the AuNPs was calculated from the measured electrophoretic mobility by applying the Henry equation using the Smoluchowski approximation ($f(\kappa a) = 1.5$).

The morphology of AuNPs was determined with TEM measurements. AuNP suspensions, diluted in a 1:1 ratio, were placed on carbon-coated TEM grids and blot dried. Grids were visualized with a Thermo Fisher Tecnai 20 transmission electron microscope operating at 120 kV acceleration voltage. Microscopic images were taken with a Gatan US 1000 CCD camera.

To verify the measured sizes and investigate the shape of the synthesized particles, visualization of AuNPs was performed with atomic force microscopy (AFM; FlexAFM atomic force microscope equipped with an Easyscan 2 controller, Nanosurf, Liestal, Switzerland). For the AFM studies, 10 μL of each AuNP suspension was placed on a silicon wafer and dried overnight at RT. Prior to use, the wafer was rinsed with deionized water and ethanol and flushed with nitrogen. Noncontact (tapping) mode with a setpoint of 60% was used for the acquisition using a silicon nitride tip (Tap300Al-G; Budgetsensors, Sofia, Bulgaria) with a radius of 10 nm and a length of 125 μm. Samples were screened with a nominal spring constant of 40 N/m and a nominal resonance frequency of 300 kHz at ambient temperature. Images were processed with Gwydion Data Processing Software (Version 2.62) [39]. Sizes of AuNPs were obtained by using ImageJ software (Version 1.53p) and analysis of 15 single particles from TEM images and 30 single particles from AFM images.

UV-Vis spectrophotometry measurements (Eppendorf BioSpectrometer® kinetic, Darmstadt, Germany) were conducted to study the SPR of AuNPs. Samples were diluted in a 1:1 ratio and the measurements were performed over a λ range from 400 to 800 nm.

2.2.3. Effect of pH and Ion Concentration on the Physicochemical and Optical Properties of AuNPs

To investigate the effect of pH and ion concentration on changes in AuNPs in terms of zeta potential, agglomeration tendency and SPR, particles were dispersed in HCl/NaCl aqueous solutions and buffer systems for wound treatment. For the HCl/NaCl aqueous solutions, pH titrations were conducted with the titrant NaOH (0.1 M) at 2 I_c and 10 and 50 mM (adjusted with HCl and NaCl). pH was recorded with a pH meter (Lab 860 pH meter; Fisher Scientific GmbH, Vienna, Austria) equipped with a combined pH electrode (SI Analytics GmbH, Mainz, Germany). pH titrations were performed in the direction from acid to base. For the adjusted buffers, LA/NaL (pH 2.8–4.8), H_3Cit/Na_3Cit buffer (pH 2.8–5.8) and PBS (pH 6.8–8.0) at two I_c, the same as for pH titrations, were prepared and the pH was recorded. 100 μL of each AuNP dispersion was incubated with 900 μL of each buffer system for 5 min, and the pH was monitored throughout the experiments. The size and zeta potential measurements were performed with the same setups as described in Section 2.2.2.

For SPR measurements, the particles were incubated with HCl/NaCl aqueous solutions and the respective buffers. To gain better insight into the optical properties of AuNPs and SPR behaviour in media with different compositions, an extended spectroscopic method was applied, which makes it possible to examine individual particles with regard to the SPR. A confocally aligned optical setup [40,41] was utilized, which featured a high numerical aperture (NA = 0.9) objective to focus a linearly polarized Gaussian beam onto the sample. The sample of interest was mounted to a 3D piezo-driven positioning stage, which provided a precision of roughly 2 nm. A 1.3 NA oil immersion objective collected the transmitted light from the sample, which was then propagated onto a photodiode. The extreme focusing, in conjunction with the precise positioning stage, enabled one to optically probe an individual NP of interest while avoiding unwanted near-field excitations. A λ sweep was performed, and the transmitted power was recorded in order to measure the resonant properties of the individual AuNPs.

2.2.4. Simulation of the Local pH-Value Distribution near the Particle Surface

The pH-value distribution near the NPs was derived from the set of spherical diffusion equations for each mobile species in the solution (Equation (1)). Due to the fast kinetics of ionic reactions in solution, all species can be assumed to be in local thermodynamic equilibrium [42]. This allows one to link the equations and express the concentration of the involved species with a pH gradient. Details of the mathematical derivation are reported elsewhere [23]. For simplicity, it is assumed that the particles are spherical.

$$\frac{\partial c_i}{\partial t} = \frac{1}{r^2}\text{div}\left(r^2 D_i \nabla c_i + r^2 \frac{D_i z_i e_0}{k_B T} c_i \nabla \phi\right), \quad (1)$$

where the index i denotes the chemical species (H^+, OH^- and the buffer ions), c the concentration, D the diffusion coefficient, r the radial distance from the centre, z the electric charge, e_0 Coulomb's constant, k_B Boltzmann's constant and ϕ the electrical potential. Due to the short diffusion pathways of only a few µm, the set of partial differential equations can be solved for steady-state conditions, as further described in [23] for the case of flat surfaces and NP geometries [24]. Given the high I_c of the solution, the gradient of the electric potential is small, and the migration term in Equation (1) can be simplified to the case of diffusion only. The effect of an electric field will lead to a transition of the pH gradient towards/away from the NPs without changing the main shape of the pH distribution.

Results are displayed as a 2-dimensional contour plot of the local pH value as a function of distance from the electrode surface (x-axis) and buffer concentration (y-axis) for a given nanoparticle size.

2.2.5. Statistics

All experiments were performed in triplicate if not stated otherwise, and mean values, including standard deviations, are presented. To evaluate the statistical significance of the size of the particles, I_c and pH values of buffers, a three-way analysis of variance (ANOVA) using the Tukey test (GraphPad Prism 8, La Jolla, CA, USA) was applied. The significance of the results is indicated according to p-values: * $p \leq 0.05$, ** $p \leq 0.01$, *** $p \leq 0.001$ and **** $p \leq 0.0001$. The p-value below 0.05, i.e., 95% of the confidence interval, was considered statistically significant.

3. Results

3.1. Design-of-Experiments (DoE) Studies

One of the most common batch synthesis methods for the preparation of AuNPs is the Turkevich method, which chemically reduces metal salts in aqueous media using suitable, mild reducing agents such as Na_3Cit [43,44] or AA [38,45]. In this process, the molar ratio of the reducing agent and metal salt is a crucial parameter that determines the particle size and PdI. Other parameters that need to be considered are pH, temperature, reaction time and the presence of other ions and molecules [46–49]. To optimize the procedure, a DoE study was performed to adjust the aforementioned parameters for both reduction agents. The obtained data, using Na_3Cit as a reduction agent for particle size, were in the range of 5.95 ± 2.26 nm and 45.91 ± 3.14 and showed PdI values between 0.07 ± 0.01 and 0.75 ± 0.29. The zeta potential values ranged from −35.53 ± 0.58 mV to 0.90 ± 0.16 mV for the prepared batches. The use of AA as a reduction agent resulted in a larger d_h with mean sizes ranging from 44.71 ± 2.99 nm to 212.33 ± 2.45 nm. DLS measurements proposed rather polydisperse samples with at least two fractions, i.e., one from 40 to 70 nm and the second between 150 and 180 nm. This may indicate agglomeration rather than the formation of larger particles. To carefully identify the most appropriate synthesis conditions yielding de-agglomerated and thus more stable batches, the focus was laid on the fraction of the smallest sizes around 40 and 70 nm by including the respective yield. The yield for the small fraction ranged from 4.5 ± 0.0% to 77.63 ± 19.43%. At specific process conditions, the small fraction disappeared and was replaced by larger agglomerate assemblies with 100% yield. The experimental data were further used for statistical data analysis. The residuals' normal probability plots were created (Figure 1a (Na_3Cit) and Figure 1b (AA)) to evaluate whether the process setup and measurement strategy were reliable and to identify potential outliners (i.e., experiments outside or close to the boundaries (dashed lines)).

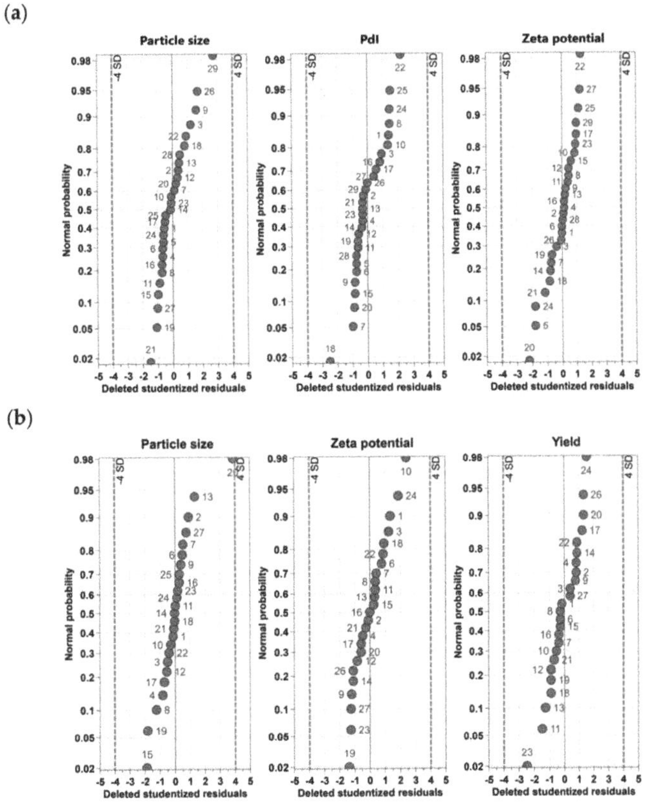

Figure 1. Normal probability plots of the standardized effects shown for (**a**) citrate-capped AuNPs and (**b**) AA-capped AuNPs.

For Na_3Cit, the results of all performed experiments were found to be normally distributed regardless of the considered response (i.e., size, PdI, zeta potential), as all data are on a straight line. By contrast, the use of AA as a reduction agent revealed some potential outliners. Considering the size of the response, experiments N5 and N26 were excluded as they were outside the boundaries. For the zeta potential and the PdI, the results of run N5 and N25 were excluded from further statistical data evaluation, as they were also outside the boundaries. Coefficients plots were created to assess the significance of the coefficients and are shown in Figure 2a (i.e., citrate-capped AuNPs) and Figure 2b (i.e., AA-capped). After excluding insignificant parameters (i.e., a small distance from y = 0 and error bars crossing y = 0) to simplify the model and to maximize the performance of prediction, all individually tested input parameters (i.e., molar ratio, pH, reaction time, stirring speed and temperature) showed a significant influence on the resulting particle sizes when using Na_3Cit (Figure 2a). In addition, the combinations of molar ratio and reaction time and reaction time and temperature also affected the sizes significantly. The PdI was mainly influenced by the molar ratio, temperature and pH. The zeta potential was affected by the molar ratio, the pH, the reaction temperature and the quadratic effects of the pH and the molar ratio. In addition, the interaction in terms of the molar ratio and reaction time, and to a lower extent, the temperature and stirring speed, as well as the pH and stirring speed, strongly influenced the resulting zeta potential. During the synthesis of AuNPs via AA, size was controlled by the quadratic effects of pH and with the combination of molar ratio and pH, molar ratio and reaction time, molar ratio and stirring speed and reaction time and stirring speed. The zeta potential was significantly influenced by the pH of the reaction media, the stirring speed and the combination of molar ratio and reaction time. Considering the yield of the smallest particle fraction in the batches, it was observed that the combinatory effect of molar ratio and stirring speed had the highest impact.

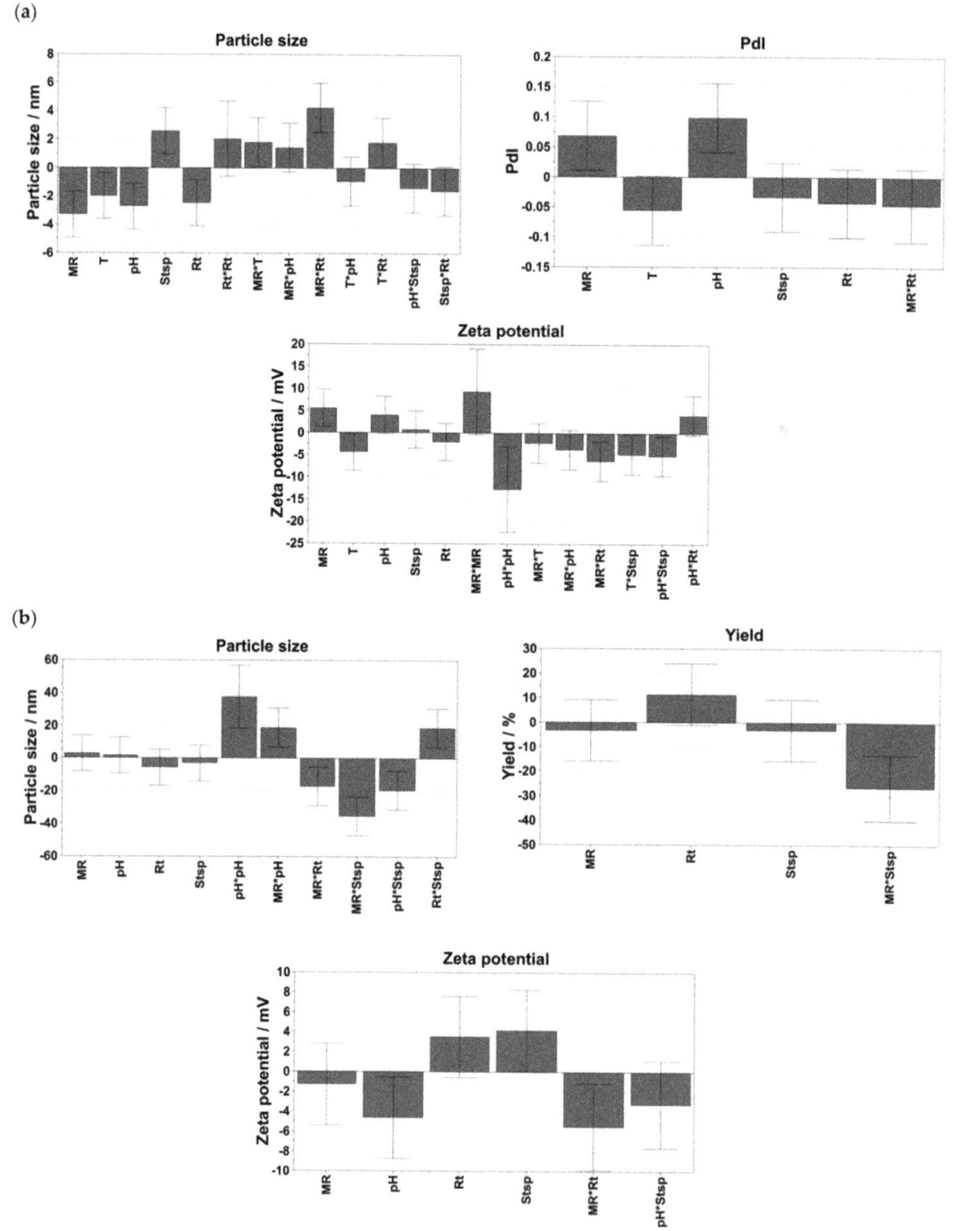

Figure 2. Coefficients plots shown for (**a**) citrate-capped AuNPs and (**b**) AA-capped AuNPs. MR = molar ratio, temperature = T, stirring speed = Stsp, reaction time = Rt. The asterisk in each case indicates the interaction between two tested parameters.

For quality assessment of the model, the obtained data were fitted using MLR, and the summary of fit was prepared considering R^2 and Q^2 values as well as the model validity and reproducibility. For both synthesis strategies, values for R^2 and Q^2 were highest for the particle sizes. With comparable high R^2 values (i.e., 0.88 for Na_3Cit and 0.89 for AA), the variability in the data can be explained by the model. As for both strategies, Q^2 values are above 0.5 (i.e., 0.57 for Na_3Cit and 0.72 for AA), and the difference between R^2 and Q^2 is less than 0.3; the precision and reliability of future prediction can be considered high. However, it must be emphasized that the reproducibility of both batch-based synthesis routes must be classified as rather low. In the synthesis with AA, obvious outliners were already identified and excluded during the statistical data analysis step. When using Na_3Cit in the synthesis, the results of the reproducibility tests already show a large variability in the results. Moreover, reliable predictions for zeta potential and PdI or yield can only be expected to a moderate extent for both synthesis strategies. Although all of the obtained R^2 values were above 0.5 (i.e., 0.57 for PdI and 0.78 for the zeta potential with Na_3Cit and 0.59 for zeta potential and 0.52 for the yield AA), which indicates an acc fit between the variability observed and the regression model; all Q^2 values were found to be below 0.5. The experimental data show a correspondingly inhomogeneous character of some batches with high PdI values (Na_3Cit) and a pronounced polydisperse particle size distribution (AA). Since the zeta potential is a size-dependent property and is therefore difficult to capture in polydisperse systems, the established models will primarily be used to produce AuNPs with the desired sizes as homogeneously as possible. With the optimized synthesis conditions, two different size ranges of AuNPs were prepared. Briefly, a molar ratio of Na_3Cit:$HAuCl_4$ of 3.7:1 at pH 3.3, a synthesis temperature of 100 °C at 250 rpm and reaction duration of 5 min, resulted in approximately 18 nm sized particles, referred here as "small particle size". The use of a 1.2:1 AA:$HAuCl_4$ molar ratio at pH 3.5 and 250 rpm and a reaction duration of 2 min resulted in about 56 nm, referred to here as "medium particle size". In particular, the results after dialysis of the nanosuspensions showed that a monomodal size distribution with a d_h of 17.65 ± 0.38 nm was achieved with the reducing agent Na_3Cit. When AA was used as the reductant, 56.02 ± 0.58 nm sized particles were obtained. As expected, the zeta potential values of produced AuNPs were −22.80 ± 2.50 mV and −37.60 ± 1.30 mV, respectively.

3.2. Physicochemical and Optical Characterization of AuNPs

The shape and size of the AuNPs were determined with TEM and AFM. Representative TEM images are shown in Figure 3a,b. Citrate-capped AuNPs were spherical and had a size of 14.57 ± 2.00 nm, while AA-capped AuNPs exhibited a size of 39.02 ± 5.36 nm. AFM images and histograms of the size distribution of the AuNPs are shown in Figure 3c–f. The fabricated small AuNPs were spherical in shape with a mean size of 15.70 ± 2.94 nm. AuNPs synthesized with AA exhibited sizes of 33.52 ± 6.42 nm. However, with increasing AuNP sizes, spherical clusters of AuNPs were noticed.

The optical properties, i.e., the changes of the SPR of the differently sized AuNPs, were studied with UV-Vis spectrophotometry (Figure 3g). The results showed that the changes in sizes correlated with a red shift of the SPR peak, i.e., from 520 nm for the small AuNPs to 535 nm for the medium-sized AuNPs' fractions.

3.3. Effect of the pH on the Physicochemical and Optical Properties of AuNPs

The changes of the zeta potential and size of the small- and medium-sized AuNPs as a function of the pH at two nearly constant I_c, i.e., 10 and 50 mM, are presented in Figure 4a,b. The results showed that, with increasing pH, the zeta potential of the small-sized AuNPs remained constant at 10 mM I_c (−28.33 ± 1.35 mV (pH = 3.0) to −28.87 ± 1.97 mV (pH = 10.2)).

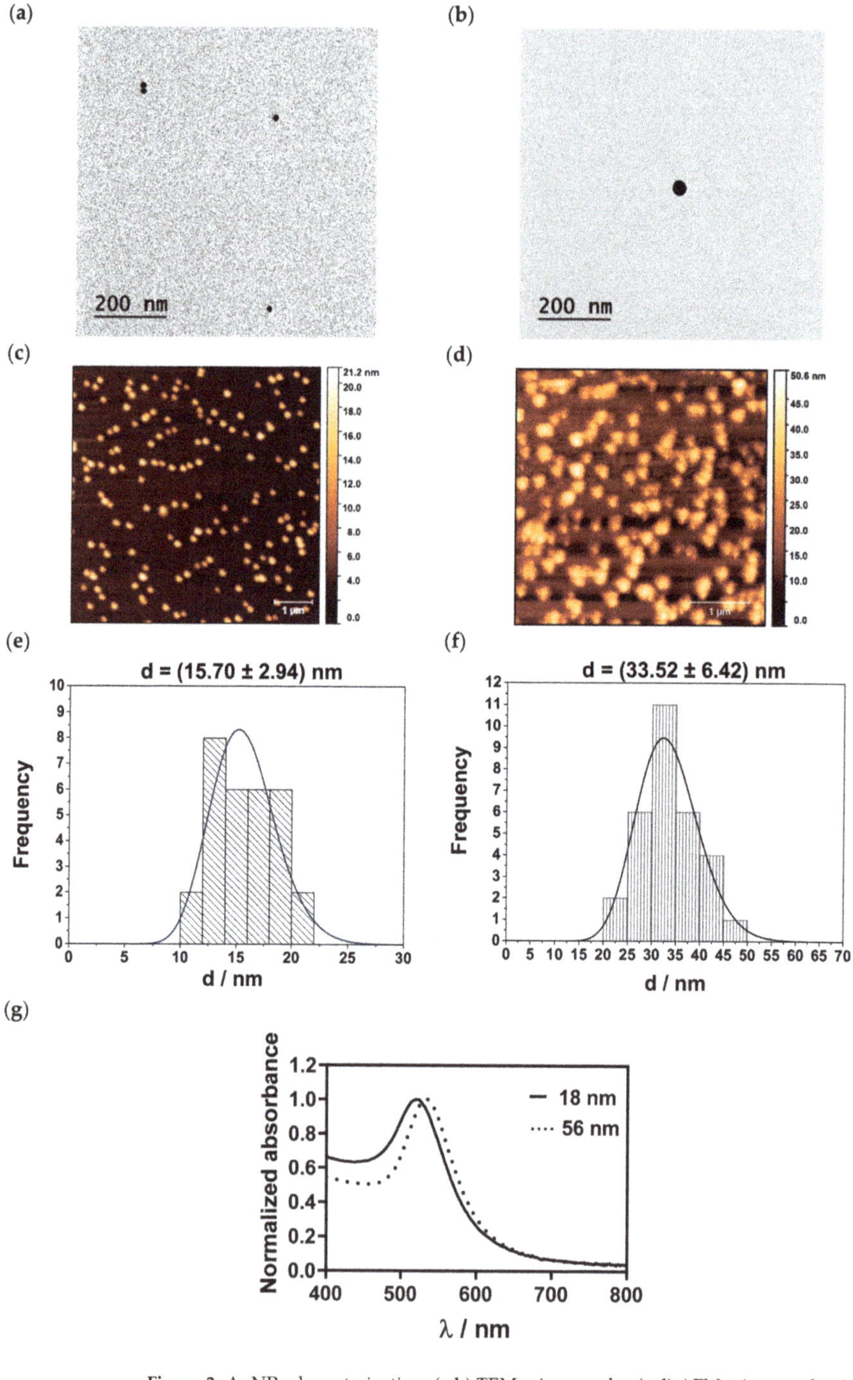

Figure 3. AuNPs characterization: (**a**,**b**) TEM micrographs, (**c**,**d**) AFM micrographs, (**e**,**f**) histograms of the AuNP size distributions obtained from AFM images and (**g**) optical properties of AuNPs of different sizes conducted with UV-Vis spectrophotometry. (**a**,**c**,**d**) 18 nm AuNPs; (**b**,**d**,**f**) 56 nm AuNPs.

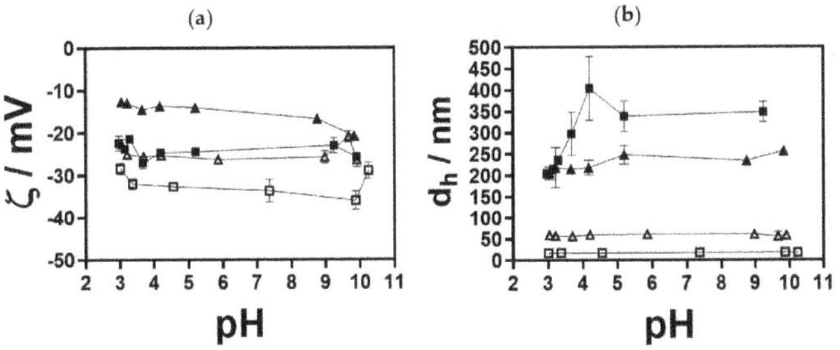

Figure 4. pH-dependent (**a**) zeta potential and (**b**) size profiles of AuNPs of different sizes in HCl/NaCl aqueous solution. pH titrations were conducted with the titrant NaOH (0.1 M) at two I_c, i.e., 10 and 50 mM (adjusted with HCl and NaCl); (□) 18 AuNPs at I_c = 10 mM, (■) 18 nm AuNPs at I_c = 50 mM, (△) 56 nm AuNPs at I_c = 10 mM and (▲) 56 nm AuNPs at I_c = 50 mM.

As expected, at higher I_c, i.e., 50 mM, the zeta potential values were lower. The isoelectric point was not noticed, and the zeta potential remained negative over the entire pH range. pH-dependent zeta potential profiles for 56 nm sized AuNPs exhibited similar behaviour, with lower zeta potential values compared to 18 nm sized AuNPs. At 10 mM, the sizes of small and medium AuNPs were similar to the AuNPs dispersed in MQ water and remained constant in the pH range tested. The increase in I_c led to an irreversible agglomeration of 18 nm and 56 nm sized AuNPs, independent of the pH. The changes in the zeta potential and size of the small- and medium-sized AuNPs in the LA/NaL-, H_3Cit/Na_3Cit buffers and PBS at 10 and 50 mM are presented in Figure 5a–f. In 10 mM LA/NaL buffer (Figure 5a,b), small-sized AuNPs retained their initial sizes regardless of the pH; however, as the I_c increased to 50 mM, the size of AuNPs increased from 48 to 290 nm with a rising pH. At 10 mM, the zeta potential was ≈ -20 mV for all pH values. In 10 and 50 mM H_3Cit/Na_3Cit buffer (Figure 5c,d), at lower pH values, the zeta potential of the small NPs changed, indicating that they exhibited less stability. Small-sized AuNPs in PBS (Figure 5e,f) were stable in terms of zeta potential and size at both investigated I_c. The medium-sized AuNPs in LA/NaL buffer (Figure 5a,b) at pH 2.8 showed a slightly lower zeta potential at both I_c. Still, the size was in the range of 50–60 nm, thus confirming a high particle stability. Incubation of the particles in 10 mM H_3Cit/Na_3Cit buffers (Figure 5c,d) exhibited the same behaviour as in the case of the 18 nm sized AuNPs. With increasing I_c, agglomeration occurred.

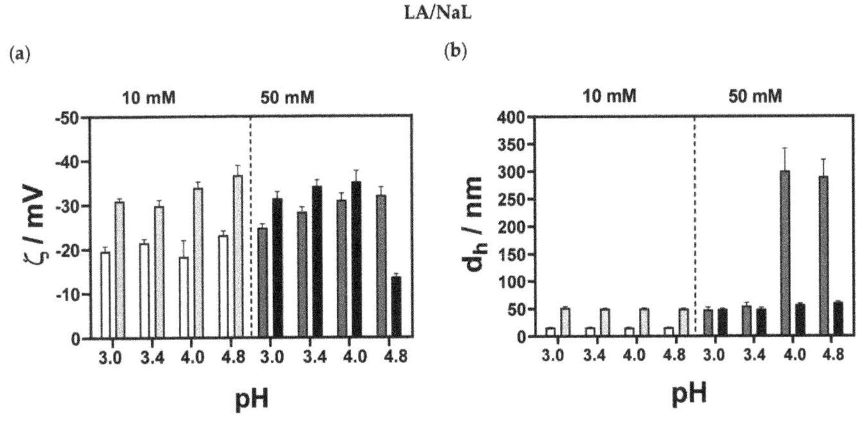

Figure 5. *Cont.*

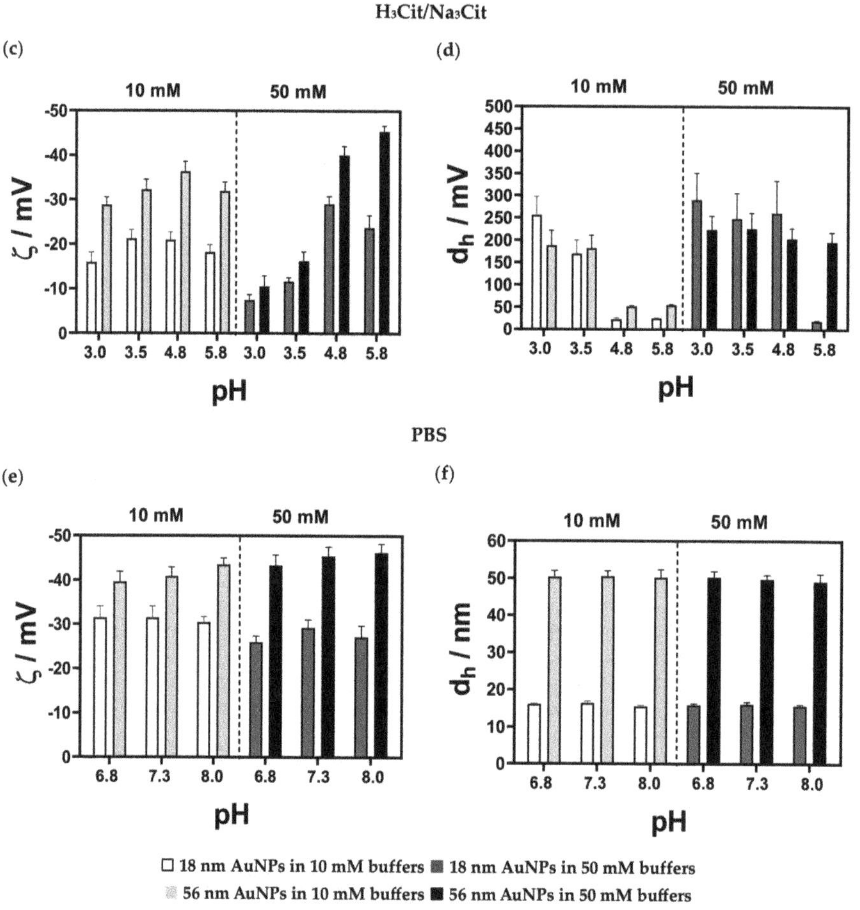

Figure 5. (**a**,**c**,**e**) Zeta potential and (**b**,**d**,**f**) size profiles of 18 nm and 56 nm sized AuNPs, in (**a**,**b**) LA/NaL buffer (pH = 2.9–4.8), (**c**,**d**) H_3Cit/Na_3Cit buffer, pH = 2.8–5.8) and (**e**,**f**) PBS (pH = 6.8–8) at two I_c, 10 mM and 50 mM. The dashed line in each case separates the datasets of 18 nm and 56 nm measured NPs in 10 mM buffer (left part) and the datasets of 18 nm and 56 nm measured NPs in 50 mM buffer (right part).

The particles dispersed in 10 and 50 mM PBS exhibited high zeta potential values ($\zeta > -40$ mV) (Figure 5e,f), and the initial particle size was preserved independent of the pH.

A three-way analysis of variance (ANOVA) was applied to test the statistical differences between size and zeta potential measurements for small- (18 nm) and medium (56 nm)-sized AuNPs at two different I_c and pH values of the buffers. There were no statistically significant differences between the AuNP sizes ($p = 0.5433$). Both I_c and buffer pH significantly affected the obtained sizes ($p \leq 0.0001$). There was a significant interaction between pairwise comparisons of I_c and buffers ($p \leq 0.0001$) and I_c and the size of AuNPs ($p \leq 0.0001$). Significance was also determined for the interaction between the size of AuNPs and I_c ($p = 0.0128$). A three-way ANOVA showed that the interaction between AuNP size, I_c and the pH value of buffers was statistically significant ($p \leq 0.0001$). A similar trend was obtained for the zeta potential. It was found that the size ($p \leq 0.0001$), I_c ($p = 0.0060$) and buffer pH ($p \leq 0.0001$) had a significant effect. For the pairwise comparison of the I_c and buffer pH, there was $p \leq 0.001$, as well as for the comparison of the AuNP size and buffer pH. The interaction among the size of AuNPs and I_c was also significant ($p = 0.0011$), as was the interaction among the size of AuNPs, I_c, and pH of buffers ($p \leq 0.0001$).

The effects of pH and I_c on the SPR peak of AuNPs were determined using UV-Vis spectrophotometry in all buffers, as shown in Figure 6. From pH 3.3 to 10.5, the small-sized AuNPs showed a red shift (Δ_{max} = 2 nm). However, the colour of the suspension remained the same. No red-wine-to-blue colour transition was noticed, suggesting that particles were stable. At 50 mM, alternations between the red and blue shift were detected ($\Delta_{max} = \pm 2$ nm), while at pH = 5.6 a red shift of 3 nm was obtained. The medium-sized AuNPs exhibited the blue shift at 10 mM in the tested pH range. At 50 mM, only at pH 3.3, a red shift of 5 nm was noticed, and the colour of the suspension changed to blue. In the pH range of 4.3–10.8, the SPR peak had the same maximum as in the MQ water (535 nm) or was blue-shifted by 2 nm.

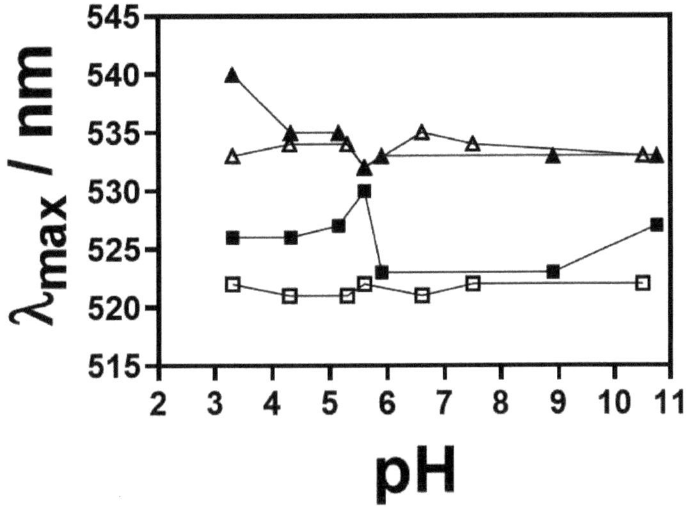

Figure 6. Maximum wavelength (λ_{max}) of the SPR peak of 18 nm and 56 nm sized AuNPs as a function of pH obtained from UV-Vis spectra at two I_c, 10 mM and 50 mM; (□) 18 AuNPs at I_c = 10 mM, (■) 18 nm AuNPs at I_c = 50 mM, (△) 56 nm AuNPs at I_c = 10 mM and (▲) 56 nm AuNPs at I_c = 50 mM.

The SPR peaks of 18 nm and 56 nm sized AuNPs were elucidated in all buffers (Figure 7a–c). When incubated in 10 mM LA/NaL buffer, small AuNPs exhibited a negligible shift of $\Delta\lambda_{max} = \pm$ 1 nm, whereas a red shift of 7–9 nm was observed at 50 mM. This indicates that AuNPs agglomerate at increased LA/NaL buffer I_c. The small-sized AuNPs showed a pronounced red shift ($\Delta\lambda_{max} \approx 100$ nm) independent of I_c at the pH range of 3.0 to 4.8. The colour of the suspension changed immediately from red wine to blue, indicating particle agglomeration. This was further confirmed with two peaks found in the UV spectra (second peak at $\lambda \approx 630$ nm). There were no changes in the SPR peaks of AuNPs in 10 and 50 mM PBS. The medium-sized AuNPs in 10 mM and 50 mM LA/NaL buffers at pH $\approx$ 3 displayed the SPR peak at λ = 535 nm, which is comparable to the results for AuNPs dispersed in MQ water. Furthermore, the pH increase led to the blue shift (2–4 nm). For both H_3Cit/Na_3Cit buffer I_c, the shift was $\pm$ 1 nm. The blue shift from 2 to 4 nm was also observed for AuNPs in PBS.

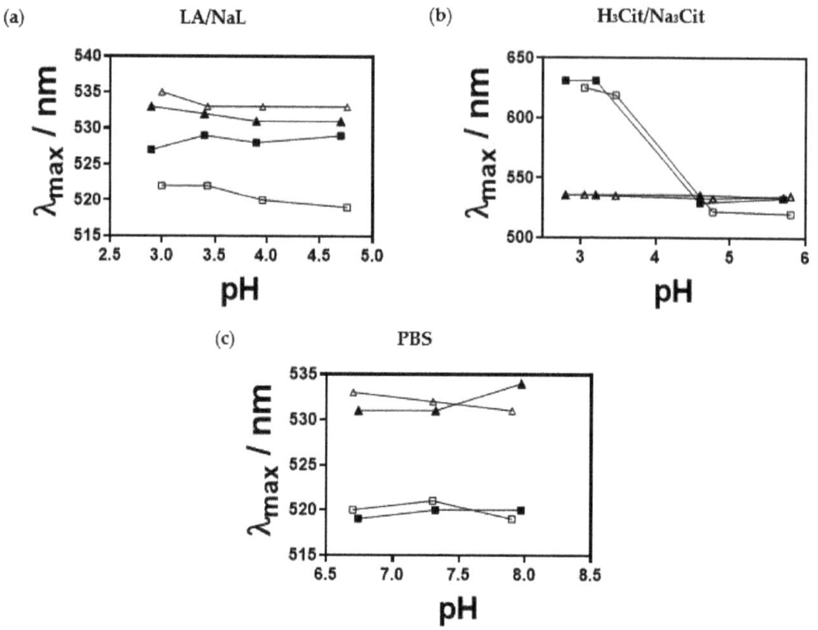

Figure 7. Maximum wavelength (λmax) of the SPR peak of 18 nm and 56 nm AuNPs as a function of pH obtained from UV-Vis spectra in (**a**) LA/NaL buffer (pH = 2.9–4.8), (**b**) H_3Cit/Na_3Cit buffer (pH = 2.8–5.8) and (**c**) PBS (pH = 6.8–8) at two I_c, 10 mM and 50 mM. (□) 18 AuNPs at I_c = 10 mM, (■) 18 nm AuNPs at I_c = 50 mM, (△) 56 nm AuNPs at I_c = 10 mM and (▲) 56 nm AuNPs at I_c = 50 mM.

The change in λ_{max} of the SPR indicates that different pH conditions affect the optical properties of the AuNPs. However, to better understand the behaviour of AuNPs in various investigated systems, more advanced experiments were conducted by individually probing single AuNPs at pH 3, 5.6 and 7.4, using buffers at high I_c following the method outlined in Section 2.2.3. Dark-field microscopy measurements were first performed to locate isolated AuNPs that would be selected for the resonance measurements. With a region of interest specified, the sample was then placed in an optical setup equipped for the single particle illumination scheme. At pH 7.4, several of the isolated AuNPs measured showed in the transmission spectra significant dips in λ between 546 and 591 nm. The variance in resonance can be associated with the standard deviation in the size of the AuNPs. At pH 5.6, no individual particles but, rather, nonspherical clusters were present. The respective transmission curves were broadened with no clear resonant wavelength; therefore, a resonance shift could not be assigned. The broadening of the resonance for this pH can be attributed to particle agglomeration, lack of spherical uniformity, and near-field interactions with neighbouring nanoclusters. For AuNPs prepared in a solution with pH 3.0, a broadened, yet identifiable, resonance occurred, which spanned 573–625 nm. The pH 3.0 AuNPs were also heavily agglomerated; however, the nanoclusters were more isolated, minimizing the effect of near-field interactions and broadening their respective resonance properties. Ultimately, the single-particle excitation schemes revealed that for a lower pH, the SPR resonance is red-shifted, primarily due to agglomeration. The significant agglomeration incurred is linked to the high buffer concentration used to prepare these particular samples.

3.4. Simulation of the Local pH-Value Distribution near the Particle Surface

The local pH values near the surface of the 18 and 56 nm sized AuNPs were determined in La/NaL, H_3Cit/Na_3Cit and PBS buffer at pH = 4. Figure 8 shows the effect of the solution pH variation and the influence of the zeta potential on the size-dependent local pH distribution at, for example, −22.8 mV (for other conditions, see Figures S1–S3 in Supplementary Materials). The simulation of the near surface pH-value distribution around 18 nm and 56 nm AuNPs showed very little difference. The same was noticed for different buffer systems (see Figure S4 in Supplementary Materials). It should be noticed that differences between the buffer systems become visible for more extreme pH values. These extreme cases, however, contradict the measured pH values and surface potentials in the experiments and have, therefore, not been considered. For the buffer variation and NPs with a size below 100 nm, a steep pH gradient was observed within the first 1 μm from the NP surface.

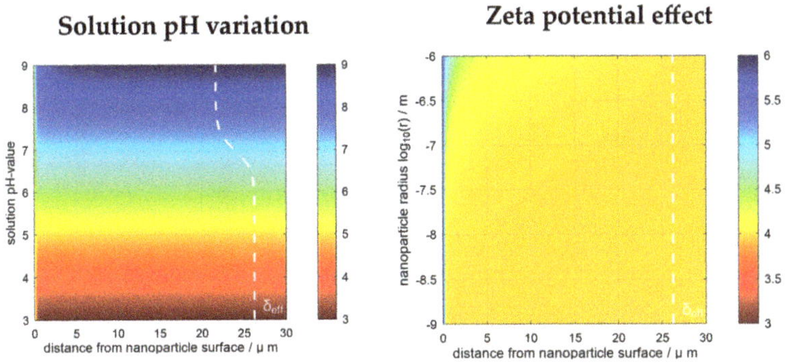

Figure 8. Calculated local pH-value distributions in the diffusion zone near the surface of 18 nm sized AuNPs. The default values for the simulations were 34 mM H_3Cit/Na_3Cit solution (solution pH = 4) and a zeta potential value of −22.8 mV (corresponding to that of the citrate capped).

Larger NPs showed a different behaviour, comprising smaller pH gradients and hence lower current densities in cyclic voltammetry curves, which confirms that smaller particles are more reactive than larger ones.

4. Discussion

One of the most widely used methods for the preparation of AuNPs is the Turkevich method, i.e., chemical reduction synthesis with the assistance of Na_3Cit, which includes two mechanistic steps: first, electron transfer, and second, reduction with acetone [50]. Once AuNPs are formed, citrate ions physically adsorb onto their surface, stabilizing the NPs. The size of AuNPs strongly depends on the molar ratio of the reactant. More precisely, in our study, monodisperse and stable 10–20 nm AuNPs could be obtained when the molar ratio was ≥2.5. However, a further increase in the molar ratio did not affect the size, as the particle was covered with citrate ions, resulting in stabilization [51]. By contrast, the reduction of metal salts with AA includes only electron transfer; therefore, Na_3Cit exhibits a higher redox potential than AA. Since Na_3Cit and AA are both weak acids, their reduction potential also depends on the pH of the reaction mixture. The increase in pH leads to subsequent deprotonation and an increase in electron density, thus increasing the redox potential. By increasing the pH in the reaction mixture, the size and morphology of the final NPs can be tailored. The synthesis conditions determined in the DoE study were chosen for the fabrication of 18 and 56 nm sized AuNPs. The particle sizes determined in MQ water with DLS agreed well with the AFM and TEM measurements. AFM and TEM measurements further showed that citrate-capped AuNPs were spherically shaped. By changing the reductant, approx. 30–40 nm sized particles were measured with AFM and TEM. However, compared to the 18 nm particles, they showed a tendency to form clusters. Regarding SPR, which depends on the size, shape and surface functionalization, a shift towards a higher λ (i.e., red shift) was detected [52,53]. In addition to the size and optical properties, the zeta potential values were above −20 mV, confirming colloidal stability in MQ water independent of the pH. However, under in vivo conditions, the situation becomes more complex, which probably affects not only biological performance but also toxicity. For example, small NPs with a larger specific surface area are more reactive because more surface area is available to interact with biological molecules and cellular components [54–59]. Furthermore, the size of metal NPs can influence the release process of ions dependent on the physiological fluid and pH [60–62]. This suggests that pH dependence profiles of NPs are important for predicting their behaviour in more complex environments. It was found that with increasing I_c, AuNPs agglomerated independently of their initial size. This is in agreement with the results of Pamies et al., who studied the stability of citrate-capped AuNPs in sodium nitrate solutions with I_c ranging from 0 to 1 M [63]. They showed that with an increasing ionic concentration, agglomeration of the particles occurred, which can be attributed to the compression of the double layer that facilitates van der Waals interactions between AuNPs [64,65]. Consequently, zeta potential values are expected to decrease at higher I_c, which coincides with our results. No isoelectric point was observed for both NP sizes, as the isoelectric point is reported to be between pH 2 to 4 [66]. In treatment-relevant buffers, AuNPs were susceptible to the buffer type and pH environment. The lower I_c did not induce agglomeration of AuNPs in LA/NaL buffer and PBS, independent of the size. However, the H_3Cit/Na_3Cit buffer had a stronger effect on the zeta potential and size and altered the stability even at low I_c. At 50 mM, agglomeration was noticed in both LA/NaL and H_3Cit/Na_3Cit buffer. Interestingly, in PBS, particles remained stable. Similar results were also found by Sangwan and Seth [67]. They monitored the stability of citrate-capped AuNPs in borate (pH = 7.5, 8.5, 9.2, 9.3), PBS (pH = 6.5, 7.5, 8.5), Tris–citrate (pH = 8.5, 9.5) and Tris–HCl (pH = 4.0, 10.5) buffer. The results confirmed that the stability of AuNPs is affected by the electrolyte composition of the buffer, pH and duration and temperature of the incubation time. To understand the behaviour of AuNPs in different buffer systems, insight into the structure of citrate and AA layers on the particle's surface is required. Both citrate and AA ligands are known to form chelate-type complexes with metallic NPs. According to Park and Shumaker-Parry, in citrate-capped AuNPs, two carboxylate groups are in direct contact with the surface, while the third terminal carboxylate group is not bound to the surface [68]. In the case of AA, the chelating is a mixed type where a monodentate and a bidentate coordination mode of AA on AuNPs are identified [69]. Since the citrate ion is more complex in terms of chelating potential compared to AA, we expected a pronounced effect in terms of the size and zeta potential in different buffers for citrate-capped AuNPs. In LA/NaL buffer, the stabilization effect at lower I_c for both AuNP sizes was achieved with the steric barrier due to the hydrogen bond network between adjacent layers of citrate or AA and lactate. The latter has an additional hydroxyl group, which increases the bonding affinity and interaction with citrate and AA. As expected, the screening effect of sodium ions at higher I_c is more pronounced for citrate-capped AuNPs. In addition, agglomeration is facilitated due to the disruption of the delicate hydrogen bonding network as steric repulsions are disabled. With increasing pH, the excess of citrate-lactate layers is removed due to deprotonation and the accumulation of negatively charged lactate ions in the vicinity of the AuNP surface, causing a negative zeta potential. Regarding the H_3Cit/Na_3Cit, the intermolecular interaction between citrate ions that coat particles and citrate ions in the bulk are more complex. Grys et al. studied the formation of AuNP aggregates in 1 M H_3Cit/Na_3Cit buffer solution at pH 3.5 [70]. The surface-enhanced Raman spectroscopy revealed differences in the coordination of AuNPs with citrate ions. Because of the protonation of nonsurface-bound terminal carboxylates, citrate ions in the vicinity formed carboxylic acid dimers, and AuNPs agglomerated at lower pH values. The pH increase enabled the deprotonation of carboxylate groups, and negatively charged citrate species around AuNPs were responsible for high negative zeta potential values. On the contrary, in PBS, both AuNPs exhibited stability independent of the I_c. The tendency of anions to interact with the NP surface is governed by the concentration and surface affinity of anions [71]. There are few studies that report on the competitive mechanism between citrate ions and ligands such as phosphate, amino acids, organothiols, halides and adenine [72–76]. White and Hjortkjaer found that phosphate anions could replace citrate ligands via oxygen atoms in phosphate anions, which resulted in phosphate-coated AuNPs [77]. The exchange between citrate and phosphate ions on the particle surface is affected by the pK values of citrate and phosphate and the ligand concentration. The concentration of phosphate species in PBS was dominant in comparison to citrate species, thus favouring the exchange between two ligands, which resulted in stable AuNPs. The obtained results suggest that the classical Derjaguin–Landau–Verwey–Overbeek (DLVO) theory, which considers the electrical double-layer interaction and attractive van der Waals force, can only partly explain our findings. Other factors that should be considered include the screening effect of monovalent cations (such as Na^+ and H^+) and the protective role of multivalent anions (which can co-adsorb or replace ligands on the surface of AuNPs). In addition, to fully comprehend the complex dynamics at the nano–bio interface, intermolecular interactions between ligands on the surface of AuNPs, and ligands from buffers such as hydrogen bonding, electrostatic attraction, and hydrophobic effects should be examined. The changes in the stability of AuNPs of different sizes considering the treatment-relevant pH values were also detected in the optical properties. The most pronounced SPR peak shift was found for 18 nm AuNPs incubated in H_3Cit/Na_3Cit buffer in the pH range 3–4. The second peak at $\approx$630 nm can be attributed to particle agglomeration, which, in turn, coincides with the size measurements [78].

The results obtained from simulations in the tested buffer systems of both AuNP sizes showed only small changes in the pH gradient, which proves the robustness of the synthesis method. The behaviour of AuNPs of both sizes in all tested buffers exhibited no significant differences, which might be due to the higher buffer concentration (mM) compared to changes of H^+ and OH^- between the surface and solution (nM or μM). The constant pH is a consequence of small changes in the protonation of the buffer species, except for in the vicinity of the surface. The steep pH gradients near the surface indicate that the interaction of (protonated) buffer species with the NPs should be considered in future studies since zeta potential measurements (see Figures 4a and 5) revealed a shift of surface pH value which the reported simulations alone cannot predict. However, simulations revealed very steep gradients less than 1 μm away from the surface, where one can assume an interaction with the outer Helmholtz plane and that van der Waals interactions are likely to have an effect. In addition, steep gradients mean a high ionic flow, so smaller NPs are much more reactive than larger NPs in the upper μm range.

5. Conclusions

The simultaneous exploitation of wound acidification and the intrinsic properties of AuNPs represent a promising approach to wound treatment. To further advance this therapeutic concept, a fundamental understanding of the interactions at the liquid–particle interface is essential. Here, we demonstrate that AuNPs dispersed in LA/NaL buffer showed the most promising results. Particles are stable at therapeutically relevant I_c regardless of their size and the reducing agent used. With an increasing I_c, the citrate-capped particles agglomerate due to the shielding effect of the sodium ions and the disruption of the hydrogen bonding network. In contrast, particles dispersed in H_3Cit/Na_3Cit buffer agglomerate at high and low I_c regardless of their size and the reducing agent used. In PBS, the particles are stable independent of their size, the reducing agent used and the I_c. PBS mimics physiological pH rather than supporting a wound-healing effect but still prevents progressive alkalinization and, thus, bacterial growth. Interestingly, the modelling studies show a steep pH gradient in close proximity (<1 μm) to the NP surface under all conditions, exclusively for particles smaller than 100 nm. Steep gradients also indicate a high ionic flow, so smaller NPs are much more reactive than larger NPs, which could further promote the healing potential due to a more acidic particle surface environment but could also be beneficial in drug loading. Thus, maintaining the colloidal stability of AuNPs in acidifying buffer systems is of utmost importance. It should be noted, however, that in the next step, the protonation level of chemical species such as amino acids and buffer ions must be quantified to understand how these molecules interact with the NP surface and possibly affect colloidal stability.

Supplementary Materials: The following supporting information can be downloaded at: https://www.mdpi.com/article/10.3390/nano13121878/s1, Figure S1. Calculated local pH-value distributions in the diffusion zone near the surface of 18 nm (top row) and 56 nm AuNPs (bottom row). The default values for the simulations were 34 mM La/NaL buffer solution (solution pH = 4) and a zeta potential value of −22.8 mV for 18 nm and −37.6 mV for 56 nm sized AuNPs, respectively; Figure S2. Calculated local pH-value distributions in the diffusion zone near the surface of 56 nm AuNPs. The default values for the simulations were 34 mM H_3Cit/Na_3Cit solution (solution pH = 4) and a zeta potential value of −37.6 mV; Figure S3. Calculated local pH-value distributions in the diffusion zone near the surface of 18 nm (top row) and 56 nm AuNPs (bottom row). The default values for the simulations were 34 mM PBS buffer solution (solution pH = 4) and a zeta potential value of −22.8 mV for 18 nm and −37.6 mV for 56 nm sized AuNPs, respectively; Figure S4. Calculated local pH-value distributions in the diffusion zone near the surface of 18 nm (top row) and 56 nm AuNPs (bottom row) with buffer concentration variation. The default values for the simulations were 34 mM buffer solutions (solution pH = 4) and a zeta potential value of −22.8 mV for 18 nm and −37.6 mV for 56 nm sized AuNPs, respectively.

Author Contributions: Conceptualization, A.S., R.J. and E.R., methodology, A.S., R.J., M.A., C.T., B.K. and G.L., writing—original draft preparation, A.S., R.J. and E.R.; writing—review and editing, A.S., R.J., M.A., C.T. and E.R.; funding acquisition, P.B., G.L. and E.R. All authors have read and agreed to the published version of the manuscript.

Funding: The Research Center Pharmaceutical Engineering (RCPE) is funded within the framework of COMET—Competence Centers for Excellent Technologies by BMK, BMDW, Land Steiermark and SFG. The COMET program is managed by the FFG. The authors acknowledge the financial support by the University of Graz.

Data Availability Statement: Not applicable.

Acknowledgments: Authors would like to thank Luca Alfred Schmid for the help with TEM measurements, Open Access Funding by the University of Graz.

Conflicts of Interest: The authors declare no conflict of interest.

References

1. Velnar, T.; Bailey, T.; Smrkolj, V. The wound healing process: An overview of the cellular and molecular mechanisms. *J. Int. Med. Res.* **2009**, *37*, 1528–1542. [CrossRef] [PubMed]
2. Lodhi, S.; Singhai, A.K. Wound healing effect of flavonoid rich fraction and luteolin isolated from Martynia annua Linn. On streptozotocin induced diabetic rats. *Asian Pac. J. Trop. Med.* **2013**, *6*, 253–259. [CrossRef] [PubMed]
3. Guo, S.; DiPietro, L.A. Factors Affecting Wound Healing. *J. Dent. Res.* **2010**, *89*, 219–229. [CrossRef]
4. Proksch, E. Buffering Capacity. In *pH of the Skin: Issues and Challenges*; Surber, C., Abels, C., Maibach, H., Eds.; Karger: Basel, Switzerland, 2018; Volume 54, pp. 11–18. [CrossRef]
5. Siddiqui, A.R.; Bernstein, J.M. Chronic wound infection: Facts and controversies. *Clin. Dermatol.* **2010**, *28*, 519–526. [CrossRef] [PubMed]
6. Aly, R.; Shirley, C.; Cunico, B.; Maibach, H.I. Effect of prolonged occlusion on the microbial flora, pH, carbon dioxide and transepidermal water loss on human skin. *J. Investig. Dermatol.* **1978**, *71*, 378–381. [CrossRef]
7. Khalilov, R. A comprehensive review of advanced nano-biomaterials in regenerative medicine and drug delivery. *Adv. Biol. Earth Sci.* **2023**, *8*, 5–18.
8. Jain, P.K.; Lee, K.S.; El-Sayed, I.H.; El-Sayed, M.A. Calculated absorption and scattering properties of gold nanoparticles of different size, shape, and composition: Applications in biological imaging and biomedicine. *J. Phys. Chem. B* **2006**, *110*, 7238–7248. [CrossRef]
9. Tiwari, P.M.; Vig, K.; Dennis, V.A.; Singh, E.R. Functionalized gold nanoparticles and their biomedical applications. *Nanomaterials* **2011**, *14*, 31–63. [CrossRef]
10. Bansal, S.A.; Kumar, V.; Karimi, J.; Singh, A.P.; Kumar, S. Role of gold nanoparticles in advanced biomedical applications. *Nanoscale Adv.* **2020**, *2*, 3764–3787. [CrossRef]
11. Akturk, O.; Kismet, K.; Yasti, A.C.; Kuru, S.; Duymus, M.E.; Kaya, F.; Caydere, M.; Hucumenoglu, S.; Keskin, D. Collagen/gold nanoparticle nanocomposites: A potential skin wound healing biomaterial. *J. Biomater. Appl.* **2016**, *31*, 283–301. [CrossRef]

12. Naraginti, S.; Kumari, P.L.; das Sivakumar, R.K.A.; Patil, S.H.; Andhalkar, V.V. Amelioration of excision wounds by topical application of green synthesized, formulated silver and gold nanoparticles in albino Wistar rats. *Mater. Sci. Eng. C* **2016**, *62*, 293–300. [CrossRef] [PubMed]
13. Hung, Y.L.; Hsiung, T.M.; Chen, Y.Y.; Huang, Y.F.; Huang, C.C. Colorimetric detection of heavy metal ions using label-free gold nanoparticles and alkanethiols. *J. Phys. Chem. C* **2010**, *114*, 16329–16334. [CrossRef]
14. Gu, H.; Ho, P.L.; Tong, E.; Wang, L.; Xu, B. Presenting Vancomycin on Nanoparticles to Enhance Antimicrobial Activities. *Nano Lett.* **2003**, *3*, 1261–1263. [CrossRef]
15. Norman, S.; Stone, J.W.; Gole, A.; Murphy, C.; Sabo-Attwood, T.L. Targeted Photothermal Lysis of the Pathogenic Bacteria, Pseudomonas aeruginosa, with Gold Nanorods. *Nano Lett.* **2008**, *8*, 302–306. [CrossRef]
16. Gil-Tomás, J.; Tubby, S.; Parkin, I.P.; Narband, N.; Dekker, L.; Nair, S.P.; Wilson, M.; Street, C. Lethal photosensitisation of Staphylococcus aureus using a toluidine blue O–tiopronin–gold nanoparticle conjugate. *J. Mater. Chem.* **2007**, *17*, 3739–3746. [CrossRef]
17. Joshi, P.; Chakraborti, S.; Ramirez-Vick, J.E.; Ansari, Z.A.; Shanker, V.; Chakrabarti, P.; Singh, S.P. The anticancer activity of chloroquine-gold nanoparticles against MCF-7 breast cancer cells. *Colloids Surf. B Biointerfaces* **2012**, *95*, 195–200. [CrossRef]
18. Schwert, G.W.; Eisenberg, M.A. The kinetics of the amidase and esterase activities of trypsin. *J. Biol. Chem.* **1949**, *179*, 665–672. [CrossRef]
19. Han, K.; Zhu, J.Y.; Wang, S.B.; Li, Z.H.; Cheng, S.X.; Zhang, X.Z. Tumor targeted gold nanoparticles for FRET-based tumor imaging and light responsive on-demand drug release. *J. Mater Chem. B.* **2015**, *3*, 8065–8069. [CrossRef]
20. Guo, B.; Zebda, R.; Drake, S.J.; Sayes, C.M. Synergistic effect of co-exposure to carbon black and Fe_2O_3 nanoparticles on oxidative stress in cultured lung epithelial cells. *Part. Fiber Toxicol.* **2009**, *6*, 4. [CrossRef]
21. Al-Jamal, W.T.; Al-Jamal, K.T.; Bomans, P.H.; Frederik, P.M.; Kostarelos, K. Functionalized-quantum-dot-liposome hybrids as multimodal nanoparticles for cancer. *Small* **2008**, *4*, 1406–1415. [CrossRef]
22. Lundqvist, M.; Stigler, J.; Elia, G.; Lynch, I.; Cedervall, T.; Dawson, K.A. Nanoparticle size and surface properties determine the protein corona with possible implications for biological impacts. *Proc. Natl. Acad. Sci. USA* **2008**, *105*, 14265–14270. [CrossRef]
23. Auinger, M.; Katsounaros, I.; Meier, J.C.; Klemm, S.O.; Ulrich Biedermann, P.; Topalov, A.A.; Rohwerdera, M.; Mayrhofer, K.J.J. Near-surface ion distribution and buffer effects during electrochemical reactions. *Phys. Chem. Chem. Phys.* **2011**, *13*, 16384–16394. [CrossRef] [PubMed]
24. Stepan, T.; Tete, L.; Laundry-Mottiar, L.; Romanovskaia, E.; Hedberg, Y.S.; Danninger, H.; Auinger, M. Effect of nanoparticle size on the near-surface pH-distribution in aqueous and carbonate buffered solutions. *Electrochim. Acta* **2022**, *409*, 139923–139932. [CrossRef]
25. Baber, R.; Mazzei, L.; Kim Thanh, N.T.; Gavriilidis, A. An engineering approach to synthesis of gold and silver nanoparticles by controlling hydrodynamics and mixing based on a coaxial flow reactor. *Nanoscale* **2017**, *9*, 14149–14161. [CrossRef]
26. Damilos, S.; Alissandratos, I.; Panariello, L.; Radhakrishnan, A.N.P.; Cao, E.; Wu, G.; Besenhard, M.O.; Kulkarni, A.A.; Makatsoris, C.; Gavriilidis, A. Continuous citrate-capped gold nanoparticle synthesis in a two-phase flow reactors. *J. Flow Chem.* **2021**, *11*, 553–567. [CrossRef]
27. Zhang, X.; Ma, S.; Li, A.; Chen, L.; Lu, J.; Geng, X.; Xie, M.; Liang, X.; Wan, Y.; Yang, P. Continuous high-flux synthesis of gold nanoparticles with controllable sizes: A simple microfluidic system. *Appl. Nanosci.* **2020**, *10*, 661–669. [CrossRef]
28. Turkevich, J.; Cooper, P.H.J. A study of the nucleation and growth process in the synthesis of colloidal gold. *Discuss. Faraday Soc.* **1951**, *55*, 55–75. [CrossRef]
29. Brust, M.; Walker, M.; Bethell, D.; Schiffrin, D.J.; Whyman, R. Synthesis of thiol-derivatised gold nanoparticles in a two-phase liquid-liquid system. *J. Chem. Soc. Chem. Commun.* **1994**, *7*, 5–7. [CrossRef]
30. Niidome, Y.; Nishioka, K.; Kawasaki, H.; Yamada, S. Rapid synthesis of gold nanorods by the combination of chemical reduction and photoirradiation processes; morphological changes depending on the growing processes. *Chem. Comm.* **2003**, *18*, 2376–2377. [CrossRef] [PubMed]
31. Kumar, S.; Gandhi, K.S.; Kumar, R. Modeling of formation of gold nanoparticles by citrate method. *Ind. Eng. Chem. Res.* **2007**, *46*, 3128–3136. [CrossRef]
32. Pal, A.; Esumi, K.; Pal, T. Preparation of nanosized gold particles in a biopolymer using UV photoactivation. *J. Colloid Interface Sci.* **2005**, *288*, 396–401. [CrossRef]
33. Wangoo, N.; Bhasin, K.K.; Mehta, S.K.; Suri, C.R. Synthesis and capping of water-dispersed gold nanoparticles by an amino acid: Bioconjugation and binding studies. *J. Colloid Interface Sci.* **2008**, *323*, 247–254. [CrossRef]
34. Sau, T.K.; Murphy, C.J. Room temperature, high-yield synthesis of multiple shapes of gold nanoparticles in aqueous solution. *J. Am. Chem. Soc.* **2004**, *126*, 9–10. [CrossRef]
35. Lee, K.X.; Shameli, K.; Yew, Y.P.; Teow, S.Y.; Jahangirian, H.; Rafiee-Moghaddam, R.; Webster, T.J. Recent developments in the facile bio-synthesis of gold nanoparticles (AuNPs) and their biomedical applications. *Int. J. Nanomed.* **2020**, *15*, 275–300. [CrossRef]
36. Wang, J.; Liu, N.; Su, Q.; Lv, Y.; Yang, C.; Zhan, H. Green Synthesis of Gold Nanoparticles and Study of Their Inhibitory Effect on Bulk Cancer Cells and Cancer Stem Cells in Breast Carcinoma. *Nanomaterials* **2022**, *12*, 3324. [CrossRef]
37. Dong, J.; Carpinone, P.L.; Pyrgiotakis, G.; Demokritou, P.; Moudgil, B.M. Synthesis of precision gold nanoparticles using Turkevich method. *Kona* **2020**, *37*, 224–232. [CrossRef]

38. Malassis, L.; Dreyfus, R.; Murphy, R.J.; Hough, L.A.; Donnio, B.; Murray, C.B. One-step green synthesis of gold and silver nanoparticles with ascorbic acid and their versatile surface post-functionalization. *RSC Adv.* **2016**, *6*, 33092–33100. [CrossRef]
39. Nečas, D.; Klapetek, P. Gwyddion: An open-source software for SPM data analysis. *Cent. Eur. J. Phys.* **2012**, *10*, 181–188. [CrossRef]
40. Woźniak, P.; Banzer, P.; Leuchs, G. Selective switching of individual multipole resonances in single dielectric nanoparticles. *Laser Photonics Rev.* **2015**, *9*, 231–240. [CrossRef]
41. Banzer, P.; Peschel, U.; Quabis, S.; Leuchs, G. On the experimental investigation of the electric and magnetic response of a single nano-structure. *Opt. Express* **2010**, *18*, 10905–10923. [CrossRef] [PubMed]
42. Eigen, M. Methods for investigation of ionic reactions in aqueous solutions with half-times as short as 10^{-9} sec. Application to neutralization and hydrolysis reactions. *Discuss. Faraday Soc.* **1954**, *17*, 194–205. [CrossRef]
43. Ji, X.; Song, X.; Li, J.; Bai, Y.; Yang, W.; Peng, X. Size control of gold nanocrystals in citrate reduction: The third role of citrate. *J. Am. Chem. Soc.* **2007**, *129*, 13939–13948. [CrossRef] [PubMed]
44. Bastús, N.G.; Comenge, J.; Puntes, V. Kinetically controlled seeded growth synthesis of citrate-stabilized gold nanoparticles of up to 200 nm: Size focusing versus Ostwald ripening. *Langmuir* **2011**, *27*, 11098–11105. [CrossRef]
45. Hussain, M.H.; Abu Bakar, N.F.; Mustapa, A.N.; Low, K.F.; Othman, N.H.; Adam, F. Synthesis of various size gold nanoparticles by chemical reduction method with different solvent polarity. *Nanoscale Res. Lett.* **2020**, *15*, 140–149. [CrossRef] [PubMed]
46. Kimling, J.; Maier, M.; Okenve, B.; Kotaidis, V.; Ballot, H.; Plech, A. Turkevich method for gold nanoparticle synthesis revisited. *J. Phys. Chem. B* **2006**, *110*, 15700–15707. [CrossRef]
47. Zhao, L.; Jiang, D.; Cai, Y.; Ji, X.; Xie, R.; Yang, W. Tuning the size of gold nanoparticles in the citrate reduction by chloride ions. *Nanoscale* **2012**, *4*, 5071–5076. [CrossRef]
48. Suchomel, P.; Kvitek, L.; Prucek, R.; Panacek, A.; Halder, A.; Vajda, S.; Zboril, R. Simple size-controlled synthesis of Au nanoparticles and their size dependent catalytic activity. *Sci. Rep.* **2018**, *8*, 4589–4599. [CrossRef] [PubMed]
49. Yazdani, S.; Daneshkhah, A.; Diwate, A.; Patel, H.; Smith, J.; Reul, O.; Cheng, R.; Izadian, A.; Hajrasouliha, A.R. Model for gold nanoparticle synthesis: Effect of pH and reaction time. *ACS Omega* **2021**, *6*, 16847–16853. [CrossRef]
50. Rodrigues, T.S.; Zhao, M.; Yang, T.H.; Gilroy, K.D.; da Silva, A.G.M.; Camargo, P.H.C.; Xia, Y. Synthesis of colloidal metal nanocrystals: A Comprehensive review on the reductants. *Chem. Eur. J.* **2018**, *24*, 16944–16963. [CrossRef]
51. Wuithschick, M.; Birnbaum, A.; Witte, S.; Sztucki, M.; Vainio, U.; Pinna, N.; Rademann, K.; Emmerling, F.; Kraehnert, R.; Polte, J. Turkevich in new robes: Key questions answered for the most common gold nanoparticle synthesis. *ACS Nano* **2015**, *9*, 7052–7071. [CrossRef]
52. Ou, J.; Zhou, Z.; Chen, Z.; Tan, H. Optical diagnostic based on functionalized gold nanoparticles. *Int. J. Mol. Sci.* **2019**, *20*, 4346. [CrossRef] [PubMed]
53. Mie, G. Contributions to the optics of turbid media, particularly of colloidal metal solutions. *Ann. Phys.* **1908**, *25*, 377–445. [CrossRef]
54. Sukhanova, A.; Bozrova, S.; Sokolov, P.; Berestovoy, M.; Karaulov, A.; Nabiev, I. Dependence of nanoparticle toxicity on their physical and chemical properties. *Nanoscale Res. Lett.* **2018**, *13*, 44. [CrossRef] [PubMed]
55. Pan, Y.; Neuss, S.; Leifert, S.A.; Fischler, M.; Wen, F.; Simon, U.; Schmid, G.; Brandau, W.; Jahnen-Dechent, W. Size-dependent cytotoxicity of gold nanoparticles. *Small* **2007**, *3*, 1941–1949. [CrossRef] [PubMed]
56. Li, X.; Hu, Z.; Ma, J.; Wang, X.; Zhang, Y.; Wang, W.; Yuana, Z. The systematic evaluation of size-dependent toxicity and multi-time biodistribution of gold nanoparticles. *Colloids Surf. B Biointerfaces* **2018**, *167*, 260–266. [CrossRef]
57. Liu, W.; Wu, Y.; Wang, C.; Li, H.C.; Wang, T.; Liao, C.Y.; Cui, L.; Zhou, Q.F.; Yan, B.; Jiang, G.B. Impact of silver nanoparticles on human cells: Effect of particle size. *Nanotoxicology* **2010**, *4*, 319–330. [CrossRef] [PubMed]
58. Yu, K.O.; Grabinski, C.M.; Schrand, A.M.; Murdock, A.M.R.C.; Wang, W.; Gu, B.; Schlager, J.J.; Hussain, S.M. Toxicity of amorphous silica nanoparticles in mouse keratinocytes. *J. Nanopart. Res.* **2009**, *11*, 15–24. [CrossRef]
59. Lin, X.; Li, J.; Ma, S.; Liu, G.; Yang, K.; Tong, M.; Lin, D. Toxicity of TiO_2 nanoparticles to escherichia coli: Effects of particle size, crystal phase and water chemistry. *PLoS ONE* **2014**, *9*, e110247. [CrossRef]
60. Misra, S.K.; Dybowska, A.; Berhanu, D.; Luoma, S.N.; Valsami-Jones, E. The complexity of nanoparticle dissolution and its importance in nanotoxicological studies. *Sci. Total Environ.* **2012**, *438*, 225–232. [CrossRef]
61. Brunner, T.J.; Wick, P.; Manser, P.; Spohn, P.; Grass, R.N.; Limbach, L.K.; Bruinink, A.; Stark, W.J. In vitro cytotoxicity of oxide nanoparticles: Comparison to asbestos, silica, and the effect of particle solubility. *Environ. Sci. Technol.* **2006**, *40*, 4374–4381. [CrossRef]
62. Bian, S.W.; Mudunkotuwa, I.A.; Rupasinghe, T.; Grassian, V.H. Aggregation and dissolution of 4 nm ZnO nanoparticles in aqueous environments: Influence of pH, ionic strength, size, and adsorption of humic acid. *Langmuir* **2011**, *27*, 6059–6068. [CrossRef]
63. Pamies, R.; Ginés Hernández Cifre, J.; Fernández Espín, V.; Collado-González, M.; Díaz Baños, F.G.; García de la Torre, J. Aggregation behaviour of gold nanoparticles in saline aqueous media. *J. Nanopart. Res.* **2014**, *16*, 2376–2386. [CrossRef]
64. Edwards, S.A.; Williams, D.R. Double layers and interparticle forces in colloid science and biology: Analytic results for the effect of ionic dispersion forces. *Phys. Rev. Lett.* **2004**, *92*, 248303. [CrossRef] [PubMed]
65. Boström, M.; Williams, D.R.; Ninham, B.W. Specific ion effects: Why DLVO theory fails for biology and colloid systems. *Phys. Rev. Lett.* **2001**, *87*, 168103. [CrossRef] [PubMed]

66. Csapó, E.; Sebők, D.; Makrai Babić, J.; Šupljika, F.; Bohus, G.; Dékány, I.; Kallay, N.; Preočanin, T. Surface and structural properties of gold nanoparticles and their biofunctionalized derivatives in aqueous electrolytes solution. *J. Dispers. Sci. Technol.* **2014**, *35*, 815–825. [CrossRef]
67. Sangwan, S.; Seth, R. Synthesis, characterization and stability of gold nanoparticles (AuNPs) in different buffer systems. *J. Clust. Sci.* **2022**, *33*, 749–764. [CrossRef]
68. Park, J.W.; Shumaker-Parry, J.S. Structural study of citrate layers on gold nanoparticles: Role of intermolecular interactions in stabilizing nanoparticles. *J. Am. Chem. Soc.* **2014**, *136*, 1907–1921. [CrossRef]
69. Zümreoglu-Karan, B. A rationale on the role of intermediate Au(III)–vitamin C complexation in the production of gold nanoparticles. *J. Nanopart. Res.* **2009**, *11*, 1099–1105. [CrossRef]
70. Grys, D.B.; de Nijs, B.; Salmon, A.R.; Huang, J.; Wang, W.; Chen, W.H.; Scherman, O.A.; Baumberg, J.J. Citrate coordination and bridging of gold nanoparticles: The role of gold adatoms in AuNP aging. *ACS Nano* **2020**, *14*, 8689–8696. [CrossRef]
71. Aubard, J.; Bagnasco, E.; Pantigny, J.; Ruasse, M.F.; Levi, G.; Wentrup-Byrne, E. An ion-exchange reaction as measured by surface-enhanced raman spectroscopy on silver colloids. *J. Phys. Chem.* **1995**, *99*, 7075–7081. [CrossRef]
72. Afshinnia, K.; Baalousha, M. Effect of phosphate buffer on aggregation kinetics of citrate-coated silver nanoparticles induced by monovalent and divalent electrolytes. *Sci. Total Environ.* **2017**, *581–582*, 268–276. [CrossRef] [PubMed]
73. Rani, M.; Moudgil, L.; Singh, B.; Kaushal, A.; Mittal, A.; Saini, G.S.S.; Tripathi, S.K.; Singhe, G.; Kaura, A. Understanding the mechanism of replacement of citrate from the surface of gold nanoparticles by amino acids: A theoretical and experimental investigation and their biological application. *RSC Adv.* **2016**, *6*, 17373–17383. [CrossRef]
74. Park, J.W.; Shumaker-Par, J.S. Strong resistance of citrate anions on metal nanoparticles to desorption under thiol functionalization. *ACS Nano* **2015**, *9*, 1665–1682. [CrossRef] [PubMed]
75. Huang, P.J.J.; Yang, J.; Chong, K.; Ma, Q.; Li, M.; Zhang, F.; Moon, W.J.; Zhang, G.; Liu, J. Good's buffers have various affinities to gold nanoparticles regulating fluorescent and colorimetric DNA sensing. *Chem. Sci.* **2020**, *11*, 6795–6804. [CrossRef]
76. Perera, G.S.; Athukorale, S.A.; Perez, F.; Pittman, C.U., Jr.; Zhang, D. Facile displacement of citrate residues from gold nanoparticle surfaces. *J. Colloid Interface Sci.* **2018**, *511*, 335–343. [CrossRef]
77. White, P.; Hjortkjaer, J. Preparation and characterisation of a stable silver colloid for SER(R)S spectroscopy. *J. Raman Spectrosc.* **2014**, *45*, 32–40. [CrossRef]
78. Li, S.; Lui, K.H.; Tsoi, T.H.; Lo, W.S.; Li, X.; Hu, X.; Tai, W.C.S.C.; Hung, H.L.; Gu, Y.J.; Wong, W.T. pH-responsive targeted gold nanoparticles for in vivo photoacoustic imaging of tumor microenvironments. *Nanoscale Adv.* **2019**, *1*, 554–564. [CrossRef]

nanomaterials

Article

Nonlinear Optical Rectification in an Inversion-Symmetry-Broken Molecule Near a Metallic Nanoparticle

Natalia Domenikou [1], Ioannis Thanopulos [1,*], Vassilios Yannopapas [2] and Emmanuel Paspalakis [1]

[1] Materials Science Department, School of Natural Sciences, University of Patras, 26504 Patras, Greece; domenikou.n@gmail.com (N.D.); paspalak@upatras.gr (E.P.)
[2] Department of Physics, National Technical University of Athens, 15780 Athens, Greece; vyannop@mail.ntua.gr
* Correspondence: ithano@upatras.gr

Abstract: We study the nonlinear optical rectification of an inversion-symmetry-broken quantum system interacting with an optical field near a metallic nanoparticle, exemplified in a polar zinc-phthalocyanine molecule in proximity to a gold nanosphere. The corresponding nonlinear optical rectification coefficient under external strong field excitation is derived using the steady-state solution of the density matrix equations. We use *ab initio* electronic structure calculations for determining the necessary spectroscopic data of the molecule under study, as well as classical electromagnetic calculations for obtaining the influence of the metallic nanoparticle to the molecular spontaneous decay rates and to the external electric field applied to the molecule. The influence of the metallic nanoparticle to the optical rectification coefficient of the molecule is investigated by varying several parameters of the system, such as the intensity and polarization of the incident field, as well as the distance of the molecule from the nanoparticle, which indirectly affects the molecular pure dephasing rate. We find that the nonlinear optical rectification coefficient can be greatly enhanced for particular incident-field configurations and at optimal distances between the molecule and the metallic nanoparticle.

Keywords: NOR; PDMs; asymmetric two-level quantum system; plasmonic nanoparticle; zinc-phalocyanine molecular complex

check for updates

Citation: Domenikou, N.; Thanopulos, I.; Yannopapas, V.; Paspalakis, E. Nonlinear Optical Rectification in an Inversion-Symmetry-Broken Molecule Near a Metallic Nanoparticle. *Nanomaterials* **2022**, *12*, 1020. https://doi.org/10.3390/nano12061020

Academic Editor: Andrés Guerrero-Martínez

Received: 16 February 2022
Accepted: 18 March 2022
Published: 21 March 2022

1. Introduction

The manipulation and tuning of the nonlinear optical properties of quantum systems (QSs) is important in various photonic applications, as it can enhance the generally weak interaction between light and quantum matter. A method recently used for the enhancement of nonlinear optical processes in QSs is the placement of the QSs near plasmonic (mainly metallic) nanostructrures. The localized surface plasmons supported by plasmonic nanostructures interact with the excitations of the QSs and enhance the nonlinear optical response of the QSs. Some examples of nonlinear optical phenomena that have been studied in coupled quantum plasmonic nanostructures are nonlinear Fano resonances in the energy absorption [1–5], optical transparency and gain without inversion [6–10], controlled four-wave mixing and Kerr nonlinearity [11–13], as well as optical bistability [14–16].

The above phenomena have been studied in QS with inversion symmetry. Asymmetric QSs that do not possess inversion symmetry give rise to second-order nonlinearities due to the occurrence of otherwise forbidden electronic transitions. At the same time, the permanent electric dipole moments (PDMs) which exist in asymmetric QSs also interact with light and can therefore modify the nonlinear optical response of the QSs. Some of the nonlinear optical phenomena that have been studied in isolated asymmetric QSs are second-order harmonic generation [17], two-photon phase conjugation [18], enhanced light emission at the terahertz [19,20], creation of high-order harmonic generation [21], saturation of the nonlinear optical response [22], and efficient generation of correlated photon pairs [23]. The modification of the population inversion as well as the nonlinear and quantum optical properties of bichromatically driven asymmetric QSs have also attracted significant attention [24–27]. Recently, work has been devoted to the influence of the PDMs on the optical response of asymmetric QSs in cavities [28,29]. Second-order nonlinear optical effects have also been studied in QSs near plasmonic nanostructures. For example, plasmon-enhanced second-harmonic generation [30–32] and difference-frequency generation [33] have been analyzed in semiconductor quantum dots coupled with metallic nanoparticles (MNPs).

An important second-order nonlinear optical effect is the nonlinear optical rectification (NOR), which has also been studied in molecules and semiconductor quantum dots near plasmonic nanostructrures [32,34,35]. Specifically, Thanopulos et al. [34] studied the enhancement of the NOR coefficient of molecules placed near a periodic plasmonic nanostructure under very weak excitation. The strong enhancement of the NOR coefficient stems from the suppression of the spontaneous decay rate of the molecules due to the Purcell effect induced by the periodic plasmonic nanostructure. Furthermore, Evangelou [32] found either suppression or enhancement for the NOR coefficient of a quantum dot near a spherical MNP under very weak excitation. In that work, the modification of the NOR coefficient was attributed to the Purcell effect. The influence of the light intensity to the NOR coefficient of a semiconductor quantum dot near a MNP, without taking into account the Purcell effect, was studied by Carreño et al. who reported saturation effects [35].

In this work, we study the NOR of an inversion-symmetry-broken QS which is placed near a MNP, both interacting with an incident electromagnetic field. More specifically, we consider a polar zinc–phthalocyanine molecule in the vicinity of a gold nanosphere. First, we derive the corresponding NOR coefficient under external strong field excitation using the steady-state solution of the relevant density matrix equations. Next, we use *ab initio* electronic structure calculations to determine the necessary spectroscopic data of the molecule under study. In addition, we perform numerical electromagnetic calculations in order to (a) calculate the modified, due to the presence of the MNP, molecular spontaneous rates and (b) the electric field acting on the molecule which is the incident field plus the field scattered off the MNP. In order to quantify the effect of the MNP on the NOR coefficient, we vary different parameters of the calculation setup, such as the polarization and intensity of the electric field, as well as the molecular pure dephasing rate. The variation of the dephasing rate is achieved by varying the distance between the molecule and the MNP. Lastly, we provide the optimal values of the above parameters for maximizing the enhancement of NOR coefficient.

The paper is organized as follows. In Section 2, we derive the equation for the NOR coefficient in the presence of the electromagnetic field using the density matrix equations for the QS under study. In Section 3, we first introduce the electronic-structure calculations for the molecule and then present the results for the NOR coefficient in the absence or presence of the MNP for various parameters of the system. Finally, in Section 4, we summarize our findings.

2. Theoretical Model

We consider a hybrid structure composed of a polar two-level QS and a MNP, as depicted in Figure 1, where the distance between the QS and the surface of the MNP is denoted by d. The ground ($|1\rangle$) and the excited state ($|2\rangle$) of the QS have energies $\hbar\omega_1$ and $\hbar\omega_2$, respectively, while the corresponding transition dipole moment is given by $\vec{\mu}_{12}$. The two states of the QS feature unequal PDMs, $\vec{\mu}_{11}$ and $\vec{\mu}_{22}$, respectively, due to the absence of the inversion symmetry in the QS. As an MNP, we consider a gold (Au) nanoparticle of radius $R = 80$ nm; the local dielectric function of Au is obtained from spectroscopic data [36]. Both the QS and the MNP are embedded in air (refractive index $n = 1$).

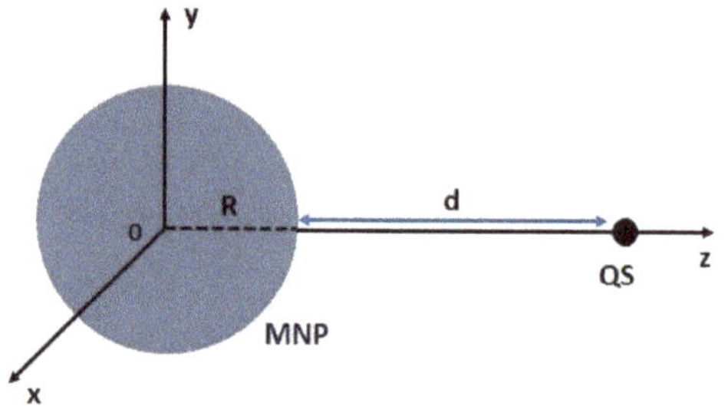

Figure 1. The quantum system (QS) at distance d from the surface of the metallic nanoparticle (MNP) with radius R.

The Hamiltonian of the hybrid system is given by

$$H(t) = \sum_{i=1,2} (\hbar\omega_i - \vec{\mu}_{ii} \cdot \vec{\mathcal{E}}(t))|i\rangle\langle i| - (\vec{\mu}_{12}\vec{\mathcal{E}}(t)|1\rangle\langle 2| + \text{H.c.}), \quad (1)$$

with $\vec{\mathcal{E}}(t) = \vec{E}_0 e^{-i\omega t} + \vec{E}_0^* e^{i\omega t}$, ω being the angular frequency of the field and $\vec{E}_0$ being the modified external field amplitude due to the presence of the MNP.

In this work, we consider the external electric field being polarized either along the z-axis or along the x-axis, corresponding to either radially or tangentially polarized external field with respect to the surface of the MNP, respectively (see Figure 1). We also assume that the dipole moments $\vec{\mu}_{12}$, $\vec{\mu}_{11}$ and $\vec{\mu}_{22}$ are always parallel to the interacting field polarization $\vec{E}_0$. Therefore, in the following, we suppress the polarization indices of the dipole moments and external electric fields, and denote $\mu_{12} = |\vec{\mu}_{12}|$ and $E_0^i = \vec{E}_0$ $(i = x, z)$.

The amplitude of the external electric field, E_0^i, at the position of the molecule is related to the amplitude of incident field, E_{0f}, by the modified field factor $|E_i| \equiv |E_0^i|/|E_{0f}|$, which is obtained by the Mie scattering method [37]. We note that E_{0f} is related to the incident irradiation intensity I by $|E_{0f}|^2 = 2I/nc\epsilon_0$.

The corresponding modified field factors are shown in panel (a) in Figure 2 as functions of the distance d between the QS and the surface of the MNP, for QS energy 1.99 eV. The $|E_z|$ factor decreases monotonically with increasing d, while the $|E_x|$ factor initially decreases below 1, assumes the minimum value 0.1 at $d \approx 10$ nm, and then starts to increase monotonically. In panel (b) of Figure 2, we present the Purcell enhancement factors $\Gamma_i = \Gamma_0^i/\Gamma_{0f}^i$, $(i = x, z)$, of a QS with transition dipole moment oriented along the i-axis, and free-space decay rate value of $\Gamma_{0f}^i = \omega_{21}^3\mu_{12}^2/3\pi c^3\hbar\epsilon_0$, with $\omega_{21} = \omega_2 - \omega_1$. These are obtained from the electromagnetic Green's tensor [37]. We observe that the Γ_z enhancement factor decreases monotonically, and, in general, it is larger than the Γ_x enhancement factor. The latter decreases up to $d \approx 34.5$ nm and increases thereafter very slowly.

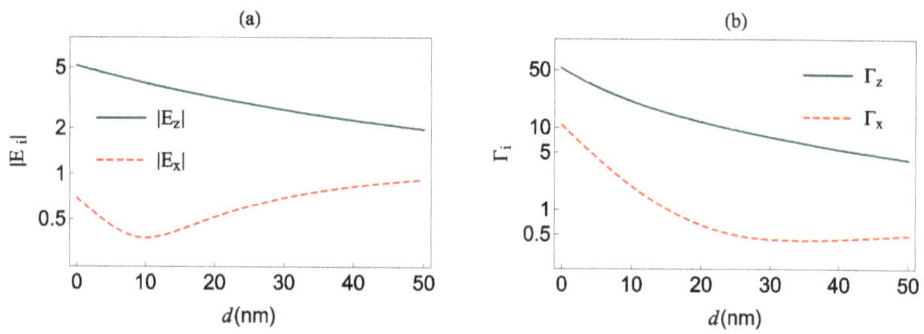

Figure 2. (**a**) The modified field factors of the field applied to the molecule and (**b**) the Purcell enhancement factors of the QS as function of the distance d of the QS to the surface of the MNP.

In order to describe the dynamics of the system within the density matrix approach, we define the population difference $\Delta(t) = \rho_{22}(t) - \rho_{11}(t)$ and the optical coherence $\sigma(t)$

$$\sigma(t) = \rho_{12}(t)e^{-i\omega t + \frac{2i|\mu|}{\hbar\omega}(E_R \sin(\omega t) - E_I \cos(\omega t))}, \tag{2}$$

where $\mu = \mu_{22} - \mu_{11}$. In Equation (2), the modified field amplitude is written as a complex number $E_0^i = E_R + iE_I$. It follows that the equations for optical coherence and population difference be given by

$$\begin{aligned}\dot{\sigma}(t) = &- \left[\frac{1}{T_2} + i\delta\right]\sigma(t) \\ &+ [i\Omega_R(A+B) + \Omega_I(A-B)]\Delta(t),\end{aligned} \tag{3}$$

$$\begin{aligned}\dot{\Delta}(t) = &- \frac{1}{T_1}[\Delta(t) + 1] \\ &- 2[i\Omega_R(A+B) + \Omega_I(A-B)]\sigma^*(t) \\ &+ 2[i\Omega_R(A^*+B^*) - \Omega_I(A^*-B^*)]\sigma(t),\end{aligned} \tag{4}$$

with $\delta = \omega - \omega_{21}$ being the field detuning, and $\Omega_m = \mu_{12}E_m/\hbar$ $(m = R, I)$ being the real and imaginary part of the complex Rabi frequency expressed as $\Omega_0 = \Omega_R + i\Omega_I = \mu_{12}E_0^i/\hbar$. In Equations (3) and (4), we also introduced

$$A = \sum_{n=-\infty}^{+\infty} (-i)^n J_n(\alpha_R)J_n(\alpha_I), \tag{5a}$$

$$B = \sum_{n=-\infty}^{+\infty} (-i)^n J_{n+2}(\alpha_R)J_n(\alpha_I), \tag{5b}$$

with $J_n(\cdot)$ denoting the nth-order ordinary Bessel function, and $\alpha_m = \frac{2|\mu|E_m}{\hbar\omega}$ $(m = R, I)$. We note that for a nonpolar system, i.e., $\mu = 0$, the parameters A and B are equal to 1 and 0, respectively [35].

We also note that T_1 and T_2 in Equations (3) and (4) are influenced by the presence of the MNP, due to the relation $\frac{1}{T_2} = \frac{1}{2T_1} + \gamma_d$, where $T_1 = 1/\Gamma_0^i$, $(i = x, z)$, is the directional relaxation time, T_2 is the corresponding dephasing relaxation time, and γ_d is the pure dephasing rate of the QS, respectively.

The steady-state solutions of Equations (3) and (4) are obtained as

$$\begin{aligned}\sigma_\infty = &-T_2[i\Omega_R(A+B) + \Omega_I(A-B)] \\ &\times \frac{1 - iT_2\delta}{1 + T_2^2\delta^2 + 4T_1T_2|\Omega_0^*A + \Omega_0 B|^2},\end{aligned} \tag{6}$$

$$\Delta_\infty = -\frac{1 + T_2^2\delta^2}{1 + T_2^2\delta^2 + 4T_1T_2|\Omega_0^*A + \Omega_0 B|^2}. \tag{7}$$

For the calculation of the NOR coefficient, we need the induced polarization P of the QS, which is given by $P = P_1 + P_2$, with

$$P_1 = \frac{1}{2}N\mu(1 + \Delta), \tag{8}$$

$$P_2 = N\mu_{12}(\rho_{12} + \rho_{21}). \tag{9}$$

where N is the effective electron volume density of the QS. The first term, P_1, which includes the PDMs, is the main contribution to P, while the second term, including the transition dipole μ_{12}, is a small contribution to P, which, of course, becomes important when the PDMs are equal [22].

The NOR coefficient can be determined using the Equations (7) and (8), in conjunction with the equation $P = \epsilon_0 \chi_0^{(2)} |E_0^i|^2/4$, due to the following identity [35]:

$$\frac{\epsilon_0 \chi_0^{(2)} |E_0^i|^2}{4} = N\mu_{12}(\sigma_\infty B + \sigma_\infty^* \mu) + \frac{N}{2}\mu(1 + \Delta_\infty), \tag{10}$$

which finally leads in our case to the $\chi_0^{(2)}$ coefficient, which is central in this work, given by

$$\begin{aligned} \chi_0^{(2)}(\delta = 0, I) = & \frac{36N\pi^2 c^6 \epsilon_0 |\mu|}{\omega_{21}^2(\Gamma_i^2 + 2\Gamma_i \bar{\gamma}_d)\mu_{12}^2} \\ & \times \frac{1}{1 + \frac{36\pi^2 c^6 \hbar^2 \epsilon_0^2 \delta^2}{\omega_{21}^2(\Gamma_i + 2\bar{\gamma}_d)^2 \mu_{12}^4} + \frac{144\pi^2 c^5 \epsilon_0 |E_i|^2 I}{\omega_{21}^6 n(\Gamma_i^2 + 2\Gamma_i \bar{\gamma}_d)\mu_{12}^2}}, \end{aligned} \tag{11}$$

with $\bar{\gamma}_d = \gamma_d/\Gamma_{0f}^i$ $(i = x, z)$. We note that the values of Γ_i and $|E_i|$ depend on the distance d between the QS and the MNP. We also note that in deriving Equation (11), we have introduced the approximation $A = 1$ and $B = 0$ in Equations (6), (7), and (10), since the numerical values of these two parameters are very close to 1 and 0, respectively. In the next section, we present the results obtained by applying Equation (11) on a realistic molecular QS.

3. Results and Discussion

3.1. *The Zinc–Phthalocyanine Molecular Complex*

We demonstrate the nonlinear optical response of a polar two-level QS, as presented in the previous section on a zinc–phthalocyanine molecular complex shown in Figure 3, which has recently been synthesized [38]. We chose this molecule because its large transition dipole moments between the ground and first excited electronic state, as well as the notable difference between the corresponding PDMs due to its inversion symmetry violation, lead to pronounced NOR effects. We stress that this is just a typical molecule that has these properties, and the effects that we describe below also apply to other molecules that possess similar properties. In fact, currently there is intensive interest in the nonlinear optical properties of molecules under the interaction with laser fields and several interesting and potentially useful experimental results have been presented, see, e.g., the works in [39,40].

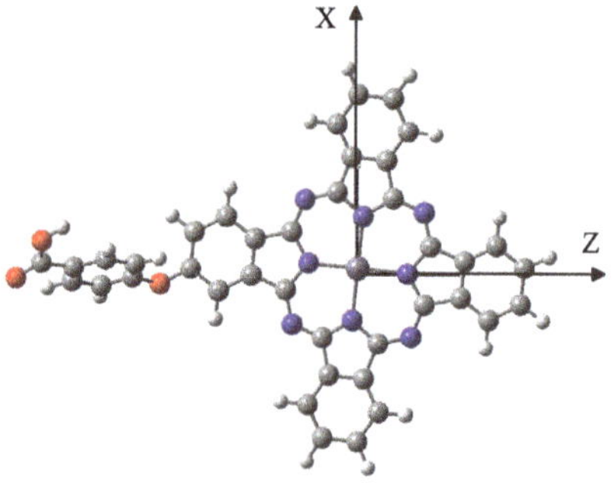

Figure 3. The polar molecular complex used: the zinc–phthalocyanine complex is composed of carbon (gray), hydrogen (white), oxygen (red), nitrogen (blue), and zinc (light blue) atoms. The complex is not planar; however, the phthalocyanine part of the complex is planar, coinciding with the zx-plane, as schematically shown.

In our calculations presented below, the ground and first singlet excited electronic states of this complex are the states $|1\rangle$ and $|2\rangle$ of the QS, respectively. The corresponding molecular spectroscopic parameters are obtained by *ab initio* electronic structure methods, after geometry optimization of the molecular structure of state $|1\rangle$ at the DFT/B3LYP/6-311+G^* and for state $|2\rangle$ at the TD-DFT/B3LYP/6-31-G^* level of theory [41].

From the *ab initio* calculations, we obtain the QS transition energy $\hbar\omega_{21} = 1.99$ eV, as well as the values of the PDMs and the transition dipole moments, as given in Table 1. Moreover, the free-space spontaneous decay widths used are $\Gamma^z_{0f} \approx 13.6$ MHz and $\Gamma^x_{0f} \approx 22.3$ kHz when the corresponding transition dipole moment is along the z-axis and the x-axis, respectively.

Table 1. *Ab initio* obtained values of dipole moments (in Debye) of the ground and first singlet excited electronic state of the molecular complex shown in Figure 3.

Dipole Moments/D	z-axis	x-axis
μ_{11}	7.2746	1.9557
μ_{22}	6.7071	1.5723
μ_{12}	−3.2476	0.1312

3.2. *NOR of the Zinc–Phthalocyanine Complex*

In Figures 4 and 5, we present the NOR coefficient $\chi_0^{(2)}$ in the absence or presence of the MNP as a function of various parameters in the case of no pure dephasing, $\gamma_d = 0$.

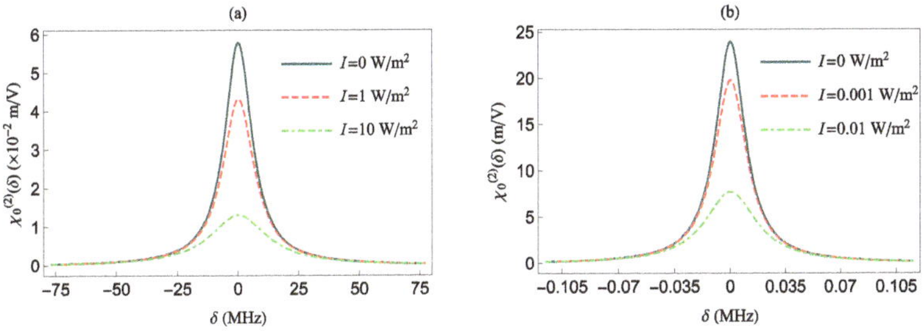

Figure 4. The $\chi_0^{(2)}$ as function of δ for various intensities in absence of the MNP and $\gamma_d = 0$. The QS transition dipole moment is along the z-axis (**a**) and along the x-axis (**b**).

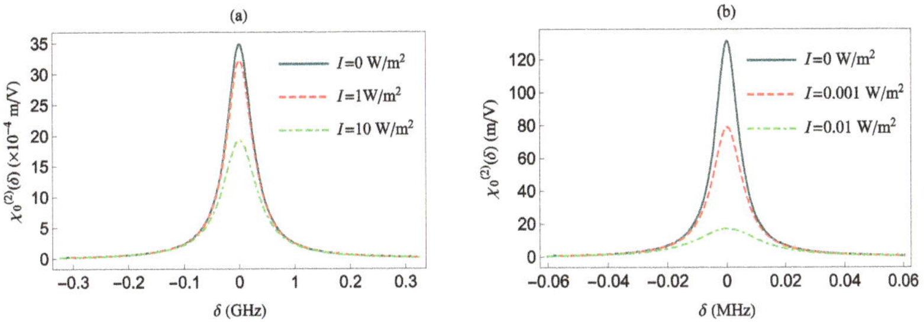

Figure 5. The $\chi_0^{(2)}(\delta)$ as function of δ for various intensities in the presence of the MNP and $\gamma_d = 0$. (**a**) The μ_{12} is along the z-axis and $d = 50$ nm. (**b**) The μ_{12} is along the x-axis and $d = 34.5$ nm.

More specifically, in Figure 4, we present the $\chi_0^{(2)}(\delta)$ at various intensities I, for a QS with $\gamma_d = 0$ and a transition dipole oriented either along the z-axis [panel (a)] or along the x-axis [panel (b)], in the absence of the MNP (i.e., $d \to \infty$). The latter means that $\tilde{\gamma}_d = 0$, $\Gamma_i = 1$, and $|E_i| = 1$. The largest values of $\chi_0^{(2)}(\delta)$, for all intensities I, are obtained for $\delta = 0$. The corresponding values for $\chi_0^{(2)}(\delta = 0)$ are given by

$$\chi_0^{(2)}(\delta = 0, I = 0; d = \infty) = \frac{36N\pi^2 c^6 \epsilon_0 |\mu|}{\omega_{21}^2 \mu_{12}^2}. \tag{12}$$

and

$$\chi_0^{(2)}(\delta = 0, I \neq 0; d = \infty) = \frac{\chi_0^{(2)}(\delta = 0, I = 0; d = \infty)}{1 + \frac{144\pi^2 c^5 \epsilon_0 I}{\omega_{21}^6 n \mu_{12}^2}}. \quad (13)$$

We observe that the $\chi_0^{(2)}(\delta = 0)$ value in each panel is maximized in the absence of the electric field. When $I \neq 0$, the system reaches saturation as the field intensity increases, resulting in a smaller value for $\chi_0^{(2)}(\delta = 0)$ due to the larger denominator in Equation (13). Moreover, since $\chi_0^{(2)}(\delta = 0)$ is inversely proportional to μ_{12}^2, the NOR coefficient assumes larger values in the bottom panel than in the top panel of Figure 4. This is because μ_{12}^2 is smaller along the x direction than in the z direction, as shown in Table 1.

In order to assess the influence of the MNP on the NOR coefficient, in Figure 5, we investigate the NOR coefficient at various intensities I, for $\gamma_d = 0$, in the presence of the MNP, i.e., for $\Gamma_i \neq 1$ and $|E_i| \neq 1$. The QS is located at $d = 50$ nm and $d = 34.5$ nm while its transition dipole moment is along the z- [panel (a)] and x-axis [panel (b)], respectively. These distance values are chosen in order to have the smallest Purcell enhancement factor in each case, according to Figure 2. Again here, the largest values of $\chi_0^{(2)}$ for all values of I are obtained at $\delta = 0$ and they are provided by

$$\chi_0^{(2)}(\delta = 0, I = 0; d) = \frac{\chi_0^{(2)}(\delta = 0, I = 0; d = \infty)}{\Gamma_i^2}, \quad (14)$$

and

$$\chi_0^{(2)}(\delta = 0, I \neq 0; d) = \frac{\chi_0^{(2)}(\delta = 0, I = 0; d)}{1 + \frac{|E_i|^2}{\Gamma_i^2} \frac{144\pi^2 c^5 \epsilon_0 I}{\omega_{21}^6 n \mu_{12}^2}}. \quad (15)$$

In Figure 5, we observe that $\chi_0^{(2)}$ (NOR) is much stronger for tangential polarization of the external field (panel (a)) than for the radial one. By comparing with the results of Figure 4 (absence of MNP), when $I = 0$, $\chi_0^{(2)}$ is suppressed by a factor equal to 0.06 for radial polarization, and enhanced by a factor equal to 5.47 for tangential polarization of the external field. This observation can be rationalized with the help of Equation (14) where it can be seen that the $\chi_0^{(2)}(\delta = 0, I = 0)$ is inversely proportional to the Γ_i^2 factor in this case. Moreover, when $I \neq 0$, the NOR suppression or enhancement, related to the radial and tangential polarization of the external field, is not as pronounced as in the cases shown in Figure 4. This is due to the presence $|E_i|^2/\Gamma_i^2$ factor in the denominator of Equation (15).

In the next figures, Figures 6 and 7, we study how the pure dephasing rate γ_d affects the NOR coefficient of the polar QS in the absence or presence of the MNP, respectively.

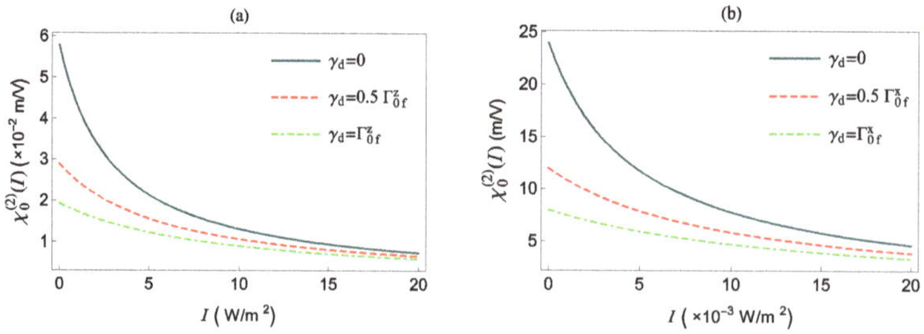

Figure 6. $\chi_0^{(2)}(\delta = 0)$ as a function of intensity I, in the absence of the MNP and for various values of γ_d. The QS transition dipole moment is along the z-axis (**a**) and x-axis (**b**), respectively.

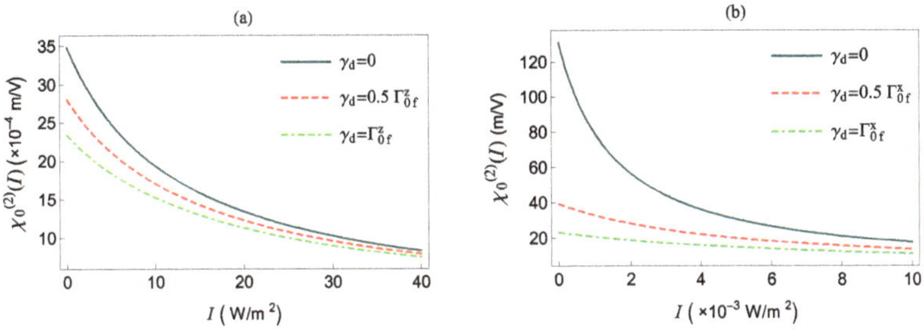

Figure 7. $\chi_0^{(2)}(\delta = 0)$ as a function of the intensity I in presence of the MNP for various γ_d. (**a**) The μ_{12} is along the z-axis and $d = 50$ nm. (**b**) The μ_{12} is along the x-axis and $d = 34.5$ nm.

In Figure 6, we present $\chi_0^{(2)}(\delta = 0)$ as a function of the intensity I, for various γ_d, in the absence of the MNP, i.e., for $\Gamma_i = 1$ and $|E_i| = 1$. We find that $\chi_0^{(2)}(\delta = 0)$ decreases monotonically as I increases due to saturation effects. We note that the field is polarized along the x-axis, the saturation requires intensity three orders of magnitude smaller than for the field being polarized along the z-axis. For $\gamma_d = 0$ and no external field, the $\chi_0^{(2)}(\delta = 0)$ coefficient assumes its largest values, according to Equation (13). For non-zero γ_d, the $\chi_0^{(2)}(\delta = 0)$ values in the absence of the MNP are given by

$$\chi_0^{(2)}(\delta = 0, I; d = \infty) = \frac{36N\pi^2c^6\epsilon_0|\mu|}{\omega_{21}^2(1 + 2\tilde{\gamma}_d)\mu_{12}^2} \cdot \frac{1}{1 + \frac{144\pi^2c^5\epsilon_0 I}{\omega_{21}^6 n(1+2\tilde{\gamma}_d)\mu_{12}^2}}; \quad (16)$$

accordingly, the nonlinear optical behavior of QS is suppressed due to the presence of the $\tilde{\gamma}_d$ factor in both factors of the product of Equation (16).

Now, in presence of the MNP, for any intensity, the $\chi_0^{(2)}(\delta = 0)$ coefficient for $\gamma_d = 0$ is given by Equation (15), while in the case of $\gamma_d \neq 0$, it is given by

$$\chi_0^{(2)}(\delta = 0, I; d) = \frac{36N\pi^2c^6\epsilon_0|\mu|}{\omega_{21}^2(\Gamma_i^2 + 2\Gamma_i\tilde{\gamma}_d)\mu_{12}^2} \cdot \frac{1}{1 + \frac{144\pi^2c^5\epsilon_0|E_i|^2 I}{\omega_{21}^6 n(\Gamma_i^2+2\Gamma_i\tilde{\gamma}_d)\mu_{12}^2}}. \quad (17)$$

We note that in Equation (17), for a given non-zero value of γ_d and I, it is the quantities $|E_i|^2$, Γ_i^2 and $\tilde{\gamma}_d$ that determine the suppression of the $\chi_0^{(2)}(\delta = 0)$, in comparison with the $\gamma_d = 0$ cases. Accordingly, in the presence of the MNP, $\chi_0^{(2)}(\delta = 0)$ in Figure 7 is suppressed (for radially polarized field (panel (a))), at twice the intensity, while it is enhanced (tangentially polarized (panel (b) field)), at half of the intensity, when we compare it with the results shown in Figure 6.

Next, in Figure 8, we investigate $\chi_0^{(2)}(\delta = 0)$ as a function of the distance d between the QS and the MNP, for z- (panel (a)) and x-polarized (panel (b)) external field, for various values of I and no pure dephasing. The values of $\chi_0^{(2)}(\delta = 0)$ for $I = 0$ are given by Equation (14), while for $I \neq 0$, they are given by Equation (15).

In the top panel of Figure 8, we observe that $\chi_0^{(2)}(\delta = 0)$ increases with increasing d, due to the decrease of Γ_z for a z-oriented transition dipole moment of the QS (see Figure 2b). On the other hand, in the bottom panel of Figure 8, for x-oriented external field, we observe that $\chi_0^{(2)}(\delta = 0)$ increases up to $d < 34.5$ nm and decreases monotonically beyond this value. For $I \neq 0$, $\chi_0^{(2)}(\delta = 0)$ is suppressed in comparison with the $I = 0$ case due to saturation effects which are evident in both panels in Figure 8. Moreover, we observe that for the tangentially polarized field, the largest value of $\chi_0^{(2)}(\delta = 0)$ occurs at a shorter distance d than for the case of the radially polarized field.

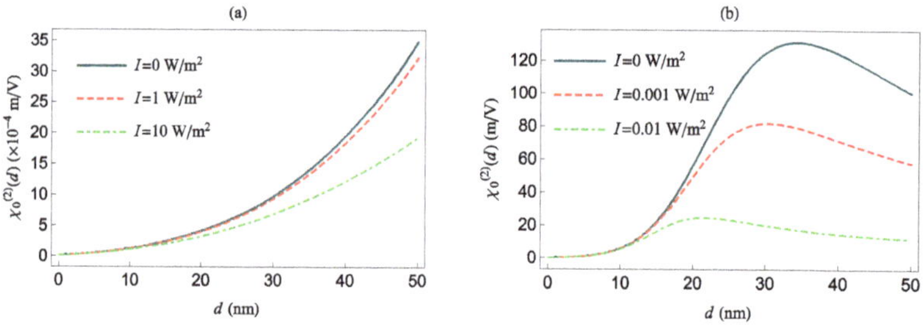

Figure 8. $\chi_0^{(2)}(\delta = 0)$ as a function of d, for $\delta = 0$ and $\gamma_d = 0$, and different values of the intensity I. The QS transition dipole moment is along the z-axis (**a**) and x-axis (**b**), respectively.

Lastly, in Figure 9, we present the nonlinear optical response of the QS as a function of d for various I and for $\gamma_d = \Gamma_{0f}^i$. For $I = 0$, $\chi_0^{(2)}(\delta = 0)$ is provided by

$$\chi_0^{(2)}(\delta = 0; d) = \frac{36 N \pi^2 c^6 \epsilon_0 |\mu|}{\omega_{21}^2 (\Gamma_i^2 + 2\Gamma_i \tilde{\gamma}_d) \mu_{12}^2}. \tag{18}$$

Accordingly, we observe that $\chi_0^{(2)}(\delta = 0)$ is suppressed by a factor of 1.49 and 5.68 for radially (panel (a)) and tangentially (panel (b)) polarized fields, respectively, when compared with the case of no pure dephasing presented in Figure 9.

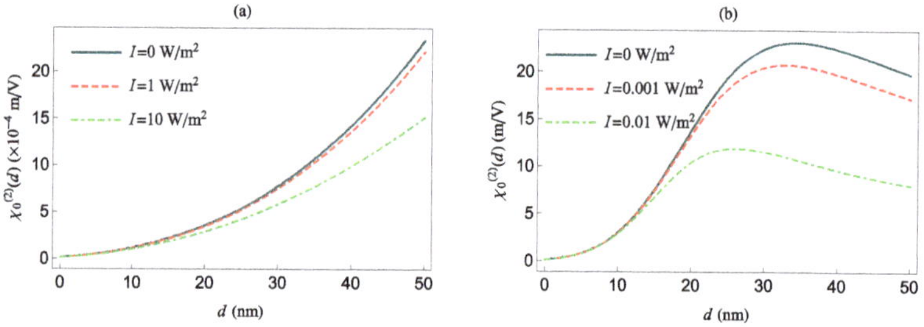

Figure 9. $\chi_0^{(2)}(\delta = 0)$ as a function of d, for $\delta = 0$ and $\gamma_d = 0$, and different values of the intensity I, but for $\gamma_d = \Gamma_{0f}^i$. The QS transition dipole moment is along the z-axis (**a**) and x-axis (**b**), respectively.

For non-zero intensity, in the case of a tangentially polarized field, we observe that as the intensity increases, the highest value of $\chi_0^{(2)}(\delta = 0)$ occurs at slightly larger d, than in the case of no pure dephasing (Figure 8b). We note that for $\gamma_d > \Gamma_{0f}^x$, we find (not shown here) that the highest value of $\chi_0^{(2)}(\delta = 0)$ occurs at about $d = 34.5$ nm; however, in this case, the NOR process becomes very weak, with an almost vanishing $\chi_0^{(2)}(\delta = 0)$.

4. Conclusions

We have investigated the influence of an MNP on the phenomenon of NOR in an inversion-symmetry-broken molecular complex, modeled as a polar two-level QS, under external light illumination. Thus, for such a system, we have derived analytically the equations of the NOR coefficient of the QS and obtained the spectroscopic parameters of the molecular complex, a polar zinc–phthalocyanine complex, via *ab initio* methods. The NOR process of an inversion-symmetry-broken QS, under weak field intensity, was found to be proportional to the difference of the PDMs and inversely proportional to the square of the transition dipole moment of the QS.

In particular, we investigated the NOR coefficient as a function of various parameters of the QS, MNP, and incident light configuration, such as the intensity and polarization of the external field, the distance of the QS to the MNP, the directional decay time of the QS, and the pure dephasing rate of the QS.

We found that in the presence of a MNP, when the electric field is polarized radially with respect to the MNP surface, the NOR is suppressed, while it is enhanced when the field is polarized tangentially. For both polarization directions, as the external field intensity increases, the NOR coefficient decreases due to saturation effects. Further, when we increase the pure dephasing rate of the QS, the NOR decreases. We also observed that, for increasing distance d, the NOR of the QS is enhanced for radial polarization; in contrast, when the polarization is tangential, NOR is enhanced up to some distance, but then it is slowly suppressed with increasing d. Lastly, we found that in presence of the MNP, for the tangential polarization of the field, the NOR process of the QS is more efficient when compared with the free-space case, for certain values of the above variables.

Our findings can be of particular interest for topical quantum technology and nanophotonic applications.

Author Contributions: Conceptualization, E.P.; methodology, N.D., I.T., V.Y. and E.P.; software, N.D. and V.Y.; visualization, N.D.; validation, N.D., I.T., V.Y. and E.P.; investigation, N.D., I.T., V.Y. and E.P.; writing—original draft preparation, N.D., I.T., V.Y. and E.P.; writing—review and editing, N.D., I.T., V.Y. and E.P.; supervision, I.T. and E.P. All authors have read and agreed to the published version of the manuscript.

Funding: This research received no external funding.

Institutional Review Board Statement: Not applicable.

Informed Consent Statement: Not applicable.

Data Availability Statement: The data presented in this study are available upon reasonable request from the corresponding author.

Acknowledgments: This research was funded in part by an Empeirikion Foundation research grant.

Conflicts of Interest: The authors declare no conflict of interest.

Abbreviations

The following abbreviations are used in this manuscript:

QS	Quantum system
MNP	Metallic nanoparticle
PDMs	Permanent electric dipole moments
NOR	Nonlinear optical rectification

References

1. Zhang, W.; Govorov, A.O.; Bryant, G.W. Semiconductor-metal nanoparticle molecules: Hybrid excitons and the nonlinear Fano effect. *Phys. Rev. Lett* **2006**, *97*, 146804. [CrossRef] [PubMed]
2. Yan, J.-Y.; Zhang, W.; Duan, S.-Q.; Zhao, X.-G.; Govorov, A.O. Optical properties of coupled metal-semiconductor and metal-molecule nanocrystal complexes: Role of multipole effects. *Phys. Rev. B* **2008**, *77*, 165301. [CrossRef]
3. Artuso, R.D.; Bryant, G.W. Strongly coupled quantum dot-metal nanoparticle systems: Exciton-induced transparency, discontinuous response, and suppression as driven quantum oscillator effects. *Phys. Rev. B* **2010**, *82*, 195419. [CrossRef]
4. Singh, M.R.; Schindel, D.G.; Hatef, A. Dipole-dipole interaction in a quantum dot and metallic nanorod hybrid system. *Appl. Phys. Lett.* **2011**, *99*, 181106. [CrossRef]
5. Kosionis, S.G.; Terzis, A.F.; Yannopapas, V.; Paspalakis, E. Nonlocal effects in energy absorption of coupled quantum dot–metal nanoparticle systems. *J. Phys. Chem. C* **2012**, *116*, 23663. [CrossRef]
6. Sadeghi, S.M. Gain without inversion in hybrid quantum dot–metallic nanoparticle systems. *Nanotechnology* **2010**, *21*, 455401. [CrossRef]
7. Kosionis, S.G.; Terzis, A.F.; Sadeghi, S.M.; Paspalakis, E. Optical response of a quantum dot–metal nanoparticle hybrid interacting with a weak probe field. *J. Phys. Condens. Matter* **2013**, *25*, 045304. [CrossRef]
8. Sadeghi, S.M. Ultrafast plasmonic field oscillations and optics of molecular resonances caused by coherent exciton-plasmon coupling. *Phys. Rev. A* **2013**, *88*, 013831. [CrossRef]

9. Zhao, D.-X.; Gu, Y.; Wu, J.; Zhang, J.-X.; Zhang, T.-C.; Gerardot, B.D.; Gong, Q.-H. Quantum-dot gain without inversion: Effects of dark plasmon-exciton hybridization. *Phys. Rev. B* **2014**, *89*, 245433. [CrossRef]
10. Li, J.-H.; Shen, S.; Ding, C.-L.; Wu, Y. Magnetically induced optical transparency in a plasmon-exciton system. *Phys. Rev. A* **2021**, *103*, 053706. [CrossRef]
11. Lu, Z.; Zhu, K.-D. Enhancing Kerr nonlinearity of a strongly coupled exciton–plasmon in hybrid nanocrystal molecules. *J. Phys. B At. Mol. Opt. Phys.* **2008**, *41*, 185503. [CrossRef]
12. Paspalakis, E.; Evangelou, S.; Kosionis, S.G.; Terzis, A.F. Strongly modified four-wave mixing in a coupled semiconductor quantum dot-metal nanoparticle system. *J. Appl. Phys.* **2014**, *115*, 083106. [CrossRef]
13. Kosionis, S.G.; Paspalakis, E. Control of self-Kerr nonlinearity in a driven coupled semiconductor quantum dot — metal nanoparticle structure. *J. Phys. Chem. C* **2019**, *123*, 7308–7317. [CrossRef]
14. Malyshev, A.V.; Malyshev, V.A. Optical bistability and hysteresis of a hybrid metal-semiconductor nanodimer. *Phys. Rev. B* **2011**, *84*, 035314. [CrossRef]
15. Nugroho, B.S.; Iskandar, A.A.; Malyshev, V.A.; Knoester, J. Bistable optical response of a nanoparticle heterodimer: Mechanism, phase diagram, and switching time. *J. Chem. Phys.* **2013**, *139*, 014303. [CrossRef]
16. Mohammadzadeh A.; Miri, M. Optical response of hybrid semiconductor quantum dot-metal nanoparticle system: Beyond the dipole approximation. *J. Appl. Phys.* **2018**, *123*, 043111. [CrossRef]
17. Bavli R.; Band, Y.B. Relationship between second-harmonic generation and electric-field-induced second-harmonic generation. *Phys. Rev. A* **1991**, *43*, 507. [CrossRef]
18. Antón, M.A.; Gonzalo, I. Two-photon phase conjugation by degenerate four-wave mixing in polar molecules. *J. Opt. Soc. Am. B* **1991**, *8*, 1035. [CrossRef]
19. Kibis, O.V.; Slepyan, G.Y.; Maksimenko, S.A.; Hoffmann, A. Matter coupling to strong electromagnetic fields in two-level quantum systems with broken inversion symmetry. *Phys. Rev. Lett.* **2009**, *102*, 023601. [CrossRef]
20. Kavokin, K.V.; Kaliteevski, M.A.; Abram, R.A.; Kavokin, A.V.; Sharkova, S.; Shelykh, I.A. Stimulated emission of terahertz radiation by exciton-polariton lasers. *Appl. Phys. Lett.* **2010**, *97*, 201111. [CrossRef]
21. Calderón, O.G.; Gutierrez-Castrejon, R.; Guerra, J.M. High harmonic generation induced by permanent dipole moments. *IEEE J. Quantum Electron.* **1999**, *35*, 47. [CrossRef]
22. Paspalakis, E.; Boviatsis, J.; Baskoutas, S. Effects of probe field intensity in nonlinear optical processes in asymmetric semiconductor quantum dots. *J. Appl. Phys.* **2013**, *114*, 153107. [CrossRef]
23. Oster, F.; Keitel, C.H.; Macovei, M. Generation of correlated photon pairs in different frequency ranges. *Phys. Rev. A* **2012**, *85*, 063814. [CrossRef]
24. Macovei, M.; Mishra, M.; Keitel, C.H. Population inversion in two-level systems possessing permanent dipoles. *Phys. Rev. A* **2015**, *92*, 013846. [CrossRef]
25. Yan, Y.; Lü, Z.; Zheng, H.; Zhao, Y. Exotic fluorescence spectrum of a superconducting qubit driven simultaneously by longitudinal and transversal fields. *Phys. Rev. A* **2016**, *93*, 033812. [CrossRef]
26. Kryuchkyan, G.Y.; Shahnazaryan, V.; Kibis, O.V.; Shelykh, I.A. Resonance fluorescence from an asymmetric quantum dot dressed by a bichromatic electromagnetic field. *Phys. Rev. A* **2017**, *95*, 013834. [CrossRef]
27. Antón, M.A.; Maede-Razavi, S.; Carreño, F.; Thanopulos, I.; Paspalakis, E. Optical and microwave control of resonance fluorescence and squeezing spectra in a polar molecule. *Phys. Rev. A* **2017**, *96*, 063812. [CrossRef]
28. Scala, G.; Slowik, K.; Facchi, P.; Pascazio, S.; Pepe, F.V. Beyond the Rabi model: Light interactions with polar atomic systems in a cavity. *Phys. Rev. A* **2021**, *104*, 013722. [CrossRef]
29. Mîrzac, A.; Carlig, S.; Macovei, M.A. Microwave multiphoton conversion via coherently driven permanent dipole systems. *Phys. Rev. A* **2021** *103*, 043719. [CrossRef]
30. Singh, M.R. Enhancement of the second-harmonic generation in a quantum dot–metallic nanoparticle hybrid system. *Nanotechnology* **2013**, *24*, 125701. [CrossRef]
31. Turkpence, D.; Akguc, G.B.; Bek, A.; Tasgin, M.E. Engineering nonlinear response of nanomaterials using Fano resonances. *J. Opt.* **2014**, *16*, 105009. [CrossRef]
32. Evangelou, S. Tailoring second-order nonlinear optical effects in coupled quantum dot-metallic nanosphere structures using the Purcell effect. *Microelectron. Eng.* **2019**, *215*, 111019. [CrossRef]
33. Yan, J.-Y.; Zhang, W.; Duan, S.-Q.; Zhao, X.-G. Plasmon-enhanced midinfrared generation from difference frequency in semiconductor quantum dots. *J. Appl. Phys.* **2008**, *103*, 104314. [CrossRef]
34. Thanopulos, I.; Paspalakis, E.; Yannopapas, V. Plasmon-induced enhancement of nonlinear optical rectification in organic materials. *Phys. Rev. B* **2012**, *85*, 035111. [CrossRef]
35. Carreño, F.; Antón, M.A.; Paspalakis, E. Nonlinear optical rectification and optical bistability in a coupled asymmetric quantum dot-metal nanoparticle hybrid. *J. Appl. Phys.* **2018**, *124*, 113107. [CrossRef]
36. Johnson, P.B.; Christy, R.W. Optical Constants of the Noble Metals. *Phys. Rev. B* **1972**, *6*, 4370. [CrossRef]
37. Yannopapas, V.; Vitanov, N.V. Electromagnetic Green's tensor and local density of states calculations for collections of spherical scatterers. *Phys. Rev. B* **2007**, *75*, 115124. [CrossRef]
38. Fashina, A.; Nyokong, T. Nonlinear optical response of tetra and mono substituted zinc phthalocyanine complexes. *J. Lumin.* **2015**, *167*, 71. [CrossRef]

39. Zhang, C.-X.; Liu, J.; Gao, Y.; Li, X.-H.; Lu, H.-B.; Wang, Y.; Feng, J.-J.; Lu, J.-B.; Ma, K.-X.; Chen, X.-H. Porous nickel oxide micron polyhedral particles for high-performance. ultrafast photonics. *Opt. Laser Tech.* **2022**, *146*, 107546. [CrossRef]
40. Li, X.-H.; Guo, Y.-X.; Ren, Y.; Peng, J.-J.; Liu, J.-S.; Wang, C.; Zhang, H. Narrow-bandgap materials for optoelectronics applications. *Front. Phys.* **2022**, *17*, 13304. [CrossRef]
41. Frisch, M.J.; Trucks, G.W.; Schlegel, H.B.; Scuseria, G.E.; Robb, M.A.; Cheeseman, J.R.; Scalmani, G.; Barone, V.; Petersson, G.A.; Nakatsuji, H.; et al. *Gaussian 09, Revision A.02*; Gaussian, Inc.: Wallingford, CT, USA, 2016.

nanomaterials

Article

High Strength Die-Attach Joint Formation by Pressureless Sintering of Organic Amine Modified Ag Nanoparticle Paste

Xingwang Shen [1,2,†], Junjie Li [2,†] and Shuang Xi [1,*]

[1] College of Mechanical and Electronic Engineering, Nanjing Forestry University, Nanjing 210037, China
[2] Shenzhen Institute of Advanced Electronic Materials, Shenzhen Institute of Advanced Technology, Chinese Academy of Sciences, Shenzhen 518100, China
* Correspondence: shuangxi@njfu.edu.cn
† These authors contributed equally to this work.

Abstract: Sintered silver (Ag) die-attach has attracted much attention in power systems with high power density and high operating temperature. In this paper, we proposed a novel surface modification method for Ag nanoparticles with organic amines as a coating agent for enhancing the pressureless sintering performance. This work systematically introduced the Ag nanoparticle modification process, Ag paste preparation, and sintering process and compared the changes in the sintering performance of Ag nanoparticles after modification with four different alkyl chain lengths of amines. The study showed that the sintered films of Ag nanoparticle pastes modified with *n*-octylamine (NOA) can achieve the lowest resistivity of the sintered film and the highest shear strength of the bonded joints. The resistivity of the sintered Ag film is affected by the grain size and microscopic morphology, and the strength of the bonded joints is also related to the sintering density and the amount of organic residues. The thermal behavior of the Ag particles coated with different amines is measured by thermal analysis. Finally, the mechanism of NOA-modified Ag nanoparticles to improve the sintering performance is proposed. This study can provide effective data and theoretical support for the further promotion and application of nano-Ag pressureless sintering.

Keywords: Ag nanoparticle paste; pressureless sintering; surface modification; high-strength joint; die-attach

check for updates

Citation: Shen, X.; Li, J.; Xi, S. High Strength Die-Attach Joint Formation by Pressureless Sintering of Organic Amine Modified Ag Nanoparticle Paste. *Nanomaterials* **2022**, *12*, 3351. https://doi.org/10.3390/nano12193351

Academic Editor: Mohd Shkir

Received: 22 August 2022
Accepted: 22 September 2022
Published: 26 September 2022

1. Introduction

With the rapid development of electric vehicles, smart grids, wireless communications, and other fields, silicon-based semiconductor devices have been unable to meet the high temperature, high power, and other service requirements in terms of its operation junction temperature which is below 200 °C and voltage blocking capabilities. Wide band gap (WBG) semiconductors, such as SiC and GaN, with the advantages of high switching frequency, high breakdown field strength, high bonding energy, high thermal conductivity, high temperature operation, and radiation resistance, have become the developing direction of new generation power devices [1–5]. At the same time, its packaging methods and interconnect materials become a real challenge in nowadays power packaging technology, for the materials used in high temperature electronic packaging must be able to withstand such conditions. Traditional tin-based solders are not suitable for the development needs of wide-band-gap semiconductor because the interconnect performance declines dramatically at service temperatures above 150 °C [6–12]. Diverse high temperature die attach materials have been studied. Gold-based and zinc-based solder alloys have limited their widespread application in power device die attachment due to their high cost, brittleness of intermetallic compounds (IMCs), and poor corrosion resistance. Although transient liquid phase bonding (TLP) can achieve bonding at lower temperatures (250–300 °C) and shorter times and obtain higher remelting temperatures, the reliability of its bonding is also vulnerable to intermetallic compound (IMCs) brittleness [13–19].

At present, Ag sintering is a promising technology for power device interconnects. Ag paste has low temperature sintering capability and its sintered structure has high temperature resistance as well as excellent electrical conductivity, thermal conductivity, and high reliability, which can meet the requirements of reliable electrical connection and heat dissipation of high power and high junction temperature chips [20–24]. According to previous reports [25,26], different sintering temperatures, pressures, heating times, and sizes and shapes of the Ag particles all affect the sintered Ag microstructure and sintering performance. The classical sintering process can be carried out at temperatures ranging from 200 to 300 °C with the sintering pressure range from 10 MPa to 40 MPa [27]. However, the pastes in which the powder particles are a few to several micrometers in size are processed at temperatures above 250 °C and at pressures higher than 10 MPa, which may increase the risk of chip damage and increase the process difficulty of chip fabrication. Therefore, the development demand for pressureless sintering Ag technology has become an increasing urgent need. According to the size effect, the surface energy of Ag particles will increase dramatically when they reach the nanometer scale, and the diffusivity of Ag atoms will be significantly enhanced during the sintering process. Therefore, nano-Ag pastes have become the primary material for the development of pressureless die-attach technology. Meanwhile, with the development of nanotechnology, the synthesis process of Ag nanoparticles has gradually matured, and the manufacturing cost has been well controlled, which provides the necessary technical support for the industrial application of nano-Ag paste. However, smaller Ag nanoparticles can easily form micron-sized agglomerates at room temperature due to their overactive surface energy, which reduces the sintering drive of the Ag paste during the sintering process and leads to degradation of the sintering performance. Therefore, how to avoid the spontaneous agglomeration of Ag nanoparticles at room temperature deserves further study. Based on the present state of research [28], the main method to prevent nano-Ag agglomeration is to add a coating agent during the synthesis of Ag monomers with Ag precursors. The types of coating agents are generally classified as polymers and small molecule compounds. Generally, polymers with carboxylate, amino, or hydroxyl functional groups possess high decomposition temperature, such as polyacrylic acid (PAA) [29,30] and polyvinylpyrrolidone (PVP) [31–33], and organic matter will remain after sintering and bonding, which influence their sintering quality. Small molecule compounds are always with long alkyl chains and polar heads, such as alkanethiols [34–36], alkylamines [37], and carboxylic acids [38], which present a relatively wide range of decomposition temperature. Therefore, choosing the small molecule compound coating agent with a suitable decomposition temperature becomes the critical issue to enhance the anti-agglomeration property and the sintering performance. At the same time, in general, the synthesis of Ag monomers also suffers from the long time required and the instability of the product.

At present, there are few studies on the surface modification of commercial Ag nanoparticles. Based on the demand, this work proposes a novel method for surface modification of commercial Ag nanopowders by organic amines, which was then prepared into nano-Ag paste for pressureless die-attach application. Four kinds of amines (*n*-octylamine, dodecylamine, hexadecylamine, and octadecylamine) were utilized to modify nano-Ag powders by washing, dispersing, and freeze-drying processes, and sintering and bonding experiments were conducted to character the microstructure, shear strength, resistivity, and cross-sectional porosity of these Ag pastes. Based on the experimental results, we found that the performance of the amine-modified Ag nanoparticles with different chain lengths differed dramatically after sintering. This is related to the difference in their boiling points. Then, the mechanism of organic amine modified Ag nanoparticles to enhance low temperature sintering properties was proposed.

2. Materials and Methods

2.1. Materials

The commercial nano-Ag powder used in this paper was purchased from Guangdong Lingguang New Material Co., Ltd. (Zhaoqing, Guangdong, China), Octylamine, dodecylamine, hexadecylamine, octadecylamine, ethanol, *tert*-butanol, and polyol ether organic solvents in Ag paste were purchased from Shanghai Aladdin Bio-Chem Technology Co., Ltd. (Shanghai, China), and all reagents were of analytical grade without further purification. The upper and lower substrates are DBC (Direct Bonded Copper) ceramic substrates with Ni/Au surface coating layer, purchased from Jiangsu Ferrotec Semiconductor Co., Ltd. (Yancheng, Jiangsu, China).

2.2. Surface Modification of Ag Nanoparticles

In this paper, four amines with different alkyl chain lengths are introduced as coating agents for Ag nanoparticles, which are *n*-octylamine ($C_8H_{19}N$, NOA, boiling point ~179 °C), dodecylamine ($C_{12}H_{27}N$, DDA, boiling point ~258 °C), hexadecylamine ($C_{16}H_{35}N$, HDA, boiling point ~322 °C), and octadecylamine ($C_{18}H_{39}N$, ODA, boiling point ~349 °C). The coating agents were used to prevent agglomeration of nanoparticles during preservation and before sintering. The detailed process of surface modification is described as follows. Firstly, 2 g of commercial Ag nanoparticles were mixed with 30 mL of ethanol and then subjected to ultrasonic dispersion and freezing centrifugation to remove organic impurities from the surface of the Ag particles. The cleaning process was conducted 3 times for better removal of surface impurities. Then, 30 mL of ethanol, and 0.04 g of organic amines were mixed with the cleaned Ag powder, and the excess organic amines were removed by ultrasonic dispersion and freezing centrifugation. Next, the nanoparticles were dispersed into a mixture of *tert*-butanol and ethanol, placed in a refrigerator for 6 h, and then freeze-dried for 10 h. Finally, the obtained surface-modified Ag nanoparticle powder was collected for further use.

2.3. Modified Nano-Ag Paste Preparation and Bonding Process

The four amine-modified Ag nanopowders were mixed with organic solvents as respective, and the homogeneous nano-Ag paste was formed by high-speed stirring. The paste had an Ag content of 85 wt.% to achieve a suitable viscosity for the dispensing process. The organic solvents in the Ag paste are ethylene glycol, terpineol and polyethylene glycol 200, which can be completely evaporated at 250 °C. The pastes prepared from unmodified Ag nanoparticles and Ag nanoparticles modified with different amines were defined as Ag-0, Ag-NOA, Ag-DDA, Ag-HDA, and Ag-ODA, respectively. The bonding substrate is DBC ceramic substrate with 5-μm thick Ni layer and 70-nm thick Au layer coated on the surface of the Cu layer of the DBC. The upper and lower substrate sizes are $3 \times 3 \times 0.3$ mm and $6 \times 6 \times 0.3$ mm to simulate practical application scenarios. The DBC substrates were removed from the surface contaminants on the substrate by sonication in ethanol for 5 min before use. The Ag paste was coated on the surface of the lower DBC substrate by a dispensing process, and then the upper substrate was placed on the Ag paste to form a sandwich bonding structure. Figure 1 shows a schematic diagram of the Ag paste preparation and bonding process. The interconnected joints in this experiment were achieved by pressureless sintering of the prepared sandwich samples under environmental conditions. During the bonding process, the sample was placed on a heating plate and heated from 30 °C to 250 °C at a steady heating rate of 16 °C/min, held for different times (10 min, 30 min, 60 min) and then cooled naturally. The heating curve of the sintering process is shown in Figure S1.

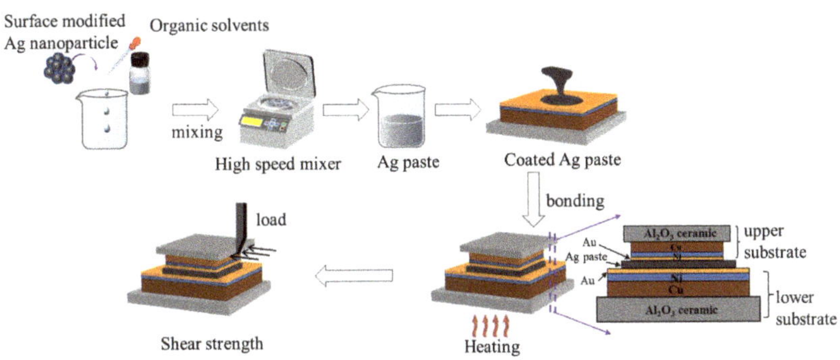

Figure 1. Nano-Ag paste preparation and pressureless sintering bonding process.

2.4. Measurement and Characterization

To examine the coating effect of Ag nanoparticles, Fourier transform infrared absorption spectroscopy (FT-IR, Invenio R, Beuker, Karlsruhe, Germany), energy dispersive X-ray energy spectroscopy (EDS, FEI Nova Nana SEM 450, FEI, Hillsboro, OR, USA), and X-ray photoelectron spectroscopy (XPS, Thermo Scientific Escalab 250xi, Thermo Fisher Scientific, Waltham, MA, USA) were used for surface composition analysis. The resistivity of the sintered Ag films was measured by a 4-point probe system (Loresta-GP MCP-T600, Mitsubishi Chemical, Kanagawa, Japan). Physical phase identification of sintered Ag films was conducted by X-ray diffractometry (XRD, D8 ADVANCE A25, Beuker, Karlsruhe, Germany). The bond strength of Ag-Au joints was evaluated by a chip shear tester (DAGE4000, Nordson DAGE, Aylesbury, UK), with a shear head speed of 100 μm/s and a shear height of 50 μm from the lower substrate surface. The average value of shear strength was calculated after testing 6 joints in each group. Clear and accurate cross-sectional structures of the bonded joints were prepared by a grinding and polishing machine (Tegramin, Struers, Ballerup, Denmark). The microscopic morphology of commercial Ag nanoparticles was observed by transmission electron microscopy (TEM, JEM-ARM200F, JOEL, Tokyo, Japan). The sizes of commercial Ag nanoparticles were measured by image processing software (Nano Measurer version 1.2, Shanghai, China). The structural features of Ag nanoparticles, sintered Ag films, and bonded joints were characterized by field emission scanning electron microscopy (FE-SEM, FEI Nova Nana SEM 450, FEI, Hillsboro, OR, USA). The porosity of the joint cross-section was calculated with the Image pro Plus (version 6.0, Media Cybernetics, Silver Spring, MD, USA) software. The thermal behavior of different amine-modified Ag nanoparticles was measured using a thermogravimetric analyzer (TGA/DSC 2, Mettler Toledo, Greifensee, Switzerland) at a holding time of 60 min at 250 °C. The measurements were conducted at an air flow rate of 100 mL/min and a heating rate of 10 °C/min.

3. Results and Discussion

3.1. Characterization of Commercial Ag Nanoparticles

Due to the problems of unstable product batches, unknown surface organics. and poor sintering performance of commercial Ag nanoparticles, we introduced a nano-Ag surface modification process to improve the stability and sintering performance of Ag nanoparticle pastes. Figure 2a shows the transmission electron micrograph (TEM) of commercial Ag nanoparticles with spherical and quasi-spherical shapes. Figure 2b shows the particle size distribution of Ag nanoparticles. The average particle size is approximately 50 nm, and the distribution is concentrated between 30 nm and 80 nm. Figure 2c shows the XRD diffraction pattern of commercial Ag nanoparticles, indicating a high purity with no other impurities.

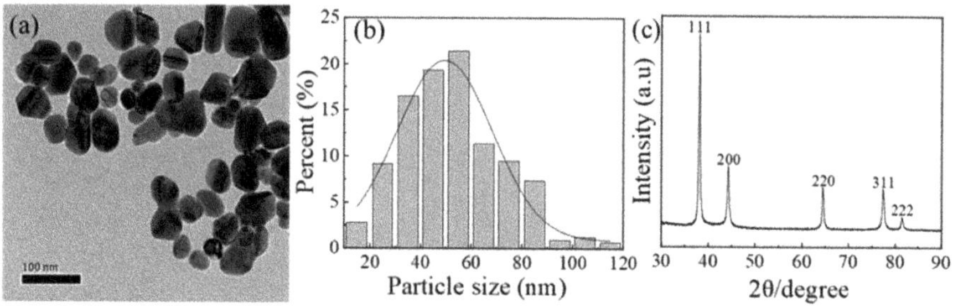

Figure 2. Morphology, size and physical phase analysis of commercial Ag nanoparticles (**a**) TEM image (**b**) Size distribution (**c**) XRD diffraction image.

3.2. Analysis of the Effect of Organic Amine Modification on the Surface of Ag Nanoparticles

Figure 3 shows the FT−IR spectra of different organic amine modified Ag nanoparticles. In fact, *n*-octylamine, dodecylamine, hexadecylamine, and octadecylamine are all primary amines and will have two moderate intensity stretching vibration peaks at 3500–3100 cm^{-1}, while no such stretching vibration peaks can be found on the amine-modified Ag powder. Combined with other related reports [35,39,40], it can be inferred that $-NH_2$ undergoes coordination reactions with Ag atoms to form Ag–N bonds, so that the characteristic peaks of N–H cannot be detected. Meanwhile, in the FT−IR spectra of NOA, DDA, HDA, and ODA treated Ag powders, we find a faint stretching vibration peak near 2750 cm^{-1}, which is a stretching belonging to the $-CH_2-$ group on the amino chain, which further proves the presence of amines. Figure 4a–d show the EDS energy spectra for NOA, DDA, HDA, and ODA modified Ag powders, respectively. The characterization results clearly show the characteristic peaks of elemental N, providing evidence that the organic amines were not removed during the cleaning process. However, we found that the proportion of N elements varied in different amine-modified Ag powders, which may be related to the number of carbon chains of the amine. Specifically, longer carbon chains will have a more significant spatial site resistance effect, which could weaken the ability to coordinate with Ag atoms [41]. In addition, in order to further observe the elemental and chemical states on the surface of organic amine-modified Ag nanoparticles, different modified Ag particles are analyzed by XPS. The binding energies in the XPS spectrum of the modified Ag particles shown in Figure 5 are calibrated by using the binding energy of C 1s (284.8 eV). Figure 5a shows the high resolution spectrum of Ag 3d, the splitting of the 3d double peak spanning 6 eV, which shows that the Ag nanoparticles are metallic in nature [42]. Meanwhile, no peaks related to Ag oxide compounds (367.3 eV for AgO and 367.7 eV for Ag_2O) are found [43]. Interestingly, the Ag 3d5/2 spectrum binding energy of Ag particles modified with NOA and DDA shifted to a higher binding energy (368.4 eV), while the Ag 3d5/2 spectrum binding energy of Ag particles modified with HDA and ODA decreased (368.0 eV), compared to the Ag 3d5/2 standard number binding energy (368.2 eV). The main reason for this occurrence is the transfer of electrons due to the interaction of organic amine and Ag nanoparticles, which is similar to other related reports [42,44–46]. Figure 5b shows the N 1s high-resolution spectrum, which provides conclusive evidence for the organic amine coating on the surface of the Ag nanoparticles, and the weaker peak intensity may be caused by the low organic amine content. In order to further confirm the absence of Ag oxide compounds in the organic amine modified Ag powder, we conducted the analysis by XRD, as shown in Figure S2. The above results show that all four different organic amines can form effective coating layers on the surface of Ag nanoparticles.

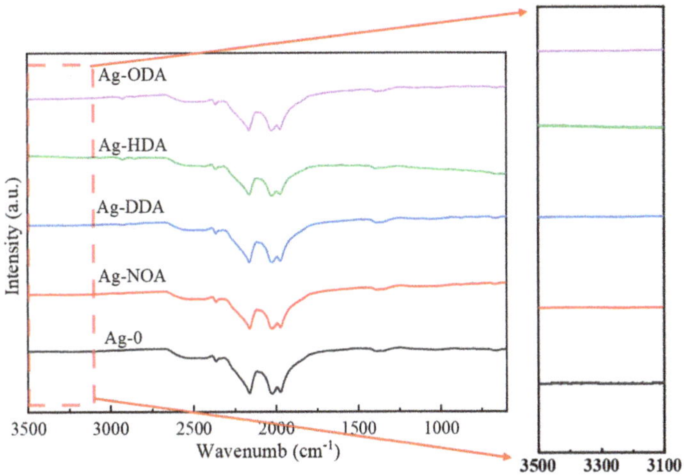

Figure 3. FT−IR spectra of different organic amine modified Ag nanoparticles.

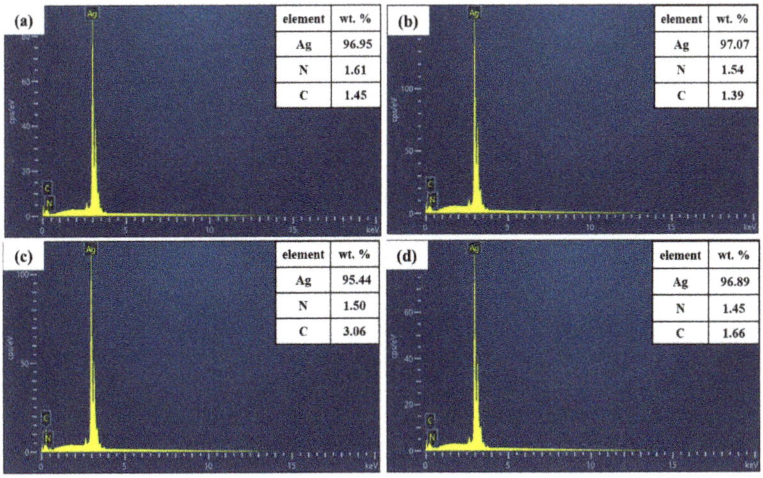

Figure 4. EDS energy spectra of Ag nanoparticles modified with different organic amines (**a**) NOA (**b**) DDA (**c**) HDA (**d**) ODA.

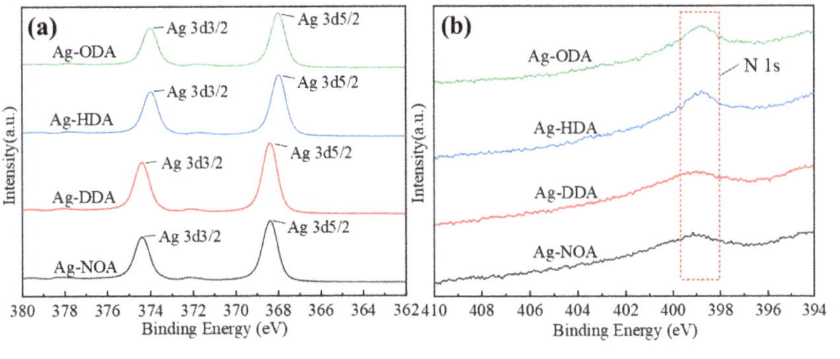

Figure 5. XPS high-resolution spectrum of Ag nanoparticles modified with different organic amines (**a**) Ag 3d spectrum (**b**) N 1s spectrum.

3.3. Resistivity of Ag Paste Sintered Films

Electrical resistivity is a critical parameter for die-attach materials and related to the current handling capability and electrical power loss [47]. Different amine modified nano-Ag pastes were sintered at 250 °C for 30 min to prepare the sintered films, and the effect of the amine type on the resistivity was studied, as shown in Figure 6. The film prepared with Ag-NOA has the lowest resistivity of only 7.31 μΩ·cm, which is approximately four times higher than that of bulk silver (1.6 μΩ·cm). The resistivity of the film prepared with Ag-DDA increased to 12.02 μΩ·cm, but it was still lower than that of the films prepared with Ag-0 (19.63 μΩ·cm). However, the resistivity of the films prepared by Ag-HDA and Ag-ODA is significantly increased to 58.66 μΩ·cm and 80.16 μΩ·cm, respectively, which are much higher than those of the films prepared by Ag-0. The results show that the type of amine has a significant effect on the film resistivity. NOA modification has the highest conductivity, while HDA and ODA modifications could weaken the conductivity.

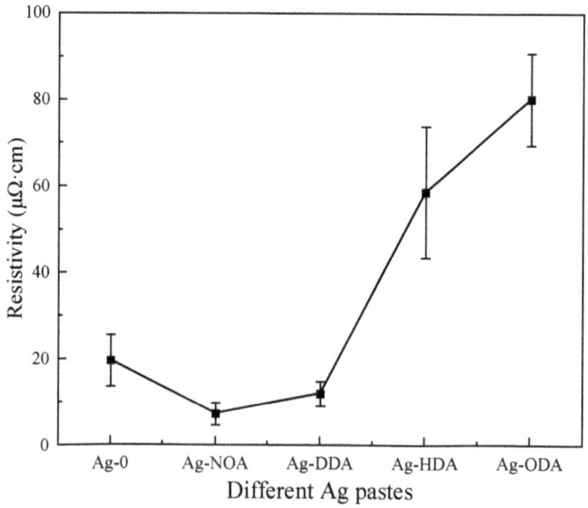

Figure 6. Resistivity of 30 min films sintered with different organic amine modified Ag pastes.

To investigate the influence mechanism on resistivity, the crystal structures and surface morphologies of the sintered films of different amine modified Ag pastes were characterized by XRD and SEM, as shown in Figures 7 and 8. There are five main diffraction peaks marked in Figure 7, namely (111), (200), (220), (311), and (222), which correspond to the crystalline plane of the pure Ag phase. These XRD diffraction peaks are consistent with the standard powder diffraction card (JCPDS No. 03-065-2871) for the Ag face-centered cubic crystal structure, and no other diffraction peaks of impurities are detected. The sintered Ag films prepared with Ag-NOA and Ag-DDA have sharp peaks, which indicates a high degree of crystallization of the film structure and the formation of well-crystallized Ag films. However, the sintered Ag films prepared with Ag-HDA and Ag-ODA show relatively weak peaks, indicating a low degree of crystallization of the film structure. In addition, we calculated the grain size of the particles in each Ag film using the Debye–Scherrer formula, as shown in Table 1. It can be clearly found that the grain size of Ag-NOA sintering (55.0 nm) is much larger and more crystalline than that of Ag-0 (35.66 nm) at a 2θ value of approximately 38°. The grain size of Ag-DDA (44.29 nm) is slightly larger than that of Ag-0. However, the grain sizes of both Ag-HDA and Ag-ODA are smaller than that of Ag-0, which further confirms the above analysis for crystallinity.

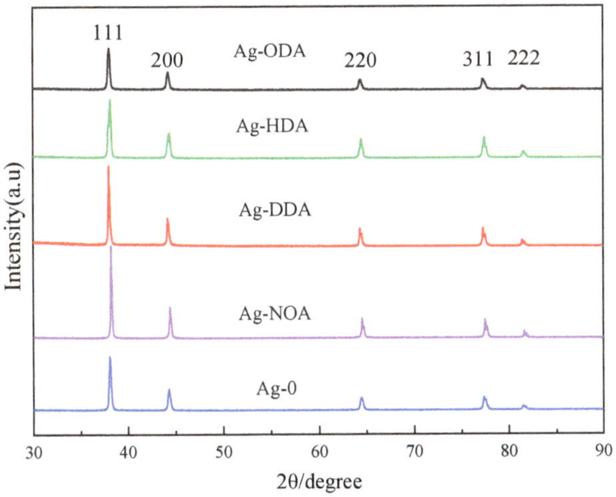

Figure 7. XRD diffractograms of 30 min films sintered with different organic amine modified Ag pastes.

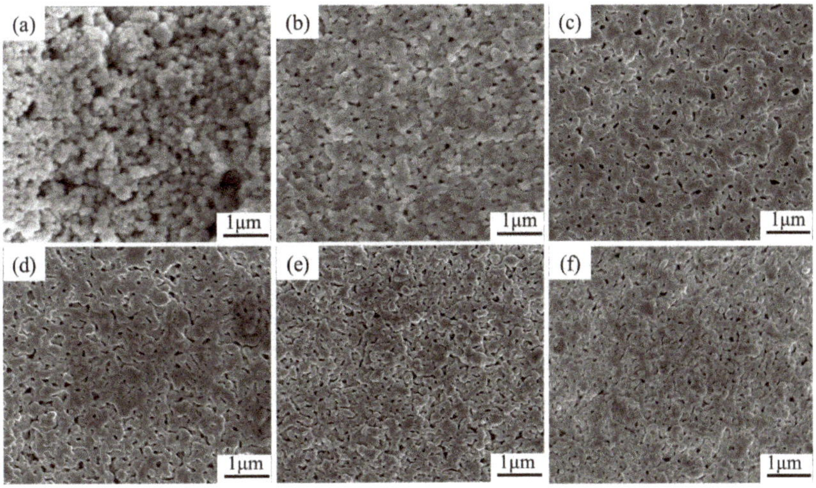

Figure 8. SEM images of commercial Ag nanoparticles at room temperature and different organic amine modified Ag pastes sintered for 30 min (**a**) commercial Ag nanoparticles (**b**) Ag-0 (**c**) Ag-NOA (**d**) Ag-DDA (**e**) Ag-HDA (**f**) Ag-ODA.

Table 1. Particle size of nano-Ag crystals in sintered films of different organic amine modified Ag pastes.

Ag Paste	2θ (°)	FWHM	Size (nm)
Ag-0	38.07556	0.24636	35.6559
Ag-NOA	38.2275	0.15977	55.00541
Ag-DDA	38.00677	0.19828	44.29276
Ag-HDA	38.1224	0.26087	33.67741
Ag-ODA	38.04315	0.36849	23.836

Figure 8a shows the SEM images of commercial Ag nanoparticles at room temperature, showing partial spontaneous agglomeration, which is a common phenomenon of metal nanoparticles due to the size effect. Figure 8b–f show SEM images of the films prepared from Ag-0, Ag-NOA, Ag-DDA, Ag-HDA, and Ag-ODA, respectively. It can be observed that the microstructure of the films has significant differences in sintering morphology, particle size, and degree of contact. The Ag-0 prepared films show an irregular crystal structure, uneven size of sintered particles, poor state of particle connection, and less sintered neck, indicating insufficient sinter formation. The film prepared by Ag-NOA formed a smooth surface and a uniform and continuous microporous structure, which means that it is already in a relatively adequate sintered state, consistent with the lower resistivity of the Ag-NOA film. Although the Ag film prepared by Ag-DDA also has a relatively dense sintering structure, the Ag particles are still not well connected in some areas, which may account for the increased resistivity. For the Ag films prepared by Ag-HDA and Ag-ODA, we can clearly observe that the degree of connection between Ag nanoparticles is significantly reduced and there are still some Ag particles in an isolated state, which indicates that sufficient sintering has not occurred, resulting in a rapid increase in resistivity.

Combining the obtained results, it can be revealed that the Ag-NOA film has the lowest resistivity and the best sintering density. Then, the film was prepared by sintering Ag-NOA at 250 °C for 10, 30, and 60 min to investigate the effect of holding time on resistivity, as shown in Figure 9. The film resistivity after 10 min of holding is 11.28 μΩ·cm, which is approximately seven times higher than that of bulk Ag. When the holding time is increased to 30 min, the resistivity decreased to 7.31 μΩ·cm. When the holding time reaches 60 min, the film resistivity continues to decrease to 5.3 μΩ·cm, which is only about three times the resistivity of bulk Ag and can meet the needs of industrial applications. The above experimental results show that the resistivity decreases with the increase of holding time, while the decrease trend slows down after 30 min sintering.

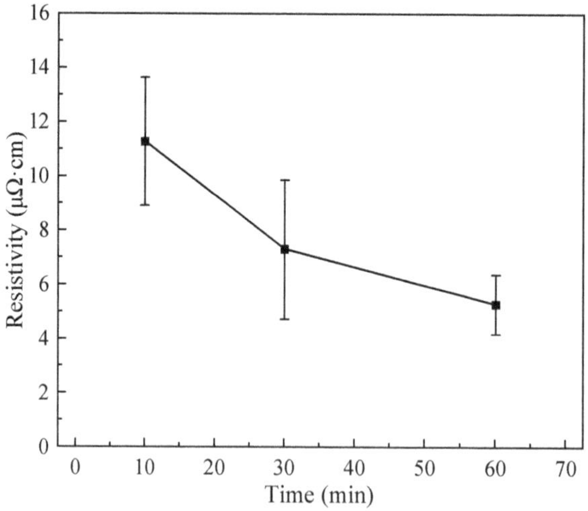

Figure 9. Resistivity of Ag-NOA films for different holding times.

In order to explore the mechanism of holding duration on resistivity of Ag-NOA film, the crystal structure and surface morphology of Ag film were characterized by XRD and SEM, as shown in Figures 10 and 11. The five main diffraction peaks marked in Figure 10, similar to the XRD characterization above, correspond to the crystalline plane of the pure Ag phase, and no impurity diffraction peaks were detected. With a continuous increase in the holding time, the diffraction peaks of the Ag films become sharper, which indicates an increasing degree of crystallization of the film structure. We calculated the particle size of Ag nanocrystals in Ag-NOA film at different holding times with the Debye–Scherrer formula, as shown in Table 2. With the increase of sintering time, we find that the particle size increases, which is consistent with the XRD diffraction peak analysis.

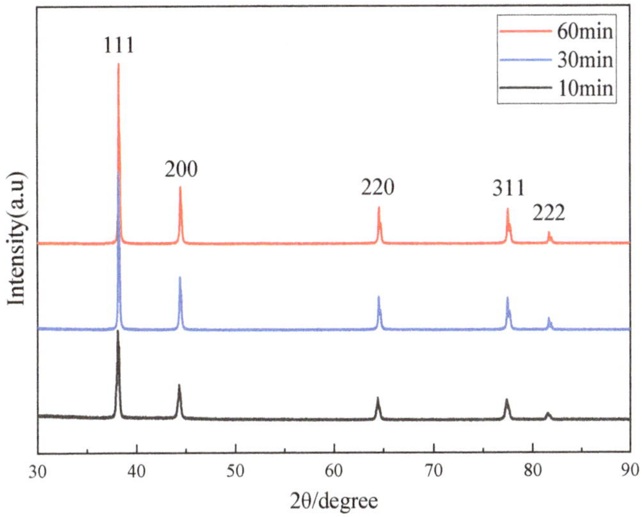

Figure 10. XRD diffraction patterns of Ag-NOA films at different holding times.

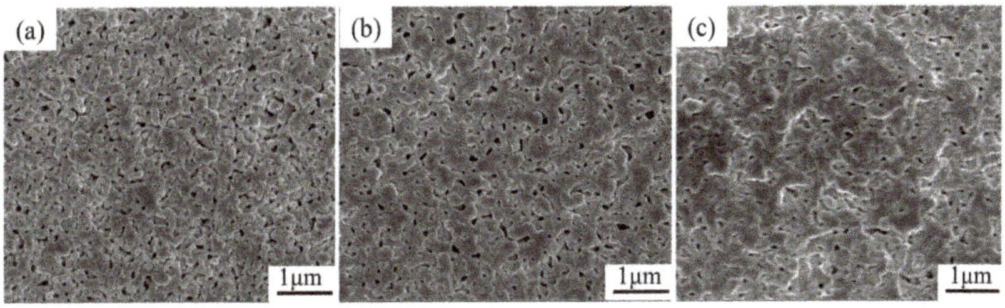

Figure 11. SEM images of Ag-NOA films with different holding times at 250 °C (**a**) 10 min (**b**) 30 min (**c**) 60 min.

Table 2. Particle size of nano-Ag crystals in Ag-NOA sintered films at different holding times.

Time (min)	2θ (°)	FWHM	Size (nm)
10	38.1475	0.25118	34.97926
30	38.2275	0.15977	55.00541
60	38.2751	0.14933	58.85945

The surface morphology of Ag-NOA sintered film with different holding times was characterized by SEM, as shown in Figure 11. With the increase of holding time, the sintering of Ag nanoparticles gradually becomes adequate. The microstructure is looser after holding for 10 min, and the Ag particles are irregular in shape and size, with a low degree of connection. As the holding time increases to 30 min, the sintered connection of Ag particles increases, and a larger sintered neck is formed. After a hold time of 60 min, the independent Ag particles have disappeared and the Ag particles are in good contact with each other, forming a denser sintered network structure. At the same time, the surface pores decreased, indicating that the Ag particles have achieved sufficient sintering. Therefore, increasing the holding time can promote the adequate sintering and diffusion of Ag particles to form a uniform and dense sintering structure, resulting in a lower resistivity of the film.

It can be concluded from the above analysis that highly conductive Ag films can be obtained by selecting suitable amine modified Ag nanoparticles for the preparation of Ag paste. The microstructure of the film prepared by Ag-NOA is uniform and dense with a large Ag crystal composition when held at 250 °C for more than 30 min, but some pores still exist, which cannot be avoided during the sintering process [48,49]. It should be noted here that the particle size of Ag nanoparticles calculated with XRD data belongs to primary particles. However, the particle size shown in SEM is formed by aggregated particles (secondary particles) [41].

3.4. Shear Strength of Bonded Joints

In order to evaluate the effect of different amine-modified Ag nanoparticles on the bond strength, a shear strength test was conducted on these samples. Figure 12 shows the shear strength of the joints after 30 min of sintered bonding with different amine modified Ag pastes, indicating that the use of the appropriate chain length amine is essential for the formation of high-strength joints. The average shear strength of Ag-0 bonded joints without the introduction of organic amines is 35.7 MPa. The average shear strength of the bonded joints with Ag-NOA is substantially increased to 61.8 MPa, indicating that the NOA-modified Ag particles can significantly enhance the bond strength of the joints. However, for the base Ag-DDA paste, the average shear strength of the bonded joints decreases to 50.78 MPa, which may be related to the increase in the number of carbon chains. The average shear strengths of Ag-HDA and Ag-ODA bonded joints are 34.98 MPa and 33.73 MPa, respectively, and this result was similar to the strength of unmodified joints. In fact, under the same conditions, the higher the number of carbon chains of the same kind of organic substance, the higher its boiling point. The boiling points of the four amines used in this paper increase with the number of carbon chains. Combined with the above analysis, we speculate that the boiling point of the amine may influence the bonding strength of the modified Ag paste, indicating that the modification of low boiling point amine can enhance the bonding performance. The thermal analysis of the modified Ag particles is further discussed in the following mechanistic explanation.

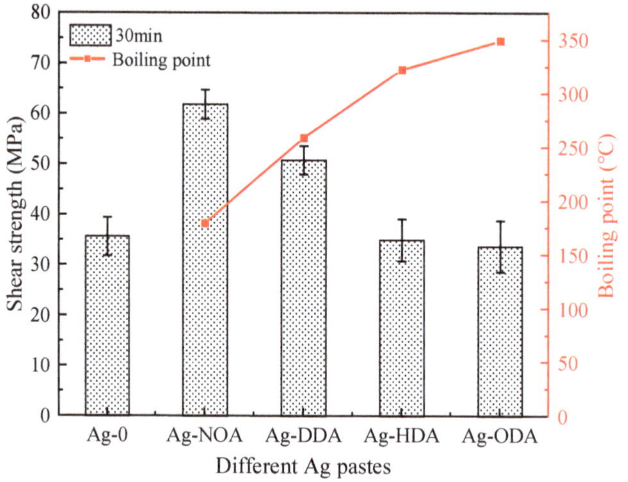

Figure 12. Shear strength of bonded joints with different organic amine modified Ag pastes sintered for 30 min.

The shear strength test above shows the optimal bonding performance of Ag-NOA joints. To investigate the effect of Ag-NOA on the bonding strength at different holding times, we added comparative experiments with holding times of 10 min, 45 min, and 60 min, as shown in Figure 13. The joint strength increases sharply when the holding time increases from 10 min to 30 min, and then increases slightly after 30 min and finally reaches the maximum value of 69.3 MPa at 60 min. The above results indicate that joint strength is positively correlated with holding time. The increase in sintering time leads to an increase in joint strength, which is consistent with existing reports [25,50,51]. Meanwhile, in order to investigate the difference of holding time on the shear strength of the four amine-modified Ag paste bonded joints, a comparison experiment was conducted, and the joint strength results are shown in Figure S3.

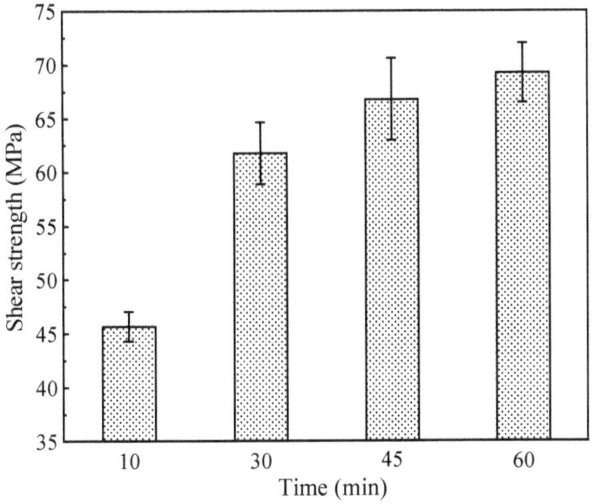

Figure 13. Strength of Ag-NOA bonded joints at different holding times.

The fracture surface morphology of the joint is further characterized to investigate the shear strength with different modification agents. Figure 14a–e show the SEM images of the fracture surfaces of bonded joints sintered with different Ag pastes for 30 min. The Ag-0 joints form fewer triangular ductile stretches of different orientations between the Ag particles. The Ag layers at both the Ag-NOA and Ag-DDA sections tilt at a large angle along one direction, forming a plastic fracture. The Ag-NOA shows greater plastic tensile deformation and a denser sintered structure, indicating a very strong bond between the sintered Ag structure and the DBC substrate. The Ag-DDA exhibits a denser porous structure and a large honeycomb-like structure with tensile deformation. However, the plastic deformation of the Ag-HDA fracture surface is significantly reduced and is no longer sharp, showing a similar morphology to that of the Ag-0 joint. This indicates that the sintering ability of the modified Ag paste is weakened after modification with HDA. The nanoparticles in the fracture surface of the Ag-ODA joint do not undergo large-scale deformation and the fracture position is close to the interface of the DBC substrate. Compared to other joints, the conversion from cohesive damage to adhesive damage is observed. This phenomenon may be related to the ODA organic residues in the bonded structure, which prevents the sintering diffusion between the Ag particles and results in weakening of bonding strength. Therefore, the results of the fracture surface analysis are consistent with the measuring results of shear strength.

To explore the reasons for the difference in strength of Ag-NOA paste with different holding times, SEM images of the fracture surface of Ag-NOA joints with holding times of 60 min are observed, as shown in Figure 14f. It is found that the sections at 60 min have a tighter sintered structure and longer plastic tensile deformation than sections at 30 min, which indicates that sufficient sintering between Ag particles has occurred to form joints with shear strengths up to 69.3 MPa. This is understandable because the increase in holding time helps the thermal decomposition of the organic matter in the Ag paste, which promotes the sintering diffusion between the Ag particles to form a high-density interconnected structure.

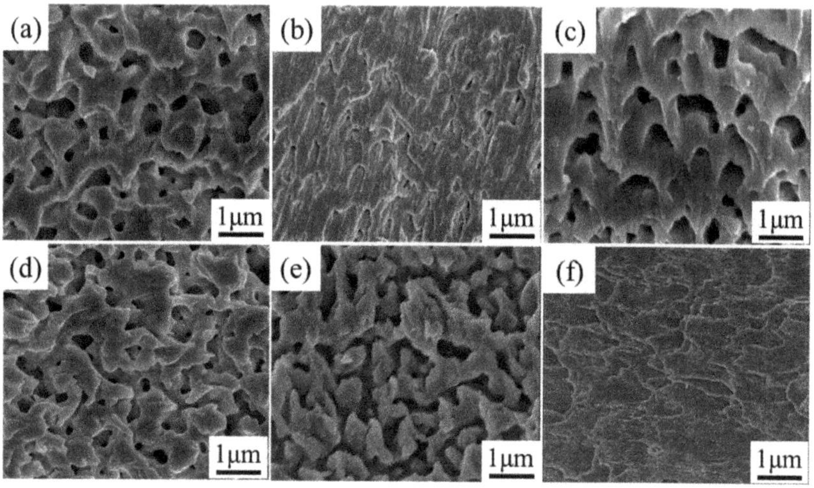

Figure 14. Fracture surface morphology of different modified Ag paste bonded joints (**a**) Ag-0 (**b**) Ag-NOA (**c**) Ag-DDA (**d**) Ag-HDA (**e**) Ag-ODA (**f**) Ag-NOA with 60 min holding time.

To investigate the relationship between the cross-sectional morphology of the joint and the bond strength, we prepared cross-sectional observation specimens by metallographic polishing method. Figure 15 shows the cross-sectional SEM characterization of the bonded specimens based on different Ag pastes. Figure 15g–i correspond to the partial high magnification features of Figure 15a–f, respectively. All joints held sintering for 30 min, except for Figure 15f,l, which present joints formed by holding under Ag-NOA sintering for 60 min. The pore size and porosity have a significant effect on the bonding performance of the joints, so we calculated the porosity of six types of joints using the Image-Pro Plus tool. In the unmodified case, there are obvious pores of different sizes in the cross-section of Ag-0 joints (Figure 15a,g) with a porosity of 11.71%, and the unevenly distributed pores lead to a reduction in shear strength. After the introduction of the lower boiling point NOA coating agent, it can be observed that large areas of Ag particles fuse together to form a uniform and continuous sintering network with a significant reduction in porosity to 5.3% and a very tightly bound Ag-Au interface (Figure 15b,h). The Ag-DDA joint also forms a denser sintered structure, but its pore size has slightly increased compared to Ag-NOA, and the porosity increases to 10.91%. Although a relatively dense sintered structure is formed in some regions of the Ag-HDA joint cross-section, large voids with 12.84% porosity are found by high magnification (Figure 15j), and the bonding line at the Ag-Au interface is also clearly visible, which indicates that the complete diffusion of nanoparticles at the binding interface is not formed. Interestingly, we find that the porosities of Ag-0, Ag-DDA, and Ag-HDA are similar, but the pore size of the latter two is significantly larger than that of the former, and these large pores can sprout and expand into cracks [52], leading to a decrease in shear strength. The Ag-ODA joint cross-section shows larger and more pores, and the sintered Ag structure is looser (Figure 15e). This may be due to the failure of the high boiling point ODA to volatilize and decompose sufficiently, and the organic residue prevents the sufficient diffusion of Ag particles. Meanwhile, the bonding connection at the Ag-Au interface is weaker near the interface (Figure 15k), and isolated Ag particles are present.

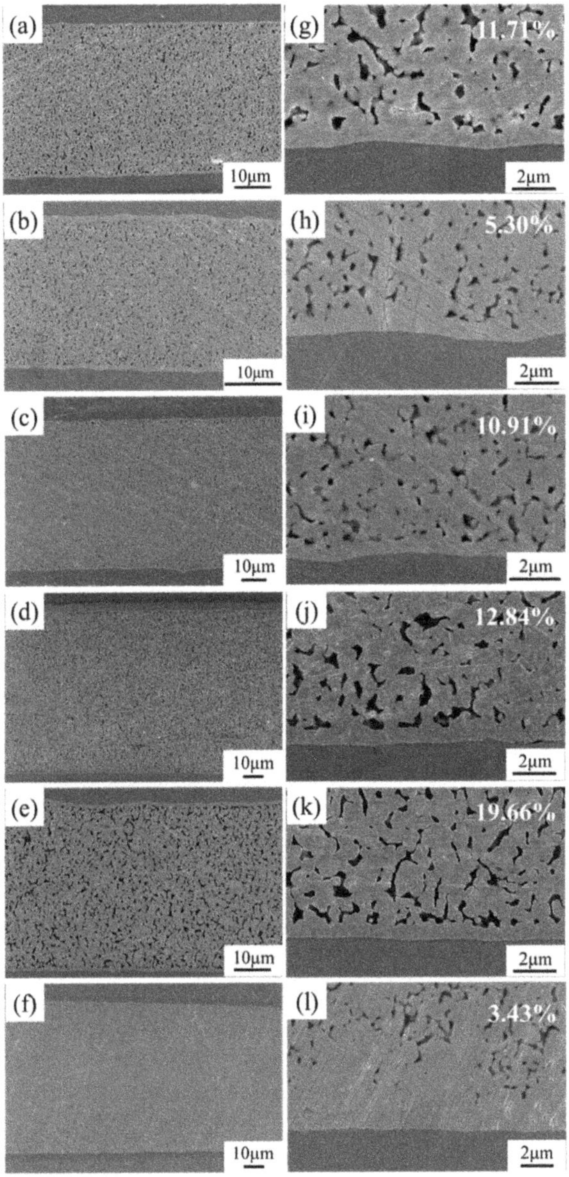

Figure 15. Cross-sections of pristine Ag pastes and different amine modified Ag paste bonded joints (**a**) Ag-0 (**b**) Ag-NOA (**c**) Ag-DDA (**d**) Ag-HDA (**e**) Ag-ODA (**f**) Ag-NOA holding for 60 min, (**g–l**) corresponding to partial enlargements of (**a–f**), respectively.

To investigate the effect of holding time on the cross-section of Ag-NOA joints, the holding time was increased to 60 min. Figure 15f,l show the cross-sectional morphology of the Ag-NOA joints at a holding time of 60 min. The morphology of the Ag particles can hardly be observed, and sufficient diffusion is achieved at the Ag–Au bonding interface. Figure 15f,l show the cross-sectional morphology of the Ag-NOA joints at a holding time of 60 min. The morphology of the Ag particles can hardly be observed, and sufficient diffusion is achieved at the Ag–Au bonding interface. Compared to the Ag-NOA joints at a holding time of 30 min, the porosity is slightly reduced to 3.43%. This can be explained here by the relationship between the diffusion coefficient and diffusion length as follows:

$$D = D_0 \exp\left(-\frac{Q_d}{RT}\right) \tag{1}$$

$$L = (2D \times t)^{\frac{1}{2}} \tag{2}$$

where D_0, Q_d, R, T, and t are the temperature-independent preexponential, diffusion activation energy, gas constant, absolute temperature, and diffusion time, respectively. According to the equations, when the sintering holding time is extended from 30 min to 60 min, the diffusion length will only increase by 1.4 times, which is not a significant effect for sintering [53,54]. This indicates that increasing the holding time can enhance the sintering density of Ag-NOA joints to a lesser degree, which is verified by the small increase in shear strength.

The chemical compositions of different Ag bonded joint cross-sections were characterized by EDS energy spectra, as shown in Figure 16. The content of Ag in the joint is compared to determine the content of organic residue in the joint. The Ag contents in the cross-section of Ag-NOA and Ag-DDA joints are relatively higher with the Ag-0 joints, reaching 97.0 wt.% and 96.4 wt.% with relatively fewer organic residues at 30 min of holding time. However, the Ag content in the Ag-HDA and Ag-ODA joints is reduced relative to the Ag-0 joints, with values of 95.1 wt.% and 93.4 wt.%, respectively. It may be that with the increase in the number of carbon chains, the higher boiling point of organic amines is not easily decomposed and volatilized, which is not beneficial to the sintering and diffusion of Ag particles. When the holding time is increased to 60 min, the Ag content in the Ag-NOA joints slightly increases to 97.4 wt.%, which is similar to that of the Ag-NOA joints at a holding time of 30 min. This indicates that most of the organic matter in the Ag-NOA joint has been volatilized and decomposed at a holding time of 30 min.

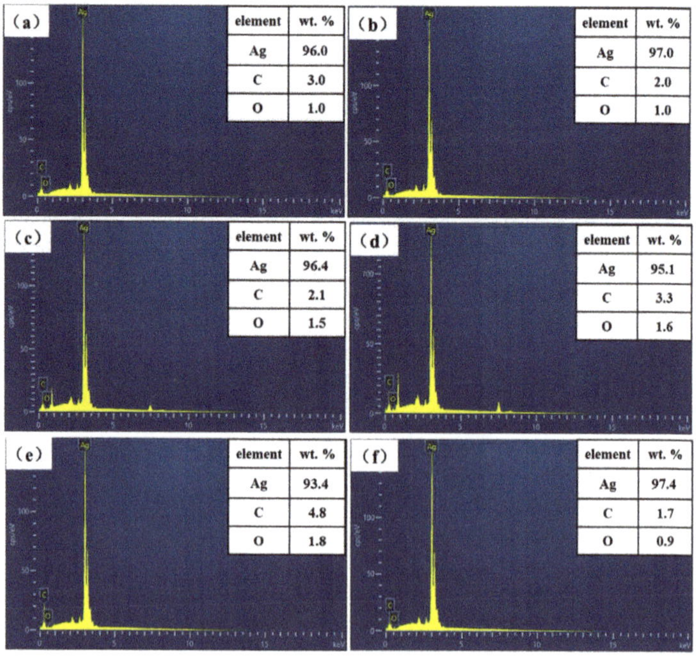

Figure 16. Cross-sectional EDS of unmodified and different amine modified Ag paste bonded joints (**a**) Ag-0, (**b**) Ag-NOA, (**c**) Ag-DDA, (**d**) Ag-HDA, (**e**) Ag-ODA and (**f**) Ag-NOA holding for 60 min.

3.5. Sintering Mechanism

Figure 17 shows the DSC curves of different amine modified Ag nanoparticles sintered at 250 °C with a holding time of 60 min, which helps to understand the sintering behavior of the Ag paste. Within the holding time of 5 min, exothermic peaks are observed for both unmodified Ag powders and different amine modified Ag powders. Interestingly, we find that the exothermic peaks of the four amine modified Ag nanoparticles appear successively with increasing boiling points. The NOA and DDA modified Ag particles show exothermic peaks at holding times of 0.8 min and 2 min, which are earlier than the appearance of exothermic peaks for unmodified Ag particles. However, the exothermic peaks of HDA and ODA modified Ag particles appear later than those of unmodified Ag particles. Based on the above analysis of the sintered Ag film and bond strength, we speculate that the earlier appearance of the exothermic peak facilitates deeper sintering diffusion between the Ag particles and improves the size and density of the sintered neck growth between the Ag particles. This is consistent with the results above that low boiling point amine modified Ag particle pastes have lower resistivity and higher joint strength.

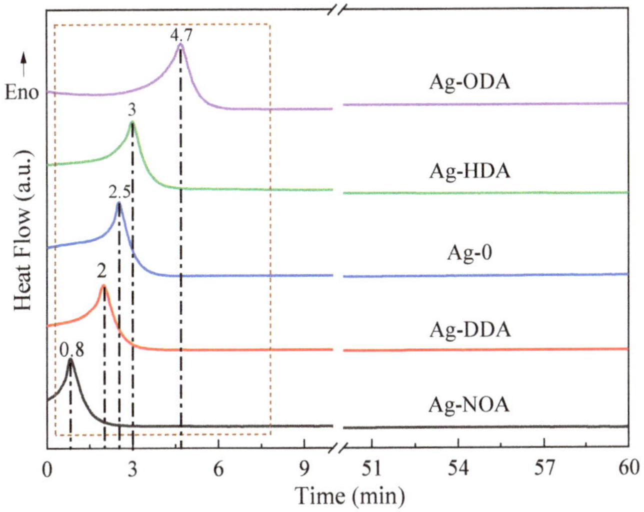

Figure 17. DSC curves of different amine modified Ag nanoparticles with holding time of 60 min at 250 °C.

The analysis based on the above results shows that the selection of suitable amine coating agents to modify the Ag nanoparticles helps to improve the sintering performance of the paste. In this paper, the sintering performance of amine modified Ag pastes with different boiling points is compared and NOA was found to have the best modification effect. Therefore, we speculate on the sintering mechanism, as shown in Figure 18. The whole sintering process can be divided into a low-temperature stage and a high-temperature stage. For the unmodified Ag particles, the exposed Ag nanoparticles with high surface activity will drive surface diffusion at room temperature or lower heating temperatures, resulting in a non-densified diffusion behavior and the formation of low-activity micron-sized agglomerated Ag structures. The non-densified diffusion behavior of nano-Ag in the low-temperature stage reduces the driving force required for densified diffusion. Therefore, it is difficult to obtain excellent sintering density when sintering in the high-temperature stage.

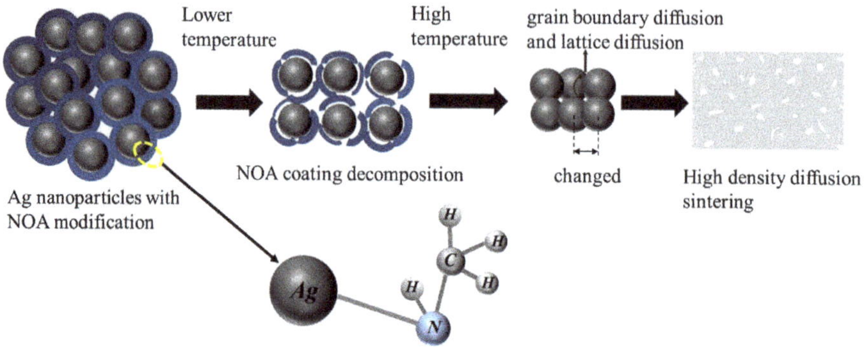

Figure 18. Sintering mechanism of NOA modified Ag nanoparticle paste.

After modification by NOA, the $-NH_2$ group undergoes a coordination reaction with Ag atoms to form Ag–N bonds attached to the surface of the particles. Due to the presence of the NOA coating layer, the Ag nanoparticles exhibit better dispersion and stability in the preparation of Ag paste and storage at room temperature. In the low-temperature stage, Ag nanoparticles do not directly trigger the sintering behavior, but first wait for the decomposition of low boiling point NOA on the particle surface. This limits the generation of low-density sintered structure formation of Ag nanoparticles at low heating temperature and promotes the main diffusion of Ag nanoparticles to change from surface diffusion to grain boundary diffusion and lattice diffusion. Smaller Ag nanoparticles with large surface-to-volume ratio can provide enough energy to stimulate grain boundaries and lattice diffusion, which can form a large, sintered grain size and high-density sintered structures at high temperature stage.

In addition, the other three higher boiling point amine modified Ag particles take longer to decompose the coating agent when sintered at 250 °C, and even too much organic matter can remain. In particular, residual organic matter can weaken the sintering diffusion between Ag particles and inhibit the densification of Ag sintering.

Based on the above analysis, it can be concluded that the commercial Ag nanoparticles can be effectively modified and the pressureless sintering performance of the Ag pastes can be enhanced by selecting the appropriate organic amine with the proper treatment process. The Ag paste prepared by using *n*-octylamine modified Ag nanoparticles has low resistivity and high bonding strength. Its surface morphology is uniform and dense, consisting of larger silver grains and fewer organic residues.

4. Conclusions

In this paper, a novel surface treatment process based on commercial Ag nanoparticles is proposed that can prevent the agglomeration of Ag nanoparticles and enhance the Ag paste sintering performance. This method is simple, efficient, and time-saving to operate. Four types of amine modified Ag particles with different boiling points are used to prepare pastes, the effect of amine type on the sintering performance is investigated, and its application in pressureless die attachment is discussed. The results show that the $-NH_2$ group in the organic amine adsorbs on the surface of Ag particles, forming an organic coating layer and preventing the agglomeration of Ag particles at room temperature. The sintering properties of Ag pastes are closely related to the amine used, since the amine influences the thermal decomposition of the Ag particles, resulting in different grain sizes, sintering structures, and organic residues. The grain size and sintering density after sintering impact the resistivity of the film, with larger grains and higher density improving the electrical conductivity of the film. Ag-NOA has the lowest resistivity of 7.31 $\mu\Omega\cdot cm$ at 250 °C with 30 min holding time, and the result is significantly better than that of the unmodified film. After increasing the holding time to 60 min, the resistivity slightly decreased to 5.3 μΩ·cm. The micro-morphology, porosity, and organic residues of sintered Ag affect the shear strength of bonded samples. The Ag-NOA joint has a bond strength of up to 61.8 MPa at 250 °C with 30 min holding time. The fracture surface shows a clearly plastic deformation structure. The cross section shows a dense sintered network with the lowest porosity and the highest interface connectivity. The EDS energy spectra showed the least organic residues after Ag-NOA sinter bonding. Increasing the holding time can promote the full diffusion sintering of Ag particles. The shear strength was slightly increased to 69.3 MPa at 250 °C with 60 min holding time. In addition, we have proposed the corresponding sintering mechanism based on the excellent sintering effect achieved by NOA-modified Ag nanoparticle paste, explaining the influence of the organic amine modification on the sinter formation of Ag nanoparticles in the low-temperature stage and high-temperature stage. The study provides effective experimental data and theoretical support for pressureless sintering technology, which has good potential for practical application in power device packaging.

Supplementary Materials: The following supporting information can be downloaded at: https://www.mdpi.com/article/10.3390/nano12193351/s1, Figure S1: Heating curve of sintering process; Figure S2: XRD diffractograms of the organic amine modified Ag nanoparticles; Figure S3: Shear strength of different organic amine modified nano-Ag pastes sintered at 250 °C for different holding times.

Author Contributions: Conceptualization, X.S. and J.L.; methodology, X.S.; Software, X.S.; validation, X.S.; formal analysis, J.L.; investigation, X.S.; resources, J.L. and S.X.; data curation, X.S.; writing—original draft preparation, X.S.; writing—review & editing, X.S. and J.L.; visualization, X.S.; supervision, J.L. and S.X.; project administration, J.L.; funding acquisition, J.L. and S.X. All authors have read and agreed to the published version of the manuscript.

Funding: This work is financially supported by National Natural Science Foundation of China with the grant number of 51805197.

Institutional Review Board Statement: Not applicable.

Informed Consent Statement: Not applicable.

Data Availability Statement: The data that support the findings of this study are available from the corresponding author upon reasonable request.

Conflicts of Interest: The authors declare no conflict of interest.

References

1. Sugiura, K.; Iwashige, T.; Tsuruta, K.; Chen, C.; Nagao, S.; Funaki, T.; Suganuma, K. Reliability Evaluation of SiC Power Module with Sintered Ag Die Attach and Stress-Relaxation Structure. *IEEE Trans. Compon. Packag. Manuf. Technol.* **2019**, *9*, 609–615. [CrossRef]
2. Villacarlos, B.J.L.; Pulutan, M.L.D. Thermomechanical Stress and Strain Distribution and Thermal Resistivity Correlation to Bondline Thickness of Ag Sinter. In Proceedings of the 2020 IEEE 22nd Electronics Packaging Technology Conference (EPTC), Singapore, 2–4 December 2020; pp. 307–311.
3. Kim, D.; Chen, C.; Nagao, S.; Suganuma, K. Mechanical Characteristics and Fracture Behavior of GaN/DBA Die-Attached during Thermal Aging: Pressure-Less Hybrid Ag Sinter Joint and Pb–5Sn Solder Joint. *J. Mater. Sci. Mater. Electron.* **2020**, *31*, 587–598. [CrossRef]
4. Jibran, M.; Sun, X.; Hua, J.; Wang, B.; Yamauchi, Y.; Da, B.; Ding, Z.J. $Cu_2Zn(Si,Ge)Se_4$ Quaternary Semiconductors as Potential Photovoltaic Materials. *Chem. Phys. Lett.* **2020**, *756*, 137820. [CrossRef]
5. Calabretta, M.; Renna, M.; Vinciguerra, V.; Messina, A.A. Power Packages Interconnections for High Reliability Automotive Applications. In Proceedings of the ESSDERC 2019—49th European Solid-State Device Research Conference (ESSDERC), Krakow, Poland, 23–26 September 2019; pp. 35–39.
6. Liu, Y.; Chen, C.; Suganuma, K.; Sakamoto, T.; Ueshima, M.; Naoe, T.; Nishikawa, H. Ag Die-Attach Paste Modified by WC Additive for High-Temperature Stability Enhancement. In Proceedings of the 2022 IEEE 72nd Electronic Components and Technology Conference (ECTC), San Diego, CA, USA, 31 May–3 June 2022; pp. 2153–2157.
7. Pietrikova, A.; Girasek, T.; Durisin, J.; Saksl, K. Pressureless Silver Sintering in Power Application. In Proceedings of the 2018 International Conference on Diagnostics in Electrical Engineering (Diagnostika), Pilsen, Czech Republic, 4–7 September 2018; pp. 1–4.

8. Guo, L.; Liu, W.; Ji, X.; Zhong, Y.; Hang, C.; Wang, C. Robust Cu–Cu Bonding with Multiscale Coralloid Nano-Cu_3 Sn Paste for High-Power Electronics Packaging. *ACS Appl. Electron. Mater.* **2022**, *4*, 3457–3469. [CrossRef]
9. Lee, H.; Smet, V.; Tummala, R. A Review of SiC Power Module Packaging Technologies: Challenges, Advances, and Emerging Issues. *IEEE J. Emerg. Sel. Top. Power Electron.* **2020**, *8*, 239–255. [CrossRef]
10. Yuan, Y.; Wu, H.; Li, J.; Zhu, P.; Sun, R. Cu-Cu Joint Formation by Low-Temperature Sintering of Self-Reducible Cu Nanoparticle Paste under Ambient Condition. *Appl. Surf. Sci.* **2021**, *570*, 151220. [CrossRef]
11. Liu, X.; Liu, L.; Sun, R.; Li, J. Low Temperature Sintering of MOD Assisted Ag Paste for Die-Attach Application. *Mater. Lett.* **2021**, *305*, 130799. [CrossRef]
12. Bhogaraju, S.K.; Hans, A.; Schmid, M.; Elger, G.; Conti, F. Evaluation of Silver and Copper Sintering of First Level Interconnects for High Power LEDs. In Proceedings of the 2018 7th Electronic System-Integration Technology Conference (ESTC), Dresden, Germany, 18–21 September 2018; pp. 1–8.
13. Mokhtari, O. A Review: Formation of Voids in Solder Joint during the Transient Liquid Phase Bonding Process—Causes and Solutions. *Microelectron. Reliab.* **2019**, *98*, 95–105. [CrossRef]
14. Jung, D.H.; Sharma, A.; Mayer, M.; Jung, J.P. A Review on Recent Advances in Transient Liquid Phase (TLP) Bonding for Thermoelectric Power Module. *Rev. Adv. Mater. Sci.* **2018**, *53*, 147–160. [CrossRef]
15. Chen, H.; Hu, T.; Li, M.; Zhao, Z. Cu@Sn Core–Shell Structure Powder Preform for High-Temperature Applications Based on Transient Liquid Phase Bonding. *IEEE Trans. Power Electron.* **2017**, *32*, 441–451. [CrossRef]
16. Yoon, J.-W.; Lee, B.-S. Initial Interfacial Reactions of Ag/In/Ag and Au/In/Au Joints during Transient Liquid Phase Bonding. *Microelectron. Eng.* **2018**, *201*, 6–9. [CrossRef]
17. Shao, H.; Wu, A.; Bao, Y.; Zhao, Y.; Zou, G.; Liu, L. Thermal Reliability Investigation of Ag-Sn TLP Bonds for High-Temperature Power Electronics Application. *Microelectron. Reliab.* **2018**, *91*, 38–45. [CrossRef]
18. Ji, H.; Zhou, J.; Liang, M.; Lu, H.; Li, M. Ultra-Low Temperature Sintering of Cu@Ag Core-Shell Nanoparticle Paste by Ultrasonic in Air for High-Temperature Power Device Packaging. *Ultrason. Sonochem.* **2018**, *41*, 375–381. [CrossRef] [PubMed]
19. Bhogaraju, S.K.; Mokhtari, O.; Pascucci, J.; Conti, F.; Kotadia, H.R.; Elger, G. A Multi-Pronged Approach to Low-Pressure Cu Sintering Using Surface-Modified Particles, Substrate and Chip Metallization. *Int. Symp. Microelectron.* **2019**, *2019*, 000387–000392. [CrossRef]
20. Wang, X.; Mei, Y.; Li, X.; Wang, M.; Cui, Z.; Lu, G.-Q. Pressureless Sintering of Nanosilver Paste as Die Attachment on Substrates with ENIG Finish for Semiconductor Applications. *J. Alloys Compd.* **2019**, *777*, 578–585. [CrossRef]
21. Li, H.; Wang, C.; Wu, L.; Chen, M.; Wu, C.; Wang, N.; Li, Z.; Tang, L.; Pang, Q. Optimization of Process Parameters, Microstructure, and Thermal Conductivity Properties of Ti-Coated Diamond/Copper Composites Prepared by Spark Plasma Sintering. *J. Mater. Sci. Mater. Electron.* **2021**, *32*, 9115–9125. [CrossRef]
22. Yan, J. A Review of Sintering-Bonding Technology Using Ag Nanoparticles for Electronic Packaging. *Nanomaterials* **2021**, *11*, 927. [CrossRef]
23. Weber, C.; Hutter, M. Ag Sintering—An Alternative Large Area Joining Technology. In Proceedings of the 2018 7th Electronic System-Integration Technology Conference (ESTC), Dresden, Germany, 18–21 September 2018; pp. 1–8.
24. Abt, M.; Roch, A.; Qayyum, J.A.; Pestotnik, S.; Stepien, L.; Abu-Ageel, A.; Wright, B.; Ulusoy, A.C.; Albrecht, J.; Harle, L.; et al. Aerosol-Printed Highly Conductive Ag Transmission Lines for Flexible Electronic Devices. *IEEE Trans. Compon. Packag. Manuf. Technol.* **2018**, *8*, 1838–1844. [CrossRef]
25. Wang, Q.; Zhang, S.; Lin, T.; Zhang, P.; He, P.; Paik, K.-W. Highly Mechanical and High-Temperature Properties of Cu–Cu Joints Using Citrate-Coated Nanosized Ag Paste in Air. *Prog. Nat. Sci. Mater. Int.* **2021**, *31*, 129–140. [CrossRef]
26. Chen, C.; Yeom, J.; Choe, C.; Liu, G.; Gao, Y.; Zhang, Z.; Zhang, B.; Kim, D.; Suganuma, K. Necking Growth and Mechanical Properties of Sintered Ag Particles with Different Shapes under Air and N_2 Atmosphere. *J. Mater. Sci.* **2019**, *54*, 13344–13357. [CrossRef]
27. Mysliwiec, M.; Kisiel, R.; Pavlov, K.; Kruszewski, M.J. Pressureless Direct Bonding of Au Metallized Substrate with Si Chips by Micro-Ag Particles. In Proceedings of the 2022 45th International Spring Seminar on Electronics Technology (ISSE), Vienna, Austria, 11–15 May 2022; pp. 1–7.
28. Mo, L.; Guo, Z.; Yang, L.; Zhang, Q.; Fang, Y.; Xin, Z.; Chen, Z.; Hu, K.; Han, L.; Li, L. Silver Nanoparticles Based Ink with Moderate Sintering in Flexible and Printed Electronics. *Int. J. Mol. Sci.* **2019**, *20*, 2124. [CrossRef] [PubMed]
29. Grouchko, M.; Kamyshny, A.; Mihailescu, C.F.; Anghel, D.F.; Magdassi, S. Conductive Inks with a "Built-In" Mechanism That Enables Sintering at Room Temperature. *ACS Nano* **2011**, *5*, 3354–3359. [CrossRef] [PubMed]
30. Lee, S.; Kim, J.H.; Wajahat, M.; Jeong, H.; Chang, W.S.; Cho, S.H.; Kim, J.T.; Seol, S.K. Three-Dimensional Printing of Silver Microarchitectures Using Newtonian Nanoparticle Inks. *ACS Appl. Mater. Interfaces* **2017**, *9*, 18918–18924. [CrossRef] [PubMed]
31. Mo, L.; Ran, J.; Yang, L.; Fang, Y.; Zhai, Q.; Li, L. Flexible Transparent Conductive Films Combining Flexographic Printed Silver Grids with CNT Coating. *Nanotechnology* **2016**, *27*, 065202. [CrossRef] [PubMed]
32. Tang, Y.; He, W.; Wang, S.; Tao, Z.; Cheng, L. New Insight into the Size-Controlled Synthesis of Silver Nanoparticles and Its Superiority in Room Temperature Sintering. *CrystEngComm* **2014**, *16*, 4431–4440. [CrossRef]
33. Khalil, A.M.; Hassan, M.L.; Ward, A.A. Novel Nanofibrillated Cellulose/Polyvinylpyrrolidone/Silver Nanoparticles Films with Electrical Conductivity Properties. *Carbohydr. Polym.* **2017**, *157*, 503–511. [CrossRef]

34. Tobjörk, D.; Aarnio, H.; Pulkkinen, P.; Bollström, R.; Määttänen, A.; Ihalainen, P.; Mäkelä, T.; Peltonen, J.; Toivakka, M.; Tenhu, H.; et al. IR-Sintering of Ink-Jet Printed Metal-Nanoparticles on Paper. *Thin Solid Films* **2012**, *520*, 2949–2955. [CrossRef]
35. Mo, L.; Liu, D.; Li, W.; Li, L.; Wang, L.; Zhou, X. Effects of Dodecylamine and Dodecanethiol on the Conductive Properties of Nano-Ag Films. *Appl. Surf. Sci.* **2011**, *257*, 5746–5753. [CrossRef]
36. Volkman, S.K.; Yin, S.; Bakhishev, T.; Puntambekar, K.; Subramanian, V.; Toney, M.F. Mechanistic Studies on Sintering of Silver Nanoparticles. *Chem. Mater.* **2011**, *23*, 4634–4640. [CrossRef]
37. Yu, B.; Wang, K.; Pang, X.; Wu, G.; Pu, J.; Zhao, H. Tribological Properties of Alkylated Reduced Graphene Oxide as Lubricant Additive. *Tribol. Int.* **2022**, *165*, 107273. [CrossRef]
38. Ankireddy, K.; Vunnam, S.; Kellar, J.; Cross, W. Highly Conductive Short Chain Carboxylic Acid Encapsulated Silver Nanoparticle Based Inks for Direct Write Technology Applications. *J. Mater. Chem. C* **2013**, *1*, 572–579. [CrossRef]
39. Shao, C.; Xiong, S.; Cao, X.; Zhang, C.; Luo, T.; Liu, G. Dithiothreitol-Capped Red Emitting Copper Nanoclusters as Highly Effective Fluorescent Nanoprobe for Cobalt (II) Ions Sensing. *Microchem. J.* **2021**, *163*, 105922. [CrossRef]
40. Tripathy, S.K.; Yu, Y.-T. Spectroscopic Investigation of S–Ag Interaction in ω-Mercaptoundecanoic Acid Capped Silver Nanoparticles. *Spectrochim. Acta A Mol. Biomol. Spectrosc.* **2009**, *72*, 841–844. [CrossRef] [PubMed]
41. Yang, W.; Wang, C.; Arrighi, V. Effects of Amine Types on the Properties of Silver Oxalate Ink and the Associated Film Morphology. *J. Mater. Sci. Mater. Electron.* **2018**, *29*, 20895–20906. [CrossRef]
42. Adhikary, J.; Das, B.; Chatterjee, S.; Kumar Dash, S.; Chattopadhyay, S.; Roy, S.; Chen, J.-W.; Chattopadhyay, T. Ag/CuO Nanoparticles Prepared from a Novel Trinuclear Compound $[Cu(Imdz)_4(Ag(CN)_2)_2]$ (Imdz = Imidazole) by a Pyrolysis Display Excellent Antimicrobial Activity. *J. Mol. Struct.* **2016**, *1113*, 9–17. [CrossRef]
43. Zhang, X.; Zhang, Y.; Wang, C.; Zhu, P.; Xiang, B.; Zhao, T.; Xu, L.; Sun, R. Exploration of Key Factors for the Sintering of Micro-Nano Silver Paste. In Proceedings of the 2022 23rd International Conference on Electronic Packaging Technology (ICEPT), Dalian, China, 10–13 August 2022; pp. 1–6.
44. Zheng, Y.; Zheng, L.; Zhan, Y.; Lin, X.; Zheng, Q.; Wei, K. Ag/ZnO Heterostructure Nanocrystals: Synthesis, Characterization, and Photocatalysis. *Inorg. Chem.* **2007**, *46*, 6980–6986. [CrossRef]
45. Xia, H.; Yijian, L. Synthesis of Ag-Coated Cu Nano Powder Applied to the Silver Paste on Front of the Solar Cell. In Proceedings of the 2016 IEEE International Conference on Manipulation, Manufacturing and Measurement on the Nanoscale (3M-NANO), Chongqing, China, 18–22 July 2016; pp. 392–395.
46. Gao, S.; Li, Z.; Jiang, K.; Zeng, H.; Li, L.; Fang, X.; Jia, X.; Chen, Y. Biomolecule-Assisted in Situ Route toward 3D Superhydrophilic Ag/CuO Micro/Nanostructures with Excellent Artificial Sunlight Self-Cleaning Performance. *J. Mater. Chem.* **2011**, *21*, 7281. [CrossRef]
47. Zhang, H.; Li, W.; Gao, Y.; Zhang, H.; Jiu, J.; Suganuma, K. Enhancing Low-Temperature and Pressureless Sintering of Micron Silver Paste Based on an Ether-Type Solvent. *J. Electron. Mater.* **2017**, *46*, 5201–5208. [CrossRef]
48. Chen, C.; Zhang, Z.; Zhang, B.; Suganuma, K. Micron-Sized Ag Flake Particles Direct Die Bonding on Electroless Ni–P-Finished DBC Substrate: Low-Temperature Pressure-Free Sintering, Bonding Mechanism and High-Temperature Aging Reliability. *J. Mater. Sci. Mater. Electron.* **2020**, *31*, 1247–1256. [CrossRef]
49. Yoon, J.-W.; Back, J.-H.; Jung, S.-B. Effect of Surface Finish Metallization on Mechanical Strength of Ag Sintered Joint. *Microelectron. Eng.* **2018**, *198*, 15–21. [CrossRef]
50. Wang, T.; Chen, X.; Lu, G.-Q.; Lei, G.-Y. Low-Temperature Sintering with Nano-Silver Paste in Die-Attached Interconnection. *J. Electron. Mater.* **2007**, *36*, 1333–1340. [CrossRef]
51. Siow, K.S. Mechanical Properties of Nano-Silver Joints as Die Attach Materials. *J. Alloys Compd.* **2012**, *514*, 6–19. [CrossRef]
52. Zhang, H.; Wang, W.; Bai, H.; Zou, G.; Liu, L.; Peng, P.; Guo, W. Microstructural and Mechanical Evolution of Silver Sintering Die Attach for SiC Power Devices during High Temperature Applications. *J. Alloys Compd.* **2019**, *774*, 487–494. [CrossRef]
53. Kwon, S.; Lee, T.-I.; Lee, H.-J.; Yoo, S. Improved Sinterability of Micro-Scale Copper Paste with a Reducing Agent. *Mater. Lett.* **2020**, *269*, 127656. [CrossRef]
54. Park, S.J.; Martin, J.M.; Guo, J.F.; Johnson, J.L.; German, R.M. Grain Growth Behavior of Tungsten Heavy Alloys Based on the Master Sintering Curve Concept. *Metall. Mater. Trans. A* **2006**, *37*, 3337–3346. [CrossRef]

nanomaterials

MDPI

Article

Synthesis of Platinum Nanoparticles Supported on Fused Nanosized Carbon Spheres Derived from Sustainable Source for Application in a Hydrogen Generation Reaction

Erik Biehler [1,2], Qui Quach [1,2] and Tarek M. Abdel-Fattah [1,2,*]

[1] Applied Research Center, Thomas Jefferson National Accelerator Facility, Newport News, VA 23606, USA; erik.biehler.16@cnu.edu (E.B.); qui.quach.13@cnu.edu (Q.Q.)
[2] Department of Molecular Biology and Chemistry, Christopher Newport University, Newport News, VA 23606, USA
* Correspondence: fattah@cnu.edu

Abstract: The dwindling supply of fossil fuels has prompted the search for an alternative energy source that could effectively replace them. Potential renewable energy sources such as solar, wind, tidal, and geothermal are all promising but each has its own drawbacks. Hydrogen gas on the other hand can be combusted to produce energy with only water as a byproduct and can be steadily generated via the aqueous media hydrolysis reaction of Sodium Borohydride ($NaBH_4$). This study successfully synthesized fused carbon spheres derived from sugar and decorated them with platinum nanoparticles to form a novel composite material (PtFCS) for catalyzing this reaction. The platinum nanoparticles were produced by reducing chloroplatinic acid in a solution with sodium borohydride and using sodium citrate as a capping agent for the nanoparticles. Transmission electron microscopy (TEM), Energy-dispersive X-ray spectroscopy (EDS), Fourier-transform infrared spectroscopy (FTIR), and X-ray diffraction (XRD) were used to characterize and determine the size and shape of the Pt nanoparticles (PtNPs) and fused carbon spheres. TEM was able to determine the average size of the fused carbon spheres to be 200 nm and the average size for the PtNPs to be 2–3 nm. The PtFCS composite was tested for its ability to catalyze the hydrolysis of $NaBH_4$ under various reaction conditions including various solution pH, various temperatures, and various dosages of sodium borohydride. The catalyst was found to perform the best under acidic solution conditions (pH 6), producing hydrogen at a rate of 0.0438 $mL/mg_{cat}{\cdot}min$. The catalyst was determined to have an activation energy of 53.0 kJ/mol and could be used multiple times in succession with no loss in the volume of hydrogen produced. This sugar-derived composite catalyst shows promise and could be implemented as a sustainable catalyst for the generation of hydrogen fuel.

Keywords: hydrogen; nanoparticles; carbon; catalysis; sustainability; energy; green chemistry; environment; metals; platinum

Citation: Biehler, E.; Quach, Q.; Abdel-Fattah, T.M. Synthesis of Platinum Nanoparticles Supported on Fused Nanosized Carbon Spheres Derived from Sustainable Source for Application in a Hydrogen Generation Reaction. *Nanomaterials* **2023**, *13*, 1994. https://doi.org/10.3390/nano13131994

Academic Editors: Andrei Honciuc and Mirela Honciuc

Received: 3 June 2023
Revised: 28 June 2023
Accepted: 28 June 2023
Published: 1 July 2023

1. Introduction

It is no secret that the world is overly reliant on fossil fuels as an energy source with an estimated 84.3% of all global energy being derived from these limited and environmentally damaging fuels [1]. The combustion of these fuels is the number one source of greenhouse gases such as carbon dioxide (CO_2), and some models predict that the reserves of coal, oil, and natural gas may be depleted in as little as 105, 35, and 37 years, respectively [2,3]. Thus, there has been a considerable global research effort that has focused on developing an alternative energy source that is not only clean but also abundant. There are multiple promising candidates for renewable energy including solar, wind, tidal, geothermal, and hydrogen, each with its own set of benefits and drawbacks [4–14]. Solar energy, for example, is a promising candidate as energy from the sun is the most abundant energy source we have currently. Direct solar energy can be used to boil water, cook food, or even dry food for storage, replacing the fossil fuels that are more typically used [5]. Solar panels have gained traction as a way to convert photovoltaic energy into electricity; however, current solar panels are often expensive to produce and lack efficiency [6]. Wind energy is another renewable source that is abundant and, through modern technology, has managed to become relatively free of pollutants [7]. Opponents of wind energy point out that the development of wind farms has led to an increase in noise pollution and that wind turbines have disrupted the flight patterns of some bird species resulting in deaths from collisions [8]. Tidal energy is a relatively new renewable energy source that is promising due to it being one of the most predictable forms of renewable energy [9]. The main concern with tidal energy is that there have not yet been enough studies on the environmental impacts. There are concerns over the impacts the construction of tidal energy devices will have on local species, as well as economic concerns over how these structures will affect shipping routes [9]. Geothermal energy is another energy source that is highly consistent like tidal energy; however, the main drawbacks of geothermal energy are inefficiency compared to other renewable energy sources and a high initial cost to implement [10,11]. Hydrogen, which has an atomic number of 1, is the most abundant element in the universe and primarily exists in a diatomic gaseous state. Hydrogen gas is unique not only in that it can be combusted to produce energy, but that there are no harmful byproducts produced via this reaction, only water. This abundance and environmentally friendly bioproduct make hydrogen gas an extremely promising choice as a renewable energy source to replace the world's reliance on fossil fuels. Hydrogen gas, however, is not without its drawbacks as the main issue standing in the way of a widely implemented hydrogen fuel economy lies in the struggle to efficiently store it. There are two main methods of hydrogen storage, first, keeping hydrogen in its diatomic gaseous state and compressing the gas into tanks, and second, cooling the gas down into a liquid to be stored in refrigerated tanks. The first method is relatively simple, however, there are safety concerns associated with compressing the highly combustible gas. The refrigeration technique also is not ideal as the energy cost to cool and store the gas would essentially offset the benefits of using the fuel. A different way to store hydrogen that has gained attention in recent years is storage within the structure of a chemical species. A class of chemicals known as metal hydrides is capable of storing large percentages of hydrogen in their structure. Sodium borohydride ($NaBH_4$), for example, contains 10.8% hydrogen by weight and when mixed in and reacted with water, produces hydrogen gas steadily over time (1) [15]. The main drawback of this reaction, however, is that the gas is produced too slowly to be effectively utilized.

$$NaBH_4 + 2H_2O \rightarrow NaBO_2 + 4H_2 \quad (1)$$

Transition metals have been extensively studied for their catalytic ability that stems from their incomplete valence shells and their ability to transfer electrons. Precious metal catalysts in particular have been studied for their catalytic ability and this team as well as others have explored catalysts made from these metals including gold, silver, palladium, and platinum to make this reaction more efficient [16–23]. Nanomaterials made using these metals and more common metals such as zinc have been used for hydrogen generation [24–29], antibacterial effects [30], gas detection [31], optoelectronic properties [32], and as catalyst [33,34]. Platinum metal, in particular, is commonly used as a catalyst in its bulk state for hydrosilation reactions [22], oxidation of methane [35], and in catalytic converters [36], while as nanoparticles (PtNPs), platinum has been used as a catalyst for the hydrogenation of alkynes [37], in hydrogen fuel cells [38], and in hydrogen generation reactions [18,39]. Metal nanoparticles make excellent catalysts due to their increased surface area over bulk materials; however, it has been noted that nanoparticles have a tendency to agglomerate in solutions, decreasing their catalytic effectiveness [40]. A way to mitigate this agglomeration is through the use of a support material, which gives the nanoparticles a surface to disperse on and bind to rather than each other. There are different support materials available, however, support materials made from carbon are an attractive option due to their high surface area and tensile strength [41]. Additionally, there is an environmentally friendly aspect to using carbon over other materials, since carbon is regularly found in the environment, and structures made of it have an easier time breaking down. In this study, we aimed to synthesize and characterize a novel catalyst comprised of platinum nanoparticles supported on a fused carbon sphere composite (PtFCS). This material was then tested for its effectiveness as a catalyst in the water-splitting reaction of sodium borohydride under a variety of different reaction conditions including different pH conditions, solution temperature conditions, and doses of $NaBH_4$ added to the solution. The activation energy of this reaction as catalyzed by PtFCS was then able to be determined and was compared to other catalysts used in the hydrolysis of sodium borohydride.

2. Experimental

2.1. *Synthesis of Platinum Nanoparticles*

The platinum nanoparticles (PtNPs) were synthesized via the reduction of chloroplatinic acid (Cl_6H_2Pt). Then, 48 mL of 10 mM beta cyclodextrin solution was mixed with 1 mM of chloroplatinic acid. The mixture was stirred for 30 min. After that, 0.25 mL of 180 mM sodium borohydride ($NaBH_4$) was added slowly to the mixture. The solution was stirred for 2 h. The nanoparticle solution was centrifuged at 10,000 rpm for 15 min to remove unreacted reactants.

2.2. *Synthesis of Fused Carbon Spheres and Nanocomposite Catalyst*

In order to synthesize the fused carbon sphere support material, dextrose was first dissolved into deionized water resulting in a solution with a 0.5 M concentration. The dextrose solution was then poured into the polytetrafluoroethylene inside the body of a stainless-steel reaction vessel so that there was a 3:2 ratio of air to dextrose solution. The stainless-steel reaction vessel containing the solution was then placed into an oven which was heated to 473 K and left overnight, which resulted in the desired fused carbon spheres (FCS). Once the fused carbon spheres had been formed, the solution was filtered via vacuum filtration to collect all solid material. The solid material was then washed several times with deionized water and left out at room temperature to dry.

The fused carbon sphere composites were produced by incipient wetness impregnation of 100 mg of fused carbon spheres by 2 mL of the PtNPs aqueous solution. First, the dried fused carbon spheres were placed into a small beaker. Next, the nanoparticle solution was poured over the top of the powdered fused carbon spheres and stirred well at room temperature. The resulting mixture was then stored at 333 K for two days to facilitate the evaporation of excess water from the composite material. After the two days had passed, the dried material was collected and stored until needed for characterization or the catalytic trials.

2.3. *Characterization*

The PtFCS material was first characterized via transmission electron microscopy (TEM, JEM-2100F) to confirm the adhesion of the nanoparticles to the surface of the spheres as well as to confirm the final morphology of the nanoparticle-coated fused carbon sphere composite.

Energy-dispersive X-ray spectroscopy (EDS, ThermoScientific UltraDry, Waltham, MA, USA) allowed us to determine what elements were present within the chemical structure of the catalyst. The surface structure of the material after reusability trials was shown by scanning electron microscopy (SEM, ThermoScientific UltraDry, Waltham, MA, USA).

Fourier-transform infrared spectroscopy (FTIR, Shimadzu IR-Tracer 100, Kyoto, Japan) allowed us to determine any functional groups present in the catalyst and supported the identification of the nanoparticles. Finally, powder X-ray diffraction (P-XRD, Rigaku MiniFlex II Benchtop X-ray Diffractometer, Tokyo, Japan) was implemented in order to further confirm any chemical species within our material.

2.4. Catalysis

The setup consisted of two vacuum flasks connected to one another by a thin plastic tube. One flask was designated as the reaction chamber inside of which the hydrolysis reaction of sodium borohydride catalyzed by platinum-decorated fused carbon sphere composites occurred. The second flask contained only deionized water, which would be displaced by the hydrogen generated in the first flask. Both flasks were sealed via rubber stoppers, however, the second flask, containing the DI water to be displaced, included a second thin plastic tube that ran through the rubber stopper that sealed the second flask. This second hose then hung over the mouth of a plastic cup that was placed on a microbalance scale. The scale was then balanced and any water that was displaced by hydrogen gas would drip into the cup so that the mass could be measured. When sodium borohydride was added to the first flask and carefully sealed, hydrogen gas would begin to fill the first flask. Once that flask was full of hydrogen gas, the gas would move through the thin plastic tubing that connected the two flasks. As the gas filled the second flask, the deionized water present in the flask would be forced up the thin plastic tube that was going through the rubber stopper and would begin filling the cup on the micro balance scale. This scale was connected to a laboratory computer and a measurement program was run, which recorded the measured mass of the water displaced every 0.25 s. This experiment was run at a variety of pH levels (6, 7, 8), temperatures (283, 288, 295, 303) K, and $NaBH_4$ doses (625, 925, 1225) μmoles to determine optimal reaction conditions. All reactions were stirred using a magnetic stir bar for the full two hours of the trial except in the cases of the temperature-controlled trials, which required insulation that interfered with the magnetic stir plate.

3. Results and Discussion

The novel PtFCS catalyst was characterized via transition electron microscopy as shown in Figure 1. Figure 1A depicts the structure of the fused carbon sphere backbone as a collection of semi-round fused objects with a diameter ranging from about 180 to 250 nm. Zooming in on these spheres revealed the presence of nanoparticles, which can be seen in Figure 1B–D. The nanoparticles can be seen spread across the fused carbon spheres in Figure 1B or in more concentrated groups as in Figure 1C,D. From these images, the average size of the nanoparticles was determined to be about 2.9 nm. Despite some grouping, it is clear from these images that the platinum nanoparticles are dispersed across the materials, showing the fused carbon sphere successfully prevented major agglomeration.

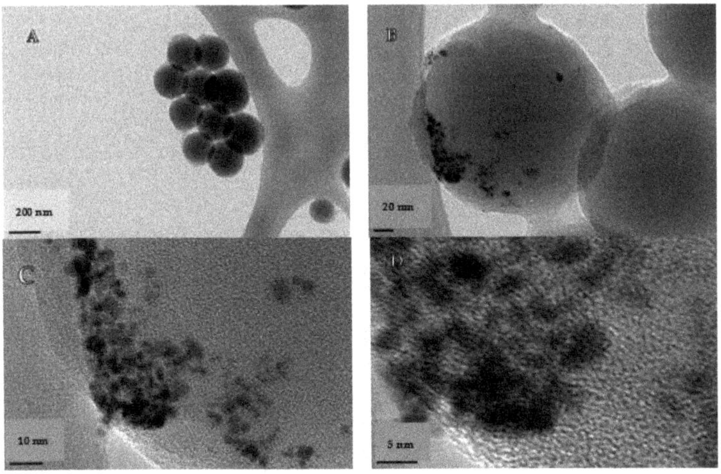

Figure 1. TEM Images of the PtFCS composite material at a scale of 200 nm (**A**), 20 nm (**B**), 10 nm (**C**), 5 nm (**D**).

After TEM, the PtFCS catalyst was characterized using EDS (Figure 2). The two main elements of the PtFCS were carbon and platinum. Carbon was the most abundant element in the composite material as the fused carbon sphere backbone is derived from carbon-based dextrose. the platinum peaks had a low number of counts; however, this is most likely due to the small nanoparticle size and the dispersion of nanoparticles across the fused carbon sphere backbone. The percentage of Pt loading is 1.57% ± 0.5 (Wt%).

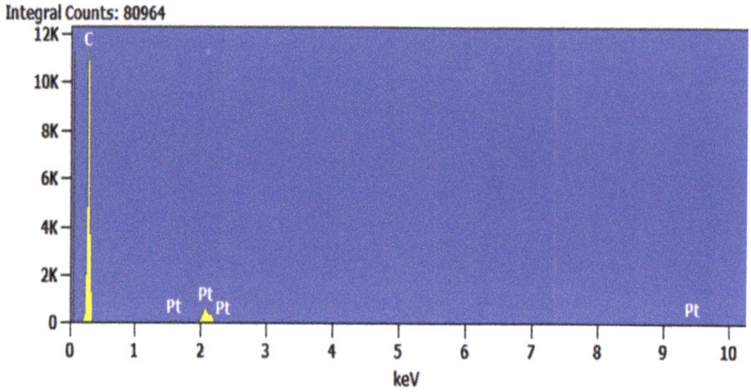

Figure 2. EDS spectra for PtFCS material.

PtFCS as well as fused carbon spheres with no nanoparticles on them were characterized via XRD analysis (Figure 3). A broad peak seen around 23 degrees was observed in the fused carbon sphere material, which is indicative of the graphitic characteristics of carbon-based materials [42]. Expectedly, this peak was also seen in the PtGLM composite since the fused carbon sphere support makes up a majority of the catalyst. The remaining peaks seen around 39.9°, 46.4°, 67.9°, and 81.6° all correspond to the (111), (200), (220), and (311) planes for the face-centered cubic structure of platinum nanoparticles (ICDD PDF 70-2431). The EDS spectra from Figure 2 confirm that there is platinum within the sample, and the TEM images from Figure 1 showed small nanoparticles; therefore, this XRD analysis further confirms that the metal present on the support material is platinum nanoparticles.

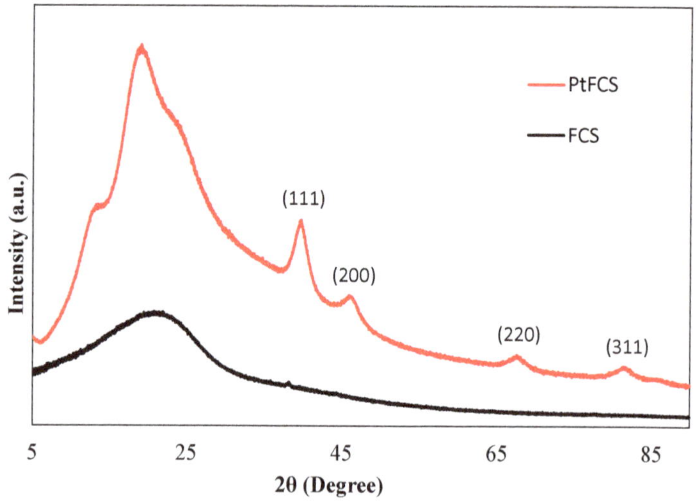

Figure 3. P-XRD analysis of PtFCS and FCS fused carbon sphere material.

The final method of characterization of the PtFCS material was through FTIR analysis (Figure 4). Fused carbon nanospheres that contained no nanoparticles were also analyzed for comparison purposes. There did not appear to be not much of a difference in the locations of the peaks seen in the two materials, which could indicate that the addition of nanoparticles did not significantly change the fused carbon sphere structure. Three major peaks were seen in the materials that are indicative of the dextrose used to prepare the fused carbon spheres. The broad stretch from 3600–3000 cm^{-1}, the small peak at 2900 cm^{-1}, and the peak at 1700 cm^{-1} represent the hydroxyl (OH), alkane (C-C), and carbonyl (C=O) functional groups of dextrose.

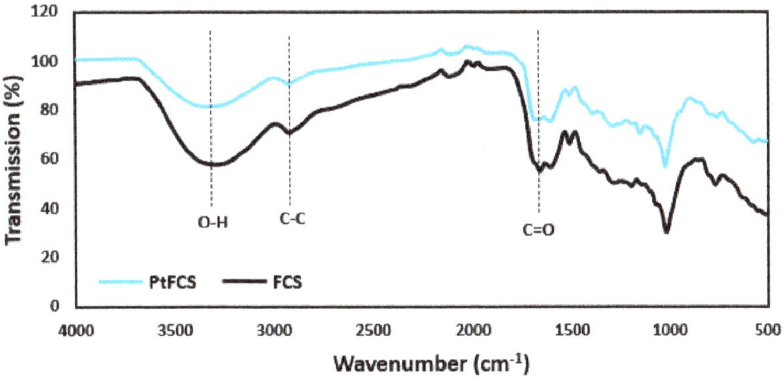

Figure 4. FTIR analysis of PtFCS and FCS fused carbon sphere material.

The novel PtFCS composite was tested for its catalytic ability at $NaBH_4$ doses of 625 µmol, 925 µmol, and 1225 µmol (Figure 5). At a dose of 625 µmol, the volume of hydrogen produced by the reaction was observed to be 10.7 mL, resulting in a generation rate of 0.0089 mL/mg_{cat}·min for the two-hour trial. The dose of the $NaBH_4$ was then raised to 925 µmol which increased the volume of hydrogen generated and the generation rate to 21.2 mL and 0.0180 mL/mg_{cat}·min, respectively. Finally, the dose was further raised to 1225 µmol the volume and rate further increased to 26.6 mL and 0.0222 mL/mg_{cat}·min. It is clear from these data that as the dosage of $NabH_4$ is increased, so did the amount of hydrogen generated from the reaction. Based on Equation (1), this result follows the equilibrium law as the increase in the dose of a reactant shifts the reaction to produce more product.

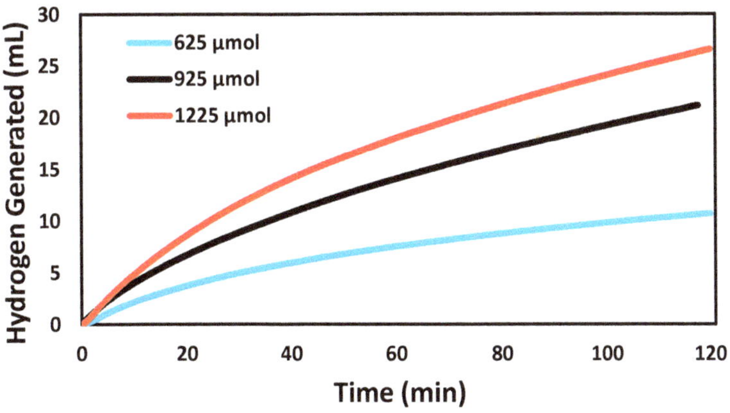

Figure 5. PtFCS catalyzed hydrolysis of $NaBH_4$ at various doses.

PtFCS was next tested for its catalytic ability under acidic conditions (pH 6), neutral conditions (pH 7), and basic conditions (pH 8) (Figure 6). The reaction under acidic conditions or pH 6 was observed to produce 52.5 mL of hydrogen gas at a rate of 0.0438 mL/mg_{cat}·min. A solution pH of 7 or neutral conditions resulted in a hydrogen generation rate of 0.0180 mL/mg_{cat}·min and a volume of 21.2 mL of hydrogen gas. When the pH of the solution was raised to a basic pH of 8, a decrease in hydrogen generation was observed, with only 4.9 mL of hydrogen gas being generated after two hours of reaction time, resulting in a reaction rate of 0.0041 mL/mg_{cat}·min. Based on these results, it is evident that the higher the pH of the reaction solution, the less hydrogen gas would be generated and that the opposite would occur for lower pHs. These results are consistent with the work of previous teams that have explored this reaction [15,35].

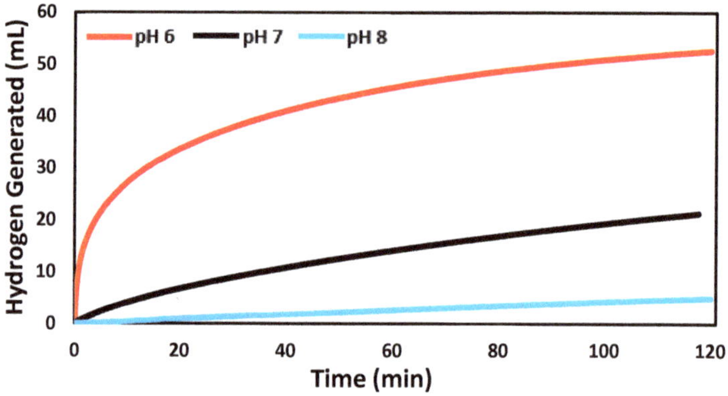

Figure 6. PtFCS catalyzed hydrolysis of $NaBH_4$ at various pH's.

The catalytic ability of PtFCS to hydrolyze $NaBH_4$ was also tested at the following temperatures: 283 K, 288 K, 295 K, and 303 K (Figure 7). The temperature of the reaction was first raised to 303 K, which resulted in a volume of hydrogen produced to be 30.8 mL at a rate of 0.0257 mL/mg_{cat}·min. At room temperature (295 K), the reaction produced 21.2 mL of hydrogen gas at a rate of 0.0180 mL/mg_{cat}·min. When cooled to 288 K, the hydrogen generation rate slowed to 0.0082 mL/mg_{cat}·min producing only 9.8 mL of hydrogen gas. Finally, when the reaction solution was further cooled to 283 K, the reaction appeared to slow even further, producing only 7.6 mL of hydrogen gas at a rate of 0.0063 mL/mg_{cat}·min. These results show a direct relationship between the temperature of the reaction solution and the volume of hydrogen gas that was produced. Since higher temperatures resulted in more hydrogen gas being produced, it was determined that this reaction is endothermic.

Once the temperature study (Figure 7) had been completed, the data could be used to find the activation energy of the reaction as catalyzed by PtFCS. Each temperature tested was entered into the Arrhenius Equation (2) where k is equal to the rate constant of the reaction at each temperature. A is equal to the pre-exponential factor, Ea is the activation energy of the reaction, R is the universal gas constant, and lastly, T represents the temperature tested.

$$k = Ae^{-\frac{Ea}{RT}} \quad (2)$$

The natural log of the rate constant (k) of the reaction at each tested temperature (T) was then plotted against 1000 divided by that temperature to create an Arrhenius plot (Figure 8). The equation of the line from this plot allowed for the activation energy of this reaction as catalyzed by PtFCS to be calculated as 53.0 kJ/mol. This activation energy was then compared to other catalysts for this reaction as seen in Table 1.

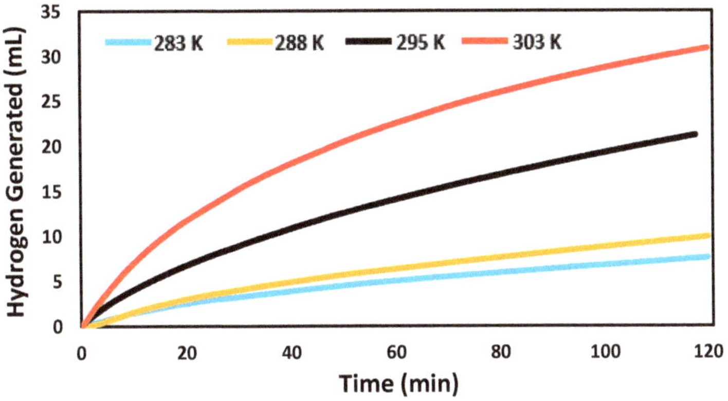

Figure 7. PtFCS catalyzed hydrolysis of $NaBH_4$ at varying temperatures.

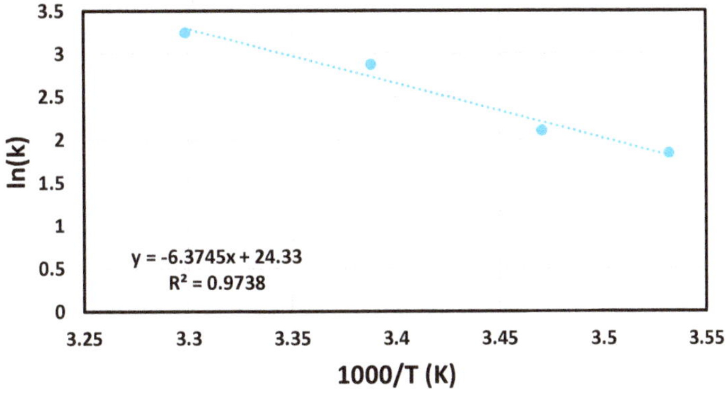

Figure 8. The Arrhenius plot created from the temperature data and the Arrhenius Equation (2).

Table 1. Comparison of Activation energies for the catalyzed hydrolysis of $NaBH_4$.

Catalyst	E_a (kJ mol^{-1})	Temperature (K)	Reference
Ni	71	273–308	[43]
Raney-Nickel	63	273–308	[43]
Co	75	273–308	[43]
Co/MWCNTs	63.8	303–318	[44]
CoB-zeolite-HCl	42.5	293–323	[45]
MCCC	64.3	303	[46]
Pt–Pd/CNTs	19	302–332	[47]
Au/MWCNTs	21.1	273–303	[48]
Ag/MWCNTs	44.5	273–303	[17]
Pd/MWCNTs	62.7	273–303	[19]
BCD-AuNP	54.7	283–303	[16]
PtNPs	39.2	283–303	[18]
Pd Nanocup	58.9	283–303	[49]
AgNPs	50.3	273–303	[50]
AgNP-FCS	37.0	273–303	[51]
Pt/C	45	298–313	[52]
PGO	N/A	298	[53]
PtFCS	53.0	283–303	This Work

When compared to other catalysts used for this hydrolysis reaction (Table 1), PtFCS shows relatively comparable activation energy. It has lower activation energy than bulk metal catalysts such as nickel, as well as certain transition metal composite catalysts such as Ag/MWCNTs, however, it is not as low as others. Despite this relatively average activation energy, many of the catalysts that have a lower activation energy use support materials that are not as sustainable as fused carbon spheres, since fused carbon spheres can be derived simply from dextrose.

The final catalytic study performed on PtFCS was its ability to be used multiple times consecutively. A standard trial was begun at pH 7, 295 K, and using 925 μmol of $NaBH_4$ which was run for a full two hours with magnetic stirring. After the first two hours had been completed, an additional 925 μmol $NaBH_4$ was added to the reaction vessel and quickly closed, marking the start of a second two-hour trial. This same method was then repeated for an additional three trials for a total of five trials (Figure 9). Across the five trials, there was an observed average volume of 30.5 mL of hydrogen gas produced per trial. It was noted that the first two trials produced similar volumes of hydrogen gas with each subsequent trial increasing amounts. One possible explanation is that the platinum nanoparticle supported on the fused carbon sphere catalyst was becoming more catalytically activated with each subsequent trial, a phenomenon previously reported by Deraedt et. al. 2014 [54]. They theorized that the BH_4^- forms strong hydridic bonds with the nanoparticles, stabilizing them and allowing for further catalyzation over the long term. After the reusability trials, the catalyst was collected and examined by P-XRD, FTIR, and SEM-EDS.

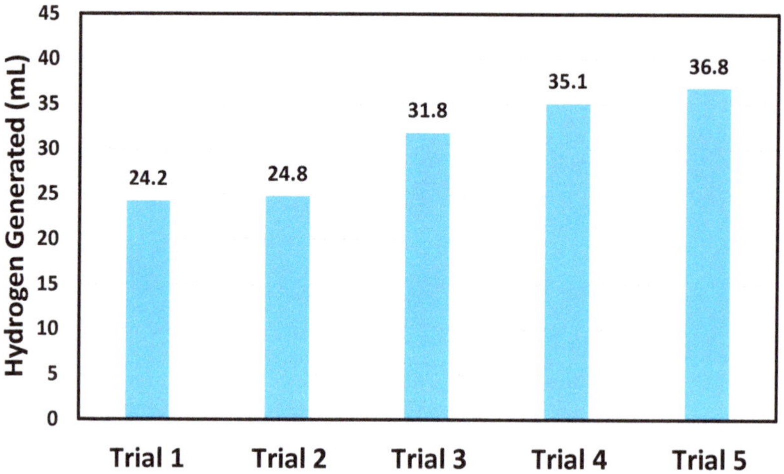

Figure 9. Catalytic reusability of PtFCS after five consecutive trials.

The P-XRD spectra of PtFCS after reusability experiments are shown in Supplementary Materials Figure S1. The broad peaks at 23° indicated the graphitic characteristics of carbon-based materials. The peaks at 39°, 46°, 67°, and 81°, which correspond to the (111), (200), (220), and (311) lattice planes of platinum nanoparticles were smaller. As discussed above, it could be due to BH_4^- forming strong bonds with nanoparticles, which affected the crystallinity structure. The inset of Figure S1 indicated the P-XRD of sodium borohydride. The signature peaks of sodium borohydride at 29°, 41°, and 48° were observed on the PtFCS after reusability trials [55].

The Supplementary Materials Figures S2 and S3 indicate the FTIR of sodium borohydride and PtFCS after reusability trials. Both B-H bending and stretching functional groups at 1327 cm^{-1}, 1095 cm^{-1}, and 2276 cm^{-1} were observed on the FTIR spectra of PtFCS. The functional groups O-H, C-C, and C=O of PtFCS were still retained after reusability trials.

The surface structure of PtFCS after reusability trials is shown in Figure S4. Through the mapping elements on the surface, it was observed that the distribution of boron was matched with that of platinum. It highly supported BH_4^- surrounding and stabilizing the platinum nanoparticles, which maintained their catalytic efficiency.

A proposed mechanism for the hydrolysis of $NaBH_4$ by the PtFCS catalyst is depicted in Scheme 1. First, a borohydride ion (BH_4^-) attaches itself to a platinum nanoparticle resting on the surface of the fused carbon sphere material. A nearby water molecule then attacks the boron from the borohydride and splits, leaving a hydroxyl group and releasing a hydrogen gas molecule. This can happen up to three more times, releasing four diatomic gas molecules in total. After the fourth hydrogen gas molecule is produced, the remaining tetrahydroxyborate [$B(OH)_4$] molecule can detach from the platinum nanoparticle and allow for another BH_4^- to take its place.

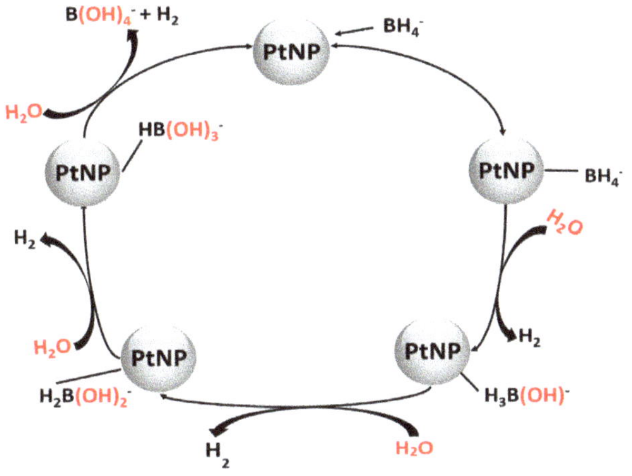

Scheme 1. Hydrolysis of $NaBH_4$ by PtFCS.

4. Conclusions

In conclusion, the successful synthesis of the novel PtFCS composite was confirmed through the use of FTIR, XRD, TEM, and EDS analysis. The catalyst was tested at various temperatures, pH conditions, and doses of $NaBH_4$ revealing that the reaction has an activation energy of 53.0 kJ/mol, which is competitive compared to similar catalysts. This catalyst produced the most hydrogen under the reaction conditions of pH 6, a temperature of 295 K, and a dosage of 925 µmol of $NaBH_4$. When tested for its structural stability, it was found that the same amount of catalyst could be used at least five times in a row with increasing volumes of hydrogen being produced with later trials. This could indicate that with each use, the catalyst becomes further activated. Since the fused carbon sphere backbone is synthesized from dextrose and the catalyst can be used multiple times without a decrease in hydrogen production, this novel catalyst shows promise as a sustainable way to produce hydrogen gas. This work can be expanded upon in a few ways. First, the conditions we already tested could be taken further, i.e., more temperatures and more pHs. Additionally, different metals such as gold, silver, and palladium could be tested to see how composite catalysts made from these metals compare to this metal and previous materials.

Supplementary Materials: The following supporting information can be downloaded at: https://www.mdpi.com/article/10.3390/nano13131994/s1, Figure S1: P-XRD of PtFCS after reusability trials. The inset showed the P-XRD of sodium borohydride. The red color showed the signature peaks of sodium borohydride. Figure S2: FTIR of sodium borohydride (NaBH4). Figure S3: FTIR of PtFCS after reusability trials. Figure S4: (a) SEM of PtFCS after reusability trials; (b) mapping of boron (B); (c) mapping of carbon (C); (d) mapping of platinum (Pt).

Author Contributions: Conceptualization, T.M.A.-F.; Methodology, T.M.A.-F.; Validation, T.M.A.-F.; Formal analysis, E.B. and Q.Q.; Resources, T.M.A.-F.; Data curation, E.B. and Q.Q.; Writing—original draft, E.B. and Q.Q.; Writing—review & editing, T.M.A.-F.; Visualization, T.M.A.-F.; Supervision, T.M.A.-F.; Funding acquisition, T.M.A.-F. All authors have read and agreed to the published version of the manuscript.

Funding: This research received no external funding.

Data Availability Statement: The data presented in this study are available on request from the corresponding author.

Acknowledgments: Corresponding Author acknowledges Lawrence J. Sacks Professorship in Chemistry.

Conflicts of Interest: The authors declare no conflict of interest.

References

1. Ritchie, H.; Roser, M. Energy-Our World in Data. 2014. Available online: https://ourworldindata.org/energy (accessed on 14 November 2022).
2. Rodhe, H.A. Comparison of the Contribution of Various Gases to the Greenhouse Effect. *Science* **1990**, *248*, 1217–1219. [CrossRef] [PubMed]
3. Shafiee, S.; Topal, E. When Will Fossil Fuel Reserves Be Diminished? *Energy Policy* **2009**, *37*, 181–189. [CrossRef]
4. Panwar, N.L.; Kaushik, S.C.; Kothari, S. Role of renewable energy sources in environmental protection: A review. *Renew. Sustain. Energy Rev.* **2011**, *15*, 1513–1524. [CrossRef]
5. Tucker, M. Can solar cooking save the forests? *Ecol. Econ.* **1999**, *31*, 77–89. [CrossRef]
6. Adamson, A.W.; Namnath, J.; Shastry, V.J.; Slawson, V. Thermodynamic inefficiency of conversion of solar energy to work. *J. Chem. Educ.* **1984**, *61*, 221. [CrossRef]
7. Ottinger, R.L. Experience with promotion of renewable energy: Successes and lessons learned. In Proceedings of the Parliamentarian Forum on Energy Legislation and Sustainable Development, Cape Town, South Africa, 5–7 October 2005.
8. US Government Accountability Office (GOA). *Wind Power: Impacts on Wildlife and Government Responsibilities for Regulating Development and Protecting Wildlife*; US Government Accountability Office: Washington, DC, USA, 2005.
9. O'Doherty, T.; O'Doherty, D.M.; Mason-Jones, A. Tidal Energy Technology. In *Wave Tidal Energy*; John Wiley & Sons Ltd.: New York, NY, USA, 2018; pp. 105–150.
10. Anderson, A.; Rezaie, B. Geothermal technology: Trends and potential role in a sustainable future. *Appl. Energy* **2019**, *248*, 18–34. [CrossRef]
11. Li, K.; Bian, H.; Liu, C.; Zhang, D.; Yang, Y. Comparison of geothermal with solar and wind power generation systems. *Renew. Sustain. Energy Rev.* **2015**, *42*, 1464–1474. [CrossRef]
12. Daut, I.; Razliana, A.R.N.; Irwan, Y.M.; Farhana, Z. A Study on the Wind as Renewable Energy in Perlis, Northern Malaysia. *Energy Procedia* **2012**, *18*, 1428–1433. [CrossRef]
13. Kabir, E.; Kumar, P.; Kumar, S.; Adelodun, A.A.; Kim, K.-K. Solar energy: Potential and future prospects. *Renew. Sustain. Energy Rev.* **2018**, *82*, 894–900. [CrossRef]
14. Bosnjakovic, M.; Sinaga, N. The Perspective of Large-Scale Production of Algae Biodiesel. *Appl. Sci.* **2020**, *10*, 8181. [CrossRef]
15. Schlesinger, H.I.; Brown, H.C.; Finholt, E.; Gilbreath, J.R.; Hoekstra, H.R.; Hyde, E.K. Sodium Borohydride, Its Hydrolysis and its Use as a Reducing Agent and in the Generation of Hydrogen. *J. Am. Chem. Soc.* **1953**, *75*, 215–219. [CrossRef]
16. Quach, Q.; Biehler, E.; Elzamzami, A.; Huff, C.; Long, J.M.; Abdel-Fattah, T.M. Catalytic Activity of Beta-Cyclodextrin-Gold Nanoparticles Network in Hydrogen Evolution Reaction. *Catalysts* **2021**, *11*, 118. [CrossRef]
17. Huff, C.; Long, J.M.; Aboulatta, A.; Heyman, A.; Abdel-Fattah, T.M. Silver Nanoparticle/Multi-Walled Carbon Nanotube Composite as Catalyst for Hydrogen Production. *ECS J. Solid State Sci. Technol.* **2017**, *6*, M115–M118. [CrossRef]
18. Huff, C.; Biehler, E.; Quach, Q.; Long, J.M.; Abdel-Fattah, T.M. Synthesis of Highly Dispersive Platinum Nanoparticles and their Application in a Hydrogen Generation Reaction. *Colloids Surf. A Physicochem. Eng. Asp.* **2021**, *610*, 125734. [CrossRef]
19. Huff, C.; Long, J.M.; Heyman, A.; Abdel-Fattah, T.M. Palladium Nanoparticle Multiwalled Carbon Nanotube Composite as Catalyst for Hydrogen Production by the Hydrolysis of Sodium Borohydride. *ACS Appl. Energy Mater.* **2018**, *1*, 4635–4640. [CrossRef]
20. Wittstock, A.; Neumann, B.; Schaefer, A.; Dumbuya, K.; Kübel, C.; Biener, M.M.; Zielasek, V.; Steinrück, H.P.; Gottfried, J.P.; Biener, J.; et al. Nanoporous Au: An Unsupported Pure Gold Catalyst? *J. Phys. Chem. C* **2009**, *113*, 5593–5600. [CrossRef]
21. Li, Z.; Fu, J.-Y.; Feng, Y.; Dong, C.-K.; Liu, H.; Du, X.-W. A silver catalyst activated by stacking faults for the hydrogen evolution reaction. *Nat. Catal.* **2019**, *2*, 1107–1114. [CrossRef]
22. Periana, R.A.; Taube, D.J.; Gamble, S.; Taube, H.; Satoh, T.; Fujii, H. Platinum Catalysts for the High-Yield Oxidation of Methane to a Methanol Derivative. *Science* **1998**, *280*, 560–564. [CrossRef]
23. Xia, R.; Zhang, S.; Ma, X.; Jiao, F. Surface-Functionalized Palladium Catalysts for Electrochemical CO_2 Reduction. *J. Mater. Chem. A* **2020**, *8*, 15884–15890. [CrossRef]
24. Dushatinski, T.; Huff, C.; Abdel-Fattah, T.M. Characterization of electrochemically deposited films from aqueous and ionic liquid cobalt precursors toward hydrogen evolution reactions. *Appl. Surf. Sci.* **2016**, *385*, 282–288. [CrossRef]

25. Osborne, J.; Horton, M.R.; Abdel-Fattah, T.M. Gold Nanoparticles Supported Over Low-Cost Supports for Hydrogen Generation from a Hydrogen Feedstock Material. *ECS J. Solid State Sci. Technol.* **2020**, *9*, 071004. [CrossRef]
26. Huff, C.; Dushatinski, T.; Barzanjii, A.; Abdel-Fattah, N.; Barzanjii, K.; Abdel-Fattah, T.M. Pretreatment of gold nanoparticle multi-walled carbon nanotube composites for catalytic activity toward hydrogen generation reaction. *ECS J. Solid State Sci. Technol.* **2017**, *6*, M69–M71. [CrossRef]
27. Nandanapalli, K.R.; Mudusu, D.; Lee, S. Defects-free single-crystalline zinc oxide nanostructures for efficient photoelectrochemical solar hydrogen generation. *Int. J. Hydrogen Energy* **2020**, *45*, 27279–27290. [CrossRef]
28. Kumar, S.; Kumar, A.; Kumar, A.; Krishnan, V. Nanoscale zinc oxide based heterojunctions as visible light active photocatalysts for hydrogen energy and environmental remediation. *Catal. Rev.-Sci. Eng.* **2019**, *62*, 346–405. [CrossRef]
29. Gao, Z.; Chen, K.; Wang, L.; Bai, B.; Liu, H.; Wang, Q. Aminated flower-like $ZnIn_2S_4$ coupled with benzoic acid modified g-C_3N_4 nanosheets via covalent bonds for ameliorated photocatalytic hydrogen generation. *Appl. Catal. B* **2020**, *268*, 118462. [CrossRef]
30. Quach, Q.; Abdel-Fattah, T.M. Silver Nanoparticles functionalized Nanoporous Silica Nanoparticle grown over Graphene Oxide for enhancing Antibacterial effect. *Nanomaterials* **2022**, *12*, 3341. [CrossRef]
31. Biehler, E.; Whiteman, R.; Lin, P.; Zhang, K.; Baumgart, H.; Abdel-Fattah, T.M. Controlled Synthesis of ZnO Nanorods Using Different Seed Layers. *ECS J. Solid State Sci. Technol.* **2020**, *9*, 121008. [CrossRef]
32. Dushatinski, T.; Abdel-Fattah, T.M. Carbon Nanotube Composite Mesh Film with Tunable Optoelectronic Performance. *ECS J. Solid State Sci. Technol.* **2015**, *4*, M1–M5. [CrossRef]
33. Abdel-Fattah, T.M.; Wixtrom, A.; Zhang, K.; Baumgart, H. Highly Uniform Self-Assembled Gold Nanoparticles over High Surface Area Dense ZnO Nanorod Arrays as Novel Surface Catalysts *ECS J. Solid State Sci. Technol.* **2014**, *3*, M61–M64. [CrossRef]
34. Abdel-Fattah, T.M.; Wixtrom, A. Catalytic Reduction of 4-Nitrophenol Using Gold Nanoparticles Supported On Carbon Nanotubes. *ECS J. Solid State Sci. Technol.* **2014**, *3*, M18–M20. [CrossRef]
35. Sabourault, N.; Mignani, G.; Wagner, A.; Mioskowski, C. Platinum Oxide (PtO_2): A Potent Hydrosilylation Catalyst. *Org. Lett.* **2002**, *4*, 2117–2119. [CrossRef]
36. Ely, J.C.; Neal, C.R.; Kulpa, C.F.; Schneegurt, M.A.; Seidler, J.A.; Jain, J.C. Implications of Platinum-Group Element Accumulation along U.S. Roads from Catalytic-Converter Attrition. *Envir. Sci. Technol.* **2001**, *35*, 3816–3822. [CrossRef] [PubMed]
37. Abu-Reziq, R.; Wang, D.; Post, M.; Alper, H. Platinum Nanoparticles Supported on Ionic Liquid-Modified Magnetic Nanoparticles: Selective Hydrogenation Catalysts. *Adv. Synth. Catal.* **2007**, *349*, 2145–2150. [CrossRef]
38. Fang, B.; Chaudhari, N.K.; Kim, M.-S.; Kim, J.H.; Yu, J.-S. Homogeneous Deposition of Platinum Nanoparticles on Carbon Black for Proton Exchange Membrane Fuel Cell. *J. Am. Chem. Soc.* **2009**, *131*, 15330–15338. [CrossRef] [PubMed]
39. Huff, C.; Quach, Q.; Long, J.M.; Abdel-Fattah, T.M. Nanocomposite Catalyst Derived from Ultrafine Platinum Nanoparticles and Carbon Nanotubes for Hydrogen Generation. *ECS J. Solid State Sci. Technol.* **2020**, *9*, 101008. [CrossRef]
40. Liu, J.; Zhou, Z.; Zhao, X.; Xin, Q.; Sun, G.; Yi, B. Studies on Performance Degradation of a Direct Methanol Fuel Cell (DMFC) in Life Test. *Phys. Chem. Chem. Phys.* **2004**, *6*, 134–137. [CrossRef]
41. Zhu, Y.; Murali, S.; Cai, W.; Li, X.; Suk, J.W.; Potts, J.R.; Ruoff, R.S. Graphene and Graphene Oxide: Synthesis, Properties, and Applications. *Adv. Mater.* **2010**, *22*, 3906–3924. [CrossRef]
42. Li, K.; Liu, Q.; Cheng, H.; Hu, M.; Zhang, S. Classification and carbon structural transformation from anthracite to natural coaly graphite by XRD, Raman spectroscopy, and HRTEM. *Spectrochim. Acta A* **2020**, *249*, 119286. [CrossRef]
43. Kaufman, C.M.; Sen, B. Hydrogen generation by hydrolysis of sodium tetrahydroborate: Effects of acids and transition metals and their salts. *J. Chem. Soc. Dalton Trans.* **1985**, *2*, 307–313. [CrossRef]
44. Narasimharao, K.; Abu-Zied, B.M.; Alfaifi, S.Y. Cobalt oxide supported multi wall carbon nanotube catalysts for hydrogen production via sodium borohydride hydrolysis. *Int. J. Hydrogen Energy* **2020**, *46*, 6404–6418. [CrossRef]
45. Saka, C.; Salih Eygi, M.; Balbay, A. CoB doped acid modified zeolite catalyst for enhanced hydrogen release from sodium borohydride hydrolysis. *Int. J. Hydrogen Energy* **2020**, *45*, 15086–15099. [CrossRef]
46. Lin, K.-Y.A.; Chang, H.-A. Efficient hydrogen production from $NaBH_4$ hydrolysis catalyzed by a magnetic cobalt/carbon composite derived from a zeolitic imidazolate framework. *Chem. Eng. J.* **2016**, *296*, 243–251. [CrossRef]
47. Peña-Alonso, R.; Sicurelli, A.; Callone, E.; Carturan, G.; Raj, R. A picoscale catalyst for hydrogen generation from $NaBH_4$ for fuel cells. *J. Power Sources* **2007**, *165*, 315–323. [CrossRef]
48. Huff, C.; Dushatinski, T.; Abdel-Fattah, T.M. Gold Nanoparticle/Multi-Walled Carbon Nanotube Composite as Novel Catalyst for Hydrogen Evolution Reactions. *Int. J. Hydrogen Energy* **2017**, *42*, 18985–18990. [CrossRef]
49. Biehler, E.; Quach, Q.; Huff, C.; Abdel-Fattah, T.M. Organo-Nanocups Assist the Formation of Ultra-Small Palladium Nanoparticle Catalysts for Hydrogen Evolution Reaction. *Materials* **2022**, *15*, 2692. [CrossRef] [PubMed]
50. Huff, C.; Long, J.M.; Abdel-Fattah, T.M. Beta-Cyclodextrin Assisted Synthesis of Silver Nanoparticle Network and its Application in a Hydrogen Generation Reaction. *Catalysts* **2020**, *10*, 1014. [CrossRef]
51. Biehler, E.; Quach, Q.; Abdel-Fattah, T.M. Silver Nanoparticle-Decorated Fused Carbon Sphere Composite as a Catalyst for Hydrogen Generation. *Energies* **2023**, *16*, 5053. [CrossRef]
52. Guella, G.; Patton, B.; Miotello, A. Kinetic Features of the Platinum Catalyzed Hydrolysis of Sodium Borohydride from 11B NMR Measurements. *J. Phys. Chem. C* **2007**, *111*, 18744–18750. [CrossRef]
53. Zhang, H.; Feng, X.; Cheng, L.; Hou, X.; Li, Y.; Han, S. Non-noble Co anchored on nanoporous graphene oxide, as an efficient and long-life catalyst for hydrogen generation from sodium borohydride. *Colloids Surf.* **2018**, *563*, 112–119. [CrossRef]

54. Deraedt, C.; Salmon, L.; Gatard, S.; Ciganda, R.; Hernandez, E.; Ruiz, J.; Astruc, D. Sodium borohydride stabilizes very active gold nanoparticle catalysts. *Chem. Commun.* **2014**, *50*, 14194–14196. [CrossRef]
55. Fang, Y.; Zhang, J.; Hua, M.Y.; Zhou, D.W. Modifying effects and mechanisms of graphene on dehydrogenation properties of sodium borohydride. *J. Mater. Sci.* **2020**, *55*, 1959–1972. [CrossRef]

nanomaterials

Article

Effect of the Elaboration Method on Structural and Optical Properties of $Zn_{1.33}Ga_{1.335}Sn_{0.33}O_4$:0.5%$Cr^{3+}$ Persistent Luminescent Nanomaterials

Guanyu Cai [1,2,†], Luidgi Giordano [1,†], Cyrille Richard [2,*] and Bruno Viana [1,*]

1 Chimie ParisTech, CNRS, Institut de Recherche de Chimie Paris (IRCP), Université PSL, 75005 Paris, France; guanyu.cai@chimieparistech.psl.eu (G.C.); luidgi.giordano@chimieparistech.psl.eu (L.G.)
2 Université Paris Cité, CNRS, INSERM, Unité de Technologies Chimiques et Biologiques pour la Santé (UTCBS), Faculté de Pharmacie, 75006 Paris, France
* Correspondence: cyrille.richard@u-paris.fr (C.R.); bruno.viana@chimieparistech.psl.eu (B.V.)
† These authors contributed equally to this work.

Abstract: Near-infrared (NIR) persistent luminescence (PersL) materials have demonstrated promising developments for applications in many advanced fields due to their unique optical properties. Both high-temperature solid-state (SS) or hydrothermal (HT) methods can successfully be used to prepare PersL materials. In this work, $Zn_{1.33}Ga_{1.34}Sn_{0.33}O_4$:0.5%$Cr^{3+}$ (ZGSO:0.5%Cr^{3+}), a newly proposed nanomaterial for bioimaging, was prepared using SS and HT methods. The results show the crystal structure, morphology and optical properties of the samples that were prepared using both methods. Briefly, the crystallite size of the ZGSO:0.5%Cr^{3+} prepared using the SS method is ~3 μm, and as expected, is larger than materials prepared using the HT method. However, the growth process used in the hydrothermal environment promotes the formation of ZGSO:0.5%Cr^{3+} with more uniform shapes and smaller sizes (less than 500 nm). Different diameter ranges of nanoparticles were obtained using HT and ball milling (BM) methods (ranging from 25–50 nm) and by using SS and BM methods (25–200 nm) as well. In addition, the SS-prepared microstructure material has stronger PersL than HT-prepared particles before they go through ball milling to create nanomaterials. On the contrary, after BM treatment, ZGSO:0.5%Cr^{3+} HT and BM NPs present higher PersL and photoluminescence (PL) properties than ZGSO:0.5%Cr^{3+} SS and BM NPs, even though both kinds of NPs present worse PersL and PL compared to the original particles before BM. To summarize: preparation methods, whether by SS or HT, with additional grinding as a second step, can have a significant impact on the morphological and luminescent features of ZGSO:0.5%Cr^{3+} PersL materials.

Keywords: persistent luminescence; phosphors; nanomaterials; $Zn_{1.33}Ga_{1.335}Cr_{0.005}Sn_{0.33}O_4$; hydrothermal; solid state

Citation: Cai, G.; Giordano, L.; Richard, C.; Viana, B. Effect of the Elaboration Method on Structural and Optical Properties of $Zn_{1.33}Ga_{1.335}Sn_{0.33}O_4$:0.5%$Cr^{3+}$ Persistent Luminescent Nanomaterials. *Nanomaterials* **2023**, *13*, 2175. https://doi.org/10.3390/nano13152175

Academic Editors: Andrei Honciuc and Mirela Honciuc

Received: 29 June 2023
Revised: 21 July 2023
Accepted: 24 July 2023
Published: 26 July 2023

1. Introduction

Persistent luminescence (PersL) refers to the phenomenon where a material continues to emit light after the excitation source has been removed [1]. PersL is applied in many fields, such as anti-counterfeiting [2,3], information storage [4–8], thermal sensors [9], lighting [10–15] and imaging [16–19]. In terms of imaging, the development of nanoprobes that can emit deep-red or near-infrared (NIR) light has attracted the attention of the scientific community [20–26]. Specifically, NIR PersL is an emerging imaging modality that is quickly gaining popularity in the field of bioimaging, due to the key advantage of PersL being able to provide a long-term, low-background signal, making it ideal for sensitive imaging applications [27–32]. PersL can be easily activated with UV/visible light and X-rays, making it a versatile tool for imaging in a range of environments and conditions [33–36].

We first reported these persistent luminescence (PersL) $Zn_{(1+x)}Ga_{(2-2x)}Sn_xO_4$:$Cr^{3+}$ spinel materials at micrometric size in 2020 [37], following our previous results obtained on

$ZnGa_2O_4:Cr^{3+}$ in 2011 [38]. Past works mainly focused on the Cr^{3+}-doped gallate materials, such as $ZnGa_2O_4$: Cr^{3+}(ZGO) [18,39–43], $Zn_{(1+x)}Ga_{(2-2x)}Ge_xO_4$: Cr^{3+}(ZGGO) [44] and $Zn_{(1+x)}Ga_{(2-2x)}Sn_xO_4:Cr^{3+}$(ZGSO) [37]. Furthermore, recent papers dealing with spinel materials doped with Ni, Yb and Er ions have recently been reported [3,8,24,45]. Important applications of PersL were documented for in vitro and in vivo imaging, and these applications were widely developed through the respective authors' research laboratories [46–48]. Cr^{3+} is an ideal candidate that provides the deep-red/near-infrared PersL at ~700 nm due to the $^2E \rightarrow {}^4A_2$ transition, with some contribution of a large band in longer wavelengths from the $^4T_2 \rightarrow {}^4A_2$ transition [49–51]. Through Sn^{4+}-doping, the persistent luminescent performance of Cr^{3+}-doped zinc gallate is improved due to the increase of spinel anti-site defects, especially for visible-light charging [37], which makes it very promising for bioimaging applications. Some researchers have worked on different synthesis methods, however, there are no reports regarding the comparison of various synthesis methods within the same paper, namely the solid-state (SS) synthesis method with ball milling (BM) and hydrothermal synthesis (HT) method with BM. Furthermore, PersL is quite interesting to study as the property is very sensitive to the defects of crystalline quality. Indeed, traps and recombination centers should be carefully controlled.

In this study, we prepare $Zn_{1.33}Ga_{1.335}Sn_{0.33}O_4:0.5\%Cr^{3+}$ (ZGSO:0.5%Cr^{3+}) PersL materials using both SS and HT methods, as shown in Scheme 1. Even though these two methods have been widely used for other materials in the industry, they have yet to be applied to ZGSO:0.5%Cr^{3+}. We then report our comparison of the structure, morphology and optical properties, and use BM to downsize these compounds to nanoscale. Following this procedure, we provide details on the optical features of ZGSO:0.5%Cr^{3+} nanomaterials. Finally, in order to move toward applications, the properties of ZGSO:0.5%Cr^{3+} while suspended in water are presented, including concentration and size effects. This paper suggests that ZGSO:0.5%Cr^{3+} nanoparticles could be valuable probes for further bioimaging applications, using the appropriate synthesis method.

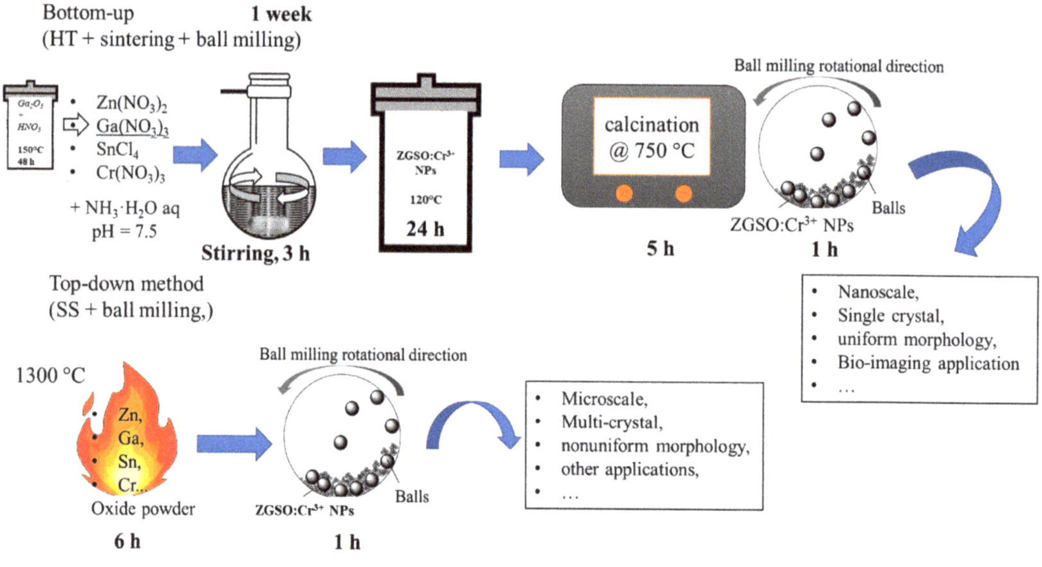

Scheme 1. $_{33}Ga_{1.335}Sn_{0.33}O_4:0.5\%Cr^{3+}$ PersL phosphors, their properties and suggested applications.

2. Materials and Methods

2.1. Materials

For the solid state method: zinc oxide (99.99%, Strem Chemicals, Newburyport, MA, USA), gallium oxide (99.99%, Strem Chemicals, Newburyport, MA, USA), tin oxide

(99.99%, Strem Chemicals, Newburyport, MA, USA), chromium oxide (99.99%, Alfa Aesar, Karlsruhe, Germany).

For the hydrothermal method: zinc nitrate hexahydrate (>99%, Alfa Aesar, Karlsruhe, Germany), gallium oxide (99.999%, Alfa Aesar, Karlsruhe, Germany), nitric acid (35 wt%), tin chloride pentahydrate (98%, Alfa Aesar, Karlsruhe, Germany), chromium (III) nitrate nonahydrate (99.9%, Alfa Aesar, Karlsruhe, Germany), ammonia solution (30 wt%), hydrochloric acid (50 mmol L^{-1}). All the chemicals were used as received without further purification.

2.2. Hydrothermal (HT) Method

The synthesis of ZGSO:0.5%Cr^{3+} refers to our previous reports [19,41], in which the tin concentration was optimized. In this method, 8.94 mmol of gallium oxide (99.999%, Alfa Aesar) was dissolved in 10 mL of concentrated nitric acid (35 wt%) under hydrothermal conditions for 48 h at 150 °C. Next, 10 mL of a solution containing 17.82 mmol of zinc nitrate hexahydrate (>99%, Fluka), 4.42 mmol tin chloride pentahydrate (98%, Alfa Aesar) and 0.07 mmol chromium (III) nitrate nonahydrate (99.9%, Alfa Aesar) was added to the vessel under vigorous stirring. All the chemicals are designed corresponding to the stoichiometric composition of $Zn_{1.33}Ga_{1.335}Cr_{0.005}Sn_{0.33}O_4$ (ZGSO:0.5%Cr^{3+}). The pH level of the solution was adjusted to 7.5 using ammonia solution (30 wt%), the mixture was stirred for 3 h at room temperature and then transferred to a stainless-steel autoclave, in which it was kept for 24 h at 120 °C. Under these conditions, the hydrothermal synthesis method ranges from about 10^5–10^7 Pa (1–100 bars) considering hydro and solvothermal methods, but at a temperature of about 120 °C the pressure does not exceed 40 bars [52,53]. The resulting solid was washed several times with water and ethanol and subsequently dried at 60 °C for 2 h. 1 g of the obtained solid was sintered in air at 750 °C for 5 h in a tubular oven.

2.3. Solid State (SS) Method

The ZGSO:0.5%Cr^{3+} powder sample was synthesized by a high-temperature solid-state method beginning with binary oxides. According to the stoichiometric ratio of the compounds, the raw material mixture was prepared. Briefly, 17.82 mmol of ZnO, 8.94 mmol of Ga_2O_3, 4.42 mmol of SnO_2 and 0.035 mmol of Cr_2O_3, corresponding to the stoichiometric composition of $Zn_{1.33}Ga_{1.335}Cr_{0.005}Sn_{0.33}O_4$ (ZGSO:0.5%Cr^{3+}), were weighed and ground homogenously in an agate mortar and the mixture was introduced into an alumina crucible. Next, the crucible was placed in a high-temperature furnace at 1300 °C for 6 h in the air to produce the final sample [38]. After cooling to room temperature, the phosphors were ground into fine powders.

2.4. BM Treatment

The ZGSO:0.5%Cr^{3+} powders (~500 mg) prepared using both methods were crushed by a Pulverisette 7 Fritsch Planetary Ball Mill, with the addition of 1 mL of 5 mM HCl solution, at the speed of 1000 rpm for 1 h. The mixture was transferred into a round-bottom flask and vigorously stirred for 24 h at room temperature. The final ZGSO:0.5%Cr^{3+} NPs were selected from the polydisperse colloidal via centrifugation on a SANYO MSE Mistral 1000 (SpectraLab Scientific Inc. 38 McPherson St. Markham, ON, Canada) at 3500 rpm for 5 min. To optimize the ball milling conditions using the Pulverisette 7 Fritsch Planetary Ball Mill program, it is possible to fine tune the speed (between 0~1000 rpm) and time (0~4 h). By changing these parameters, we found the best conditions (at a speed of 1000 rpm for 1 h), as proposed in the manuscript.

2.5. Pellets Preparation for Spectroscopy Measurement

For all dry powder luminescence measurements via spectroscopy, a pellet containing 60 mg of the prepared material and 180 mg of KBr was prepared. We compacted the powder to form a whole pellet body with ϕ10 mm, with 2 mm thickness under uniaxial pressure of 5 MPa. Pellet samples are further used for optical spectroscopy measurements.

2.6. Characterization

2.6.1. X-ray Diffraction (XRD)

XRD of ZGSO:0.5%Cr^{3+} was performed on an X-ray diffractometer (XPert PRO, PANalytical, Malvern Panalytical Ltd., Malvern, UK) equipped with a Ge111 single crystal monochromator selecting the $K_{\alpha 1}$ radiation wavelength of the Cu X-ray tube (0.15405 nm).

2.6.2. Transmission Electron Microscopy (TEM)

Observations of ZGSO:0.5%Cr^{3+} particles were carried out on a FEI® Tecnai Spirit G2 TEM (ThermoFisher Scientific Inc., Hillsboro, OH, USA) working with an acceleration voltage of 120 kV. For the analysis, one drop of particle suspension is deposited on a carbon film-coated copper grid.

2.6.3. Photoluminescence (PL)

Visible and deep red (or NIR-I) photoluminescence (PL) measurements of dry ZGSO:0.5%Cr^{3+} microstructured materials and NPs were performed using a CCD camera cooled at −65 °C and coupled to a monochromator (Acton Spectra Pro 2500, Princeton Instruments, Trenton, NJ, USA), at 300 grooves per mm, centered at 500 nm.

2.6.4. PL Excitation Spectroscopy

PL Excitation measurements of the dry ZGSO:0.5%Cr^{3+} microstructured materials and NPs were carried out using an Agilent Cary Eclipse UV/Vis spectrophotometer.

2.6.5. Persistent Luminescence (PersL)

The dry ZGSO:0.5%Cr^{3+} microstructured materials and NPs samples were thermally de-trapped at 70 °C before each experiment and then kept in the dark. The samples were excited with a UV lamp for 5 min at 290 K, and after cutting off the excitation, the afterglow was recorded for up to 60 min at the same temperature. The signal was followed with the same camera as in the PL experiment. Afterglow decay curves were obtained by integrating the intensity of the PersL spectra as a function of time.

2.6.6. Photoluminescence Lifetime (PL Lifetime)

To measure the PL lifetime (PL decay curves), we use a pulse EKSPLA laser (8 ns, 10 Hz) with tunable wavelengths from UV to NIR for the excitation, and for the detection, an intensified CCD camera (ICCD) cooled at −65 °C, coupled to a monochromator (Acton Spectra Pro, Princeton Instruments), with a grating of 300 grooves per mm, centered at 500 nm. The data of the decay curve were collected using Winspec software V2.6A for some PersL decay profiles we used either the Imax Roper Pixis camera with an integration time of 1 s (and measured the decay just after stopping the excitation) or the Optima Biospace Lab camera described below.

2.6.7. Thermoluminescence (TL)

Thermoluminescence (TL) measurements were performed using a closed-cycle He-flow cryostat (Sumitomo Cryogenics HC-4E) attached to a Lakeshore 340 temperature controller. The samples were cooled down to 10 K and irradiated with a UV lamp for 10 min. Next, the TL curves were recorded at a 10 K/min incline while heating up to 470 K. The signal was recorded using an ICCD camera (Roper Pixis 100) coupled to a visible monochromator (Acton Spectra Pro, Princeton Instruments), at 300 grooves per mm, centered at 500 nm).

2.6.8. Persistence Luminescence of ZGSO:0.5%Cr^{3+} NPs Suspension

Dry ZGSO:0.5%Cr^{3+} NPs were weighted and dispersed into DI water filling 2 mL Ependorf, (weighing, according to the targeted concentration, from 0.2 mg/mL to 2 mg/mL), via ultrasonic dispersion. ZGSO:0.5%Cr^{3+} NPs in suspension (series of concentration from 0.2 to 2 mg/mL) are transferred into a 96-well plate (0.3 mL for each suspension hole), and irradiated under a UV lamp for 2 min. The excited suspension was then observed using an Optima camera (Biospace Lab, Nesles-la-Vallée, France) for 5 min of acquisition in bioluminescence mode. After collection, M3-vision software was employed for data treatment.

3. Results and Discussion

3.1. Crystal Structure

ZGSO:0.5%Cr^{3+} possesses a normal spinel structure belonging to the cubic lattice system with two tetrahedral and four octahedral sites per formula unit, as shown in Figure S2 in the Supplementary Materials. The ZGSO matrix is comprised of cubic, close-packed oxides with space group *Fd-3m* [37]. Ga^{3+} cation size (0.76 Å, CN = VI) is smaller than the Zn^{2+} (0.88 Å, CN = VI) and Sn^{4+} (0.83 Å, CN = VI) cations, respectively. The radius of trivalent cations in octahedral site Cr^{3+} (0.755 Å) is close to that of Ga^{3+}, and therefore Cr^{3+} is inclined to occupy the octahedral Ga^{3+} sites. In this structure, a part of the Zn occupies the tetrahedral sites, while the rest of the Zn^{2+}, as well as all Ga^{3+}, and Sn^{4+} are expected to occupy the octahedral sites [41]. To confirm the crystal structure of our materials, Powder XRD measurements were conducted and analyzed, as shown in Figure 1. The XRD results show that ZGSO:0.5%Cr^{3+} can be successfully prepared using both HT and SS methods. All of the recorded diffraction peaks in the XRD pattern are basically consistent with the standard $ZnGa_2O_4$ card (ICSD No. 81105) with a slight shift in the diffraction peaks due to the presence of Sn^{4+} inside the structure with a larger radius than Ga^{3+}.

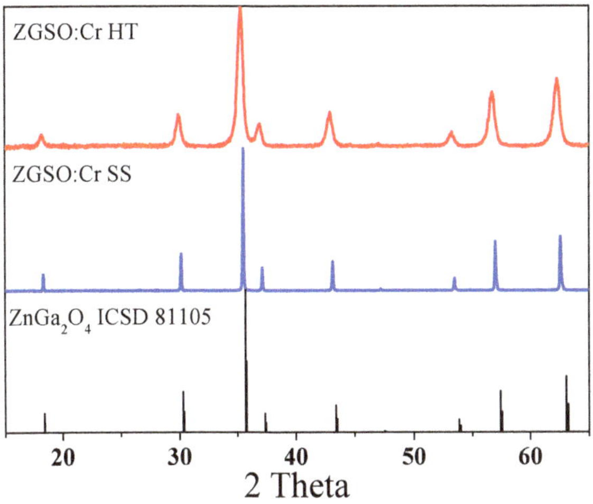

Figure 1. XRD patterns of $Zn_{1.33}Ga_{1.335}Sn_{0.33}O_4$:0.5%$Cr^{3+}$ (ZGSO:0.5%Cr^{3+}) prepared using solid-state (SS) or hydrothermal (HT) methods.

3.2. Microscopic Investigation

Transmission electron microscopy (TEM) was used to determine the microstructure of the ZGSO:0.5%Cr^{3+} samples (see Figure 2), prepared using both methods, ZGSO:0.5%Cr^{3+} SS (a–c) and ZGSO: Cr^{3+} HT (d–f). These microscopy images show that particle sizes of ZGSO:0.5%Cr^{3+} prepared using the solid-state method are larger than 1 μm. In fact, most of the particles are about 3 μm (see Figure 2a). Their irregular shape is due to a higher degree of agglomeration, and the formation of crystalline structures, as in the electron diffraction (ED) results of the polycrystal, could be approximated by an annular diffraction spot (in Figure 2c). On the contrary, the growth process used in the hydrothermal environment promotes the formation of ZGSO:0.5%Cr^{3+} HT materials with smaller sizes (less than 500 nm, not shown here), and more uniform shapes (in Figure 2d) when compared to the ones obtained using the SS method. The ED of the particles prepared using the HT method also shows the formation of single crystal-like nanostructures with regular shapes (in Figure 2f).

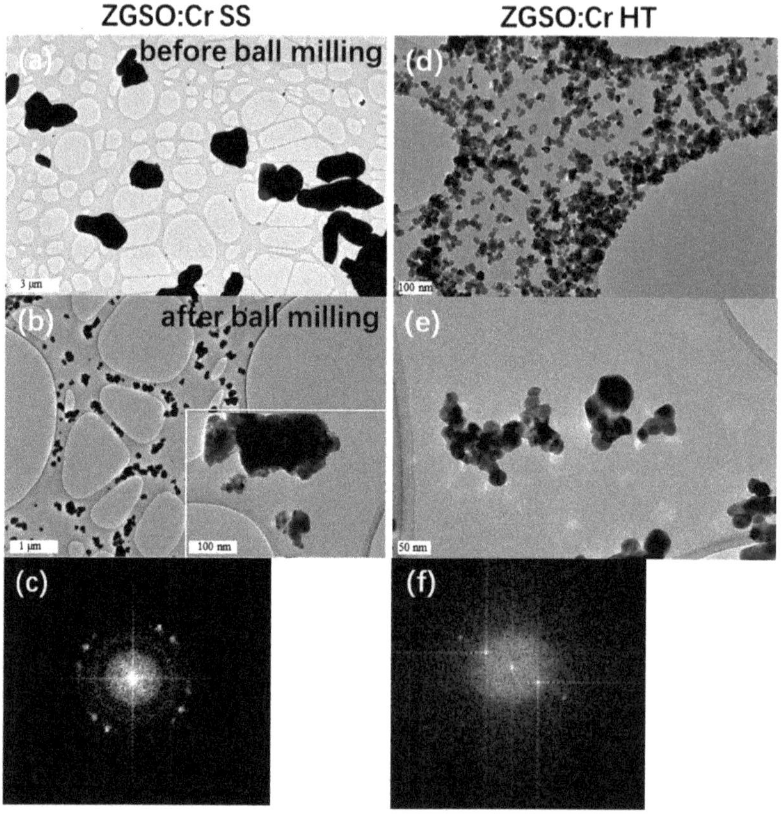

Figure 2. TEM images of ZGSO:0.5%Cr^{3+} (**a**) before and (**b**) after a BM treatment after SS method preparation, insert of (**b**) is part enlargement. (**c**) Electron diffraction (ED) of the ZGSO:0.5%Cr^{3+} SS. (**d**,**e**) TEM images of ZGSO:0.5%Cr^{3+} prepared by HT method after BM treatment. (**f**) Electron diffraction (ED) of the ZGSO:0.5%Cr^{3+} HT phosphors after BM, respectively.

To obtain nanoparticles from both methods, it is necessary to grind the particles a second time, to reduce the size of the grains. This leads to size dispersion, as seen in Figure 2b. Nanoparticles with different diameter ranges could be further extracted using SS and BM (sizes in the range of 25–200 nm) as seen in Figure 2b, and ranging from 25–50 nm for HT and BM, as seen in Figure 2d,e. Monodisperse and smaller average-sized nanoparticles are obtained for ZGSO:0.5%Cr^{3+} HT and BM in comparison to the ones prepared using solid-state and ball milling. Particles are selected via high-speed centrifugation (speed of 1000 rpm for 1 h). For example, ZGSO:0.5%Cr^{3+} NPs with average sizes of ~25, ~35 and ~50 nm can be separately selected for further measurements in Section 3.4. Usually, there is less than a 10% margin of error in determining the NPs sizes (for example, NPs diameter = 25 nm ± 2 nm; 50 nm ± 5 nm).

3.3. Optical Properties

The effect of the preparation method on the optical properties of the $Zn_{1.33}Ga_{1.335}Sn_{0.33}O_4$: 0.5%$Cr^{3+}$ (ZGSO:0.5%Cr^{3+}) was also analyzed. Normalized PL excitation spectra monitoring emission at 700 nm (shown in Figure 3a) presents large UV excitation peaks (at wavelengths below 370 nm) for all samples. This can be attributed to the band gap energy of these materials. The other two excitation peaks located at ~425 nm and ~580 nm, respectively, typical of the trivalent chromium ($^4A_2 \rightarrow ^4T_1$ and $^4A_2 \rightarrow ^4T_2$ transitions, respectively) can be observed as well. Throughout both synthesis methods and comparisons of excitation bands before and after BM, no band shift could be observed. This suggests that, in all cases, Cr^{3+} is under the same crystal field and environment. This indicates that Sn^{4+} content remains at the same level in both cases, as the introduction of Sn^{4+} instead of Ga^{3+} in this ZGSO matrix drastically modifies the crystal field around Cr^{3+}, and therefore its excitation wavelength, as reported by Pan et al. [37]. The presence of these bands also indicates that Cr^{3+} can be directly excited through the $^4A_2 \rightarrow ^4T_1$ (^4F) and $^4A_2 \rightarrow ^4T_2$ (^4F) transitions. After BM treatment, the excitation peaks of both ZGSO:0.5%Cr^{3+} samples do not move, which means that the decrease in size to nanoscale does not change the Cr^{3+} crystal field.

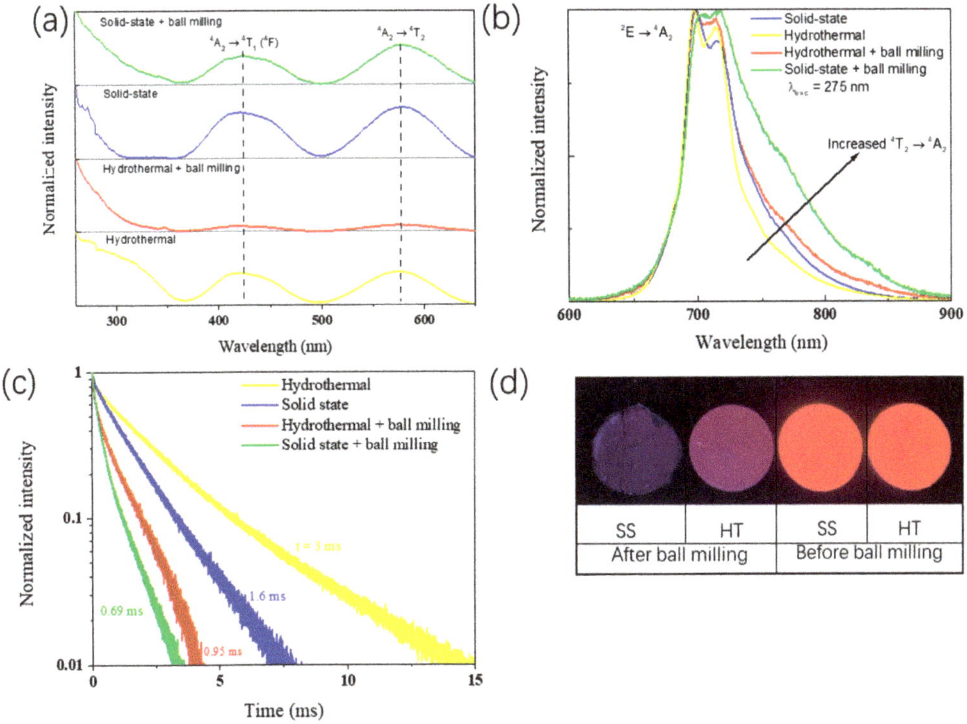

Figure 3. (**a**) PL Excitation with λ_{em} = 700 nm, (**b**) emission spectra with λ_{exc} = 275 nm, (**c**) lifetime curve (λ_{exc} = 450 nm), (**d**) PL digital images of the ZGSO:0.5%Cr^{3+} phosphors prepared using solid-state or hydrothermal methods, before and after BM treatment, in dry powder form during excitation with a UV lamp (λ_{exc} = 254 nm, room temperature).

As can be seen from the emission spectra obtained under excitation at 275 nm (UV excitation) in Figure 3b, both synthesized samples before BM have a strong emission peak at ~700 nm in the deep red (or NIR-I) range. While the emission peak is mainly caused by the $^2E \rightarrow ^4A_2$ transition of Cr^{3+}, there is a small contribution from the $^4T_2 \rightarrow ^4A_2$ transition, as these levels are in thermal equilibrium [38]. In relation to their emission bands, there is a higher PL intensity ratio of Cr^{3+} $^4T_2 \rightarrow ^4A_2$ transition over $^2E \rightarrow ^4A_2$ transition in the solid-state samples compared to the samples obtained using the HT method [54,55]. When samples were treated by BM for HT and SS, an increased contribution of the $^4T_2 \rightarrow ^4A_2$ transition was found. The increase of $^4T_2 \rightarrow ^4A_2$ to $^2E \rightarrow ^4A_2$ ratio after BM suggests the creation of new non-radiative energy loss paths after obtaining the nanoparticles, possibly due to the formation of surface defects, which is a side effect of the BM procedure and also explains the poor signal to noise ratio in samples after BM. In Figure 3d, the image obtained during UV excitation of the samples clearly shows that after BM there is a significant decrease in the PL intensity.

Lifetime was used for the evaluation of the optical properties of materials, with a focus on their efficiencies. For the ^2E emission level of Cr^{3+} in ZGSO, a lifetime value of around 3 ms was found for the material prepared using the HT method, while for the material prepared using the SS method, its lifetime value is around 1.6 ms when excited at 450 nm. Even after changing the excitation wavelengths through a large excitation energy range from 254 nm to 550 nm, Cr^{3+} ^2E manifold lifetime does not change significantly (Figure S1). It is at first quite surprising to find higher lifetime values for the material prepared using the HT method, as the calcination temperature is lower, which usually leads to lower crystallinity and the existence of additional non-radiative paths, decreasing the average lifetime. However, in PersL phosphors, as trapping is the most relevant effect, this suggests that during excitation, traps are being filled, which would reduce the lifetime of the manifold. This will be investigated in more detail in the following section (see Figure 4d) with thermoluminescence glow curves. In the case of ZGSO:0.5%Cr^{3+}, one must also consider that in the solid-state elaboration method, there is also a higher contribution of the $^4T_2 \rightarrow {}^4A_2$ transition, which also explains the lower lifetime value when compared to the samples prepared using the HT method. After BM, lifetime values for both preparation methods decrease. ^2E manifold lifetimes decrease from 3 ms to 0.95 ms for hydrothermal method and from 1.69 ms to 0.69 ms for solid-state one, respectively, corroborating the BM effect on the PersL intensity. As for the PL intensity, lifetime measurements show that the BM creates alternative, non-radiative pathways, as the ^2E manifold's lifetime decreases drastically following the formation of nanoparticles.

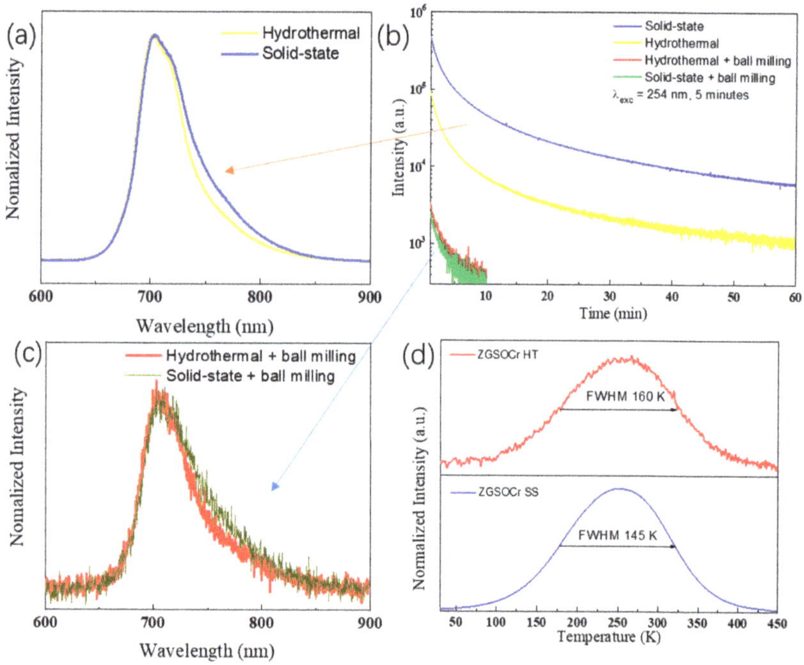

Figure 4. (**a**,**c**) PersL glow curves 5 s after ceasing UV excitation and (**b**) decay curves of ZGSO:0.5%Cr^{3+} phosphors prepared using solid-state or hydrothermal methods, before and after BM treatment. (**d**) thermoluminescence (TL) spectra of microscale ZGSO:0.5%Cr^{3+} phosphors prepared using solid-state or hydrothermal methods. All collections of PersL glow curves are started 5 s after stopping the 5 min UV excitation.

To evaluate the PersL properties of ZGSO:0.5%Cr^{3+} prepared using solid-state (SS) or hydrothermal (HT) methods, before and after BM treatment, their decay curves in a longer time range were collected (in Figure 4b) by monitoring Cr^{3+} emission at ~700 nm after 5 min of UV excitation. The nature of the antisite defects and mechanism of persistent luminescence for this family of materials is well known [36,37] and will not be discussed in detail in this work. When comparing the PersL of the four samples, samples prepared using the solid-state method appear to be the most effective. When comparing both materials before BM, the higher synthesis temperature of the material prepared using the solid-state method is responsible for a higher crystallinity, and due to the thermodynamic nature of the defect formation, there is also a higher concentration of controlled defects useful for PersL using this synthesis method. Controlled defects mean that the defect's depth is well adapted to the persistent luminescence, namely at about 0.5–0.7 eV below the conduction band. However, after BM and nanoparticles extraction, both materials have lost significant performance, with the one prepared using the HT method remaining slightly superior. The crucial effects of BM leading to performance losses can be attributed to the formation of quenching defects and non-radiative pathways, interfering with energy storage in controlled defects and antisites, for instance, favorable to PersL, thus decreasing their PersL performance [56]. In terms of emission profiles (see Figure 4a,c), the samples prepared by the solid-state method have a higher relative contribution of the $^4T_2 \rightarrow {}^4A_2$ emission, as also recognized from photoluminescence features.

As PersL is closely related to the existence of traps, the traps depth profile was studied by their thermoluminescence glow curves (λ_{exc} = 254 nm, Figure 4d). Notice that traps depths about 0.5–0.7 eV below the conduction band roughly correspond to peaks in the thermoluminescence glow curves around 250 to 350 K. For both materials prior to BM, the thermoluminescence glow curve shows maximum emission around 255 K. This, as shown previously for ZGSO [19,37], suggests that an important part of ZGSO's traps are outside the optimal temperature range for long persistent phosphors, being quickly thermalized at room temperature. However, this is indeed favorable for bio-imaging, as the intensity in the first minutes will therefore be enhanced [37]. In both cases, the bands are large, with full width at half maximum (FWHM) of 145 K and 160 K for solid state and hydrothermal methods, respectively. The broadness of this band is a consequence of the formation of $Sn_{Ga}{}^{\circ}$ and $Zn_{Ga}{}'$ defects acting as electron and hole traps, respectively, when Sn^{4+} is added to $ZnGa_2O_4$ [37]. In terms of absolute intensity, through signal-to-noise ratio comparison, the sample prepared by solid-state has a much higher thermoluminescence intensity. This is in correlation with the higher PersL of this sample, and is likely related, as stated previously, to the thermodynamics of defect formation at higher temperatures, allowing more energy storage.

3.4. Colloidal Stability and PersL Properties of NPs in Suspension

Taking into account further applications of ZGSO:0.5%Cr^{3+} PersL NPs for bioimaging as presented previously [57], circulating NPs could have a long journey via blood circulation (pH $\approx$ 7.4) until they reach their target or are trapped by the liver, for instance. Therefore, suspension in water is an important step for the stability characteristic of the ZGSO:0.5%Cr^{3+} PersL NPs, even though suspension in water is different from the real in vivo system and environment. Respectively, ZGSO:0.5%Cr^{3+} SS and BM NPs and ZGSO:0.5%Cr^{3+} HT and BM NPs are washed via DI water twice and then dispersed into DI water via ultrasonic dispersion into. Both kinds of NP's suspension samples with a concentration ~0.2 mg/mL are pipetted into a cave (3 mL). The stability of the NP's suspension is characterized by the sedimentation behavior of NPs in water suspension. Figure 5a shows the colloidal stability of ZGSO:0.5%Cr^{3+} NPs dispersed in water after BM treatment of materials prepared using both the solid-state (SS) and hydrothermal (HT) methods. As can be seen in Figure 5a, up to at least 6 h, there are no apparent changes in the stability or decantation of the solid. Subsequently, the PersL NPs in suspension were excited by UV (254 nm) for 2 min, as shown in Figure 5b. The PersL decay curves of the suspension samples are reported in Figure 5c. PersL NPs in suspension with ZGSO:0.5%Cr^{3+} prepared using the HT method present higher intensity and longer durations than those prepared using the SS method, which is correlated to the PersL results obtained from the dry NPs after BM. This may be related to the controlled growth conditions and the smaller particles formed during the HT synthesis (as seen in TEM) and slower reaction rates in the hydrothermal environment. The HT reaction condition can help preserve the luminescent centers and reduce the formation of quenching defects even after BM. Thus, ZGSO:0.5%Cr^{3+} NPs can present more effective optical properties when made using the HT method. Furthermore, ZGSO:0.5%Cr^{3+} HT and BM NPs also show remarkable optical stimulated luminescence (OSL) properties (Figure 5d) when excited by a NIR laser diode. This corresponds to light-induced trap redistribution. The suspension of ZGSO:0.5%Cr^{3+} HT and BM NPs presents desirable optical stimulated PersL property over several cycles after excitation under 980 nm laser (see Figure 5e).

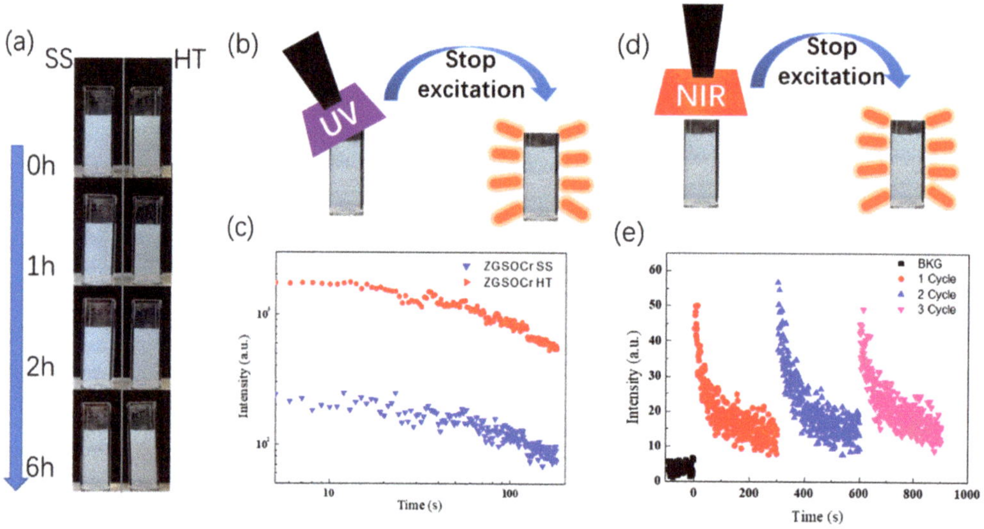

Figure 5. (**a**) Suspension stability measurement of the NPs obtained using the solid-state or hydrothermal methods, ZGSO:0.5%Cr^{3+} NPs dispersed in water for up to 6 h, (**b**) scheme of UV excitation for PersL of ZGSO:0.5%Cr^{3+} NPs suspension, (**c**) PersL decay curves of the suspension (0.2 mg/mL dispersed ZGSO:0.5%Cr^{3+}NPs prepared using both methods). (**d**) scheme and (**e**) decay curve of NIR optical stimulated luminescence of the ZGSO:0.5%Cr^{3+} HT NPs with ~25 nm diameter in 0.2 mg/mL water suspension. (Excitation by 980 nm laser diode with three cycles).

To better understand the nanoparticles' size effect on PersL properties in suspension, suspension samples of ZGSO:0.5%Cr^{3+} NPs (0.2 mg/mL in DI water) were prepared, for both the SS and HT methods, using NPs with average sizes of 25, 35 and 50 nm (with <10% size errors, when determining average particle size under TEM). NPs were selected via careful centrifugation. Figure 6a shows the total counts of PersL intensity (5 min acquisition time), monitored at 700 nm, for the three ZGSO:0.5%Cr^{3+} NP sizes obtained using the SS method after BM treatment. Meanwhile, their decay curves are shown in Figure 6c. The same was repeated for ZGSO:0.5%Cr^{3+} NPs obtained using the HT method after BM, as can be seen in Figure 6b,d. All signal acquisitions of the suspension samples started 1 min after a 2 min irradiation with UV light. Higher average particle sizes yield higher persistent luminescence signal intensity, as they have a lower surface-to-volume ratio. Due to this lower ratio, less energy is lost through surface ligands and defects. When comparing both methods, one can observe that the PersL intensity of nanoparticles prepared using the HT method is about 50 times more intense than the one prepared using the SS method with similar particle sizes.

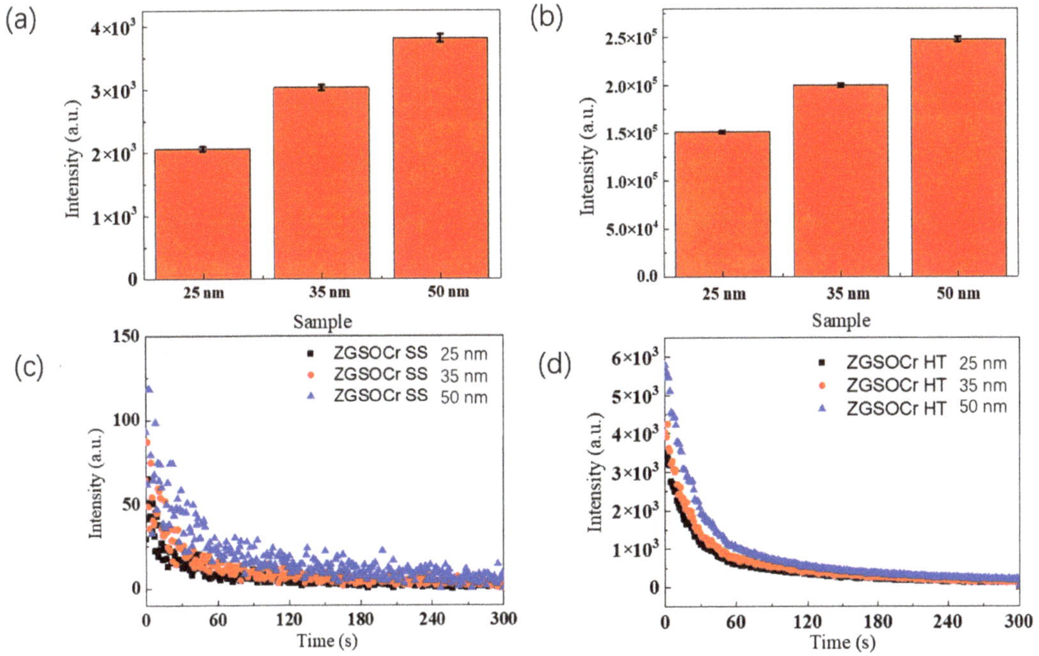

Figure 6. (**a**,**b**) Total counts of PersL of ZGSO:0.5%Cr^{3+} NPs prepared using solid-state and hydrothermal methods, with different diameters in suspension (in water) with 0.2 mg/mL and (**c**,**d**) their respective decay curves (acquisition time of 5 min, starting 1 min after stopping 254 nm excitation).

To understand the relationship between the PersL and the ZGSO:0.5%Cr^{3+}NPs' mass concentration in suspension, the suspension samples with a series of concentration (0.5 mg/mL, 1 mg/mL and 2 mg/mL) were prepared for both SS and HT methods, using nanoparticles of average size (~25 nm, as measured by TEM). Figure 7a shows the total counts of PersL (5 min. acquisition) monitored at 700 nm NPs obtained using the SS method, after BM treatment. Their respective decay curves are presented in Figure 7c. Similarly, the total counts of the ZGSO:0.5%Cr^{3+} suspensions prepared using the HT method and their respective decay curves are shown in Figure 7b,d. As expected, a concentration increase is responsible for a significant increase in collected PersL signals. The signal intensity collected from ZGSO:0.5%Cr^{3+} NPs obtained using the HT method is around 50 times higher than that of the NPs obtained using the SS method for a given concentration.

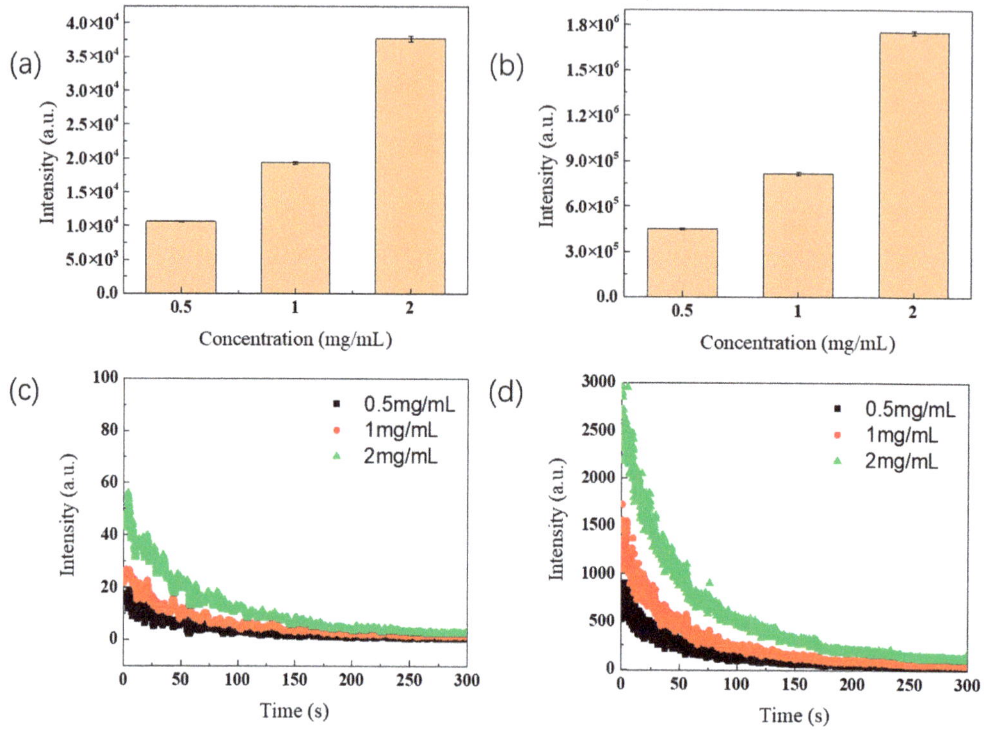

Figure 7. (**a**,**b**) Total counts of PersL decay of ZGSO:0.5%Cr^{3+} NPs prepared using solid-state or hydrothermal methods (~25 nm average diameter determined by TEM) with different concentrations in water suspension and (**c**,**d**) their decay curves profiles within 5 min, starting 1 min after the end of the 254 nm of UV excitation.

4. Conclusions

In this work, $Zn_{1.33}Ga_{1.335}Sn_{0.33}O_4$:0.5%$Cr^{3+}$ (ZGSO:0.5%Cr^{3+}) deep-red persistent luminescence phosphors were successfully prepared using solid-state (SS) and hydrothermal (HT) methods. The characterization results clearly show the significant impacts of the grinding method on the structural, morphological and luminescent properties of ZGSO:0.5%Cr^{3+}. Briefly, the SS method, without further grinding, leads to irregularly shaped particles, and a crystallite size of the ZGSO:0.5%Cr^{3+} phosphor over 1 μm. Furthermore, SS-prepared micrometric phosphors exhibit high efficiency in terms of persistent luminescence (PersL) duration, which is correlated to the high-temperature reaction. This method leads to the narrowing of defects' energy measured by thermoluminescence (TL) and their localization below the conduction band, as evidenced by the TL glow curve observed between 250–300 K. Even after mechanical ball milling (BM) treatment, the size of the NPs is hardly reduced to reach 25 nm, while their PersL intensity drastically decreases.

On the other hand, for those prepared using the HT method, the growth process of the nanoparticles in the hydrothermal environment promotes the formation of ZGSO:0.5%Cr^{3+} with more uniform shapes and smaller sizes, thus even after a BM treatment, the size of NP products are easily reduced, and their PersL is proportionally less reduced than NPs obtained from the top-down solid state process. For both preparation methods, BM creates new non-radiative energy loss paths, possibly due to the formation of surface defects, which leads to a significant decrease in the PL intensity, a shorter PL lifetime, and weaker PersL performance.

To conclude, on one hand, for applications where particle size is not critical, such as anti-counterfeiting and lighting, ZGSO:0.5%Cr^{3+} prepared using the SS method shows more effective persistent luminescence properties than those prepared using the HT method, and is therefore the optimal choice. On the other hand, when nanoparticles are required, such as in bioimaging and theranostics [58–60], ZGSO:0.5%Cr^{3+} HT and BM NPs are indeed more interesting, presenting higher luminescent intensity and longer PersL duration compared to ZGSO:0.5%Cr^{3+} SS and BM NPs. This is observed in both dry powder and suspension.

Supplementary Materials: The following supporting information can be downloaded at: https://www.mdpi.com/article/10.3390/nano13152175/s1, Figure S1. Lifetime decay profiles of $Zn_{1.33}Ga_{1.335}Sn_{0.33}O_4$:$Cr^{3+}$ (ZGSO:0.5%Cr^{3+}) phosphors prepared by (a) solid-state and (b) hydrothermal methods before BM, under the different wavelengths of the laser excitation. Figure S2. (a) Crystal structure of the $ZnGa_2O_4$ normal spinel. The blue spheres represent Zn, red spheres represent O, and yellow spheres represent Ga. (b) Polyhedral view of a complex spinel. Tetrahedral sites are expected to be filled by Zn^{2+} ions, the octahedral sites are filled by the remaining Zn^{2+} ions and all Sn^{4+} and Ga^{3+} ions.

Author Contributions: Conceptualization, B.V.; Methodology, G.C. and L.G.; Investigation, C.R.; Writing—original draft, G.C. and L.G.; Writing—review & editing, C.R. and B.V.; Supervision, C.R. and B.V. All authors have read and agreed to the published version of the manuscript.

Funding: This research was funded by Agence Nationale de la Recherche (ANR-22-CE09-0029-01 PLEaSe) and the CSC Grant program.

Data Availability Statement: The data presented in this study are available on request from the corresponding author.

Conflicts of Interest: The authors declare no conflict of interest.

References

1. Xu, J.; Tanabe, S. Persistent luminescence instead of phosphorescence: History, mechanism, and perspective. *J. Lumin.* **2019**, *205*, 581–620. [CrossRef]
2. Cai, G.Y.; Delgado, T.; Richard, C.; Viana, B. ZGSO Spinel Nanoparticles with Dual Emission of NIR Persistent Luminescence for Anti-Counterfeiting Applications. *Materials* **2023**, *16*, 1132. [CrossRef] [PubMed]
3. Ma, C.Q.; Liu, H.H.; Ren, F.; Liu, Z.; Sun, Q.; Zhao, C.J.; Li, Z. The Second Near-Infrared Window Persistent Luminescence for Anti-Counterfeiting Application. *Cryst. Growth Des.* **2020**, *20*, 1859–1867. [CrossRef]
4. Du, J.; Lyu, S.; Jiang, K.; Huang, D.; Li, J.; Van Deun, R.; Poelman, D.; Lin, H. Deep-level trap formation in Si-substituted Sr_2SnO_4:Sm^{3+} for rewritable optical information storage. *Mater. Today Chem.* **2022**, *24*, 100906. [CrossRef]
5. Lai, X.H.; Fang, Z.Y.; Zhang, J.; Wang, B.; Zhu, W.F.; Zhang, R. Structure and luminescence properties of Ce^{3+}-activated $BaLu_2Al_2Ga_2SiO_{12}$ persistent phosphors for optical information storage. *Opt. Mater.* **2021**, *120*, 111391. [CrossRef]
6. Li, W.H.; Zhuang, Y.X.; Zheng, P.; Zhou, T.L.; Xu, J.; Ueda, J.; Tanabe, S.; Wang, L.; Xie, R.J. Tailoring Trap Depth and Emission Wavelength in $Y_3Al_{5-x}GaxO_{12}$:Ce^{3+},V^{3+} Phosphor-in-Glass Films for Optical Information Storage. *ACS Appl. Mater. Interfaces* **2018**, *10*, 27150–27159. [CrossRef]
7. Zhuang, Y.X.; Chen, D.R.; Xie, R.J. Persistent Luminescent Materials with Deep Traps for Optical Information Storage. *Laser Optoelectron. Prog.* **2021**, *58*, 151600.
8. Castaing, V.; Giordano, L.; Richard, C.; Gourier, D.; Allix, M.; Viana, B. Photochromism and Persistent Luminescence in Ni-Doped $ZnGa_2O_4$ Transparent Glass-Ceramics: Toward Optical Memory Applications. *J. Phys. Chem. C* **2021**, *125*, 10110–10120. [CrossRef]
9. Back, M.; Ueda, J.; Xu, J.; Asami, K.; Brik, M.G.; Tanabe, S. Effective Ratiometric Luminescent Thermal Sensor by Cr^{3+}-Doped Mullite $Bi_2Al_4O_9$ with Robust and Reliable Performances. *Adv. Opt. Mater.* **2020**, *8*, 2000124. [CrossRef]
10. Castaing, V.; Monteiro, C.; Sontakke, A.D.; Asami, K.; Xu, J.; Fernandez-Carrion, A.J.; Brik, M.G.; Tanabe, S.; Allix, M.; Viana, B. Hexagonal $Sr_{1-x/2}Al_{2-x}Si_xO_4$:$Eu^{2+}$,$Dy^{3+}$ transparent ceramics with tuneable persistent luminescence properties. *Dalton Trans.* **2020**, *49*, 16849–16859. [CrossRef]
11. Chen, W.W.; Huang, X.J.; Dong, Q.; Zhou, Z.H.; Xiong, P.X.; Le, Y.K.; Song, E.H.; Qiu, J.R.; Yang, Z.M.; Dong, G.P. Thermally stable and tunable broadband near-infrared emission from NIR-I to NIR-II in Bi-doped germanate glass for smart light sources. *J. Mater. Chem. C* **2023**, *11*, 953–962. [CrossRef]
12. Fujita, S.; Sakamoto, A.; Tanabe, S. Luminescence Characteristics of YAG Glass-Ceramic Phosphor for White LED. *IEEE J. Sel. Top. Quantum Electron.* **2008**, *14*, 1387–1391. [CrossRef]
13. Guimaraes, V.F.; Salaun, M.; Burner, P.; Maia, L.J.Q.; Ferrier, A.; Viana, B.; Gautier-Luneau, I.; Ibanez, A. Controlled preparation of aluminum borate powders for the development of defect-related phosphors for warm white LED lighting. *Solid State Sci.* **2017**, *65*, 6–14. [CrossRef]

14. Nakanishi, T.; Tanabe, S. Novel Eu^{2+}-Activated Glass Ceramics Precipitated with Green and Red Phosphors for High-Power White LED. *IEEE J. Sel. Top. Quantum Electron.* **2009**, *15*, 1171–1176. [CrossRef]
15. Nishiura, S.; Tanabe, S.; Fujioka, K.; Fujimoto, Y. Transparent Ce^{3+}:GdYAG ceramic phosphors for white LED. In Proceedings of the Conference on Optical Components and Materials VIII, San Francisco, CA, USA, 25–26 January 2011.
16. Maldiney, T.; Lecointre, A.; Viana, B.; Bessiere, A.; Gourier, D.; Bessodes, M.; Richard, C.; Scherman, D. Trap depth optimization to improve optical properties of diopside-based nanophosphors for medical imaging. In Proceedings of the Conference on Oxide-Based Materials and Devices III, San Francisco, CA, USA, 22–25 January 2012.
17. Maldiney, T.; Scherman, D.; Richard, C. Persistent Luminescence Nanoparticles for Diagnostics and Imaging. In *Functional Nanoparticles for Bioanalysis, Nanomedicine, and Bioelectronic Devices Volume 2*; Hepel, M., Zhong, C.J., Eds.; American Chemical Society: Washington, DC, USA, 2012; Volume 1113, pp. 1–25.
18. Richard, C.; Maldiney, T.; de Chermont, Q.L.; Seguin, J.; Wattier, N.; Courties, G.; Apparailly, F.; Bessodes, M.; Scherman, D. Persistent Luminescence Nanoparticles for Bioimaging. In Proceedings of the Workshop on Bio-Imaging Technologies, Biopolis, Singapore, January 2012; pp. 37–51.
19. Giordano, L.; Cai, G.Y.; Seguin, J.; Liu, J.H.; Richard, C.; Rodrigues, L.C.V.; Viana, B. Persistent Luminescence Induced by Upconversion: An Alternative Approach for Rechargeable Bio-Emitters. *Adv. Opt. Mater.* **2023**, *11*, 2201468. [CrossRef]
20. Arras, J.; Brase, S. The World Needs New Colors: Cutting Edge Mobility Focusing on Long Persistent Luminescence Materials. *Chemphotochem* **2018**, *2*, 55–66. [CrossRef]
21. Xu, J.; Murata, D.; Ueda, J.; Viana, B.; Tanabe, S. Toward Rechargeable Persistent Luminescence for the First and Third Biological Windows via Persistent Energy Transfer and Electron Trap Redistribution. *Inorg. Chem.* **2018**, *57*, 5194–5203. [CrossRef]
22. Xu, J.; Cherepy, N.J.; Ueda, J.; Tanabe, S. Red persistent luminescence in rare earth-free $AlN:Mn^{2+}$ phosphor. *Mater. Lett.* **2017**, *206*, 175–177. [CrossRef]
23. Katayama, Y.; Viana, B.; Gourier, D.; Xu, J.; Tanabe, S. Photostimulation induced persistent luminescence in $Y_3Al_2Ga_3O_{12}:Cr^{3+}$. *Opt. Mater. Express* **2016**, *6*, 1405–1413. [CrossRef]
24. Pellerin, M.; Castaing, V.; Gourier, D.; Chaneac, C.; Viana, B. Persistent luminescence of transition metal (Co, Ni center dot center dot center dot)-doped $ZnGa_2O_4$ phosphors for applications in the near-infrared range. In *Oxide-Based Materials and Devices Ix*; Rogers, D.J., Look, D.C., Teherani, F.H., Eds.; SPIE: San Francisco, CA, USA, 2018; Volume 10533.
25. Adachi, S. Review-Mn^{4+} vs Cr^{3+}: A Comparative Study as Activator Ions in Red and Deep Red-Emitting Phosphors. *ECS J. Solid State Sci. Technol.* **2020**, *9*, 026003. [CrossRef]
26. Gupta, I.; Singh, S.; Bhagwan, S.; Singh, D. Rare earth (RE) doped phosphors and their emerging applications: A review. *Ceram. Int.* **2021**, *47*, 19282–19303. [CrossRef]
27. Algar, W.R.; Massey, M.; Rees, K.; Higgins, R.; Krause, K.D.; Darwish, G.H.; Peveler, W.J.; Xiao, Z.J.; Tsai, H.Y.; Gupta, R.; et al. Photoluminescent Nanoparticles for Chemical and Biological Analysis and Imaging. *Chem. Rev.* **2021**, *121*, 9243–9358. [CrossRef]
28. Jiang, W.J.; Huang, L.; Mo, F.; Zhong, Y.Y.; Xu, L.J.; Fu, F.F. Persistent luminescent multifunctional drug delivery nano-platform based on nanomaterial $ZnGa_2O_4:Cr^{3+},Sn^{4+}$ for imaging-guided cancer chemotherapy. *J. Mater. Chem. B* **2019**, *7*, 3019–3026. [CrossRef]
29. Katayama, Y.; Kobayashi, H.; Tanabe, S. Deep-red persistent luminescence in Cr^{3+}-doped $LaAlO_3$ perovskite phosphor for in vivo imaging. *Appl. Phys. Express* **2015**, *8*, 012102. [CrossRef]
30. Li, Z.J.; Shi, J.P.; Zhang, H.W.; Sun, M. Highly controllable synthesis of near-infrared persistent luminescence $SiO_2/CaMgSi_2O_6$ composite nanospheres for imaging in vivo. *Opt. Express* **2014**, *22*, 10509–10518. [CrossRef]
31. Li, Y.; Zhou, S.F.; Li, Y.Y.; Sharafudeen, K.; Ma, Z.J.; Dong, G.P.; Peng, M.Y.; Qiu, J.R. Long persistent and photo-stimulated luminescence in Cr^{3+}-doped Zn-Ga-Sn-O phosphors for deep and reproducible tissue imaging. *J. Mater. Chem. C* **2014**, *2*, 2657–2663. [CrossRef]
32. Liu, H.H.; Ren, F.; Zhang, H.; Han, Y.B.; Qin, H.Z.; Zeng, J.F.; Wang, Y.; Sun, Q.; Li, Z.; Gao, M.Y. Oral administration of highly bright Cr^{3+} doped $ZnGa_2O_4$ nanocrystals for in vivo targeted imaging of orthotopic breast cancer. *J. Mater. Chem. B* **2018**, *6*, 1508–1518. [CrossRef]
33. Richard, C.; Viana, B. Persistent X-ray-activated phosphors: Mechanisms and applications. *Light-Sci. Appl.* **2022**, *11*, 123. [CrossRef]
34. Wei, J.J.; Liu, Y.Y.; Zhang, M.R.; Zheng, W.; Huang, P.; Gong, Z.L.; Li, R.F.; Chen, X.Y. Blue-LED-excitable NIR-II luminescent lanthanide-doped SrS nanoprobes for ratiometric thermal sensing. *Sci. China-Mater.* **2022**, *65*, 1094–1102. [CrossRef]
35. Zhang, M.R.; Zheng, W.; Liu, Y.; Huang, P.; Gong, Z.L.; Wei, J.J.; Gao, Y.; Zhou, S.Y.; Li, X.J.; Chen, X.Y. A New Class of Blue-LED-Excitable NIR-II Luminescent Nanoprobes Based on Lanthanide-Doped CaS Nanoparticles. *Angew. Chem.-Int. Ed.* **2019**, *58*, 9556–9560. [CrossRef]
36. Bessiere, A.; Sharma, S.K.; Basavaraju, N.; Priolkar, K.R.; Binet, L.; Viana, B.; Bos, A.J.J.; Maldiney, T.; Richard, C.; Scherman, D.; et al. Storage of Visible Light for Long-Lasting Phosphorescence in Chromium-Doped Zinc Gallate. *Chem. Mater.* **2014**, *26*, 1365–1373. [CrossRef]
37. Pan, Z.F.; Castaing, V.; Yan, L.P.; Zhang, L.L.; Zhang, C.; Shao, K.; Zheng, Y.F.; Duan, C.K.; Liu, J.H.; Richard, C.; et al. Facilitating Low-Energy Activation in the Near-Infrared Persistent Luminescent Phosphor $Zn_{1+x}Ga_{2-2x}Sn_xO_4:Cr^{3+}$ via Crystal Field Strength Modulations. *J. Phys. Chem. C* **2020**, *124*, 8347–8358. [CrossRef]

38. Bessiere, A.; Jacquart, S.; Priolkar, K.; Lecointre, A.; Viana, B.; Gourier, D. $ZnGa_2O_4:Cr^{3+}$: A new red long-lasting phosphor with high brightness. *Opt. Express* **2011**, *19*, 10131–10137. [CrossRef]
39. Lecuyer, T.; Teston, E.; Ramirez-Garcia, G.; Maldiney, T.; Viana, B.; Seguin, J.; Mignet, N.; Scherman, D.; Richard, C. Chemically engineered persistent luminescence nanoprobes for bioimaging. *Theranostics* **2016**, *6*, 2488–2524. [CrossRef]
40. Maldiney, T.; Richard, C.; Seguin, J.; Wattier, N.; Bessodes, M.; Scherman, D. Effect of Core Diameter, Surface Coating, and PEG Chain Length on the Biodistribution of Persistent Luminescence Nanoparticles in Mice. *ACS Nano* **2011**, *5*, 854–862. [CrossRef]
41. Maldiney, T.; Byk, G.; Wattier, N.; Seguin, J.; Khandadash, R.; Bessodes, M.; Richard, C.; Scherman, D. Synthesis and functionalization of persistent luminescence nanoparticles with small molecules and evaluation of their targeting ability. *Int. J. Pharm.* **2012**, *423*, 102–107. [CrossRef]
42. Maldiney, T.; Ballet, B.; Bessodes, M.; Scherman, D.; Richard, C. Mesoporous persistent nanophosphors for in vivo optical bioimaging and drug-delivery. *Nanoscale* **2014**, *6*, 13970–13976. [CrossRef]
43. Maldiney, T.; Bessiere, A.; Seguin, J.; Teston, E.; Sharma, S.K.; Viana, B.; Bos, A.J.J.; Dorenbos, P.; Bessodes, M.; Gourier, D.; et al. The in vivo activation of persistent nanophosphors for optical imaging of vascularization, tumours and grafted cells. *Nat. Mater.* **2014**, *13*, 418–426. [CrossRef]
44. Pan, Z.W.; Lu, Y.Y.; Liu, F. Sunlight-activated long-persistent luminescence in the near-infrared from Cr^{3+}-doped zinc gallogermanates. *Nat. Mater.* **2012**, *11*, 58–63. [CrossRef]
45. Ge, P.H.; Chen, S.; Tian, Y.X.; Liu, S.K.; Yue, X.T.; Wang, L.; Xu, C.H.; Sun, K.N. Upconverted persistent luminescent $Zn_3Ga_2SnO_8$: Cr^{3+}, Yb^{3+}, Er^{3+} phosphor for composite anti-counterfeiting ink. *Appl. Opt.* **2022**, *61*, 5681–5685. [CrossRef]
46. Suarez, P.L.; Garcia-Cortes, M.; Fernandez-Arguelles, M.T.; Encinar, J.R.; Valledor, M.; Ferrero, F.J.; Campo, J.C.; Costa-Fernandez, J.M. Functionalized phosphorescent nanoparticles in (bio)chemical sensing and imaging—A review. *Anal. Chim. Acta* **2019**, *1046*, 16–31. [CrossRef] [PubMed]
47. Viana, B.; Richard, C.; Castaing, V.; Glais, E.; Pellerin, M.; Liu, J.; Chanéac, C. NIR-Persistent Luminescence Nanoparticles for Bioimaging, Principle and Perspectives. In *Near Infrared-Emitting Nanoparticles for Biomedical Applications*; Benayas, A., Hemmer, E., Hong, G., Jaque, D., Eds.; Springer International Publishing: Cham, Switzerland, 2020; pp. 163–197.
48. Qiu, J.; Li, Y.; Jia, Y. 7—Applications. In *Persistent Phosphors*; Qiu, J., Li, Y., Jia, Y., Eds.; Woodhead Publishing: Cambridge, UK, 2021; pp. 245–287.
49. Back, M.; Ueda, J.; Brik, M.G.; Lesniewki, T.; Grinberg, M.; Tanabe, S. Revisiting Cr^{3+}-Doped $Bi_2Ga_4O_9$ Spectroscopy: Crystal Field Effect and Optical Thermometric Behavior of Near-Infrared-Emitting Singly-Activated Phosphors. *Acs Appl. Mater. Interfaces* **2018**, *10*, 41512–41524. [CrossRef] [PubMed]
50. Back, M.; Trave, E.; Ueda, J.; Tanabe, S. Ratiometric Optical Thermometer Based on Dual Near-Infrared Emission in Cr^{3+}-Doped Bismuth-Based Gallate Host. *Chem. Mater.* **2016**, *28*, 8347–8356. [CrossRef]
51. Back, M.; Ueda, J.; Brik, M.G.; Tanabe, S. Pushing the Limit of Boltzmann Distribution in Cr^{3+}-Doped $CaHfO_3$ for Cryogenic Thermometry. *Acs Appl. Mater. Interfaces* **2020**, *12*, 38325–38332. [CrossRef]
52. Byrappa, K.; Adschiri, T. Hydrothermal technology for nanotechnology. *Prog. Cryst. Growth Charact. Mater.* **2007**, *53*, 117–166. [CrossRef]
53. Kolb, E.D.; Key, P.L.; Laudise, R.A.; Simpson, E.E. Pressure-Volume-Temperature Behavior in the System H_2O-NaOH-SiO_2 and its Relationship to the Hydrothermal Growth of Quartz. *Bell Syst. Tech. J.* **1983**, *62*, 639–656. [CrossRef]
54. Castaing, V.; Sontakke, A.D.; Fernandez-Carrion, A.J.; Touati, N.; Binet, L.; Allix, M.; Gourier, D.; Viana, B. Persistent Luminescence of $ZnGa_2O_4:Cr^{3+}$ Transparent Glass Ceramics: Effects of Excitation Wavelength and Excitation Power. *Eur. J. Inorg. Chem.* **2017**, *2017*, 5114–5120. [CrossRef]
55. Castaing, V.; Sontakke, A.D.; Xu, J.; Fernandez-Carrion, A.J.; Genevois, C.; Tanabe, S.; Allix, M.; Viana, B. Persistent energy transfer in ZGO:Cr^{3+},Yb^{3+}: A new strategy to design nano glass-ceramics featuring deep red and near infrared persistent luminescence. *Phys. Chem. Chem. Phys.* **2019**, *21*, 19458–19468. [CrossRef]
56. Hai, O.; Jiang, H.Y.; Xu, D.; Li, M. The effect of grain surface on the long afterglow properties of $Sr_2MgSi_2O_7$: Eu^{2+}, Dy^{3+}. *Mater. Res. Bull.* **2016**, *76*, 358–364. [CrossRef]
57. Maldiney, T.; Scherman, D.; Richard, C. B1-2 Persistent luminescence nanoparticles for in vivo imaging: Characteristics and targeting. In *The CliniBook*; EDP Sciences: Les Ulis, France, 2012; pp. 386–393.
58. Jiang, Y.; Li, Y.; Richard, C.; Scherman, D.; Liu, Y.S. Hemocompatibility investigation and improvement of near- infrared persistent luminescent nanoparticle $ZnGa_2O_4:Cr^{3+}$ by surface PEGylation. *J. Mater. Chem. B* **2019**, *7*, 3796–3803. [CrossRef]
59. Teston, E.; Maldiney, T.; Marangon, I.; Volatron, J.; Lalatonne, Y.; Motte, L.; Boisson-Vidal, C.; Autret, G.; Clement, O.; Scherman, D.; et al. Nanohybrids with Magnetic and Persistent Luminescence Properties for Cell Labeling, Tracking, In Vivo Real-Time Imaging, and Magnetic Vectorization. *Small* **2018**, *14*, e1800020. [CrossRef] [PubMed]
60. Lecuyer, T.; Seguin, J.; Balfourier, A.; Delagrange, M.; Burckel, P.; Lai-Kuen, R.; Mignon, V.; Ducos, B.; Tharaud, M.; Saubamea, B.; et al. Fate and biological impact of persistent luminescence nanoparticles after injection in mice: A one-year follow-up. *Nanoscale* **2022**, *14*, 15760–15771. [CrossRef] [PubMed]

nanomaterials

MDPI

Article

Extraction of Metal Ions by Interfacially Active Janus Nanoparticles Supported by Wax Colloidosomes Obtained from Pickering Emulsions

Oliver Pauli [1] and Andrei Honciuc [2,*]

1 Institute of Chemistry and Biotechnology, Zurich University of Applied Sciences, Einsiedlerstrasse 31, 8820 Wädenswil, Switzerland
2 Electroactive Polymers and Plasmochemistry Laboratory, "Petru Poni" Institute of Macromolecular Chemistry, Aleea Gr. Ghica Voda 41A, 700487 Iasi, Romania
* Correspondence: honciuc.andrei@icmpp.ro

Abstract: Most common wastewater treatment technologies for ion extraction and recovery rely on pumping wastewater through ion-exchange columns, filled with surface-functionalized polymer microspheres. To avoid the energetically intensive process of pumping large quantities of water through ion-exchange columns, alternative technologies are being developed, such as water-floating membranes containing ligands. In this context, innovative materials could be deployed. Here, we report nanostructured paraffine wax microspheres capable of floating on water, a design based on Pickering emulsion technology, where Janus nanoparticles act both as emulsion stabilizers and as ligand carriers. In the process of emulsification of molten wax in water, followed by cooling, the branched polyethylenimine (bPEI) carrying Janus nanoparticles are trapped at the molten wax/water interface, forming spherical microspheres or colloidosomes decorated with nanoparticles. The paraffine wax colloidosomes stabilized by ligand-carrying Janus nanoparticles are capable of floating on water and show high metal ion extraction capacities towards Cr(VI), Co(II), Ni(II), Cu(II) and Zn(II). In addition, we demonstrate that the ions can be recovered from the colloidosomes and that the colloidosomes can withstand several extraction/recovery cycles with little or no loss in the absorption capacity.

Keywords: Janus nanoparticles; colloidosomes; metal ion extraction; wastewater treatment; Pickering emulsions

Citation: Pauli, O.; Honciuc, A. Extraction of Metal Ions by Interfacially Active Janus Nanoparticles Supported by Wax Colloidosomes Obtained from Pickering Emulsions. *Nanomaterials* **2022**, *12*, 3738. https://doi.org/10.3390/nano12213738

Academic Editor: Rajinder Pal

Received: 6 September 2022
Accepted: 21 October 2022
Published: 25 October 2022

1. Introduction

The development of energetically efficient wastewater treatment technologies for the removal of small organic pollutants or heavy metal ions is of paramount importance for maintaining a clean environment and mitigating current water pollution problems. The advantages and drawbacks of various existing technologies for metal ion removal from wastewater, such as chemical precipitation, ion exchange, adsorption, membrane filtration, coagulation and flocculation, flotation, electrochemical treatment, etc., have been extensively reviewed [1]. Finding alternative technologies to the currently existing methods that avoid secondary pollution, generation of solid waste, are energetically efficient and have low operation costs could be achieved by deployment of advanced materials, and this represents a forthcoming challenge for fundamental science and engineering. In this context, for extraction and removal of metal ion pollutants or hydrological mining of noble metals, liquid membranes technologies such as bulk liquid membranes (BLMs), emulsion liquid membranes (ELMs) [2], supported liquid membrane (SLMs) [3] supported ionic liquid membranes (SILM) [4], or polymer inclusion membranes (PIM) [5,6], nanoparticles incorporated into an absorptive film [7], have been proposed as energetically efficient alternatives to classical methods. All these alternative methods involve the interfacial transfer of the metal ions from the water phase into a liquid organic phase or a solid-state polymer containing chelating agents. For example, in ELMs methods, the oil-in-water (o/w) emulsion, with the organic phase containing a chelating agent, is stirred in a reservoir until the ion extraction is completed. For a large interfacial area and efficient extraction, small emulsion droplets are required, however, droplet coalescence and emulsion stability are the limiting factors of the method, which must be resolved [8]. One possible solution to solve the emulsion stability in ELMs is to use the more stable Pickering emulsions, which are emulsions stabilized by nanoparticles or amphiphilic Janus nanoparticles (JNPs) [9,10]. In this case, amphiphilic nanoparticles such as JNPs can play a double role, as emulsion stabilizers in ELMs but also as ligand carriers, whereas ligands can be immobilized on the surface of the nanoparticle by specific surface grafting techniques. In this work we synthesize amphiphilic JNPs, which carry ligands and are also interfacially active, being capable of partitioning at the oil-water interface and are thus able to stabilize Pickering emulsions. JNPs are asymmetric nanoparticles consisting of at least two lobes differing in their chemical composition or surface properties. The contrasting properties between lobes give rise to an intrinsic amphiphilic property resembling molecular surfactants [11,12]. With the help of these ligand-carrying amphiphilic JNPs, we emulsify molten paraffin wax to create o/w emulsions. The proof of concept of this technology is illustrated in the cartoon of Scheme 1. By emulsifying the molten wax in water, colloidosomes are obtained, which have a monolayer of amphiphilic JNPs at their surface. Amphiphilicity is key to the emulsification ability of the JNPs [13]. Colloidosomes can be defined as microcapsules with a shell of nanoparticles, which are obtained by self-assembly of nanoparticles at the interface between two immiscible liquids, most commonly water and oil [14]; these colloidosomes can be employed directly for ion extraction. Although the molten wax colloidosomes cannot preserve their integrity in ion extraction and recovery cycles, upon cooling, the liquid molten paraffin droplets solidify, generating surface nanostructured microspheres by trapping the Janus nanoparticles at the interface. The aim of this work is to show that water-floating wax microspheres decorated with ligand-carrying JNPs, which result from the solidification of wax-in-water Pickering emulsions, can be successfully employed for metal ion extraction, as depicted in Scheme 1. Furthermore, the resulting solid-state colloidosomes/microspheres can be easily regenerated and re-used in many metal ion extraction cycles. The great advantage of using wax colloidosomes over liquid emulsions is that the wax colloidosomes, in absence of stirring, float on the surface of the water which greatly simplifies the collection process, eliminating the need for filtration and further processing steps. The novelty of this work consists in obtaining paraffin microspheres decorated with amphiphilic JNPs that carry ligands and employing them in ion extraction technologies. In addition, their ability to float on water surfaces makes them attractive for the fact that they can be easily collected by water-sweeping barriers and thus minimize the microparticle loss in the water and water pollution with microplastics. This could be a great advantage for the treatment of wastewater, especially in comparison to more energetically intensive techniques based on pumping water through ion exchange columns and can be more practical than the above-mentioned liquid-based floating membrane technologies SLM, ELM and PIM. The collected microspheres from the water surface can be regenerated by the recovery of the absorbed ions in acidic water and can be reutilized in ion extraction, see Scheme 1. To demonstrate this, we use amphiphilic Janus nanoparticles and solid Pickering emulsion for the recovery of Cr(VI), Co(II), Ni(II), Cu(II), Zn(II), which are among the most common heavy metal ions found in wastewater.

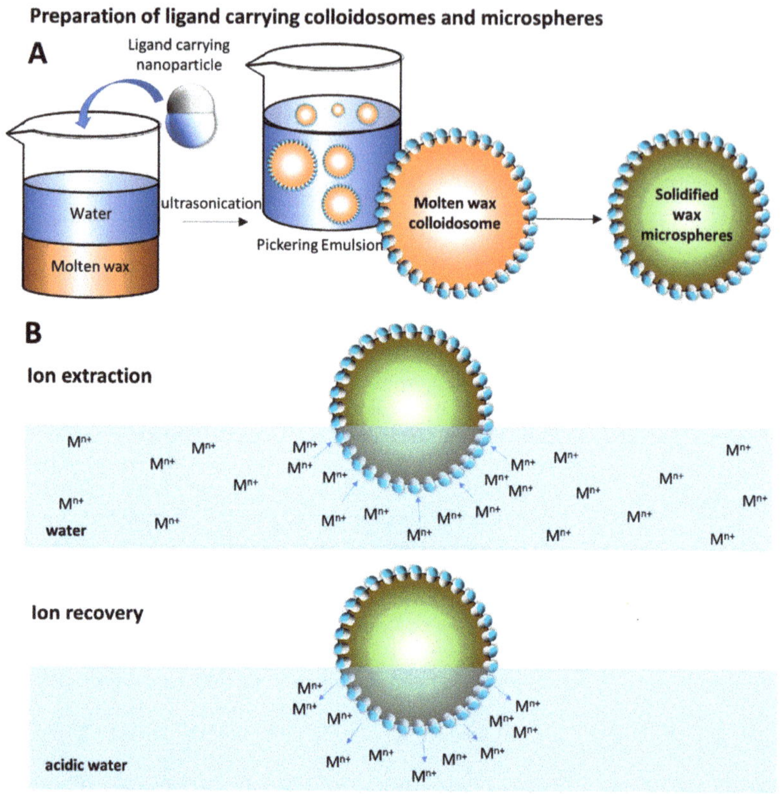

Scheme 1. (**A**) Illustration of the methodology proposed for the synthesis of molten wax colloidosomes, and microspheres decorated with ligand-carrying JNPs; (**B**) Principles of the method for employing water-floating ligand-carrying microspheres in ion extraction from wastewater at normal pH and ion recovery in acidic water.

2. Materials and Methods

2.1. Materials

Styrene (St) (>99%), divinylbenzene (DVB) (80%), sodium 4-vinylbenzenesulfonate (NaVBS) (>90%), ammonium persulfate $(NH_4)_2S_2O_8$ (APS) (>98%), 2,2′-azobis(2-methylpropionitrile) (AIBN) (>98%), ammonium hydroxide solution ($NH_4 \cdot OH$) (28%) and basic alumina (Al_2O_3) (≥98%), 3-(triethoxysilyl)propionitrile (TESPN) (97%), 3-(triethoxysilyl)propyl-methacrylate (TSPM) (99%), sulfuric acid (H_2SO_4) (95.0–98.0%), branched *polyethyleneimine* (bPEI, $M_n \approx$ 10,000 by GPC, $M_w \approx$ 25,000 by LS) and paraffin wax (mp 53–58 °C) were purchased from Sigma-Aldrich (Buchs, Switzerland). *N,N′*-diisopropylcarbodiimide (DIC) (99%) was purchased from Acros Organics (Basel, Switzerland). St and DVB were passed through basic alumina to remove the stabilizer before usage. AIBN was purified by re-crystallization twice from methanol and stored at −20 °C before usage. Other reagents were used as received. Ultrapure water (UPW; conductivity c = 0.055 μS/cm and resistivity, ρ = 18.2 MΩ cm at 298 K) was obtained from an Arium 611 VF water purification system (Startorius stedim biotech, Aubagne, France), and it was used as the aqueous medium in all experiments.

2.2. Synthesis of PS Seed Nanoparticles

The surfactant-free emulsion *co*-polymerization of styrene (St), divinylbenzene (DVB) and sodium vinylbenzenesulfonate (NaVBS) was performed according to a procedure we have previously reported [10,15].

2.3. Synthesis and Surface Modification of Janus Nanoparticles

To differentiate between the multiple types of particles and lobe size ratios, the notation JNP-X-Y was adopted, where X is the predominant surface group on the P(3-TSPM) lobe (e.g., CN) and Y is the volume of 3-TSPM/3-TSPCN mixture in mL per 1 g of PS seed NPs.

The synthesis procedure is described in the example of JNP-CN 2 mL. Reagent quantities used for the synthesis of the homologous series (JNP-CN 2 to JNP-CN 4 mL) are given in the Supporting Information (SI).

A suspension of 2 g PS seed NPs in UPW was deoxygenated by bubbling Ar gas under stirring at room temperature (RT). 3 mL 3-TSPM, 1 mL 3-TSPCN and 40 mg AIBN were mixed in a scintillator vial and sonicated for a few seconds to dissolve the AIBN. A total of 11 mL of argonated UPW was added to the mixture, which was then cooled to approx. −40 °C using an ice/acetone bath. The mixture was emulsified by ultrasonication (Branson Sonifier 450, ½ inch processing horn, Branson Ultrasonics Dietzenbach, Dietzenbach, Germany, 2 min at 50% intensity). The monomer-in-water emulsion (which is an o/w emulsion) was added to the dispersion of PS seed NPs using a syringe. The reaction mixture was left to stir at RT for approx. 3 h. 15 droplets of NH_4OH (30% aq. sol.) were then added to adjust the pH to 9. The polymerization was carried out under an Ar atmosphere while stirring at 70 °C for approx. 12 h. The JNPs were purified by centrifugation/resuspension in ethanol (EtOH) and UPW (three cycles each).

The JNP-CN were then hydrolyzed to JNP-COOH in HCl 2.5 M under reflux for approx. 12 h. The particles were washed by centrifugation/resuspension in UPW until the supernatant had a neutral pH.

2.4. Grafting of bPEI to the Janus Nanoparticles

The JNP-bPEI used for the metal ion extraction studies was synthesized using an excess of bPEI and DIC. The quantities of the reagents used for the bPEI loading study are given in the Supplementary Materials.

To a suspension of 2 g JNP-COOH in dry *N,N*-dimethylformamide (DMF), 10 mL DIC was added. The reaction mixture was left to stir under an Ar atmosphere for 15 min. 200 mg bPEI (10% m/m in relation to JNP-COOH) was dissolved in a small volume of DMF and added dropwise. The mixture was left to stir at RT for approx. 12 h. The particles were washed by centrifugation/resuspension, once in a solution of DMF/H_2O (9:1), twice in EtOH and three times in UPW.

2.5. Preparation of Wax Colloidosomes

2500 mg paraffin wax with a melting point of 53–58 °C (C_nH_{2n+2}, n = 24–36 estimated from melting point [16]) was added to a suspension of 250 mg JNP-bPEI in 50 mL UPW. The mixture was heated to 80 °C using a water bath. After the wax was completely molten, the phase-separated mixture was emulsified by sonication (Branson Sonifier 450, ½ inch processing horn, 40 s at 50% intensity). The emulsion was then cooled rapidly using an ice/acetone bath. The solidified wax colloidosomes were filtered off, washed with copious amounts of UPW, and left to dry at RT. The dry colloidosomes were Au-sputtered (Quorum Q150 RS Plus, 20 mA for 30 s), (Quorumtech, Laughton, UK) for characterization by SEM.

2.6. Metal Ion Extraction and Recovery

To test the extraction capacity of the synthesized materials, 50 mg JNP-bPEI or 250 mg colloidosomes were suspended in 5 mL of a metal salt solution with c = 10 mmol/L at the natural pH of UPW (pH 5.5 ± 0.5). After mechanical shaking at 5000 rpm for 10 min, the samples were left undisturbed for approx. 12 h. The JNP suspensions were centrifuged at 10,000 rpm for 20 min and the supernatant was collected for analysis. The particles were washed through three cycles of centrifugation/resuspension in UPW. The colloidosome samples were filtered (10 μm pore size) and the filtrate was collected for analysis. The colloidosomes were washed with copious amounts of UPW. The supernatant and filtrate obtained from experiments using JNPs and colloidosomes, respectively, were diluted to a metal ion concentration of approx. 50 mg/L and acidified to prevent the formation of metal hydroxides during analysis. Standard solutions for calibration were prepared from the corresponding metal salts in UPW.

The metal ion concentration in the diluted supernatant and filtrate were analyzed using Inductively coupled plasma-optical emission spectrometry (ICP-OES) (ICP-OES 5100, Agilent Technologies, Basel, Switzerland) and compared to the initial concentration. The metal ion extraction capacity q_e (mg/g) was calculated by:

$$q_e = \frac{(c_i - c_e)\ V}{m_P} \tag{1}$$

where c_i (mg/L) is the initial concentration, c_e (mg/L) is the extracted concentration, V (L) is the volume of the sample, and m_P (g) is the dry mass of the sorbent.

The washed JNPs and colloidosomes were then resuspended in 5 mL 0.5% H_2SO_4 by means of mechanical shaking at 5000 rpm for 10 min. To aid redispersion, the JNP samples were also sonicated for 10 min. The samples were then left undisturbed for approx. 12 h. The particles were separated from the aqueous medium as described previously. Supernatant and filtrate were analyzed using ICP-OES, and the metal ion recovery capacity q_r (mg/g) was calculated by:

$$q_r = \frac{c_r\ V}{m_P} \tag{2}$$

where c_r (mg/L) is the concentration of metal ions recovered from the particles, V (L) is the volume of the sample, and m_P (mg) is the dry mass of the particles.

The JNPs and colloidosomes were washed as described previously, and the entire procedure was repeated at least three times. All ICP-OES measurements were performed in triplicate.

2.7. Statistical Analysis of the Data

To have a proper understanding of the reproducibility of the extraction and recovery of the ions by both the JNPs and wax microspheres, we have repeated all ICP-OES measurements at least three times. For each extraction step, we have repeated the procedure three times; in this work, the average value of the obtained extraction capacity q_e is given in the graphics together with the standard deviation obtained for each measurement. The same statistical treatment was done for measurements of the ion recovery capacity q_r of the materials tested, i.e., JNPs and microspheres. The total error of the measurement σ_{total} was considered to be the total standard deviation from two sources, namely the error in the ICP-OES measurements $\sigma_{measurement}$ and that of extraction or recovery experiments $\sigma_{experiment}$:

$$\sigma_{total} = \sqrt{(\sigma_{measurement})^2 + (\sigma_{experiment})^2}$$

To determine if the means of the two groups are significantly different, a standard independent (two-sample) *t*-test was used.

3. Results and Discussions

3.1. Synthesis and Functionalization of Janus Nanoparticles (JNPs)

Polystyrene seed PS seed-NPs with an average diameter of 305 ± 3 nm were synthesized using surfactant-free emulsion polymerization, according to Scheme 2. Onto these PS seed NPs, a second lobe was grown through a seeded emulsion co-polymerization of 3-(trimethoxysilyl)propyl-methacrylate (TSPM) and 3-cyanopropyltriethoxysilane (TESPN) monomers followed by phase separation, see Scheme 2, resulting in Janus nanoparticles bearing -CN groups on the second lobe (JNPs-CN) according to previously reported methods published by our group [10,15]. The size of the second lobe relative to the PS lobe could be adjusted by varying the volume of TSPM and TESPN monomers in relation to a reference weight of 1 g of PS NPs. Thus, a homologous series of JNPs-CN was created by using 2 mL, 3 mL and 4 mL of monomers per 1 g of PS NPs. The nitrile groups on the surface of the second Janus lobe were hydrolyzed under reflux in hydrochloric acid, resulting in Janus nanoparticles bearing carboxylic acid surface functional groups (JNPs-COOH), as shown in Scheme 2. The conversion was confirmed through IR spectrometry (Figure S1) and the pH-dependent measurement of the surface zeta potential in Figure S2 and the comparative values of the homologous series of JNPs in Table S1. Through SEM imaging it was confirmed that the appearance of the particles was not altered by the acid treatment (Figure S3), a testament to the good chemical stability of polymeric JNPs.

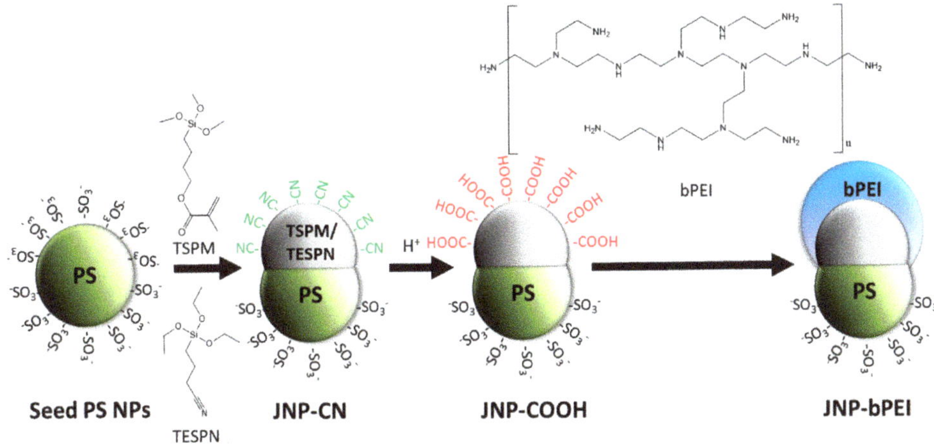

Scheme 2. Reaction scheme depicting the synthesis of seed PS JNPs, synthesis of JNP-CN and their conversion into JNP-COOH and JNP-bPEI.

3.2. *Selective Grafting of bPEI on One Janus Lobe*

N,N'-Diisopropylcarbodiimide (DIC), a coupling agent commonly employed in peptide chemistry, was used to selectively graft branched polyethylenimine (bPEI, $M_n \approx 10{,}000$ by GPC) on the second Janus lobe bearing carboxyl functional groups JNPs-COOH, according to Scheme 2. This was the only reaction where the main solvent was not water. The success of the reaction was confirmed through IR spectrometry (Figure S1) and the pH-dependent measurement of the surface zeta potential (Figure S4). The appearance of the JNPs under the SEM remained unchanged (Figure S5) and a corona of bPEI on the TSPM/TESPN Janus lobe is too thin to be visible.

3.3. *Interfacial Activity of JNP-bPEI Homologous Series*

The interfacial activity of JNP-bPEI at the heptane/water interface was determined using the pendant drop method (OCA 25, DataPhysics). A droplet of an aqueous suspension of JNPs was formed in n-heptane and the curvature of the droplet fitted to the Young–Laplace equation. The interfacial tension (IFT) of a pure heptane/water interface at 25 °C is $\gamma = 50.71$ mN·m^{-1} [17]. In the presence of the JNPs-bPEI, the IFT decreases vs. time due to interfacial adsorption. In Figure S6 it is shown that the interfacial activity, as judged by the lowest values of the IFT reached in the plateau, is significantly larger at lower pH values, which can be explained by the protonation of the bPEI grafted on the second Janus lobe. With the protonation of the bPEI, the surface of the second Janus lobe becomes more polar and produces a good amphiphilic contrast to the less polar PS lobe.

Furthermore, we investigated the influence of the Janus lobe ratio on the interfacial activity of the JNP-bPEI. The influence of the lobe size ratio on the interfacial activity of the particles is demonstrated in Figure 1. The IFT of all suspensions starts out ± 1 mN·m^{-1} of the pristine heptane/water IFT value and then slowly declines as the particles reach the interface. This also indicates the absence of fast-acting molecular surfactants. It is apparent that increasing the size of the hydrophilic bPEI-modified P(3-TSPM) lobe increases the interfacial activity of the particles (compare curves A, B and C in Figure 1). A similar observation has been reported for similar JNPs by Wu et al. [10,18] and was attributed to the hydrophilic-lyophilic balance of the JNPs, which is also called the Janus balance [19].

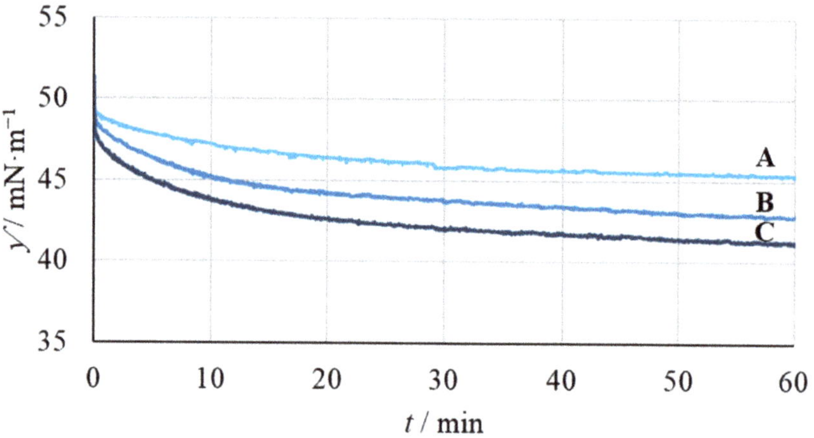

Figure 1. The IFT evolution vs. time of the heptane/water interface in the presence of the Janus nanoparticles (50 mg/mL in the aqueous phase) of the homologous series (A) JNP-bPEI 2 mL, (B) JNP-bPEI 3 mL, (C) JNP-bPEI 4 mL.

Furthermore, the IFT is greatly affected by JNP-bPEI concentration, with a minimum value of $\approx$ 33 mN/m for a concentration of 300 mg/mL, see Figure S7.

3.4. Preparation of Wax Colloidosomes

Colloidosomes consisting of paraffin wax (melting temperature, T_m = 53–57 °C) and 10% m/m bPEI-modified JNPs were prepared by emulsifying molten wax in an aqueous suspension of JNPs through sonication, followed by rapid cooling. The washed colloidosomes were characterized using SEM (Figure 2). The diameters of the colloidosomes obtained by using the homologous series of JNPs (JNP-bPEI 2 mL to JNP-bPEI 4 mL) were not significantly different (Figure S8). On average, the diameter of the colloidosomes is d = 15.8 ± 0.3 μm. Figure S9 also shows the qualitative difference in the packing densities of the JNP-bPEI 4 mL vs. JNP-bPEI 2 mL on the surface of the wax colloidosome.

3.5. Extraction of Metal Ions by JNPs

JNP-bPEI 2 mL was employed in the extraction of metal ions: Cr (VI), Co(II), Ni(II), Cu(II) and Zn(II). The procedure consisted of adding 10 mg/mL JNP-bPEI to a metal ion solution with a concentration of 10 μmol/mL, the conditions were kept constant throughout all ion extractions as described in the experimental section. The photograph in Figure 3 shows the color of JNP-bPEI 2 mL, after extraction/absorption of Cr (VI), Co(II), Ni(II), Cu(II) and Zn(II). Furthermore, to quantitatively determine the absorption capacity for each metal ion, i.e., metal ion load, the concentrations of metal ions in the supernatant were analyzed using Inductively Coupled Plasma-Optical Emission Spectrometry (ICP-OES) and compared to the initial concentration. The metal ion extraction capacity q_e (mg/g) was calculated with the formula given by the Equation (1) [20].

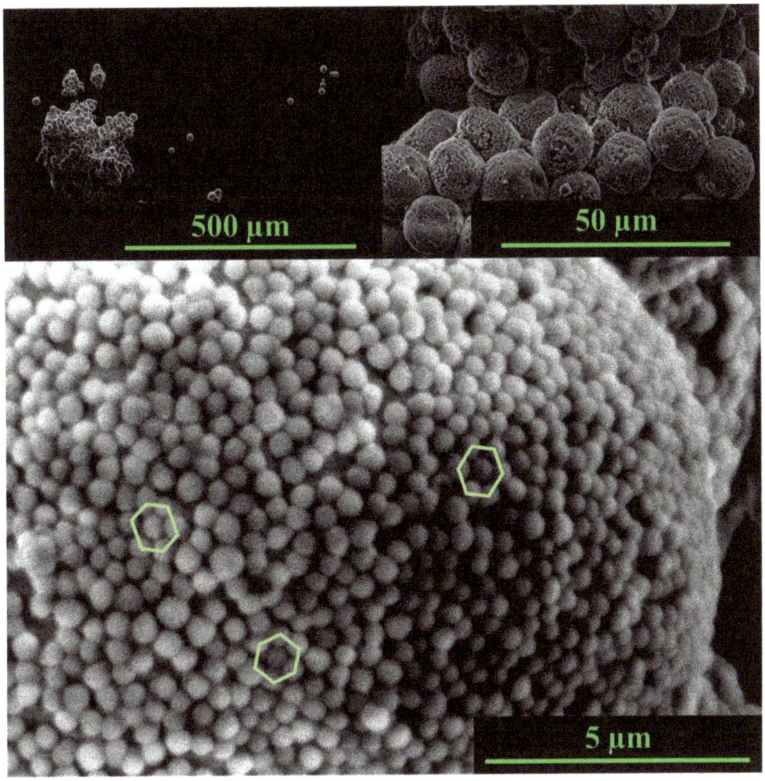

Figure 2. Wax colloidosomes obtained with JNP-bPEI 4 mL.

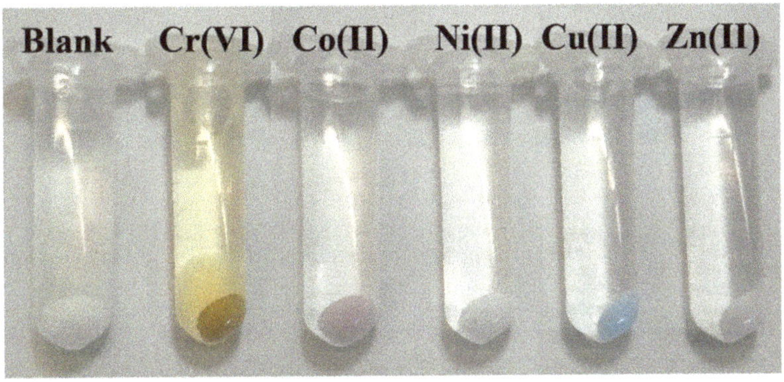

Figure 3. Photograph of the JNP-bPEI 2 mL after absorption of Cr (VI), Co(II), Ni(II), Cu(II), Zn(II) and centrifugation.

Figure 3 shows the color change of the particles occurring upon the absorption of certain metal ions, and Figure 4 shows the measured metal ion extraction capacity q_e of the JNP-bPEI 2 mL. Furthermore, we have determined the number of metal ions that can be recovered, extracted from the JNP-bPEI 2 mL. The recovery was performed in acidic conditions according to the procedures described in the experimental section. The supernatant was analyzed using ICP-OES, and the metal ion recovery capacity q_r (mg/g) was calculated with formula given by the Equation (2).

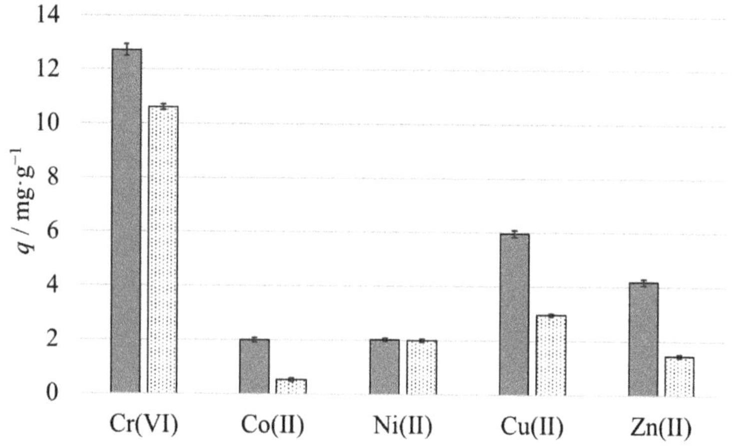

Figure 4. Graph showing the metal ion extraction capacity q_e (solid bars) and the metal ion recovery capacity q_r (dotted bars). Error bars represent $\pm$ 1 σ.

To test the metal ion absorption capacity and recovery by JNPs with multiple duty cycles, the JNPs were washed as described previously, and the entire procedure was repeated up two or five times for Cu(II). After an initial decrease of roughly 25% after the first cycle, the extraction/recovery capacity of JNP-bPEI 2 mL remains unchanged for at least four subsequent cycles, Figure S10.

3.6. Metal Ion Extraction Performance in the Homologous Series of JNPs

The metal ion extraction and recovery performance in the homologous series of JNPs, with the varying Janus lobe size was investigated for Cr(IV) and Cu(II). One way to indirectly prove the presence of bPEI on the P(3-TSPM) lobe is by increasing its surface area and observing the effect on the metal ion extraction capacity. In the homologous series of JNP-bPEI 2 mL, 3 mL, 4 mL, by increasing the TSPM/TESPN lobe size while the PS lobe remains unchanged, the surface area of the bPEI covered TSPM/TESPN lobe also increases, thus the extraction capacity of the JNPs should also increase. To quantitatively assess this, due to the fact that in the extraction/recovery experiments we have used the same mass of particles, see Table S2, we must be careful to consider the mass of a single particle also increases with an increase in the TSPM/TESPN lobe, which means that fewer nanoparticles will be present in a given mass of JNPs used for extraction. A normalization procedure was devised to account for this effect. Since the average volumes of the individual lobes V_{PS}, V_{TSPM} and their average densities ρ_{PS} = 1.03 g/mL, $\rho_{TSPM/TESPN}$ = 1.07 g/mL are known, the average mass of a single particle can be calculated according to the following equation:

$$m_{JNP} = V_{PS} \times \rho_{PS} + V_{TSPM/TESPN} \times \rho_{TSPM/TESPN} \quad (3)$$

From this, the approximate number of JNPs in a given dry mass of particles m_w can be calculated using the following equation:

$$N_{JNP} = \frac{m_w}{m_{JNP}} \quad (4)$$

To make the trends of q_e and q_r of the homologous series visible, the m_w values were normalized for an arbitrary particle number of 10^{12} and by plugging in the obtained values in Equations (1) and (2) the normalized metal ion extraction q_{Ne} and recovery q_{Nr} efficiencies are obtained. The un-normalized and normalized values with Cr(VI) and Cu(II) are depicted in Figure S11 and Figure S12. Normalized values indicate a significant increase in the extraction and recovery capacities with increasing TSPM/TESPN lobe size in the homologous series JNP-bPEI 2 mL, 3 mL and 4 mL. The extraction and recovery capacities of the tested homologous series are listed in Table S2. In comparison, Tan et al.[16] reported q_e $\approx$ 32 mg/g for both Cu(II) and Cr(VI) for their PEI-PS nanoparticles. This is a factor of 5.3 and 2.1 higher than the values found during this study for Cu(II) and Cr(VI), respectively. One possible explanation could be the smaller size of PEI-PS as compared to JNP-bPEI, which results in a higher surface area. Furthermore, PEI-PS is completely covered in bPEI, while only half of JNP-bPEI is covered.

It is instructive to compare the maximum ion uptake obtained for the JNP-bPEI (where the bPEI is a multidentate ligand brush or corona), with the maximum ion load of other micro- and nanoparticles carrying other types of ligands, immobilized in different ways, capable of forming chemical or physical bonds with the metal ion. Among these, most noteworthy are ion imprinted polymer (IIP) micro- and nanoparticles (with ligands immobilized in the bulk of the polymer, capable of chemical bonding with the metal ion), core-shell micro- and nanoparticles whereas the shell is constituted of IIPs, micro- and nanoparticles carrying surface ligands, micro- and nanoparticles carrying anionic functional groups such as ion-exchangers, or agro-based biomasses (capable of interacting with metal ions via physical bonds), etc. Judging purely by the number of ligands a particle can carry, the metal ion uptake capacity for each technology is also expected to decrease with the decrease in ligand carrying capacity of the particulate material, in the following order: IIPs > core-shell IIPs surface immobilized ligands > surface immobilized ligands and ligand brushes > ion-exchangers and biomasses.

For example, the reported ion uptake capacity by JNPs for Zn(II), 4 mg/g, see Figure 4, is higher than the non-chelating physical adsorbing biomasses, whereas the reported maximum ion uptake capacity of biomasses by biosorption of Zn(II) was found to be around 1.688 mg/g for eucalyptus bark, 1.028 mg/g for mango bark and 0.45 mg/g for pineapple fruit peel at a particle size of 0.5 mm [21]. On the other hand, it is expected that the ion-imprinted polymer nanoparticles (IIPs-NPs) offer the best ion uptake capacity, whereas, for the Co(II) for example, the maximum loading capacities reported ranged between 78.3 to 96.6 mg/g by core-shell nanoparticles obtained by constructing an IIP shell of Co-polyacrylamide onto SiO_2 nanoparticles [22] and 74 mg/g by an IIP shell of Co-dithizone/poly(methacrylic)acid [23] onto magnetic Fe_3O_4 core nanoparticles. On the other hand composite chelating IIPs-NPs constituted by a shell of polyamidoxime and a silica core PAO/SiO_2, showed a maximum Cu(II) ion uptake of 10 mg/g [24], which is comparable to the Cu(II) ion uptake by JNP-bPEI of 6 mg/g, see Figure 4. Even so, the Ni(II) ion absorption capacity of an ion-imprinted polymer (IIP) obtained by copolymerization of a Ni-dithizone complex with 4-vinylpyridine and ethyleneglycoldimethacrylate (EGDMA) is 1.3 mg/g [25] The value found for Ni(II) for JNP-bPEI in this study is higher by a factor 1.5, see Figure 4. Tan et al. [20] have reported ion extraction capacities between 20 and 40 mg/g for Cu(II), Co(II), Ni(II) and Cr(VI) for polystyrene nanoparticles modified as in the current case with a bPEI ligand brush or corona Thus, it can be concluded that the JNP-bPEI exhibit a good ion extraction or ion-uptake capacity, comparable to the extraction capacity of other ligand carrying micro- and nanoparticles obtained with other technologies, in the expected performance range.

3.7. Metal Ion Extraction by JNPs Supported by Wax Colloidosomes

The wax colloidosomes obtained by stabilizing the molten wax with JNP-bPEI were also employed in the metal ion extraction and recovery cycles, see Figure 5.

As already mentioned, the great advantage of these materials, JNPs and JNP decorated colloidosomes, is the ability to employ them in interfacial extraction technologies, with the main benefit of ease of deployment, collection and recovery from the water's surface and re-deployment on the surface of the water for the subsequent metal ion extraction/recovery cycles. To test the versatility of these JNPs-decorated wax colloidosomes in such technologies, we have deployed them for Cu(II) extraction. The graph in Figure 6A shows the extraction efficiencies q_e for three consecutive cycles from fresh 10 mM Cu(II) solutions, as described in the experimental section. In a different experiment, two portions of 6 g of colloidosomes were used to extract Cu(II) from the same 10 mL solution with an initial concentration of 10 mM. After the first extraction cycle, the concentration of copper in the solution was reduced to 4 mM and after the second extraction cycle, the concentration of Cu(II) decreased further to 0.3 mM, see Figure 6B,C. The numbers are in good agreement with the extraction capacities q_e calculated by the "differential concentration" method: (q_e $\approx$ 0.55 mg/g Colloidosome $\approx$ 8.7 µmol/g $\approx$ 52 µmol Cu(II)/6 g Colloidosome; 10 mL soln. with c = 10 mM $\rightarrow$ 100 µmol Cu(II)). The photographs in Figure 6B,C clearly demonstrate that a floating layer of 6 g of wax colloidosomes decorated with JNP-bPEI 3 mL, placed above a water solution of Cu(II) ions, and after brief shaking to ensure the wetting of the nanoparticles, is capable of almost complete interfacial extraction of the ions overnight. Next, the metal-loaded colloidosomes can be easily collected from the water's surface and dried; see the color change from white to blue of colloidosomes, before and after the Cu(II) metal ion extraction in the photographs of Figure 6B,C. The experiments were limited only to extraction, to demonstrate the working principle and versatility of wax colloidosomes and a recovery cycle for Cu(II) was not made.

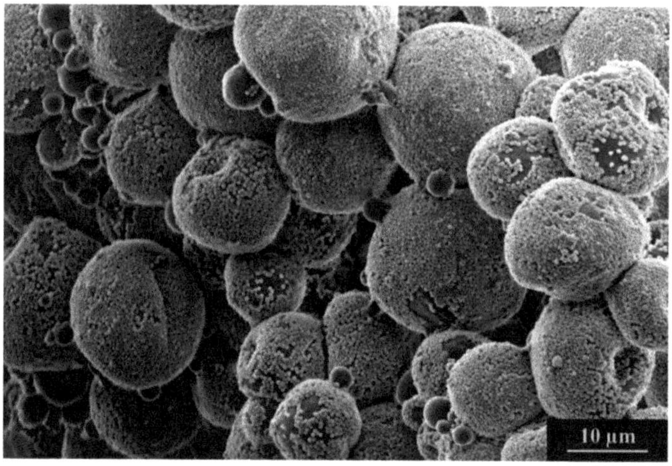

Figure 5. SEM image of the wax colloidosomes decorated with JNP-bPEI 3 mL, employed in interfacial extraction of metal ions.

It is important to note that the amphiphilicity and the interfacial activity of JNP-bPEI are important for both emulsification and to ensure the proper orientation of the Janus lobes at the surface of wax colloidosomes. In this case, good amphiphilic behavior ensures that the hydrophobic PS Janus lobe is oriented toward the wax surface, while the hydrophilic lobe covered with bPEI provides a large surface area for capturing metal ions. Therefore, this aspect is crucial in designing water-floating colloidosomes and employing them in the interfacial extraction of metal ions.

Furthermore, for a quantitative interpretation of the absorption capacity, both JNP-bPEI and JNP-bPEI-covered wax colloidosomes were employed in the extraction and recovery of the Cr(VI) comparative study. Thus, for the preparation of wax colloidosomes, 50 mg JNP-bPEI and 500 mg wax were used. Since not all colloidosome surfaces are fully covered it is reasonable to assume that the entire quantity of JNPs is incorporated into the surface of the wax, which for colloidosomes with an average radius of 15 µm, results in a mass fraction of $m_{JNP}/m_{Colloidosome}$ = 10%. Calculating q_e and q_r based on that assumption, it becomes apparent that incorporation of the JNP-bPEI into wax does not hinder their ability to interact with the ions, see Figure 7. It is important to note that the colloidosome samples were left on a mechanical shaker at 150 rpm overnight to ensure the complete wetting of the surfaces. The low density and slight hydrophobicity of the colloidosomes would otherwise prevent the wetting of the entire material and falsify the results. For demonstrative purposes, Figure S13 shows a sequence of images taken at 10 min intervals from a sample of 200 mg colloidosomes shaken in 5 mL aqueous solution containing Cr(VI) with a concentration of 10 mmol/L, and left to sit undisturbed. After 30 min, the wax colloidosomes can be found floating on the solution surface with complete separation from the liquid phase.

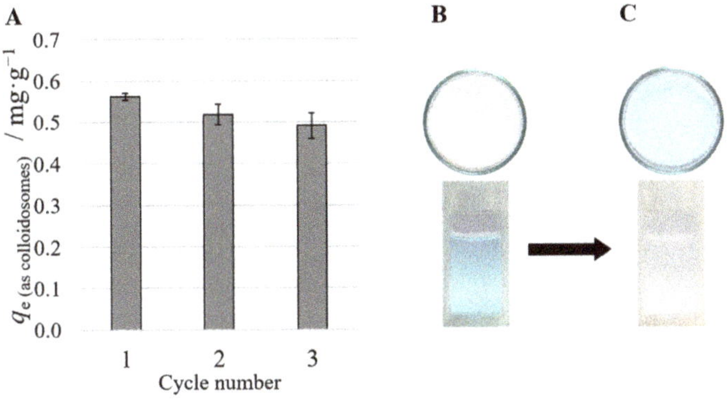

Figure 6. (**A**) Extraction performance of Cu(II) by wax colloidosomes stabilized with JNP-bPEI 3 mL after 3 consecutive extraction/recovery cycles. (B,C) Qualitative proof of the interfacial absorption/extraction of Cu(II) ions from water by floating JNP-bPEI decorated wax colloidosomes. To accurately represent the colors of the solution and the colloidosomes, they were photographed separately. (**B**) A layer of white, pristine (unwetted) colloidosomes with a mass of 6 g is floating on the surface of 10 mL of a blue 10 mM Cu(II) sulfate solution. (**C**) The sample was mechanically shaken and then left to separate overnight. After extraction of the solution with an appropriate amount of colloidosomes, the former becomes colorless while the latter appears bright blue.

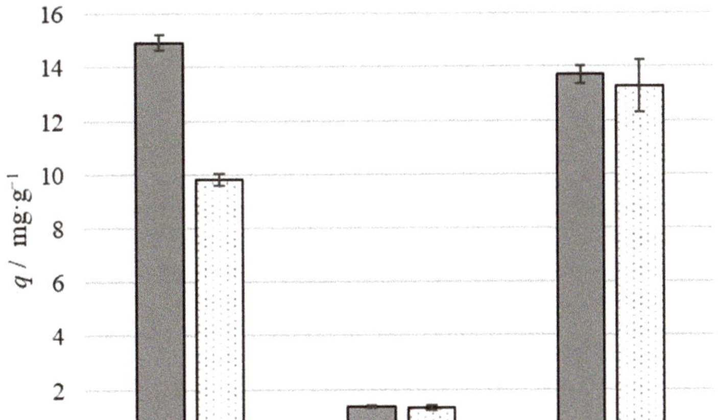

Figure 7. Comparison of q_e (filled bars) and q_r (dotted bars) between JNP-bPEI 3 mL and the colloidosomes that were prepared with it. q of Colloidosomes (**A**) was calculated using the total mass of the colloidosomes. q of Colloidosomes (**B**) was calculated using the approx. mass of JNP-bPEI in the colloidosomes.

4. Conclusions

It was shown that JNP-bPEI exhibits both of the desired characteristics: amphiphilicity and capability to absorb metal ions. We have demonstrated that the amphiphilicity of JNPs is key to enabling new ion extraction and recovery technologies. For example, we have demonstrated that amphiphilic JNP-bPEI can operate as (a) standalone ion-extraction agents, capable of absorbing metal ions from water and then transporting these to the surface where they could be collected and (b) as emulsifiers of molten wax to generate a new type of carriers capable of floating on the surface of water but also of metal ion extraction. There are, however, still multiple aspects left to explore. The kinetics of the extraction and recovery processes could be assessed by analyzing the metal content of the solutions as a function of time. With the optimized extraction times, the recyclability experiments using Cr(VI) could be repeated to see if the capacity still degrades. Different metal ions such as Ni(II) or Co(II) could be used to gather more data concerning the extraction and recovery capacities. The selectivity towards certain ions could also be determined—an important parameter when it comes to real-world applications. In future works, the long-term stability of JNP-bPEI-covered wax colloidosomes should be tested. The colloidosomes are only viable carriers if the nanoparticles stay attached to their surfaces even after prolonged agitation. A scale-up of the preparation procedure should also be attempted, with regard to manufacturing at an industrial scale.

Supplementary Materials: The following supporting information can be downloaded at: https://www.mdpi.com/article/10.3390/nano12213738/s1. Figure S1: FTIR spectra of polystyrene seed nanoparticles, JNP-CN, JNP-COOH and JNP-bPEI, Figure S2: Zeta potential with pH for JNP-CN and JNP-COOH, Figure S3: SEM images of JNP-CN and JNP-COOH, Figure S4: Zeta potential with pH for JNP-bPEI, Figure S5: SEM images of the homologous series of JNPs funcionalized with bPEI, Table S1: Zeta potential of the homologous series of JNP-CN, JNP-COOH and JNP-bPEI, Figure S6: Evolution of the interfactial tension of heptane/water in time in the presence of JNP-bPEI with pH, Figure S7: Interfacial tension of heptane/water in time as a function of JNP-bPEI concentration in water, Figure S8: SEM images of wax colloidosomes prepared with JNP-bPEI 2–4 mL, Figure S9: SEM images of colloidosomes, Figure S10: Performance of JNP-bPEI 2 mL after five consecutive extraction and recovery cycles for Cu(II), Figure S11: Regular and number-normalized Cr(VI) extraction and recovery values for JNP-bPEI 2–4 mL, Figure S12: Regular and number-normalized Cu(II) extraction and recovery values for JNP-bPEI 2–4 mL, Table S2: Metal ion extraction and recovery capacities of the homologous series of JNP-bPEI, Figure S13: Image sequence of colloidosome flotation and images of filtered colloidosomes.

Author Contributions: Conceptualization, O.P. and A.H.; methodology, O.P.; software, O.P.; validation, A.H.; formal analysis, O.P. and A.H.; investigation, O.P. and A.H.; resources, A.H.; data curation, O.P.; writing—original draft preparation, O.P. and A.H.; writing—review and editing, A.H.; visualization, O.P. and A.H.; supervision, A.H.; project administration, A.H.; funding acquisition, A.H. All authors have read and agreed to the published version of the manuscript.

Funding: This work was supported by a grant of the Ministry of Research, Innovation and Digitization, CNCS/CCCDI-UEFISCDI, project number PN-III-P4-PCE-2021-0306 (Contract Nr. PCE62/2022).

Institutional Review Board Statement: Not applicable.

Informed Consent Statement: Not applicable.

Data Availability Statement: All data is available upon request from the corresponding author.

Conflicts of Interest: The authors declare no conflict of interest.

References

1. Fu, F.; Wang, Q. Removal of Heavy Metal Ions from Wastewaters: A Review. *J. Environ. Manag.* **2011**, *92*, 407–418. [CrossRef]
2. Sulaiman, R.N.R.; Othman, N.; Amin, N.A.S. Emulsion Liquid Membrane Stability in the Extraction of Ionized Nanosilver from Wash Water. *J. Ind. Eng. Chem.* **2014**, *20*, 3243–3250. [CrossRef]
3. Kocherginsky, N.M.; Yang, Q.; Seelam, L. Recent Advances in Supported Liquid Membrane Technology. *Sep. Purif. Technol.* **2007**, *53*, 171–177. [CrossRef]

4. Wang, J.; Luo, J.; Feng, S.; Li, H.; Wan, Y.; Zhang, X. Recent Development of Ionic Liquid Membranes. *Green Energy Environ.* **2016**, *1*, 43–61. [CrossRef]
5. Fontàs, C.; Tayeb, R.; Dhahbi, M.; Gaudichet, E.; Thominette, F.; Roy, P.; Steenkeste, K.; Fontaine-Aupart, M.-P.; Tingry, S.; Tronel-Peyroz, E. Polymer Inclusion Membranes: The Concept of Fixed Sites Membrane Revised. *J. Membr. Sci.* **2007**, *290*, 62–72. [CrossRef]
6. Nghiem, L.; Mornane, P.; Potter, I.; Perera, J.; Cattrall, R.; Kolev, S. Extraction and Transport of Metal Ions and Small Organic Compounds Using Polymer Inclusion Membranes (PIMs). *J. Membr. Sci.* **2006**, *281*, 7–41. [CrossRef]
7. Ghaemi, N.; Daraei, P. Enhancement in Copper Ion Removal by PPy@Al2O3 Polymeric Nanocomposite Membrane. *J. Ind. Eng. Chem.* **2016**, *40*, 26–33. [CrossRef]
8. Ahmad, A.L.; Kusumastuti, A.; Derek, C.J.C.; Ooi, B.S. Emulsion Liquid Membrane for Heavy Metal Removal: An Overview on Emulsion Stabilization and Destabilization. *Chem. Eng. J.* **2011**, *171*, 870–882. [CrossRef]
9. Wu, D.; Honciuc, A. Design of Janus Nanoparticles with PH-Triggered Switchable Amphiphilicity for Interfacial Applications. *ACS Appl. Nano Mater.* **2018**, *1*, 471–482. [CrossRef]
10. Wu, D.; Chew, J.W.; Honciuc, A. Polarity Reversal in Homologous Series of Surfactant-Free Janus Nanoparticles: Toward the Next Generation of Amphiphiles. *Langmuir* **2016**, *32*, 6376–6386. [CrossRef]
11. Honciuc, A. Amphiphilic Janus Particles at Interfaces. In *Flowing Matter*; Toschi, F., Sega, M., Eds.; Springer International Publishing: Cham, Switzerland, 2019; pp. 95–136. ISBN 978-3-030-23369-3.
12. Honciuc, A. *Chemistry of Functional Materials Surfaces and Interfaces: Fundamentals and Applications*, 1st ed.; Elsevier: Amsterdam, The Netherlands, 2021; ISBN 978-0-12-821059-8.
13. Honciuc, A.; Negru, O.-I. Role of Surface Energy of Nanoparticle Stabilizers in the Synthesis of Microspheres via Pickering Emulsion Polymerization. *Nanomaterials* **2022**, *12*, 995. [CrossRef] [PubMed]
14. Thompson, K.L.; Williams, M.; Armes, S.P. Colloidosomes: Synthesis, Properties and Applications. *J. Colloid Interface Sci.* **2015**, *447*, 217–228. [CrossRef]
15. Mihali, V.; Honciuc, A. Self-Assembly of Strongly Amphiphilic Janus Nanoparticles into Freestanding Membranes. *Adv. Mater. Interfaces* **2021**, *9*, 2101713. [CrossRef]
16. Csikós, R.; Kestzhelyi, S.; Mózes, G. *Paraffin Products*; Mózes, G., Ed.; Elsevier Science: Amsterdam, The Netherlands, 1983.
17. Zeppieri, S.; Rodríguez, J.; de Ramos, A.L.L. Interfacial Tension of Alkane + Water Systems. *J. Chem. Eng. Data* **2001**, *46*, 1086–1088. [CrossRef]
18. Wu, D.; Binks, B.P.; Honciuc, A. Modeling the Interfacial Energy of Surfactant-Free Amphiphilic Janus Nanoparticles from Phase Inversion in Pickering Emulsions. *Langmuir* **2018**, *34*, 1225–1233. [CrossRef] [PubMed]
19. Jiang, S.; Granick, S. Janus Balance of Amphiphilic Colloidal Particles. *J. Chem. Phys.* **2007**, *127*, 161102. [CrossRef] [PubMed]
20. Tan, Y.Z.; Wu, D.; Lee, H.T.; Wang, H.; Honciuc, A.; Chew, J.W. Synthesis of Ligand-Carrying Polymeric Nanoparticles for Use in Extraction and Recovery of Metal Ions. *Colloids Surf. A: Physicochem. Eng. Asp.* **2017**, *533*, 179–186. [CrossRef]
21. Mishra, V.; Balomajumder, C.; Agarwal, V.K. Biosorption of Zn (II) onto the Surface of Non-Living Biomasses: A Comparative Study of Adsorbent Particle Size and Removal Capacity of Three Different Biomasses. *Water Air Soil Pollut.* **2010**, *211*, 489–500. [CrossRef]
22. Liu, Y.; Zhong, G.; Liu, Z.; Meng, M.; Jiang, Y.; Ni, L.; Guo, W.; Liu, F. Preparation of Core–Shell Ion Imprinted Nanoparticles via Photoinitiated Polymerization at Ambient Temperature for Dynamic Removal of Cobalt in Aqueous Solution. *RSC Adv.* **2015**, *5*, 85691–85704. [CrossRef]
23. Adibmehr, Z.; Faghihian, H. Preparation of Highly Selective Magnetic Cobalt Ion-Imprinted Polymer Based on Functionalized SBA-15 for Removal Co2+ from Aqueous Solutions. *J. Environ. Heal. Sci. Eng.* **2019**, *17*, 1213–1225. [CrossRef]
24. Gao, B.; Gao, Y.; Li, Y. Preparation and Chelation Adsorption Property of Composite Chelating Material Poly (Amidoxime)/SiO2 towards Heavy Metal Ions. *Chem. Eng. J.* **2010**, *158*, 542–549. [CrossRef]
25. Saraji, M.; Yousefi, H. Selective Solid-Phase Extraction of Ni (II) by an Ion-Imprinted Polymer from Water Samples. *J. Hazard. Mater.* **2009**, *167*, 1152–1157. [CrossRef] [PubMed]

nanomaterials

Article

Water-Floating Hydrogel Polymer Microsphere Composites for Application in Hydrological Mining of Cu(II) Ions

Andrei Honciuc *, Ana-Maria Solonaru and Mirela Honciuc *

"Petru Poni" Institute of Macromolecular Chemistry, Gr. Ghica Voda Alley 41A, 700487 Iasi, Romania; solonaru.anamaria@icmpp.ro
* Correspondence: honciuc.andrei@icmpp.ro (A.H.); teodorescu.mirela@icmpp.ro (M.H.)

Abstract: Innovative materials and technologies capable of extraction and recovery of technologically relevant metal ions from various water sources, such as lakes, oceans, ponds, or wastewater reservoirs, are in great demand. Polymer beads are among the most well-known solid-phase adsorbents and ion exchangers employed in metal ion recovery. On the other hand, hydrogels are an emerging platform for producing innovative adsorbents, which are environmentally friendly and biocompatible materials. In this work, we take advantage of both technologies and produce a new type of material by loading nanostructured polymer microsphere adsorbent into a PVA matrix to obtain a hydrogel polymer microsphere (HPM) composite in the form of a block. The main role of the poly(4-vinylpyrridine-co-methacrylic acid) microspheres is to adsorb metal ions, such as Cu(II), from model water samples. The secondary role of these microspheres in the hydrogel is to change the hydrogel morphology by softening it and stabilizing it under a foam-like morphology. The foam-like morphology endows these composites with the capability of floating on water surfaces. In this work, we report, for the first time, an HPM composite capable of floating on water surfaces and extracting Cu(II) ions from model water samples. This could enable more environmentally friendly hydrological mining technologies by simply deploying adsorbents on water surfaces for metal ion extraction and recovery, thus eliminating the need for water pumping and mechanical processing steps.

Keywords: metal ion extraction and recovery; hydrological mining; polymer adsorbents; Pickering emulsion polymerization technology; hydrogel polymer composites

check for updates

Citation: Honciuc, A.; Solonaru, A.-M.; Honciuc, M. Water-Floating Hydrogel Polymer Microsphere Composites for Application in Hydrological Mining of Cu(II) Ions. *Nanomaterials* **2023**, *13*, 2619. https://doi.org/10.3390/nano13192619

Academic Editor: George Z. Kyzas

Received: 1 September 2023
Revised: 19 September 2023
Accepted: 21 September 2023
Published: 22 September 2023

1. Introduction

Water and aqueous reserves from any source, such as oceans, lakes, or used waters from domestic, industrial, commercial, or agricultural activities, can be a valuable secondary resource for raw materials [1,2]. Innovative materials and technologies that can be deployed in hydrological mining aimed at recovering technologically relevant metal ions are in great demand. Synergistically, the same materials and technologies can be deployed for the removal of toxic metal ions from contaminated wastewater. However, recovery of metal ions must also be economically feasible; thus, materials and technologies that rely on minimal energy consumption are desirable. The current materials and technologies for wastewater treatment and purification have been extensively reviewed [3–5]. Thus far, solid-phase adsorbents and ion exchangers are deployed on a large scale for wastewater treatment and purification [4]. These could be cheap adsorbents that come from agricultural waste, food waste, biomass, inorganic materials, natural or synthetic polymers, etc. [6–8]. Among these, engineered nanomaterials could play a significant role [3,4,9], for example, micro or nanostructured polymers, such as microporous monoliths [10]. Recently, Pickering emulsion technology [10,11] and hydrogel-based composites [12] have been considered viable green water-based platforms for the preparation of solid- and semi-solid-phase adsorbents for the removal of metal ions from water and soil. As already alluded, future solid-phase adsorbent materials must also address the issue of energy efficiency and play the role of enablers for green technologies. For example, one way the energy consumption can be decreased is to eliminate the need for the energy-intensive process of pumping water through columns filled with ion exchangers. Instead, water-floating materials could be deployed on the surface of waters and ponds to adsorb polluting heavy metal ions or organic pollutants. Currently, such innovative materials with self-floating capabilities are being developed, among which we mention functionalized hollow glass microspheres [13,14], chitosan-based aerogels [15], graphene oxide aerogels [16,17], self-separating polymers [18], covalent-organic frameworks (COFs) [19], etc. On the other hand, the synthesis of hydrogels as adsorbents has been reported with good adsorption capacities [7,12,20,21]. Unfortunately, none of these studies focus on developing water-floating hydrogel or hydrogel polymer composites for the metal-ion extractions from water. Water-floating capability can be a great advantage for novel technologies in hydrological mining, as it minimizes the need for pumping large amounts of water into ion-exchange columns or other energy-intensive technologies requiring a great number of mechanical operations. In this work, we address exactly this aspect, and we show that such materials can be prepared and deployed in the extraction and recovery of some technologically relevant metal ions, such as Cu(II) metal ions [6]. The material consists of polymer microspheres synthesized via Pickering emulsion polymerization technology (PEmPTech) [22]. Due to the unique surface nanostructuring with the silica nanoparticles, these microspheres are perfectly dispersible in aqueous solutions and hydrophilic hydrogel precursor solutions, thus enabling their utilization in hydrogel matrices. Thus, we take advantage of both technologies and produce a new type of material by loading nanostructured polymer microsphere adsorbents into a PVA/Glycerol hydrogel to obtain a hydrogel polymer microsphere (HPM) composite in the form of a block. Upon inclusion in the hydrogel, these polymer microspheres can aid in foaming the hydrogel and stabilizing this foam, endowing the composite with a solid foam-like structure and water-floating capabilities. Therefore, we have prepared two classes of HPM composites: (i) hydrogel polymer microsphere blocks (HAM) and (ii) hydrogel polymer microsphere foams that can float on the surface of water (FAM). These were then deployed for the extraction and recovery of Cu(II) ions from model water samples. From a mass transfer perspective, materials that float on the surface may exhibit different ion adsorption capacities than those that are completely submersible in water. Therefore, in this work, we analyze, for the first time, this aspect of the capacity of extraction of metal ions between adsorbents that float and those that are completely submersible in water.

2. Materials and Methods

2.1. Materials

Tetraethylorthosilicate (TEOS) 99%, (3-glycidoxypropyl)trimethoxysilane (Gly) 98%, divinylbenzene (DVB) technical grade 80%, containing monomethyl ether hydroquinone as inhibitor, 4-vinylpiridine (4-VP) 95%, containing 100 ppm hydroquinone as inhibitor, methacrylic acid, (MA) 99% stabilized with 250 ppm 4-methoxyphenol, aluminium oxide (Al_2O_3), and 2,2′-Azobis(2-methylpropionitrile) (AIBN) 98% were purchased from Sigma-Aldrich (Merck, KGaA, Darmstadt, Germany). Poly(vinyl alcohol) (PVA) granules with an average molecular weight (Mw) of 12.4×10^4 g/mol and a 99–100% degree of hydrolysis and Glycerol (Gly) (99.6%) were purchased from Acros. Organics (Geel, Belgium). Copper chloride (II) dihydrate pure p.a. ($CuCl_2{\cdot}2H_2O$) was purchased from Chempur (GmbH Rueppurrer, Karlsruhe, Germany); hydrochloric acid (HCl) $\geq$ 37% was purchased from Fluka (Honeywell Specialty Chemicals, Seelze, Germany); ethanol absolute (EtOH), 99.3%, toluene, and n-hexane were purchased from Chemical Company; and ammonium hydroxide solution (28–30%) was purchased from analysis EMSURE ACS. Reag. Ph Eur. Supelco. All reagent-containing inhibitors were passed through aluminium oxide to remove the stabilizer before usage. All the aqueous solutions were prepared in freshly distilled water.

2.2. Synthesis and Functionalization of Silica Nanoparticles and Polymer Microspheres

The preparation procedure for silica nanoparticles (NP) and silica nanoparticles functionalized with epoxy (NP-Gly) by reaction with Gly was previously reported [23]. Briefly, in a 1000 mL round-bottom flask, 9 mL TEOS, 300 mL EtOH, 33 mL H_2O, and 27.7 mL NH_4OH were mixed at room temperature and 1000 rpm. Next, 54 mL TEOS dissolved in 200 mL EtOH was slowly added via separatory funnel (for 3 h), and the final reaction mixture was left for 24 h at room temperature. After this, the mixture was neutralized with 18 mL of HCl. The obtained nanoparticles were separated by centrifugation and washed three times with EtOH and three times with water. The surface functionalization reaction proceeded by dispersing 1.2 g of silica NPs in 30 mL of EtOH and then pouring in a 250 mL flask containing 10 mL EtOH, which was purged under Ar atmosphere. The reaction mixture was stirred at 1000 rpm. Subsequently, 2 mL of Gly was added dropwise. At the end of the addition time, the reaction mixture was heated and maintained to 60 °C for 24 h. The functionalized nanoparticles were washed three times with EtOH and another three times with water before being finally redispersed in water.

For the preparation of the polymer microspheres (PMs) via PEmPTech, two vinyl-bearing monomers, 4-VP and MA, having different polarities, and DVB as crosslinker, were used for the preparation of three batches of Pickering emulsions, PM1, PM2, and PM3. These were produced by first adding 30 mg of AIBN radical initiator to a 20 mL glass scintillator vial, followed by 2.5 mL of equimolar mixture of monomers (MA or 4-VP), 0.5 mL of crosslinker (DVB), and 0.75 mL of porogen solvent—and in our case, toluene. Next, 5 mg of colloidal particles NP-Gly and 12 mL of water were added. The glass scintillator vials with Pickering emulsion were then sonicated with a Vortex mixer LLG (Lab Logistics Group GmbH, Meckenheim, Germany) for 60 s at 3000 rpm, and every Pickering emulsion was then polymerized in an oil bath for 24 h at 70 °C. After the polymerization, the products were filtered and thoroughly washed with ethanol to remove the unreacted monomers and were dried at room temperature.

2.3. Preparation of Hydrogel—Polymer Microspheres Composites

First, a homogenous PVA solution with a concentration of 3% was prepared by dissolving the required amount of polymer in distilled water at 90 °C and vigorously stirring for 3 h. After the polymer was completely dissolved, Glycerol was added to obtain a mixture of 1/2 ratio of PVA/Glycerol and further stirred until complete homogenization. The HPM composites were prepared by mixing the previously obtained polymer solution with different quantities of PMs (i.e., 0.45 g and 1 g) so that they resulted in samples with two ratios of PVA/PMs. These were stirred for five minutes, with 200 rpm, at room temperature, and after that, were subjected to 10 subsequent cycles of freezing (at −20 °C) and thawing (at room temperature for 8 h). Samples thus obtained were named HAM-1 and HAM-2, respectively. A different series of samples (i.e., FAM-1 and FAM-2, respectively) were prepared by following the same recipes as for HAM-1 and HAM-2, with the difference that after obtaining the solutions, these were foamed and immediately frozen in liquid nitrogen, then subjected to 10 freezing/thawing cycles. Also, a blank sample (without PMs) of PVA/Glycerol was obtained by 10 freezing/thawing cycles.

2.4. Measurement of Ion Extraction and Recovery Capacity of HPM Composite

The Cu(II) ion concentration in the diluted supernatant and filtrate were analyzed using a UV–vis spectrophotometer (DLAB Scientific Co., Ltd., Beijing, China). First, calibration curves were generated corresponding to maximum absorption wavelength λ_{max} = 810 nm for $CuCl_2 \cdot 2H_2O$; see Figure S1 in the Supplementary Materials.

For the ion extraction, which refers to the extraction of metal ions from a stock solution, a weighted amount of HPM was immersed in 100 mL stock solution with a 5×10^{-2} M concentration.

The metal ion extraction capacity q_e (mg/g) was calculated with the formula:

$$q_e = \frac{(c_i - c_e)\ V}{m_P} \tag{1}$$

where c_i (mg/L) is the initial concentration of a stock solution or the contact solution, c_e (mg/L) is the extracted concentration, V (L) is the volume of the sample, typically 100 mL, and m_P (g) is the dry mass of the HPM (see Table S1).

For the ion recovery, which refers to the recovery of metal ions from the polymer adsorbent, the HPM composite was immersed in 50 mL of 5% HCl. The samples were then left in this condition for approx. 12 h. Supernatant and filtrate were analyzed using UV–vis, and the metal ion recovery capacity q_r (mg/g) was calculated by

$$q_r = \frac{c_r\ V}{m_P} \tag{2}$$

where c_r (mg/L) is the concentration of metal ions recovered from the HPM composite, V (L) is the volume of the sample, and m_P (mg) is the dry mass of the HPM composite (see Table S1).

The procedure of extraction–recovery was repeated five times unless otherwise specified.

2.5. *Material Characterization*

2.5.1. Scanning Electron Microscopy

The materials were investigated with a Verios G4 UC (Thermo Fischer Scientific Inc., Eindhoven, The Netherlands) scanning electron microscope (SEM), with a 5 keV beam energy, using an Everhart–Thornley detector, beam spot 50 pA.

2.5.2. Optical Microscopy

Microspheres and HPM composites were characterized with an IM-5FLD inverted fluorescence microscope (Optika Srl, Ponteranica, Italy) equipped with (i) an 8W XLED illumination source for sample analysis under transmitted light; (ii) 5W LED excitation illumination sources at 470, 560, and 385 nm and blue, green, and UV filter sets for sample analysis in fluorescence mode; (iii) color digital Camera Optika C-P6, 6.3 MP; and (iv) OPTIKA PRO VIEW (Optika Srl, Ponteranica, Italy) software for image acquisition and processing. Samples were characterized with 10× magnification objectives in transmitted illumination mode.

2.5.3. Contact Angle—Washburn Method

Water contact angle of the nanostructured microspheres obtained by PEmPTech was determined via the capillary rise method, using the DCAT 15 Tensiometer balance (DataPhysics Instruments GmbH, Filderstadt, Germany), equipped with the DCATS 32 software module for calculating the contact angle via Washburn method. The capillary constant of the PM samples packed in the glass capillary was first determined using hexane. For comparison, the contact angles of marine sand, which was sieved to a granulation of <250 μm and calcinated at 850 °C, were also measured. After the determination of the capillary constant, the polymer microsphere was loaded in special glass capillaries with a porous glass bottom (DataPhysics Instruments GmbH, Filderstadt, Germany), with an inner diameter of 9 mm, outer diameter of 11 mm, and height of 62 mm, and was filled with powder up to 22.5 mL dry volume. After filling with powder, the capillary was gently knocked with a wooden popsicle stick to achieve a compact packing of the powder and removal of the packing voids. Then, the Washburn capillary was lowered gently until it touched the water surface; once it touched the water surface, the capillary stopped, and the water started rising into the capillary packed with powder. The raw data consisted of the recorded weight of the water intake vs. time. The total duration of the experiment was 40 s.

2.5.4. Penetration Experiments

Penetration experiments of the HPM composites were conducted with a DCAT 15 Tensiometer balance (DataPhysics Instruments GmbH, Filderstadt, Germany), equipped with the DCATS 35 software module for penetration experiments. The penetration experiments were conducted with a metal penetration cone as the penetrometer probe. Upon free hanging of the metal cone from the piezoelectric weighing sensor, the tensiometer balance registered 20 g, and the software automatically tarred to zero before the penetration experiments. This cone was lowered slowly onto the soft HPM composite, and upon contacting the HPM surface, a negative weight due to the opposing force was recorded, which increased in absolute value with the cone immersion depth. The total penetration depth of the cone into the HPM was 4 mm. Thus, the raw data recorded by the balance was the negative weight due to the opposing force to penetration of the cone vs. time. The total time of the penetration experiments was 30 s.

3. Results and Discussion

3.1. Preparation via PEmPTech and Characterization of Polymer Microspheres

Pickering emulsion polymerization technology (PEmPTech) is a recently developed green, water-based method, also developed by our group and by others [24,25], for the facile synthesis of polymer microspheres with nanostructured surfaces. The method utilizes oil-in-water (*o*/*w*) Pickering emulsions, emulsions that are stabilized by silica nanoparticles, and the dispersed phase is a water-immiscible vinyl-bearing monomer. As a side note, other groups utilize the same technology for producing microporous polymer monoliths [10] and asymmetrically structured Janus membranes [26]. The silica nanoparticles utilized in the current work are 500 $\pm$ 10 nm diameter silica nanoparticles modified with glycidyl functional groups on the surface (see Figure 1A), which have been previously shown to produce preferentially *o*/*w* emulsions [27]. The Pickering emulsions stabilization mechanism by nanoparticles has been previously described, and it is mainly due to the interfacial adsorption of the silica nanoparticles at the oil–water interface with the formation of a self-assembled monolayer which acts like a shield preventing the coalescence of the oil droplets [28].

We have prepared *o*/*w* Pickering emulsion, where the oil phase contains MA and 4-VP, DVB, and a common water-immiscible solvent, toluene. The polymerization mechanism resembles suspension polymerization. Interestingly, the PEmPTech is extremely versatile, allowing for a broad spectrum of monomer composition cocktails to be used; for example, partially water-immiscible vinyl-bearing monomers, as well as completely water-insoluble monomers, can be used if they have a common solvent or they are miscible with one another. Upon polymerization of the *o*/*w* Pickering emulsion, the dispersed phase, the oil droplets are converted into [25] poly(4-vinyl pyridine-co-methacrylic acid) polymer microspheres (see Figure 1B,C), whereas the self-assembled monolayer is now trapped and gives the microspheres the typical nanostructuring (see Figure 1D), as observed for the microspheres produced by this technology (PEmPTech). The implications of nanostructuring in the water-wetting of the polymer microspheres are significant. We have previously demonstrated that for a hydrophobic polymer whose typical water contact angles are around 80°, due to the nanostructuring from a self-assembled monolayer of NP-Gly on the surface of the polymer, the wettability increases significantly, lowering the water contact angles to values below 60° characteristic for a hydrophilic surface [26]. This improvement in water wettability due to nanostructuring has significant implications for the extraction of metal ions from aqueous solutions by these polymer microspheres.

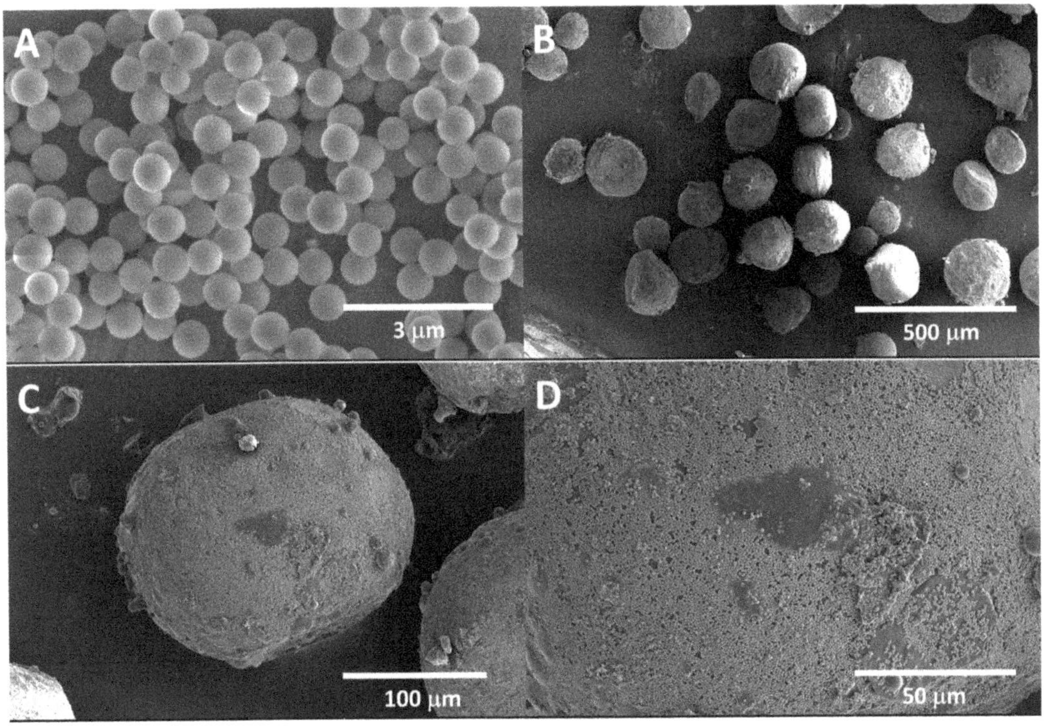

Figure 1. SEM images of (**A**) silica nanoparticles functionalized with glycidyl, (**B**) polymer microspheres obtained via PEmPTech, exhibiting nanostructured surface (**C**,**D**) due to trapping at the oil/water interface of a self-assembled monolayer of silica nanoparticles, the Pickering emulsion stabilizing nanoparticles.

In the current case, for the polymer microspheres synthesized, we have measured the water wettability of the polymer microsphere powder using the Washburn method [29], which is based on monitoring the weight of the liquid intake of a powder due to capillary forces at constant temperature:

$$\cos\theta = \frac{\mu}{C \cdot \rho^2_{Liquid} \cdot \gamma_{Liquid}} \cdot \frac{m}{t} \tag{3}$$

where θ is the contact angle (wettability); μ is the viscosity of the contacting liquid; ρ is the density; γ is the surface tension; m is the mass intake of the liquid; t is the time; and C is the packing constant dependent on the capillarity, with the capillaries being formed between the powder grains or nanoparticles. In the current case, the capillary constant for the polymer microspheres was determined to be C = 1.179 mm^5 with hexane (see Figure 2).

From the capillary rise data presented in Figure 2, we can see that the water wettability of powder consisting of PMs from three different batches is consistent and is distributed around an average value of the contact angle of $56^\circ \pm 5^\circ$. Based on this value, we can draw the conclusion that the PMs can be dispersed reasonably well into an aqueous system, such as the PVA solution, for the generation of the HPM composites.

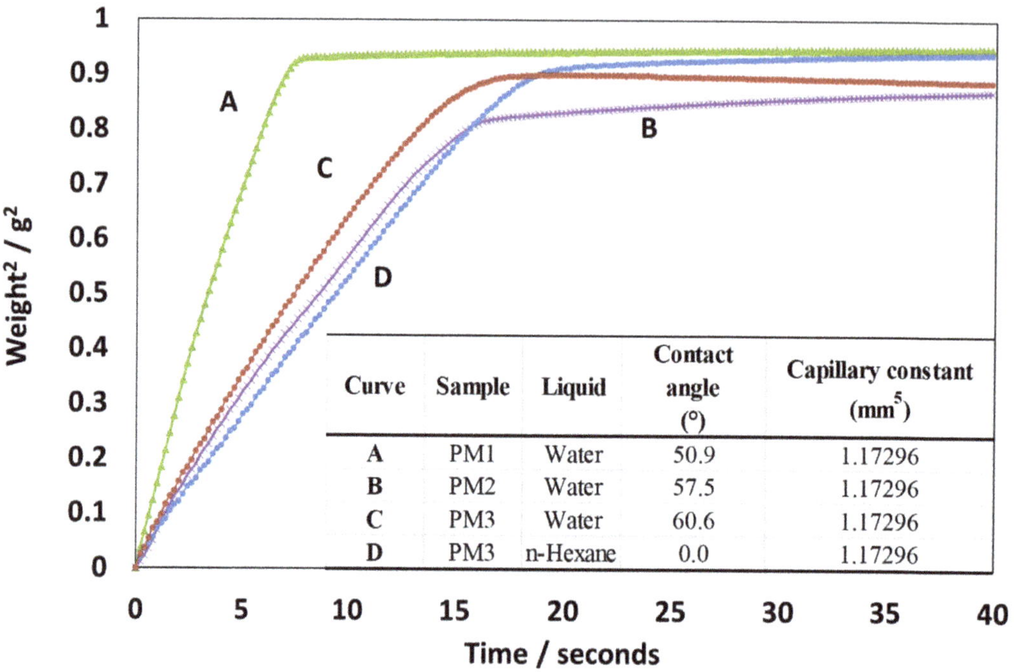

Curve	Sample	Liquid	Contact angle (°)	Capillary constant (mm^5)
A	PM1	Water	50.9	1.17296
B	PM2	Water	57.5	1.17296
C	PM3	Water	60.6	1.17296
D	PM3	n-Hexane	0.0	1.17296

Figure 2. Graph of the liquid intake weight squared vs. time by the capillary packed with powder consisting of polymer microspheres (PMs). The curve (A) represents batch PM3 wetted by the hexane, from which the packing constant C was calculated (see Equation (3)). Curves (B), (C), and (D) represent the water intake with time by the powder consisting of PMs manufactured in three different batches, from which the water contact value was calculated (PM1, PM2, and PM3, respectively). Note that a steeper slope of liquid intake corresponds to a lower contact angle with the liquid. The table inset shows the value of the contact angles with the liquid obtained for each corresponding curve and PM batch with the corresponding capillary constant.

The interaction of the poly(4-vinyl pyridine-co-methacrylic acid) PM with the metal ion can be evidenced from the FTIR-ATR spectra of the PMs before and after the adsorption of Cu(II) ions. Figure S2 shows the IR spectra of the 4-VP-co-MA copolymer—Cu(II) complex. The absorption bands at 1606 (C=N stretching), 1541, 1463, and 1398 cm^{-1} are assigned to the characteristic vibration of the pyridine ring [30–32]. The absorption bands at 1076 and 948 cm^{-1} could be assigned to the in-plane and out-of-plane C–H bending of the pyridine ring [30], while the vibration at 1163 and 1076 cm^{-1} could be assigned to the single-bond C-O stretch in the carboxyl. The effect of the Cu(II) adsorption by the PMs can be best observed in the region 1200–1800 cm^{-1}. For example, the peak at 1614 cm^{-1} is a new peak, strongly enhanced in the presence of Cu(II) ions and representing the fraction of the coordinated vinyl pyridine rings, i.e., ascribed to the pyridine ring–Cu^{2+} bond vibration [30,32]. On the other hand, after Cu(II) coordination, the characteristic vibrations of the pyridine ring mentioned are blue-shifted to 1517, 1452, and 1384 cm^{-1} [30]. At the same time, the peak at 1705 cm^{-1}, corresponding to the stretching modes of the carbonyl groups, is strongly enhanced in the presence of the Cu(II) ions, presumably due to complexation [33], or as Lee et al. ascribe, a liberation of the carbonyl stretching due to the complexation of the acid hydroxyl group [31]. The characteristic stretching vibration for the carboxylate group, usually at 1600 cm^{-1}, overlaps with the characteristic bands of pyridine; therefore, an enhancement of the band at 1600 cm^{-1} can be coming from both units, due to pyridinium coordination with the Cu(II) as well as the electrostatic interaction of the carboxylic group with the Cu(II) [32,33]. Thus, the data indicate strong interaction and coordination of the Cu(II) ions by the PMs.

3.2. Preparation, Characterization, and Morphology of HPM Composites

In this work, a series of HPM composites have been prepared, where both their compositions vary in terms of microparticle content and preparation method. For example, two categories of HPM composites have been created: (i) simple HPMs with increasing amounts of microsphere content HAM-1 and HAM-2 and (ii) foamed HPM series FAM-1 and FAM-2 with increasing amounts of polymer microspheres that are capable of floating on the water surface (see Table S1).

Photographs of the obtained HPMs and the PVA hydrogel are given in Figure S3 in the Supplementary Materials (SM). From these images (Figure S3), the physical dimensions of the HPMs vary as a function of composition. Although they were prepared in a silicon form that had a diameter of 40 mm, the PVA hydrogel containing no microspheres shrunk by a few millimeters after preparation as compared to the container. The foamed and non-foamed HPM composite FAM-1 and HAM-1 have retained the lateral dimensions of the container, but each has a different thickness, albeit the composition stays the same. The fact that FAM-1 and FAM-2 are thicker than HAM-1 and HAM-2, respectively, is due to the foaming of the former sample. The foaming in the case of the former sample was preserved by instantaneous freezing with liquid nitrogen of the hydrogel after mechanical agitation (shaking). Thus, we can note a difference in morphology between the HAM- and FAM-type samples, while the PVA hydrogel samples are compact and homogeneous. The HAM- and FAM-type samples are heterogeneous, and this was evidenced by cutting the hydrogels and imaging them in the cross-section after the adsorption of Cu(II) ions. Thus, it can be seen from Figure 3 that the HAM-1 and HAM-2 samples are only colored in intense blue in the bottom part of the sample (see Figure 3B,C for the bottom layer), where the polymer microsphere adsorbents have accumulated due to sedimentation during the gelation time. This is explained by the fact that only the PMs are capable of chemically binding Cu(II) ions, while the gel itself can only weakly physically adsorb these ions; thus, the middle of the sample remains white, as with the PVA hydrogel Figure 3A.

The FAM-1 and FAM-2 HPM composite samples (Figure 3D,E) appear structured similar to a sponge, are more voluminous than the HAM-type samples, and the adsorption of Cu(II) evidences a heterogeneous distribution of PM, mostly in the bottom part of the sample, as it can be clearly seen in Figure 3F. The sponge-like structure of the FAM-type sample endows them with the capacity to float on the surface of the water (see Figure 3G), while the HAM-type samples are not capable of floating. This capability is the property we were looking for in HPM composite that could be deployed on the surface of the water to extract metal ions from waters for hydrological mining and provide a valuable secondary source for raw materials, such as metal ions.

Further, the softness of the HPM composites was evaluated for the top and bottom of the samples using the penetration probe, as described in the Materials and Methods section. The experimental results are presented in Figure 4. Also, penetration experiments were carried out both at the top and at the bottom of the sample to see if there were differences. From the penetration experimental data, we observe three major trends, namely (i) the increase in the softness of the material with the addition and an increasing amount of polymer microparticles added in the HPM composition; (ii) the samples with microparticles are anisotropic, they are softer on the top than on the bottom; and (iii) the foamed samples are softer than the non-foamed samples. For the first case, (i) the addition of polymer microparticles appears to break the cohesion of the hydrogel and soften it considerably; the softness increases in the order PVA-Hydrogel < HAM-1 < HAM-2 < FAM-1 < FAM-2 (see Figure 4). In addition, it is also obvious that the addition of polymer microparticles causes the volume of the sample to change considerably, compared with the dimensions of the reference PVA-Hydrogel sample with the HAM-1, where the reference sample PVA-Hydrogel seems rather hard (Figure 4) and compact (Figure S3). This change in volume from PVA-Hydrogel to HAM-1 can be attributed to the formation of foam during shaking, whereas particles are known to stabilize foams. The increase in the degree of softness from HAM-1 to HAM-2, which differ only in the number of microspheres, appears to support the hypothesis that additional microspheres cause more foam bubbles. For the second case (ii), anisotropy of the sample arises during the gelation process, whereas the microparticles sediment on the bottom part of the sample, while some degree of foam is preserved on top. For the last case, (iii) the freezing with liquid N_2 of the freshly shaken sample preserves the foam bubbles, while clearly the microspheres sediment on the bottom of the sample. The degree of softness of the FAM-1 and FAM-2 at the top is comparable, while FAM-1 is harder than FAM-2. This is probably due to variations in sample preparation and particle sedimentation.

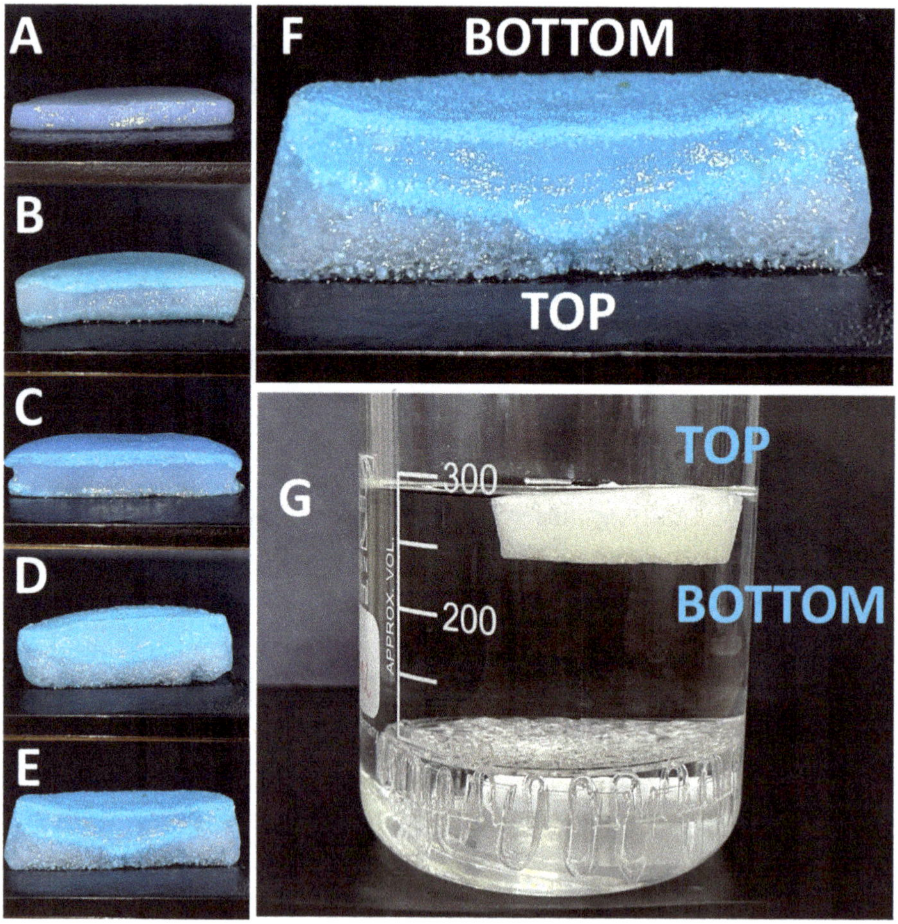

Figure 3. Photographs of the cross-section of the PVA hydrogel (**A**) and of HPM composites (**B**) HAM-1, (**C**) HAM-2, (**D**) FAM-1, and (**E**) FAM-2, having all the same orientation with the bottom of the HPM, upwards, and the top, downwards. The images were taken after HPM composite exposure to a solution of 5×10^{-2} M $CuCl_2 \cdot 2H_2O$. The reference of top and down for the HPM is taken as the position in which the HPM sat during the preparation (gelation). Image (**F**) is a close-up photograph of FAM-2 and (**G**) of the floating FAM-2 HPM composite on the surface of the water.

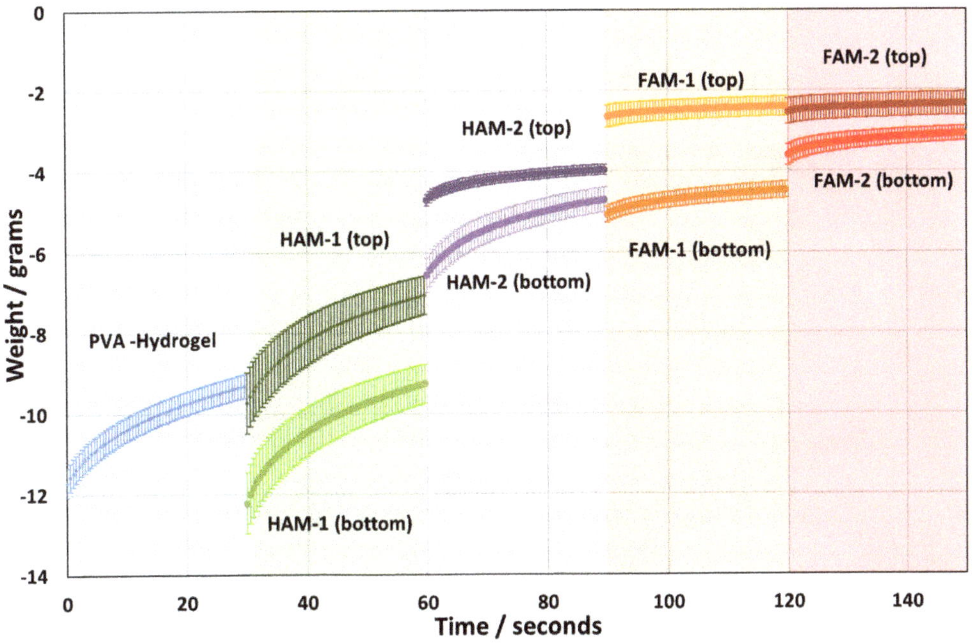

Figure 4. Resistance weight of the HPM to cone penetration vs. time for the series of HPM. The penetration experiment spanned the duration of 30 s for an immersion depth of the cone of 4 mm into the sample. The penetration tests were executed both to the top and to the bottom of the samples. The increase in softness of the material is evidenced by a lower absolute weight resistance to cone penetration. The error bars represent the standard deviation from at least three measurements.

3.3. Role of Morphology of the HPM Composites in Cu(II) Adsorption and Water-Floating Ability

The HPMs were further employed in Cu(II) ion adsorption studies. The experimental procedures for adsorption/extraction of Cu(II) from model water samples with a 5×10^{-2} M concentration were kept the same for the control sample, the PVA hydrogel, the HAM-1 and -2, and the FAM-1 and -2 samples, as described in the experimental procedures. Similarly, the Cu(II) ion desorption/recovery studies were kept the same for all samples and were proceeded by treatment with a 5% HCl solution for a period of 12 h, after which the concentration of the Cu(II) ion recovered from the material was measured. Both types of experiments were carried out under gentle stirring of 200 rpm. It can be noted that FAM-1 and FAM-2 samples were floating on the surface of the water (see Figure 3G) both during extraction and during metal ion recovery experiments. The PVA, HAM-1, and HAM-2 samples stayed submerged in the water, at the bottom, all the time; thus, in the case of these samples, an enclosing plastic cage with holes to allow water diffusion was manufactured to isolate the stirrer and avoid the magnetic stirrer physically hitting the sample.

The HAM-1 and HAM-2 samples had a drastic change in color, from white to deep blue, upon absorption of the Cu(II) ions, even compared to the PVA control hydrogel samples (Figure S4). Initially, the HPM composites are white in color, as shown in the image in Figure 5A, and in the optical microscope image (Figure 5A), it can be seen that the embedded polymer microspheres are rather colorless. When the HPM has been exposed for 12 h to a 5×10^{-2} M $CuCl_2 \cdot 2H_2O$ solution, its color changes to deep blue (Figure 5C), and the microspheres become intensely blue-colored (Figure 5D). Upon removal from the Cu(II) ion solution and treatment with a 5% solution of HCl, the HAM-2 changes color again (Figure 5E), and the microspheres become colorless, as shown in the optical microscope images in Figure 5F. The same is true for FAM-1 and FAM-2 HPM composites (see Figure S5). The difference in morphology between the HAM- and FAM-type samples noted above is becoming evident after the adsorption of Cu(II) ions. Upon Cu(II) ion adsorption, the HAM-type non-floating submersible samples and the FAM-type floating samples are colored in the bottom part of the sample, where the polymer microsphere mostly accumulated due to sedimentation during preparation (Figure 3B–E).

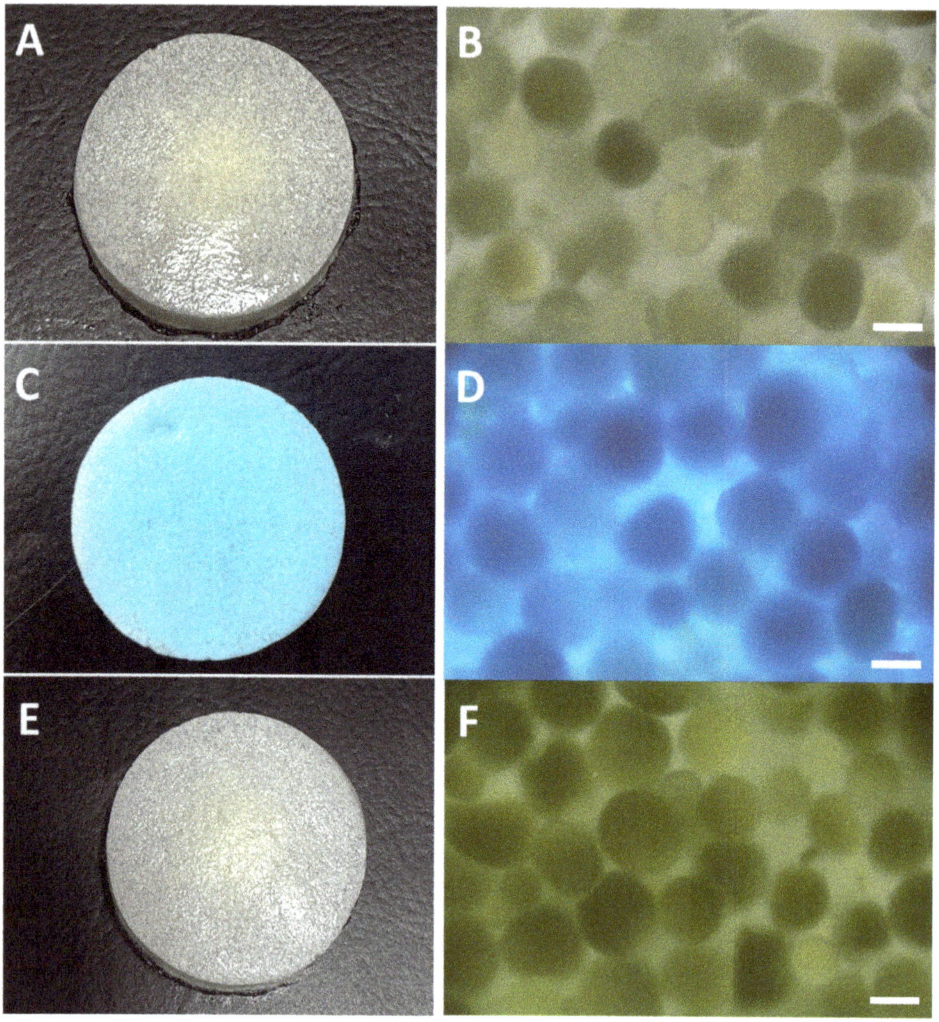

Figure 5. (**A**) Photograph of the as-prepared HPM composite, HAM-2, with the corresponding optical microscope image (**B**) taken with a 10× magnification showing the embedded polymer microparticles. (**C**) Photograph of the same hydrogel after exposure for 12 h to a 5×10^{-2} M $CuCl_2 \cdot 2H_2O$ solution and the corresponding optical microscope image (**D**) at 10× magnification, showing a strong blue coloration of the microspheres. (**E**) Photograph of the same hydrogel after being kept in a 5% solution of HCl solution and the corresponding microscope (**F**) at 10× showing a discoloration of the microspheres. The scale bar in the microscope images is 100 µm.

Thus, due to their unique morphology, the FAM-type samples are capable of floating due to the existence of air bubbles in the foamed part of the sample. However, in contrast to the HAM-type samples, the FAM-type samples show, at least qualitatively, adsorption in the bottom part of the samples, in other words, in the part of the sample with fewer bubbles, where the polymer microsphere adsorbents are concentrated. Next, we will analyze in quantitative terms how these two types of samples perform with the given difference in morphologies and the same composition.

3.4. Capacity of Ion Extraction and Recovery of HPM Composites

The capacity for Cu(II) ion extraction q_e refers to the adsorption or removal of metal ions from model water samples, stock solutions of 5×10^{-2} M concentration. However, because for these materials, we are also interested in the ability to recover the metal ions from the material, we have also measured the Cu(II) recovery capacity q_r, which was achieved by treating the HPM composites with an acidic solution of 5% HCl. This could demonstrate that these materials are indeed feasible to be deployed in technologies interested in recovering raw materials by hydrological mining. Thus, each sample from both classes of composites, HAM type and FAM type, as well as control samples, the polymer microsphere adsorbents and PVA hydrogel control samples, have been employed in at least four cycles of extraction and recovery of Cu(II) ions. The results are presented in Figure 6. Here, we note that the adsorption capacities calculated with Equations (1) and (2) were performed for the entire mass of the hydrogel, and Figure 6 shows the effective q_r and q_e capacities, meaning that the adsorption capacity of the PVA control sample, q_r = 15.7 mg/g and q_e = 11.6 mg/g, accounting for physical adsorption of Cu(II) ion, have been subtracted from each value. From the results presented, it can be immediately noted that the capacities of the HAM-type samples are only slightly less than that of FAM-type samples. Furthermore, while HAM-2 and FAM-2 both contain a double amount of polymer microsphere adsorbents, no significant change in the adsorption capacity within the experimental error for HAM-2 is observed; both the q_e and q_r are comparable to those of HAM-1, while for FAM-2, both the q_e and q_r are slightly less than that of FAM-1. Further, the only parameter that changes between the two sets of samples, HAM and FAM, is their morphology; thus, the effect of the morphology on the adsorption capacity can be understood by comparing HAM-1 to FAM-1 and HAM-2 to FAM-2. Here, we note a decrease in the adsorption performance for the Cu(II) ions of the FAM-2 type samples (q_e = 7.7 mg/g and q_r = 5.6 mg/g) compared to the HAM-2 (q_e = 11.1 mg/g and q_r = 11.5 mg/g) samples of about 31% for q_e and 51% for q_r, due to their unique morphology, whereas we hypothesize that the part containing air bubbles causing their floating ability, contributes less to ion adsorption. Also, a slight change in adsorption capacity can be noted for the HAM-1 (q_e = 12.0 mg/g and q_r = 11.8 mg/g) vs. FAM-1 (q_e = 9.9 mg/g and q_r = 11.2 mg/g) samples of about 18% for q_e and 5% for q_r, we believe also due to their different morphology. By a more careful analysis, however, it can be noted that the adsorption capacity data in Figure 6 are inversely correlated with the softness of the sample in Figure 4. In other words, the softer the sample, which is equivalent to saying the more foamed the sample is, the less adsorption capacity for Cu(II) ions. On the other hand, only the foamed samples are capable of floating on water. Thus, we conclude the tradeoff for having water-floating adsorbent samples is only a slight decrease in the adsorption capacity on the part of the FAM-1 and FAM-2 samples.

In absolute terms, while the difference in adsorption capacity of HAM- and FAM-type materials with the same chemical composition is only a reflection of material morphology, the total mass of the Cu(II) ions adsorbed by each HPM composite is due to their chemical composition. The total mass of Cu(II) ions adsorbed by HPM composite series and the reference PVA hydrogel is shown in Figure 7, where a clear change in the total mass of adsorbed metal ions can be seen with the load in the PM adsorbents in the HPM, decreasing in the order HAM-2 (q_e = 42.70 mg/g and q_r = 62.3 mg/g) > FAM-2 (q_e = 36.12 mg/g and q_r = 50.1 mg/g) > HAM-1 (q_e = 34.05 mg/g and q_r = 48.8 mg/g) > FAM-1 (q_e = 33.44 mg/g and q_r = 48.0 mg/g). HAM-2 and FAM-2 have the same amount of polymer microspheres, of 1 g, while the HAM-1 and FAM-1 each have an amount of 0.45 g of polymer microsphere load. Thus, the total mass of adsorbed Cu(II) ions reflects the load of the HPM composite with the PM adsorbent. It is important to note that for FAM-1 and FAM-2 samples, there were no negative effects on their floating capability observed with the PM loading amount.

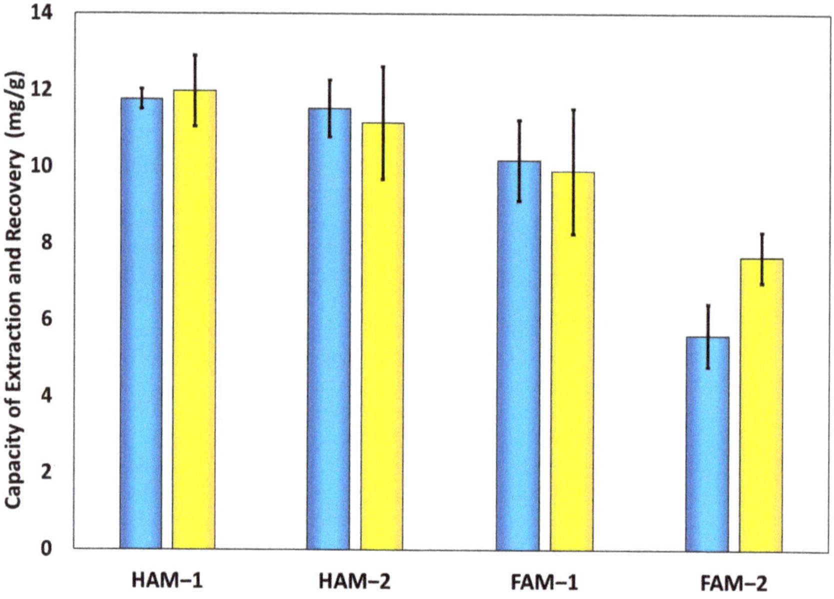

Figure 6. Histogram showing the effective recovery capacity q_r (blue bars) and extraction capacity q_e (yellow bars) data for the HAM- and FAM-type samples, from which the corresponding capacities of the control PVA hydrogel sample have been extracted.

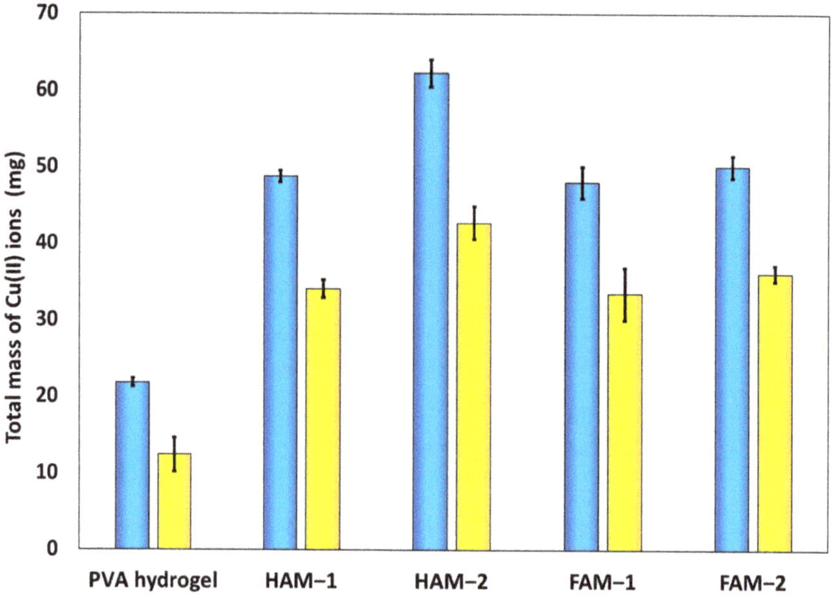

Figure 7. Mass of Cu(II) ions recovered (blue) and extracted (yellow) by the HPM composites.

In addition, the q_e and q_r capacities have been monitored with the metal ion extraction and recovery cycle number, as with HAM-1, for example (see Figure 8). The mass of Cu(II) ions adsorbed or desorbed from the HPM composite shows no change or loss in capacity up to the fifth cycle of extraction and recovery. These data strongly indicate that these HPM composite materials, especially those of FAM-type morphology, are a new technology that can be successfully deployed on the surface of the water in extraction and recovery of the metal ions from various water sources, lakes, oceans, ponds, etc., for hydrological mining of technological relevant metal ions.

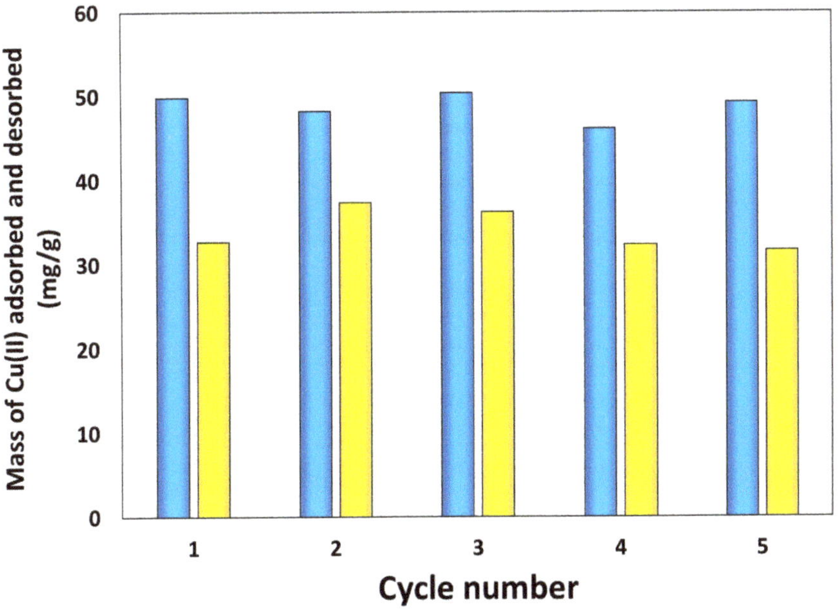

Figure 8. Mass of Cu(II) recovered (blue) and extracted (yellow) by HAM-1 with the cycle number, showing no loss in adsorption and desorption capacities.

At this point, it is important to compare the current results with the other results in literature. Functional hydrogels generated from polymers with functionality, such as amine, amide functional groups, or carboxylic groups, capable of binding metal ions exhibit excellent adsorption capacities for Cu(II) ions. This should obviously be due to the high functional group density provided by these polymer chains. For example, for the recently reported hydrogels obtained from poly(acrylic acid-co-acrylamide), the Cu(II) ion adsorption capacity was 211.7 mg/g [34], and for the poly(acrylamide-co-sodium methacrylate) hydrogel the Cu(II) ion adsorption capacity was 24.05 mg/g [35]. The former is significantly larger than the one reported here, while the q_e of the latter is comparable to that of the HPM composites. On the other hand, chitosan hydrogel beads, with chitosan being a well-known natural polymer with a high density of glucosamine groups capable of binding metal ions, have a q_e of 130 mg/g. Finally, hydrogel–particle composites, where both particles and the hydrogel are capable of binding the Cu(II) ions, such as hydrogel–clay nanocomposites, have shown a q_e of 68 mg/g [36], while the hydrogel–graphene oxide composite has shown a much-reduced q_e of 5.99 mg/g for Cu(II) ions [37]. Thus, we conclude that the Cu(II) ion adsorption capacities obtained in the current work, q_e ranging from 7.7 to 12.0 mg/g, fall in the middle of the high and low range of the values reported in the literature.

In addition, we have studied the adsorption kinetics of the HAM-2 and FAM-2 materials, the data presented in Figure 9, and we fitted the adsorption data to three different adsorption kinetic models: (i) pseudo-first-order kinetics, described by the equation $q(t) = q_e\left(1 - e^{-k_1 t}\right)$, where k_1 is the rate constant of the first-order adsorption process, *q(t)* is the amount of metal ion adsorbed at any given time, *t*, and q_e at equilibrium [38]; (ii) pseudo-second-order rate expression $q(t) = \frac{q_e^2 k_2 t}{1+q_e k_2 t}$, where k_1 is the rate constant of the second-order adsorption process [38]; and (iii) intraparticle diffusion model, which indicates that the intraparticle diffusion is the rate-limiting step $q(t) = k_d t^{0.5} + C$, where the k_d is the intraparticle metal ion diffusion constant, and C is an arbitrary constant [39]. The fit parameters are given in Table S2. It can be seen that the best fit to the data was obtained for the pseudo-second-order kinetics model for HAM-2, followed by the intraparticle diffusion model for FAM-2. This suggests that the pseudo-second-order kinetics model dominates the adsorption characteristic for the HAM-2, and the intraparticle diffusion model is characteristic of the FAM-2 material. In reality, the adsorption phenomenon in the HPM composite cannot be described purely by a single model, but its adsorption characteristic can be best modeled by a combination of different components kinetics of the adsorption and diffusion. Thus, it can be said that FAM-2 has a stronger diffusive component than the HAM-2 composite. Further, judging by the lower value of the rate constants for FAM-2 in Table S2, but also by the fact that the saturation plateau in Cu(II) ion intake is reached much later than that of HAM-2 (see Figure 9), we conclude that the adsorption in FAM-2 is about twice as slow as the adsorption of Cu(II) ions in HAM-2. This can only be explained by the difference in the morphology of the material, whereas the foamed FAM-2 composite lengthens the diffusion path of the Cu(II) ions inside the material to the adsorption sites. Other authors have also related the pseudo-second-order kinetics to chemisorption rather than physisorption [40].

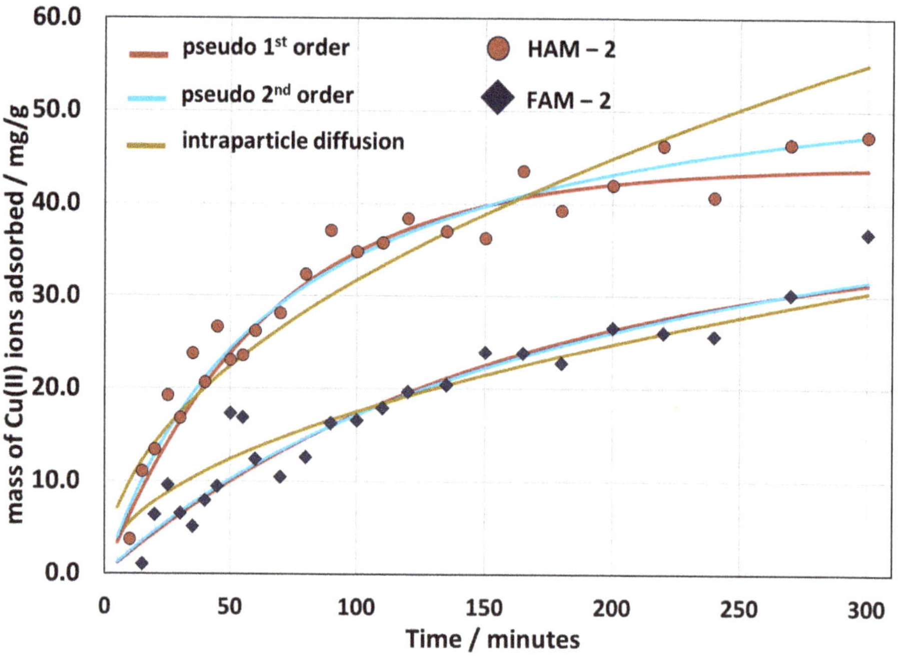

Figure 9. Experimental data of the mass intake of the Cu(II) ions by HAM-2 and FAM-2 with time. Each data set was fitted to a pseudo-first-order kinetic equation, a pseudo-second-order kinetic equation, and an intraparticle diffusion model equation, as indicated in the legend of the graph.

4. Conclusions

In this work, we have synthesized HPM composite materials with different morphologies, capable of floating on the surface of water and carrying polymer microsphere adsorbents for Cu(II) metal ions extraction from water samples. The HPMs have shown a good adsorption capacity for these metal ions from water samples, and we have shown that the Cu(II) ions can be easily recovered from these materials. We have demonstrated that the HPMs in different morphologies, such as water-floating FAM types (foamed hydrogels) exhibit only a small loss in the extraction capacity in comparison to the non-foamed hydrogel HAM types, proving the feasibility of these water-floating materials. Further work should focus on loading the floating FAM-type HPM composites with additional amounts of polymer microsphere adsorbents and testing their adsorption performance in complex matrices of ions, samples of different ionic strengths, competitive adsorption studies, and even real marine, lake, or wastewater samples spiked with ions of interest. We believe, in fact, that the HAM and FAM samples have a high technical readiness level to be tested on real water samples. Thus, future work shall also focus on deploying these materials on real water samples either in the laboratory or in a water purification station, at least after the water has been treated with flocculation agents. While these materials are useful in the recovery of Cu(II) metal ions present in the wastewater produced in various industrial activities, such as mining, smelters, foundries, electroplating, batteries manufacturing, etc., it is also conceivable that these could also be deployed in hydro-mining applications. In the hydro-mining applications of water-floating adsorbents, the presence of free copper ions has to be evaluated because the existence of ion species in the given environmental conditions or the presence of bio-produced ligands can lower the concentration of Cu(II) to insignificant levels [41,42]. Nevertheless, areas with excess Cu(II) ions can be identified that produce stress to phytoplankton and aquatic life, and such adsorbents could be involved in environmental remediation and hydro-mining. Further, it may also be of interest to use HAM-type materials in soil remediation applications.

Supplementary Materials: The following are available online at https://www.mdpi.com/article/10.3390/nano13192619/s1, Figure S1. UV–vis absorption spectrum of 5×10^{-2} M $CuCl_2{\cdot}2H_2O$ solution; Figure S2. ATR-FTIR spectra of the PMs before and after adsorption of Cu(II) ions; Figure S3. Photograph of the PVA hydrogel (A), the hydrogel polymer microsphere (HPM) composite HAM-2 (B), and the foamed HPM composite FAM-2 (C); Figure S4. Photograph of the HAM-1 (left) and PVA hydrogel control sample without adsorbent microspheres (right) after being in contact with a stock solution of 5×10^{-2} M $CuCl_2{\cdot}2H_2O$ for 12 h, showing a significant change in color from white to deep blue; Figure S5. Photograph of the FAM-2 HPM composites, before and after adsorption of Cu(II) ions from a 5×10^{-2} M solution of $CuCl_2{\cdot}2H_2O$; Table S1. Composition of the HPM composite series in the form of hydrogel and foam; Table S2. The fit parameters of the kinetic adsorption of Cu(II) ions by the HAM-2 and FAM-2 to three different models, as indicated by the equations in the text, and r represents the goodness-of-fit parameter.

Author Contributions: Conceptualization, A.H.; methodology, A.H., A.-M.S. and M.H.; validation, A.H., A.-M.S. and M.H.; formal analysis, A.H.; investigation, A.H., A.-M.S. and M.H.; resources, A.H.; data curation, A.H.; writing—original draft preparation, A.H.; writing—review and editing, A.H., A.-M.S. and M.H.; visualization, A.H., A.-M.S. and M.H.; supervision, A.H.; project administration, A.H.; funding acquisition, A.H. All authors have read and agreed to the published version of the manuscript.

Funding: This work was supported by a grant from the Ministry of Research, Innovation and Digitization of Romania, CNCS/CCCDI-UEFISCDI, project number PN-III-P4-PCE-2021-0306 (Contract Nr. PCE62/2022).

Data Availability Statement: Data is available at https://osf.io/m8uva/?view_only=4c737cad1ee748dd8922358d5099bca5.

Conflicts of Interest: The authors declare no conflict of interest.

References

1. Dutta, D.; Arya, S.; Kumar, S. Industrial Wastewater Treatment: Current Trends, Bottlenecks, and Best Practices. *Chemosphere* **2021**, *285*, 131245. [CrossRef] [PubMed]
2. Gadipelly, C.; Pérez-González, A.; Yadav, G.D.; Ortiz, I.; Ibáñez, R.; Rathod, V.K.; Marathe, K.V. Pharmaceutical Industry Wastewater: Review of the Technologies for Water Treatment and Reuse. *Ind. Eng. Chem. Res.* **2014**, *53*, 11571–11592. [CrossRef]
3. Nnaji, C.O.; Jeevanandam, J.; Chan, Y.S.; Danquah, M.K.; Pan, S.; Barhoum, A. Chapter 6—Engineered Nanomaterials for Wastewater Treatment: Current and Future Trends. In *Fundamentals of Nanoparticles*; Barhoum, A., Hamdy Makhlouf, A.S., Eds.; Elsevier: Amsterdam, The Netherlands, 2018; pp. 129–168, ISBN 978-0-323-51255-8.
4. Sharma, Y.C.; Srivastava, V.; Singh, V.K.; Kaul, S.N.; Weng, C.H. Nano-adsorbents for the Removal of Metallic Pollutants from Water and Wastewater. *Environ. Technol.* **2009**, *30*, 583–609. [CrossRef] [PubMed]
5. Fu, F.; Wang, Q. Removal of Heavy Metal Ions from Wastewaters: A Review. *J. Environ. Manag.* **2011**, *92*, 407–418. [CrossRef]
6. Al-Saydeh, S.A.; El-Naas, M.H.; Zaidi, S.J. Copper Removal from Industrial Wastewater: A Comprehensive Review. *J. Ind. Eng. Chem.* **2017**, *56*, 35–44. [CrossRef]
7. Zhou, T.; Zhao, M.; Zhao, X.; Guo, Y.; Zhao, Y. Simultaneous Remediation and Fertility Improvement of Heavy Metals Contaminated Soil by a Novel Composite Hydrogel Synthesized from Food Waste. *Chemosphere* **2021**, *275*, 129984. [CrossRef]
8. Andreazza, R.; Morales, A.; Pieniz, S.; Labidi, J. Gelatin-Based Hydrogels: Potential Biomaterials for Remediation. *Polymers* **2023**, *15*, 1026. [CrossRef]
9. Wang, X. Nanomaterials as Sorbents to Remove Heavy Metal Ions in Wastewater Treatment. *J. Environ. Anal. Toxicol.* **2012**, *2*, 154–158. [CrossRef]
10. Zhu, Y.; Wang, W.; Yu, H.; Wang, A. Preparation of Porous Adsorbent via Pickering Emulsion Template for Water Treatment: A Review. *J. Environ. Sci.* **2020**, *88*, 217–236. [CrossRef]
11. Han, J.; Du, Z.; Zou, W.; Li, H.; Zhang, C. Fabrication of Interfacial Functionalized Porous Polymer Monolith and Its Adsorption Properties of Copper Ions. *J. Hazard. Mater.* **2014**, *276*, 225–231. [CrossRef]
12. Shan, S.; Sun, X.-F.; Xie, Y.; Li, W.; Ji, T. High-Performance Hydrogel Adsorbent Based on Cellulose, Hemicellulose, and Lignin for Copper(II) Ion Removal. *Polymers* **2021**, *13*, 3063. [CrossRef] [PubMed]
13. Wang, T.; An, Y.; Sun, J.; Yang, H.; Huang, Y.; Zheng, H. Hollow Self-Floating Microspheres Capture Cobalt (Co^{2+})/Nickel (Ni^{2+}) Ions from the Acidic Leachate of Spent Lithium-Ion Battery Cathodes. *Chem. Eng. J.* **2023**, *465*, 142950. [CrossRef]
14. Tang, Q.; Zhang, F.; Chen, W.; Ma, D.; Du, B.; Zhang, K.; Huang, X.; Luo, H.; Fan, L.; An, X.; et al. Floating-Separation Adsorbent for Methylene Blue and Pb(II) Removal: Structure Construction and Adsorption Mechanism. *Sep. Purif. Technol.* **2022**, *295*, 121332. [CrossRef]
15. Li, S.; Li, Y.; Fu, Z.; Lu, L.; Cheng, J.; Fei, Y. A 'Top Modification' Strategy for Enhancing the Ability of a Chitosan Aerogel to Efficiently Capture Heavy Metal Ions. *J. Colloid Interface Sci.* **2021**, *594*, 141–149. [CrossRef] [PubMed]
16. Mi, X.; Huang, G.; Xie, W.; Wang, W.; Liu, Y.; Gao, J. Preparation of Graphene Oxide Aerogel and Its Adsorption for Cu^{2+} Ions. *Carbon* **2012**, *50*, 4856–4864. [CrossRef]
17. Bloor, J.M.; Handy, R.D.; Awan, S.A.; Jenkins, D.F.L. Graphene Oxide Biopolymer Aerogels for the Removal of Lead from Drinking Water Using a Novel Nano-Enhanced Ion Exchange Cascade. *Ecotoxicol. Environ. Saf.* **2021**, *208*, 111422. [CrossRef]
18. Mir, N.; Castano, C.E.; Rojas, J.V.; Norouzi, N.; Esmaeili, A.R.; Mohammadi, R. Self-Separation of the Adsorbent after Recovery of Rare-Earth Metals: Designing a Novel Non-Wettable Polymer. *Sep. Purif. Technol.* **2021**, *259*, 118152. [CrossRef]
19. Liu, R.; Yan, Q.; Tang, Y.; Liu, R.; Huang, L.; Shuai, Q. NaCl Template-Assisted Synthesis of Self-Floating COFs Foams for the Efficient Removal of Sulfamerazine. *J. Hazard. Mater.* **2022**, *421*, 126702. [CrossRef]
20. Jing, G.; Wang, L.; Yu, H.; Amer, W.A.; Zhang, L. Recent Progress on Study of Hybrid Hydrogels for Water Treatment. *Colloids Surf. A Physicochem. Eng. Asp.* **2013**, *416*, 86–94. [CrossRef]
21. Shalla, A.H.; Yaseen, Z.; Bhat, M.A.; Rangreez, T.A.; Maswal, M. Recent Review for Removal of Metal Ions by Hydrogels. *Sep. Sci. Technol.* **2019**, *54*, 89–100. [CrossRef]
22. Honciuc, A.; Solonaru, A.-M.; Honciuc, M. Pickering Emulsion Polymerization Technology—Toward Nanostructured Materials for Applications in Metal Ion Extractions from Wastewaters. *ACS Appl. Polym. Mater.* **2023**. [CrossRef]
23. Honciuc, A.; Negru, O.-I. NanoTraPPED—A New Method for Determining the Surface Energy of Nanoparticles via Pickering Emulsion Polymerization. *Nanomaterials* **2021**, *11*, 3200. [CrossRef]
24. Yang, J.; Li, Y.; Wang, J.; Sun, X.; Cao, R.; Sun, H.; Huang, C.; Chen, J. Molecularly Imprinted Polymer Microspheres Prepared by Pickering Emulsion Polymerization for Selective Solid-Phase Extraction of Eight Bisphenols from Human Urine Samples. *Anal. Chim. Acta* **2015**, *872*, 35–45. [CrossRef] [PubMed]
25. Ma, H.; Luo, M.; Sanyal, S.; Rege, K.; Dai, L. The One-Step Pickering Emulsion Polymerization Route for Synthesizing Organic-Inorganic Nanocomposite Particles. *Materials* **2010**, *3*, 1186–1202. [CrossRef]
26. Honciuc, A.; Negru, O.-I. Asymmetrically Nanostructured 2D Janus Films Obtained from Pickering Emulsions Polymerized in a Langmuir–Blodgett Trough. *Micromachines* **2023**, *14*, 1459. [CrossRef] [PubMed]
27. Honciuc, A.; Negru, O.-I. Role of Surface Energy of Nanoparticle Stabilizers in the Synthesis of Microspheres via Pickering Emulsion Polymerization. *Nanomaterials* **2022**, *12*, 995. [CrossRef] [PubMed]
28. Honciuc, A. *Chemistry of Functional Materials Surfaces and Interfaces: Fundamentals and Applications*, 1st ed.; Elsevier: Amsterdam, The Netherlands, 2021; ISBN 978-0-12-821059-8.

29. Kirdponpattara, S.; Phisalaphong, M.; Newby, B.Z. Applicability of Washburn Capillary Rise for Determining Contact Angles of Powders/Porous Materials. *J. Colloid Interface Sci.* **2013**, *397*, 169–176. [CrossRef]
30. Wu, K.H.; Wang, Y.R.; Hwu, W.H. FTIR and TGA Studies of Poly(4-Vinylpyridine-Co-Divinylbenzene)–Cu(II) Complex. *Polym. Degrad. Stab.* **2003**, *79*, 195–200. [CrossRef]
31. Lee, J.Y.; Painter, P.C.; Coleman, M.M. Hydrogen Bonding in Polymer Blends. 4. Blends Involving Polymers Containing Methacrylic Acid and Vinylpyridine Groups. *Macromolecules* **1988**, *21*, 954–960. [CrossRef]
32. Zhou, X.; Goh, S.H.; Lee, S.Y.; Tan, K.L. XPS and FTi.r. Studies of Interactions in Poly(Carboxylic Acid)/Poly(Vinylpyridine) Complexes. *Polymer* **1998**, *39*, 3631–3640. [CrossRef]
33. Inai, Y.; Kato, S.-I.; Hirabayashi, T.; Yokota, K. Complexation of Sequence-Ordered Methacrylic Acid Copolymers with Poly(4-Vinylpyridine). *J. Polym. Sci. A Polym. Chem.* **1996**, *34*, 2341–2348. [CrossRef]
34. Morán-Quiroz, J.L.; Orozco-Guareño, E.; Manríquez, R.; Carbajal-Arízaga, G.G.; de la Cruz, W.; Gomez-Salazar, S. Polymeric Hydrogels Obtained Using a Redox Initiator: Application in Cu(II) Ions Removal from Aqueous Solutions. *J. Appl. Polym. Sci.* **2014**, *131*, 39933. [CrossRef]
35. Milosavljević, N.; Debeljković, A.; Krušić, M.K.; Milašinović, N.; Üzüm, Ö.B.; Karadağ, E. Application of Poly(Acrlymide-*Co*-Sodium Methacrylate) Hydrogels in Copper and Cadmium Removal from Aqueous Solution. *Environ. Prog. Sustain. Energy* **2014**, *33*, 824–834. [CrossRef]
36. Kaşgöz, H.; Durmuş, A.; Kaşgöz, A. Enhanced Swelling and Adsorption Properties of AAm-AMPSNa/Clay Hydrogel Nanocomposites for Heavy Metal Ion Removal. *Polym. Adv. Technol.* **2008**, *19*, 213–220. [CrossRef]
37. Choi, J.-W.; Kim, H.J.; Ryu, H.; Oh, S.; Choi, S.-J. Three-Dimensional Double-Network Hydrogels of Graphene Oxide, Alginate, and Polyacrylonitrile for Copper Removal from Aqueous Solution. *Environ. Eng. Res.* **2019**, *25*, 924–929. [CrossRef]
38. Revellame, E.D.; Fortela, D.L.; Sharp, W.; Hernandez, R.; Zappi, M.E. Adsorption Kinetic Modeling Using Pseudo-First Order and Pseudo-Second Order Rate Laws: A Review. *Clean. Eng. Technol.* **2020**, *1*, 100032. [CrossRef]
39. Yan, H.; Dai, J.; Yang, Z.; Yang, H.; Cheng, R. Enhanced and Selective Adsorption of Copper(II) Ions on Surface Carboxymethylated Chitosan Hydrogel Beads. *Chem. Eng. J.* **2011**, *174*, 586–594. [CrossRef]
40. Ho, Y.S.; McKay, G. Pseudo-Second Order Model for Sorption Processes. *Process Biochem.* **1999**, *34*, 451–465. [CrossRef]
41. Heller, M.I.; Croot, P.L. Copper Speciation and Distribution in the Atlantic Sector of the Southern Ocean. *Mar. Chem.* **2015**, *173*, 253–268. [CrossRef]
42. Karavoltsos, S.; Kalambokis, E.; Sakellari, A.; Plavšić, M.; Dotsika, E.; Karalis, P.; Leondiadis, L.; Dassenakis, M.; Scoullos, M. Organic Matter Characterization and Copper Complexing Capacity in the Sea Surface Microlayer of Coastal Areas of the Eastern Mediterranean. *Mar. Chem.* **2015**, *173*, 234–243. [CrossRef]

nanomaterials

MDPI

Article

Antifungal Activity and Molecular Mechanisms of Copper Nanoforms against *Colletotrichum gloeosporioides*

Mun'delanji C. Vestergaard [1,*], Yuki Nishida [1], Lihn T. T. Tran [1], Neha Sharma [2], Xiaoxiao Zhang [1], Masayuki Nakamura [1], Auriane F. Oussou-Azo [3] and Tomoki Nakama [1]

1 Faculty of Agriculture, Kagoshima University, Kagoshima City 890-0065, Japan
2 Faculty of Fiber Science and Engineering, Kyoto Institute of Technology, Matsugasaki, Sakyo, Kyoto 606-8585, Japan
3 United Graduate School of Agricultural Sciences, Kagoshima University, Kagoshima City 890-0065, Japan
* Correspondence: munde@agri.kagoshima-u.ac.jp

Abstract: In this work, we have synthesized copper nanoforms (Cu NFs) using ascorbic acid as a reducing agent and polyvinylpyrrolidone as a stabilizer. Elemental characterization using EDS has shown the nanostructure to be of high purity and compare well with commercially sourced nanoforms. SEM images of both Cu NFs show some agglomeration. The in-house NFs had a better even distribution and size of the nanostructures. The XRD peaks represented a face-centered cubic structure of Cu_2O. The commercially sourced Cu NFs were found to be a mixture of Cu and Cu_2O. Both forms had a crystalline structure. Using these two types of Cu NFs, an antimicrobial study against *Colletotrichum gloeosporioides*, a devastating plant pathogen, showed the in-house Cu NFs to be most effective at inhibiting growth of the pathogen. Interestingly, at low concentrations, both Cu NFs increased fungal growth, although the mycelia appeared thin and less dense than in the control. SEM macrographs showed that the in-house Cu NFs inhibited the fungus by flattening the mycelia and busting some of them. In contrast, the mycelia were short and appeared clustered when exposed to commercial Cu NFs. The difference in effect was related to the size and/or oxidation state of the Cu NFs. Furthermore, the fungus produced a defense mechanism in response to the NFs. The fungus produced melanin, with the degree of melanization directly corresponding to the concentration of the Cu NFs. Localization of aggregated Cu NFs could be clearly observed outside of the model membranes. The large agglomerates may only contribute indirectly by a hit-and-bounce-off effect, while small structures may adhere to the membrane surface and/or internalize. Spatio-temporal membrane dynamics were captured in real time. The dominant dynamics culminated into large fluctuations. Some of the large fluctuations resulted in vesicular transformation. The major transformation was exo-bud/exo-cytosis, which may be a way to excrete the foreign object (Cu NFs).

Keywords: copper nanoforms (Cu NFs); antimicrobial; *Colletotrichum gloeosporioides*; giant unilamellar vesicles (GUVs); melanin production

Citation: Vestergaard, M.C.; Nishida, Y.; Tran, L.T.T.; Sharma, N.; Zhang, X.; Nakamura, M.; Oussou-Azo, A.F.; Nakama, T. Antifungal Activity and Molecular Mechanisms of Copper Nanoforms against *Colletotrichum gloeosporioides*. *Nanomaterials* **2023**, *13*, 2990. https://doi.org/10.3390/nano13232990

Academic Editors: Andrei Honciuc and Mirela Honciuc

Received: 19 October 2023
Revised: 16 November 2023
Accepted: 17 November 2023
Published: 22 November 2023

1. Introduction

Metals have been exploited for their antimicrobial properties for thousands of years. Since the era of Persian kings, copper and silver have been used for food preservation and water sanitization. Copper compounds have been used in agriculture as a fungistatic agent on potatoes and grapes [1]. Nanoforms of metal ions have been reported to increase their antimicrobial properties [2]. These nanoforms possess significant cytotoxicity activity against viruses, fungi, and bacteria. Metal-based nanoparticles (NPs) have the potential to be effective as antimicrobial agents through different mechanisms with respect to the classical treatments and have the potential to be able to target multiple microbes and biomolecules compromising the development of resistant strains [3]. For example, silver, gold, zinc oxide, copper, and copper oxide nanoparticles are commonly used in antibiotherapy. Abbaszadegan et al. demonstrated that positively charged Ag NPs are strongly attracted to the surface of bacteria, resulting in significant antibacterial activity [4]. ZnO NPs and Au NPs possess antimicrobial activity against different Gram-positive (*S. aureus* and many others) and Gram-negative (namely *P. aeruginosa*, *E. coli*, etc.) bacteria [5].

In the synthesis of metal NPs, natural plant extracts can be used both as reductants and stabilizers. They are easily accessible, biocompatible, and eco-friendly, which may synergistically increase the antimicrobial performances of Cu NPs [6]. Furthermore, Cu NPs dissolve faster than other noble metals by releasing ions into the surroundings [7]. Thus, they have found numerous applications in several areas, such as additives in metal coatings, inks, skin products, and plastics for food packaging [8]. Concerning their antimicrobial activity, Cu NPs generate oxidative stress, cause the disassembly of viruses or bacterial membranes, and can interfere with virus activity [9]. At the same time, Cu ions increase the generation of compounds that are toxic for microbes. Copper oxide nanoparticles (CuO NPs) stand out by possessing very well-known antibacterial properties, being active against a varied range of pathogenic bacteria [10]. The integration of metal oxide NPs improves its antimicrobial as well as physical, chemical, and mechanical properties [11]. Cu_2O NPs showed strong antibacterial properties and effectively damaged the bacterial cell membranes [12].

Metal ions can create strong coordination bonds with atoms (N, O, or S) that are abundantly present in biomolecules and organic compounds. Generally, the bond between metal ions and biomolecules is non-specific, which exhibits a broad-spectrum activity to the metal-based NPs [13].

Multiple studies have investigated the different actions that copper nanoforms (Cu NFs), that is, nanostructured copper (Cu NPs) and its oxidized forms (CuO NPs, Cu_2O NPs) exert on microbes [14,15]. In general, Cu NPs are more prone to interact with the bacterial membrane, compromising its integrity [14]. Meanwhile, CuO NPs, depending on the shape of the NP, tend to penetrate the bacterial membrane and release ions within the cell [15]. The basic process that takes place for Cu NPs is based on the redox equation, which involves the energetically favorable oxidation of copper to Cu(II) ($Cu(s) + O_2 + 2H^+ = Cu^{2+} + H_2O_2$). Thus, the local production of hydrogen peroxide (H_2O_2) causes membrane damage [16]. This was in accordance with the cellular membrane study, where exposure to Cu ions caused a minor impact [17]. Another research has suggested that CuO NPs can also produce reactive oxygen species (ROS). The authors showed that *E. coli* is more affected by CuO NPs than *S. aureus*, suggesting that the membrane differences between Gram-negative and Gram-positive bacteria can influence the resistance to ROS [18]. The authors also evidenced a correlation between the amount of killed bacteria and the size of the NP and highlighted that smaller NPs are more effective against bacteria. Furthermore, smaller NPs are associated with an increased amount of superoxide anions, generating more intense oxidative stress. Another study demonstrated that the shape of the NP significantly influences the overall antibacterial activity by comparing the minimum bactericidal concentration and minimum inhibitory concentration of nanosheets of CuO and spherical CuO NPs against *Proteus vulgaris*, *E. coli*, *Bacillus subtilis*, and *Micrococcus luteous* [19]. When tested in mice, Cu nanocrystals were reported to catalyze the production of the hydroxyl radical to kill the bacteria and hinder the further production of biofilms [20]. Cu NPs and CuO NPs have shown antimicrobial effects, but the mechanism behind the activity of these NPs is not well elucidated yet. However, it is believed that it comprises bacterial cell wall adhesion facilitated by electrostatic interactions. Previously, we have shown the antifungal properties of Cu and its oxide forms [21]. As a follow-up study, in our current work, we investigated some of the mechanisms involved behind the antimicrobial action of Cu NFs. We used commercial as well as Cu NFs synthesized in our laboratory. In this way, we were able to also observe the effect of the nanostructures depending on the concentration and size of the Cu NFs.

Structural organization such as the molecular packing of lipids are intrinsic properties of cell membranes and contribute to biophysical processes, including endocytosis, exocytosis, and autophagy. These are important for cellular function. Introduction of some substances such as NPs may interfere with these vital functions by changing lipid structural organization, thus disrupting normal/expected membrane dynamics. Therefore, studying these structural dynamics provides insight into the physical basis of changes in cellular geometry, which occur during biophysical processes such as membrane invagination. Indeed, different types of stimuli, including physical (heat, light), biomolecules (amyloid beta), and antioxidants (polyphenols) introduced to lipid vesicles, have been shown to induce morphological transformations [22,23]. Giant unilamellar vesicles (GUVs), that is, model membranes with diameters greater than 10 μm, have been used for investigations of the physical and biological properties of vesicle membranes. The cell-sized vesicles allow for the real-time observation of dynamic morphological changes to be clearly visualized without compromising on the "controllable" analytical advantage [24]. They also mimic the physiological environment closely in terms of the spatio-temporal scale at which interactions take place, thus allowing for the understanding and clarification of mechanisms. They have been extensively used in research [25,26]. In our current work, we utilized GUVs to study, in real-time, the spatio-temporal membrane dynamics and changes induced by the presence of Cu NFs. Bhat et al. have studied the role of gold and silver NPs with GUVs [27] in cancer therapies.

Herein, besides demonstrating that our in-house synthesized Cu NFs were antimicrobial and performed better than the commercially soured Cu NFs, we investigated the molecular mechanisms underlying the anti-fungal activities of nanostructured copper and copper oxide. Using biomimetic model membrane systems, we studied in real time the spatio-temporal membrane dynamics induced by the Cu NFs. We observed dynamics with a dominance of large fluctuations, some of which culminated into membrane transformations. Exo-bud/-cytosis was the major membrane change induced by both Cu NFs. Chemical characterization of the functional groups using FTIR revealed that both Cu NFs interfered/interacted with the functional groups in all major biomolecules (lipids, proteins, and carbohydrates). Furthermore, we could visually see the production of a dark pigment. We imagine this to be the production of melanin due to the stress caused by the Cu NFs on the fungus. Melanin production was confirmed and semi-quantified.

2. Materials and Methods

2.1. Materials

Copper nanoforms (Cu NFs), copper acetate monohydrate, melanin, phosphate buffer saline (PBS) tablets, polyvinyl pyrrolidone (MW 10,000, PVP-10), potato dextrose agar (PDA), and potato dextrose broth (PDB) were obtained from Sigma Aldrich Corporation (Tokyo, Japan) and were used without further purification. Dimethyl sulfoxide (DMSO) was purchased from Wako Pure Chemical Industries, Ltd., Osaka, Japan. Chloroform and methanol were purchased from Kanto-Chemical Co. Inc. (Tokyo, Japan) and Nacalai Tesque Inc. (Kyoto, Japan), respectively. *Colletotrichum gloeosporioides* was a gift from the Plant Pathology Laboratory, Faculty of Agriculture, Kagoshima University, Japan. Membrane lipid 1,2-dioleoyl-*sn*-glycero-3-phosphocholine (DOPC) was purchased from Avanti Polar Lipids, Inc. (Alabaster, AL, USA). All chemicals were of analytical grade and used as received. Deionized water was obtained using a Millipore MilliQ (MilliQ Water, Millipore S.A.S., Molsheim, France) purification system at conductivity 18.2 MΩ cm^{-1}. Unless otherwise stated, all analyses were conducted at room temperature (RT).

2.2. Synthesis of and Preparation of Copper Nanoforms

Cu NFs were synthesized using a bottom-up approach with copper acetate as a precursor, ascorbic acid as a reducing agent, and PVP-10 as a stabilizer. PVP has been used as a stabilizer in the synthesis of metal NPs including palladium [28]. Briefly, copper acetate (300 mM) and PVP-40 (5% (v/v)) in MilliQ water were mixed using a magnetic stirrer at ~40 °C for 10 min. Thereafter, 1M of ascorbic acid was slowly added to the solution at 10 μL per min with gentle stirring until the solution changed from bluish to dark red. The dark red precipitate was collected and washed with MilliQ water. This process was repeated twice. Then, the sample was dried in a convection oven at 80 °C for over 12 h. Lastly, the dried sample was ground slightly before analysis. We refer to this sample as the in-house Cu NFs.

2.3. *Characterization of Synthesized Copper Nanoforms*

2.3.1. X-ray Diffraction

Cu NFs were characterized using XRD. The dried sample was fixed on a slide glass for measurement. X-ray diffractograms were obtained using a PANalytical X'PERT PRO MPD diffractometer (Malvern Panalytical, Kobe-shi, Japan) operating at 45.0 kV and 40 mA; the Cu-Kα radiation (λ = 1.54060 Å) was measured at a scan speed of 0.1°/s, step size of 0.01°, and from the 2θ: 20° to 80° range. The commercial Cu NFs were similarly analysed.

2.3.2. Field-Emission Scanning Electron Microscopy (FE-SEM): Energy-Dispersive X-ray Spectroscopy (SEM-EDS)

First, a carbon tape was pasted on the sample holder, and the dried sample was placed on the holder. Excess sample was blown off with a blower, and a coater was used to coat the sample surface with Au. The coating was performed once from directly above the sample and from the oblique direction at 10 mA × 90 s × once using a sputter coater (Polaron SC7610, Fisons Instruments, Saitama-shi, Japan). The synthesized Cu NFs were imaged using FE-SEM (FE-SEM: Hitachi High-Tech Corporation, Model: SU-70, Tokyo, Japan), at an accelerating voltage of 5.0 kV.

Energy-dispersive X-ray spectroscopy (EDS) is a technique used to detect elements in a sample. It is used in conjunction with SEM or TEM. In our study, it was used with SEM. For EDS measurement, pellets (3 mm) were prepared from the dried samples and fixed on a sample holder using carbon tape. Afterwards, the surface of the sample was coated with carbon using a carbon coater (VC-100, Vacuum Device, Ibaraki, Japan). EDS analysis was conducted at an accelerating voltage of 25.0 kV and magnification of 1000× using an Octane Elect Super instrument (AMETEK, Tokyo, Japan).

2.3.3. Dynamic Light Scattering

Immediately after sonication of the Cu NFs, the size distribution and zeta potential of the NPs were analyzed using the DLS technique using a Zetasizer Nano ZS90 (Malvern Panalytical, Japan). All measurements were conducted in water at room temperature (25 °C) and at an approximately 200 kcps count rate.

2.4. *Preparation of Fungal Culture*

A culture of *C. gloeosporioides* was grown on PDA. Fungal isolates were first routinely cultured in PDA in agar plates to build a stock of culture. Agar plates were incubated at 25 °C for 14 d to obtain a good culture. Once the fungi were at an active growth stage, agar blocks were cut and transferred into PDA petri dishes and PDB tubes.

2.5. *Effect of Copper Nanoforms on Fungal Growth and Structure*

The PDA and PDB media were treated with Cu NFs at concentrations ranging from 0 to 1.0 mg/mL. The growth of fungus was measured daily for a period of 14 d using a homemade calibrated ruler as previously reported [24].

For imaging of fungal microstructure, mycelia were harvested after 14 days of incubation (with and without treatment with Cu NFs) from the PDB media and gently collected on filter paper. Thereafter, the mycelia were washed three times with MilliQ water to remove residual PDB. Subsequently, the mycelia were dried at 80 °C for 24 h then ground into fine powder using a mortar and pestle. The sample was fixed on a sample holder using carbon tape. A Quanta 400 instrument (FEI, Thermo Fisher Scientific, Tokyo, Japan) operating in low-vacuum mode was used with water vapor and an accelerating voltage of 15.0 kV.

2.6. Chemical Composition of Fungus

To understand how the Cu NFs interacted with the fungus at the chemical level, we used Fourier transform infrared spectroscopy (FTIR) to elucidate the types of bonds involved and any disruptions in the presence of the nanoparticles. FTIR has been used to even discriminate between genus, species, and strain level [29,30].

Mycelia were harvested as in Section 2.5. However, they were subsequently ground after drying and were analyzed using a JASCO FT/IR-4200 instrument (JASCO International Co. Ltd., Tokyo, Japan) with an attenuated total reflection set-up (ATR PRO 410-S). The samples were deposited on the ATR diamond prism (45° angle of incidence), yielding one reflection. Absorption spectra at a resolution of one datapoint for every 1 cm^{-1} were obtained in the region between 4000 and 400 cm^{-1}.

2.7. Preparation of Cell-Sized Lipid Vesicles

DOPC lipid vesicles were prepared following the natural swelling method [31]. The lipid mixture was dissolved in chloroform:methanol (2:1; (*v*/*v*)) in a Durham glass bottle to a final concentration of 0.2 mM. Glucose in methanol was added to the lipid to give a final concentration of 0.2 mM. The solvent was subsequently removed by evaporating the tube under a gentle nitrogen stream and drying it in a desiccator for 3 h, resulting in a thin lipid film at the bottom of tube. The film was rehydrated with 3 μL of MilliQ water and incubated at 55 °C for 10 min. Afterwards, 197 μL of MilliQ water was added, and the swollen lipid solution was left on a dry heating block overnight at 37 °C for the self-assembly of lipid vesicles.

2.8. Observation of Membrane Dynamics

A liposome solution (5 μL) prepared above was placed in silicon well (0.2 mm) on a slide glass and covered with a small cover slip. To study the interaction of NPs with the lipid membrane, 5 μL of the liposome solution and 5 μL of Cu NF prepared in MilliQ water (Cu NPs and Cu_2O NPs; (1 mg/mL)) were poured into a test tube, gently mixed by soft tapping, and observed soon after. Observation of the vesicular dynamics was within 2 min of the solution's introduction to the lipid vesicles. As time affects the interaction between the membrane and NPs, we carefully followed the exact same experimental conditions and procedures for each analysis and conducted at least 20 replicates for each type of interaction. Microscope observation was carried out within 3 min of sample mixing. We observed changes in membrane morphology using a phase-contrast microscope (Olympus AX80 and IX70; Olympus, Tokyo, Japan) at RT. The silicon well and cover slip ensured that evaporation of the solution did not occur over the duration of the experiment. Images were recorded on a hard disc.

2.9. Detection of Melanin Production

We used a fluorescence-based assay for the detection of melanin [32]. First, the melanin had to be extracted from the fungal culture grown in-PDA incubated (with and without Cu NFs) Petri dishes for a period of 20 days. The fungi, together with the PDA media, were transferred to PBS (pH 7.4) solution and mixed thoroughly. The mixture was heated with constant stirring at approximately 95 ± 5 °C for 20 min and centrifuged at 2000× *g* for 10 min to separate the fungus from the media. The precipitate was then dissolved in 10% (*v*/*v*) DMSO of 1 M NaOH solution using an ultrasonic wave water bath at 75 °C for 4 h. This solution was referred to as "extractable" melanin. After centrifugation at 14,000 rpm for 15 min, 35% (*v*/*v*) H_2O_2 solution was added to the supernatant and kept in the dark at RT for 4 h. Fluorescence was measured at excitation 470 nm and emission 550 nm using an Infinite 200 PRO Mplex spectrometer, TECAN, Mannedorf, Switzerland. Standard solutions of melanin were similarly detected, and a standard curve was drawn. Concentrations of extractable melanin were henceforth calculated using the standard curve.

3. Results and Discussion

3.1. Characterization of Copper Nanoforms

3.1.1. XRD

The Cu NFs were characterized using XRD spectroscopy. Major sharp diffraction peaks were obtained at 2θ 30°, 36°, 42°, 53°, 63°, 73.5°, and 77° for the in-house synthesized Cu NFs (Figure 1A). These peaks were assignable to lattice planes (110), (111), (200), (211), (220), (310), and (222), respectively, of crystalline Cu_2O (JCPDS Card No. 04-003-6433). The peaks represent a face-centered cubic structure of Cu_2O [33]. The diffraction patterns for the commercial Cu NFs were very similar to the in-house ones (Figure 1B) with the exception of an additional not fully resolved diffraction peak at 43°. This corresponds to (111) of cubic symmetry Cu^0 (JCPDS Card No. 01-070-3039). Another exception is a barely resolved peak (shoulder to peak at 74.1°). This corresponds to (310) of Cu_2O (JCPDS Card No. 04-003-6433) and the main resolved peak at 74.1° (220) of cubic symmetry Cu^0 (JCPDS No. 01-070-3039).

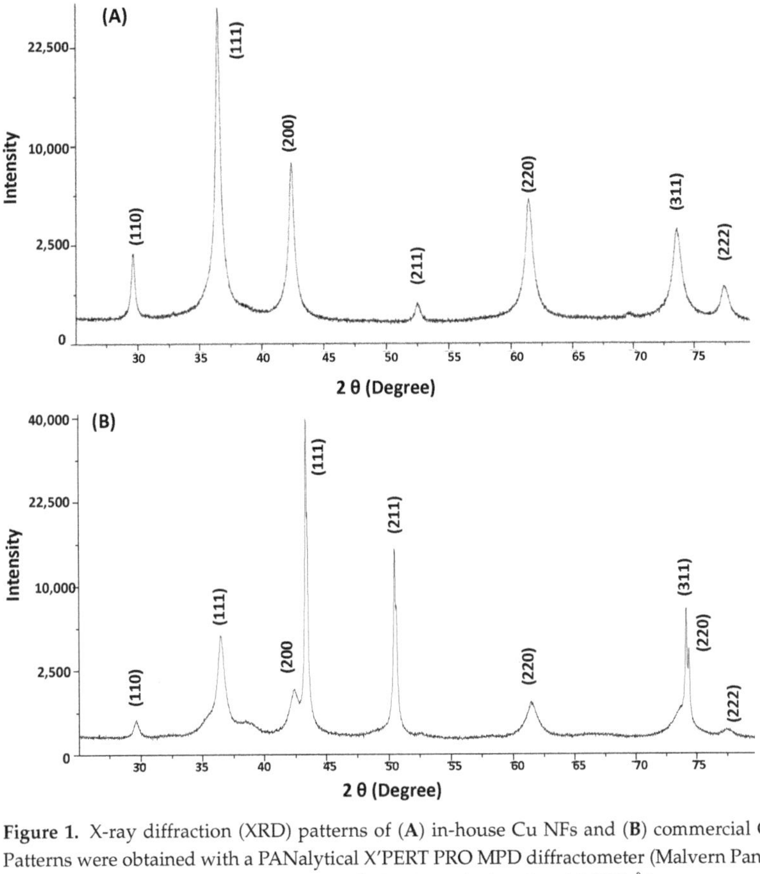

Figure 1. X-ray diffraction (XRD) patterns of (**A**) in-house Cu NFs and (**B**) commercial Cu NFs. Patterns were obtained with a PANalytical X'PERT PRO MPD diffractometer (Malvern Panalytical, Japan) operating at 45 kV and 40 mA with Cu-Kα radiation (λ = 1.54060 Å) measurement being performed at a scan speed of 0.01(units) from the 2θ 20° to 80° range.

Thus, the synthesized Cu NF sample was characterized as a crystalline pure Cu_2O, and the commercial Cu NF could be identified as a mixture of Cu_2O and Cu^0, both of which are crystalline cubic structures.

3.1.2. Energy Dispersive X-ray Spectroscopy (EDS)

Energy-dispersive X-ray spectroscopy (EDS) was used in conjunction with SEM to analyze elements in the Cu NFs. EDS measurement is based on the principle that when the elements are subjected to excitation upon being shot by an electron beam, the elements emit/release a specific amount of X-ray energy that is particular to each element. The energy emitted for each element is also related to how far excited electrons in the element fall from upper orbitals to lower ones to fill the vacancy of displaced electrons. The purity of the synthesized Cu NFs is very high indeed, and only three elements were detected. As shown in Figure 2, the spectra of the synthesized and commercial Cu NFs are very similar. They both display typical emission energies for Cu at ~0.93 keV, which corresponds to the fall of electrons from an upper orbital M to L shell (L_α). The energy spikes at ~8.04 keV and ~8.91 keV correspond to X-ray emissions K_α and K_β, respectively, which were released by electrons as they fell from the L to K electron orbitals. The amount (counts) of released energy is not too different between the two Cu NFs. The commercial Cu NFs had a slightly higher count at 69.7% compared with the in-house Cu NFs (63.3%). In both spectra, carbon was detected. Oxygen element was also detected. We imagine that the O comes from O in the Cu_2O nanostructures. We imagine that this is contamination from the carbon tape that was used to fix the sample to the sample holder [34]. The micrograph of the synthesized Cu NFs (A) shows that their structures have more evenly distributed surfaces than the commercial nanoforms (B).

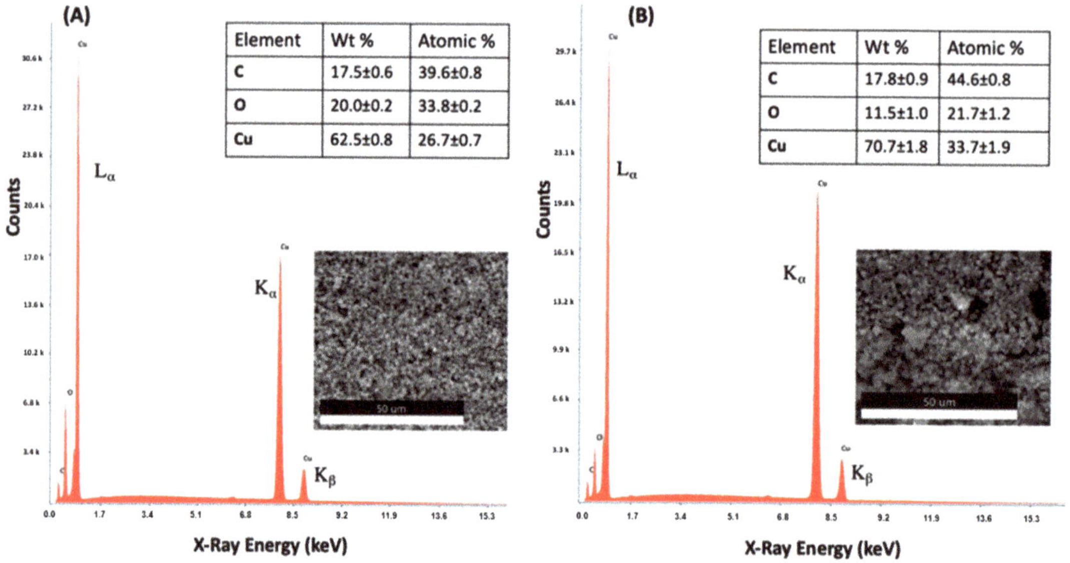

Figure 2. Dispersive energy X-ray spectra of (**A**) Cu NFs synthesized in house using ascorbic acid and polyvinyl pyrrolidone and (**B**) commercial Cu NFs. The elemental (% weight and atomic %), values and SEM images of the analysis area can be seen (inset). The analysis was conducted in triplicate at an accelerating voltage of 25.0 kV and magnification of 1000×. Elemental analysis data are the mean of three replicates.

3.1.3. Field-Emission Scanning Electron Microscope (FE-SEM)

The structure of Cu NFs was analyzed using field-emission scanning electron microscopy (FE-SEM). The images in Figure 3 show an even size and surface distribution of particles for the synthesized Cu NFs (A) compared with the commercial Cu NFs (B). A closer look at the insets shows smooth, spherically shaped nanoparticles, which are agglomerated into bigger structures (the self-evolution of the agglomerates into grapes-on-stick structures are also interesting in their own right. A study of structural evolution under different conditions is underway and will be reported separately). The nanoparticle agglomeration in both samples is due to the presence of high surface energy. Agglomeration of Cu NFs has been studied. It has been reported that commercially purchased Cu NFs of 40, 60, and 80 nm resulted in agglomerates of 335, 360, and 365 nm in water respectively. The agglomerates were even larger in cell media [35].

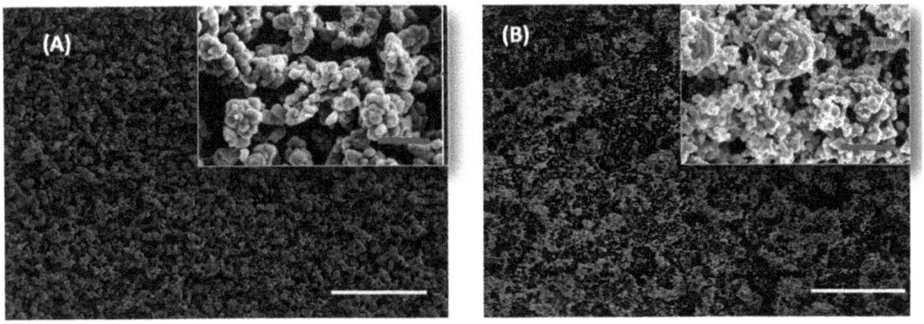

Figure 3. Typical SEM images of Cu NFs (**A**) synthesized in house using ascorbic acid and polyvinylpyrrolidone and (**B**) commercially sourced Cu NFs. Bars indicate 20 µM and 2 µM (insets). The white bars represent 100 µm size; the red bars (inset images) represent 10 µm size Samples were imaged using an accelerating voltage of 5.0 kV.

3.2. *Effect of Copper Nanoforms on Fungal Growth and Microstructure of Mycelia*

3.2.1. Effect of Copper Nanoforms on Fungal Growth

We exposed *C. gloeosporioides* to the in-house Cu NFs and commercial Cu NFs at different concentrations. As can be seen in Figure 4A, both types had the effect of increasing the growth of the fungus at low concentrations (less than or equal to 0.25 mg/mL). In fact, commercial NFs increased fungal growth even at 0.5 mg/mL at this concentration; we could begin to observe a decrease in growth when the fungus was exposed to in-house NFs. It was only at a higher concentration of commercial NFs (1 mg/mL) that there was a decrease in fungal growth (Figure 4A). Our inhouse Cu NFs were characterized as crystalline Cu_2O NFs that were center-cubic faced. The commercial NFs were characterized as a combination of crystalline Cu_2O NFs and Cu NFs. Another difference between the two Cu NFs are their respective sizes. The shape and size of NFs influence how they interact with biological molecules, including cells and organisms (bacteria, fungi, etc.) [36]. At this point, we cannot yet confirm which, if not both, of the parameters (oxidation state and size of NFs) is most effective at inhibiting the growth of the fungus. However, we can note that in our previous work [21], we reported that CuO NPs were more effective at inhibiting the same fungus compared with Cu NFs.

The increase in fungal growth at a low concentration of Cu NFs was very interesting. The same effect has been reported in some bacterial and fungal species in the presence of low concentrations of Cu [37]. Therefore, we decided to examine the overall structure of the mycelium using a simple optical microscope (BA210EINT (Shimadzu, Tokyo, Japan)). Compared with the control, we observed thinner, longer, and less dense mycelia at low concentrations (Figure S1A). These appeared darker in color, and the changes in fungal structure could be seen by the naked eye (Figure S1B). Furthermore, we used a scanning electron microscope to zoom in on the microstructure of the fungus.

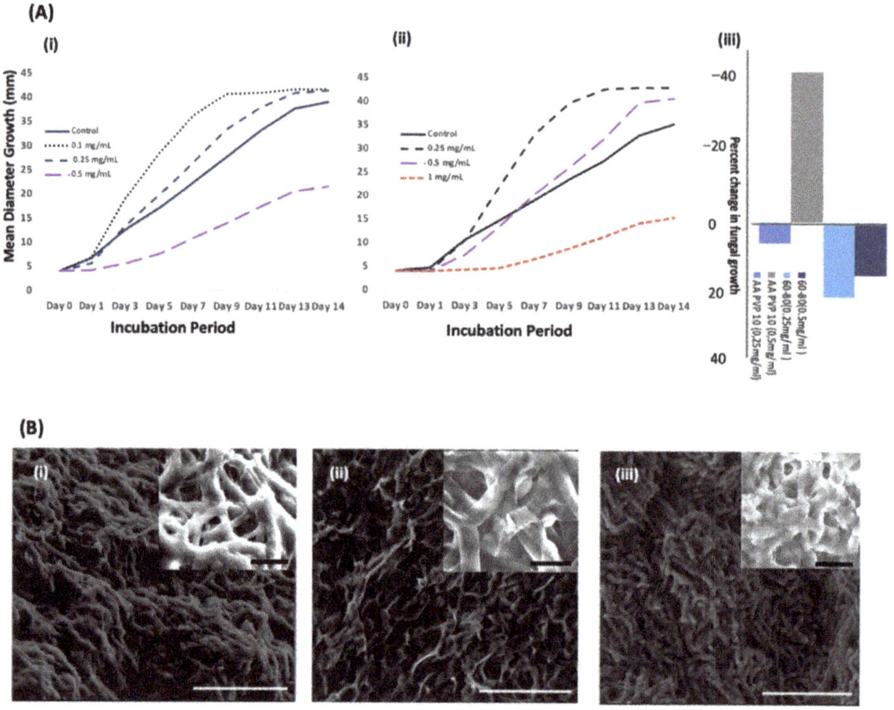

Figure 4. Effect of Cu NFs on *C. gloeosporioides*, a fungus that attacks food crops. (**A**) Diameter growth of the fungus in the presence and absence (control) of (**i**) in-house Cu NFs and (**ii**) commercial Cu NFs; Percent inhibition in fungal growth in the presence of Cu NSFs (**iii**); (**B**) SEM images of the of the fungus (i) alone; in the presence of (ii) inhouse Cu NFs and (iii) commercial Cu NFs. The white bars represent 50 µm size; the white bars (inset images) represent 10 µm size. Images were obtained in low-vacuum mode with water vapor and an accelerating voltage of 15.0 kV.

3.2.2. Micrographs Fungus after Exposure to Copper Nanoforms

The mycelia of healthy fungus can be seen as a long tubular structure (Figure 4(Bi)). This structure has been flattened, and some tears (broken mycelia) could be clearly observed when the fungus was exposed to in-house Cu NFs (Figure 4(Bii)). Interestingly, treatment of the fungus with commercial Cu NFs resulted in a different macrostructure of the fungus. The mycelia look very short and are somehow clustered in bunches (Figure 4(Biii)). The two forms of Cu NFs were different in their size and chemical structure. Analysis of the in-house Cu NFs by XRD showed that its peaks represented a face-centered cubic structure of Cu_2O, while the commercial Cu NFs could be identified as a mixture of Cu_2O and Cu^0. The SEM images showed both Cu NFs to have agglomerated, and the agglomeration with individual particles were slightly smaller for the commercial sample compared with the in-house sample. Because the EDS data showed both forms to be of similarly good purity with slight contaminations from carbon, we can propose that the differences in the macrostructure of the fungus after treatment with the two Cu NFs could only be due to the size difference and/or the oxidation state of the copper nanostructures.

3.3. Interaction of Copper Nanoforms with Fungus: Mechanisms

3.3.1. FTIR Analysis of the Interaction of Copper Nanoforms with Fungus

Salman and colleagues provided a thorough characterization of fungal mycelia using FTIR [38]. We have analyzed our results based on that characterization. As can be seen in Figure 5, we observed four major absorbance peaks in all three samples (mycelium only, mycelium treated with Cu NFs, and mycelium treated with Cu_2O NFs) and minor ones in the fingerprint region. The peak at ~3280 cm^{-1} corresponds to an O–H stretch band from water [38]. Absorbance peaks at 2923 and 2855 cm^{-1} are C–H stretching bands (and a barely visible band at 1748 cm^{-1} due to a C=O stretch) attributed to lipid molecules. At lower wavenumbers within the borders of the diagnostic and fingerprint regions, we can observe absorption bands corresponding to amide (1636 cm^{-1}, 1555 cm^{-1}) stretches caused by proteins in the sample and chitin. The amidic C–N stretch (1372 cm^{-1}) is from chitin and proteins in the sample [39]. In the fingerprint region, we see a small band at 1239 cm^{-1} corresponding to tertiary amine, N–H bending [40], with two bands corresponding to carbohydrate C–O stretches at 1147 cm^{-1} and 1026 cm^{-1} and a barely visible PO_2 stretch (1075 cm^{-1}) occurring due to nucleic acids.

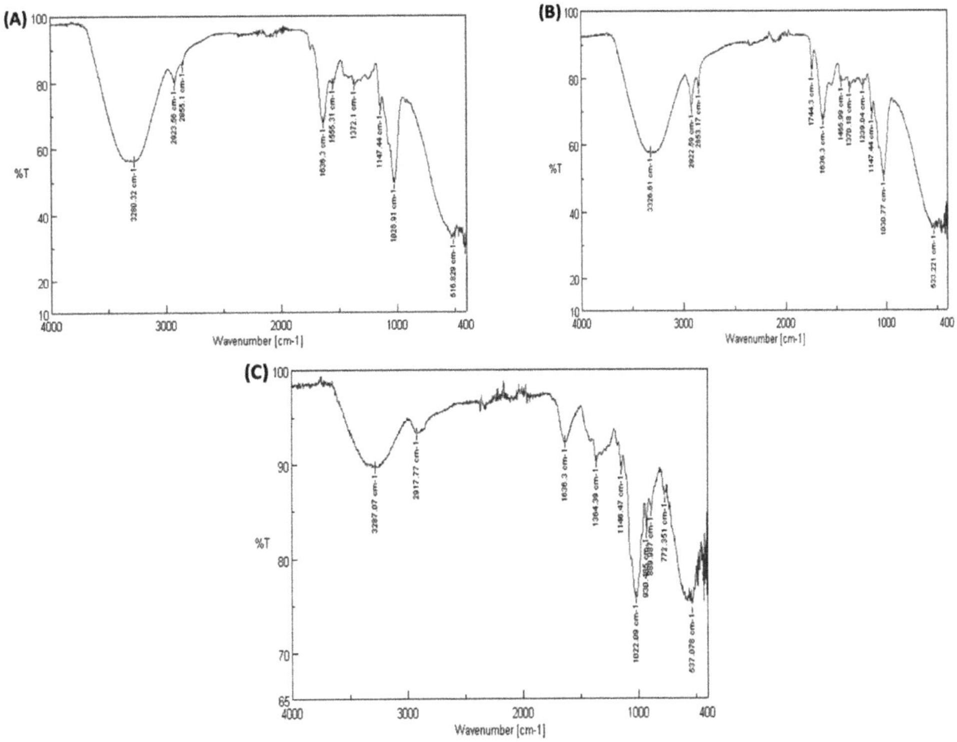

Figure 5. Interaction of *C. gloeosporioides* with copper nanoforms analyzed using FTIR. FTIR spectra of fungal mycelium (**A**) alone (control), (**B**) with Cu NPs, (**C**) with Cu_2O NPs (FTIR analyses were conducted using a JASCO FT/IR-4200 instrument with an attenuated total reflection (ATR) set-up). Absorption spectra were obtained in the region between 4000 and 400 cm^{-1}.

In the presence of Cu NFs, the FTIR spectra appear to be very similar to the control with only a small peak, possibly due to methylene scissoring vibrations from the proteins in the solution, and the peak appears at a 1455 cm^{-1} wavenumber [41]. Cu NPs have been shown to be antibacterial due to their affinity for highly abundant amines and C=O on the cellular surface [42]. Thus, another possibility is that the strong interaction of Cu NPs and amide may have caused a band displacement. This minor vibration was not automatically detected in the control sample but can be observed by the eye similarly to the Cu NFs-treated sample. Peaks at 1744 cm^{-1} and 1239 cm^{-1} are more prominent here than that in the control. It indicates a decrease in the concentration of C=O of lipids showing affinity towards Cu NPs.

Of further interest is the total absence of an absorbance band at 1555 cm^{-1} due to amides from proteins [39,41] and chitin [39]. In addition, there are three more bands in the fingerprint region at 930 cm^{-1}, 889 cm^{-1} (P–O stretching from polyphosphates in the cell wall), and 772 cm^{-1}. Here, an increase in the number of bands shows the masking of all of the biomolecules by Cu_2O NFs, which indicates a decrease in the concentration of the biomolecules after interaction with Cu_2O NFs.

Using a different fungal strain, we could still observe the same four major absorption bands in fungal mycelium only and in mycelium treated with Cu NFs, as well as minor ones in the fingerprint region (Figure S2 in Supplementary Materials). The peak at ~3320 cm^{-1} corresponds to an O–H stretch band from water [38]. Absorbance peaks at 3008 and 2922 cm^{-1} and 2853 cm^{-1} are C–H stretching bands from Csp2-H, Csp3-H, and Csp3-H, respectively; a barely visible band at 1745 cm^{-1} is due to C=O stretch attributed to lipid molecules. In the fingerprint region, we see two bands corresponding to carbohydrate C-O stretches at 1148 cm^{-1} and 1026 cm^{-1}. In the presence of Cu NFs, the FTIR spectra appears to be very similar to the control.

3.3.2. Aggregation and Localization of Copper Nanoforms in Lipid Vesicles

First, we introduced 1.0 mg/mL of Cu NFs and Cu_2O NFs into separate DOPC GUVs to provide final metal and lipid concentrations of 0.5 mg/mL and 0.2 mM, respectively. Upon observation under phase contrast microscopy, we could clearly observe Cu NFs with sizes well above what was supplied (quoted by supplier as 60 nm) (Figure 6A). This shows that the Cu NFs were not stable and had aggregated in the biological environment. These observations are in agreement with previous reports and that the NPs aggregated in the aqueous environment, as the zeta potential of both Cu NFs and Cu_2O NFs samples were very low at 0.05 mV and −0.05 mV [41]. Furthermore, the size distribution of the "aggregated" Cu and CuO NFs averaged 175.6 nm and 254.8 nm, respectively, with a wide spread between 80 and 850 nm (Figure S3). In this experiment with lipid vesicles, some of the observed aggregated Cu NFs were larger than 1 µm. Indeed, agglomeration of Cu NFs in cells and cell media have been reported to be higher than in water [35].

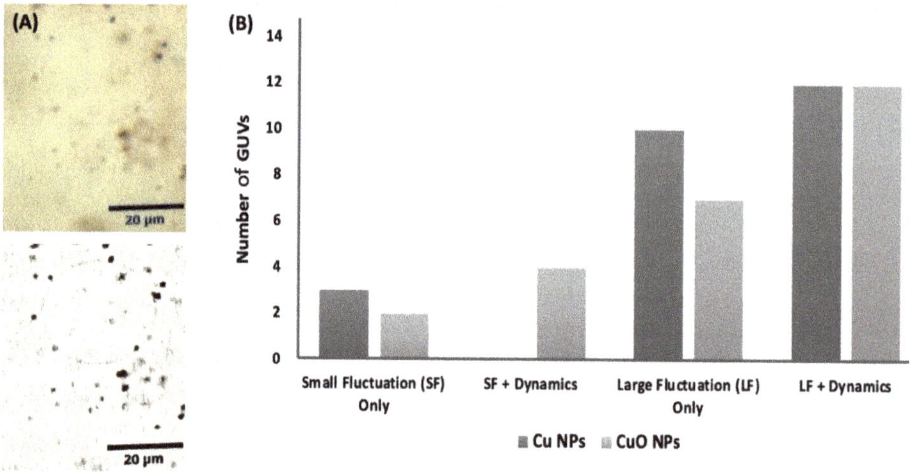

Figure 6. (**A**) Aggregated Cu NFs in aqueous lipid media (black dots). The outline of a lipid vesicle can be seen. Giant unilamellar vesicles were prepared using the 1,2-dioleoyl-*sn*-glycero-3-phosphocholine (DOPC, 0.2 mM) by the natural swelling method. Top: original image; Bottom: image after artistic modification to enhance contrast; (**B**) membrane dynamics induced by 0.5 mg/mL of Cu NPs and Cu_2O NPs. Vesicles were prepared as in (**A**).

Both forms of Cu were observed outside of the lipid vesicles. In real time, the aggregated nanoforms were captured bouncing on and off of lipid vesicles. We imagine that there was a repulsion between the hydrophilic head groups of the lipid molecules and the aggregated copper nanoforms through sheer steric repulsion. As there is only a slight negative charge on the phosphate group, a very slight negative charge (−0.05 mV) on the surface of the Cu_2O NFs, and no charge on the Cu NFs, it seems unlikely that the effect was due to charge. Hydrophobicity of the copper nanoforms, especially in aggregated forms, may also have hindered a direct interaction with the lipid headgroups (the nanoparticles were prepared as a suspension in MilliQ water. That is, they did not dissolve). What we were able to observe using a simple optical microscope was mainly relatively large particles. A focus on the lipid vesicles themselves showed that the vesicles were destabilized in the presence of the nanoparticles. This has been observed with amyloid beta fibrils upon interaction with lipid vesicles. The fibrils were observed to not closely associate with lipid vesicles. Localization of labelled oligomeric amyloid beta species could be seen on the vesicular surface [23]. This could be due to the discussed possible stearic repulsion between the vesicles and the vesicles and/or direct interaction between the vesicles and smaller (<100 nm) nanoparticles not visible under the microscope used.

3.3.3. Spatio-Temporal Membrane Dynamics Induced by Copper Nanoforms

There are two oppositely acting major forces existing among the phospholipids in the interfacial region of the lipid bilayer that control the surface area (A) per molecule. There is a negative pressure (also known as interfacial tension) from the hydrophilic–hydrophobic interface as a result of the hydrophobic effect; the positive pressure is due to repulsion between the phospholipid headgroups. The former force, which is generated from interfacial tension among identical molecules, causes the molecules to assemble and decreases the surface area A. In contrast, the latter, which includes steric repulsive interaction, hydration force, and electrostatic double-layer contribution, tends to disaggregate the molecules and increases A [43]. Changes in the balance between the two forces significantly influence membrane area and dynamics.

We investigated the effect of Cu NFs and Cu_2O NFs upon interaction with DOPC lipid vesicles by observing, in real time, the membrane instability characterized by fluctuation (undulation) and/or membrane transformation. Both Cu and Cu_2O nanoforms induced membrane fluctuation for all vesicles (n = 25). This suggests that there was an increase in the surface area A to inner volume V ratio [23]. The fluctuations were initially small. Eighty-eight percent (n = 22) of total lipid vesicles increased the intensity of the undulations in the presence of both copper forms (Figure 6B). Although only marginally different, Cu_2O NFs seemed to have had a higher effect on membrane dynamics than the Cu NFs. They also agree with the effect on fungal growth and observed images of *C. gloeosporioides* after exposure to the coppers (Figure 5).

Some of the lipid vesicles exhibited membrane transformations, forming exo-buds via large membrane fluctuations as the most dominant pathway (Figure 7). Exocytosis and product endo-bud formation by membranes are normal biological processes, such as the release of large molecules from within the cell into the extracellular space. It can also be a way of eliminating of unwanted molecules adsorbed or attached to the membrane surface. It is a form of transport. In this study, the majority of the membrane dynamics were exo-bud/-cytosis. In biophysical terms, we imagine two possible scenarios: (i) adsorption of small-sized Cu NFs and Cu_2O NFs to the membrane surface; and/or (ii) membrane penetration of even smaller-sized NPs. Upon these events, the lipid bilayer is destabilized as observed by the membrane fluctuations and ensuing dynamics. These membrane dynamics are a result of the increase in the surface area to volume ratio. The membrane tries to regain equilibrium by reducing its effective surface area. Indeed, the opposite dynamics (endo-bud/-cytosis) would also similarly reduce the effective surface area and enable equilibrium to be re-attained. In biological terms, we begin to imagine how the presence of Cu and Cu_2O NFs, especially the very small (non-aggregated forms or small aggregates) forms, can cause membrane exocytosis as a means to eliminate any adsorbed and penetrated materials on the cell surface. Exocytosis of NPs has been reported [44]. Although not a major dynamic, endo-bud/cytosis was also observed. Similarly, we can imagine that this would most likely be a means by which small nanoparticles would gain entry into the cell through clathrin-mediated endocytosis. Our future research will focus on (i) measuring the size distribution and zeta potential of the copper and copper nanoforms in aqueous solutions containing lipid vesicles; (ii) studying their localization in lipid vesicle solutions and using higher-resolution microscopy; and (iii) labelling the vesicles to investigate any phase changes and quantify membrane fluidity. Furthermore, the observed *hit-and-bounce-off* effect of the aggregated copper forms on the membrane surface could be studied deeply to understand their immediate effect on membrane curvature.

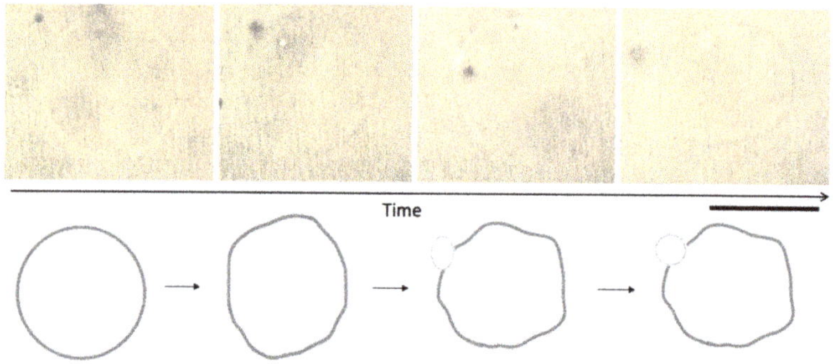

Figure 7. Endo-bud formation. The most dominant lipid vesicular transformation pathway induced by Cu NFs. Top: Original image; Bottom: stylized image to clearly show what was observable under the objective lens. Giant unilamellar vesicles were prepared using 1,2-dioleoyl-*sn*-glycero-3-phosphocholine (DOPC, 0.2 mM) by the natural swelling method. Size bar = 20 μm.

3.3.4. Production of Melanin by Fungus

In this section, we discuss the fungal response upon exposure to the Cu NFs. This was very interesting for us as we did not set up our experiment to investigate melanin production. During the experiments on the effect of Cu NFs on growth of *C. gloeosporioides*, we observed pigment formation. We believed it to be melanin, as melanin has been reported to be produced by organisms including fungus [45] in response to stress. The exposure to Cu NFs must have induced a defense mechanism in the fungus. The degree of melanization was determined using a hydrogen peroxide-based assay [32]. The results show that melanin was produced in a concentration-dependent manner, that is, the higher the concentration of Cu NFs, the bigger the amount of melanin produced (Figure 8A). An estimate of the melanin produced was calculated from a standard curve of melanin (Figure 8B). In the absence of Cu NFs, the fungus produced 0.057 mg/g of melanin. This amount changed to 0.041 mg/g, 0.166 mg/g, and 0.199 mg/g in the presence of 0.1 mg/mL, 0.25 mg/mL, and 0.5 mg/mL of Cu NFs.

At a low concentration of Cu NFs, melanization was lower than the control. This is a similar trend to the effect on fungal growth. We can imagine that the low concentration of Cu NFs is in fact a stimulant to fungal growth rather than a stress factor. This observation is in agreement with a report on how some species of bacteria and fungus perform better in the presence of low levels of copper [37].

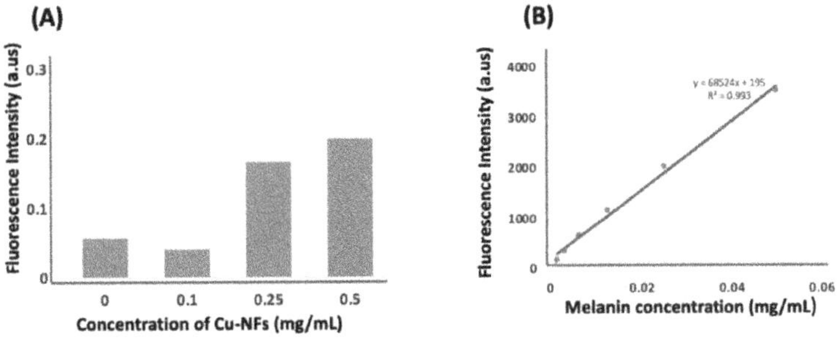

Figure 8. Melanization of *C. gloeosporioides* in the presence of in-house Cu NFs (**A**). *C. gloeosporioides* was cultured in potato dextrose agar for 14 d at 25 °C in the presence of in-house Cu NFs. Standard curve of melanin (**B**).

4. General Summary

In this research, we investigated the interaction of copper nanoforms (Cu NFs) with a plant-devastating pathogen *C. gloeosporioides*. The details and discussion of the findings are in the previous section. To summarize, we synthesized Cu NFs from copper acetate using a top-down approach, with ascorbic acid acting as a reducing agent and polyvinylpyrrolidone 10,000 (PVP-10) acting as the stabilizing agent. We also purchased Cu NFs. We characterized both Cu NFs using EDS, and the results of the in-house and commercial Cu NFs were very similar in terms of purity (Figure 2). Both were of very high purity. Only carbon was noted as a contaminant in both samples, most likely originating from the carbon tape used to fix the sample onto the sample holder. Similar sources of contamination have been reported [34]. This shows that the protocol that we used to synthesis our Cu NFs was very robust. Furthermore, the in-house Cu NFs showed a better even distribution of structures compared with the commercial Cu NFs. In both nanomaterials, there was a degree of agglomeration. The agglomerates, as well as individual particles, were smaller for commercial Cu NFs, as can be seen in the SEM images (Figure 3). XRD was used to understand their chemical properties. Both Cu NFs were characterized as having a cubic crystalline structure. However, the in-house Cu NFs were a cubic structure of Cu_2O, while the commercial Cu NFs could be identified as a mixture of Cu_2O and Cu^0. In short, we could synthesize high-purity Cu NFs using ascorbic acid as a reductant and PVP-10 as a stabilizer.

We studied the antimicrobial activity of the Cu NFs on *C. gloeosporioides*. The growth of the fungus in the presence and absence of the NFs was measured daily for 2 weeks. In addition, simple optical microscopic and SEM images of the fungus were obtained at the end of the incubation period. The following are some of the notable findings: (i) The in-house Cu NFs were more effective antimicrobial agents than commercial NFs. The difference could be attributed to the differences in size and/or oxidation states of the Cu NFs; (ii) At low concentrations of Cu NFs (0.25 mg/mL for in-house and up to 0.5 mg/mL for commercial NFs, the fungal growth was faster, but the mycelial density was less than the control. In the near future, we will investigate why the low concentration of Cu NFs boost the growth of fungus, albeit seemingly having a weaker vegetative form; (iii) Melanization of the fungus, in response to the presence of Cu NFs was observed and quantified. It corresponded to the concentration of the added Cu NFs.

Lastly, we studied some of the possible mechanisms behind the antimicrobial activity of Cu NFs. The localization of aggregated Cu NFs could be clearly observed outside the model membranes. Spatio-temporal membrane dynamics (fluctuation and membrane transformation) were captured in real-time. The dominant dynamics culminated into large fluctuations. Some of the large fluctuations resulted in vesicular transformation. The major transformation was exo-bud/exo-cytosis.

Supplementary Materials: The following supporting information can be downloaded at: https://www.mdpi.com/article/10.3390/nano13232990/s1, Figure S1: Effect of Cu-NFs on *C. gloeosporioides*, a fungal pathogen that attacks food crops (A) image of fungus (i) alone (Control); and in the presence of (ii) in-house Cu-NFs, (iii) commercial Cu-NFs, taken using an optical microscope at ×25 magnification; (B) Photos of the fungus (i) Control; and in the presence of (ii) inhouse Cu-NFs (0.5 mg/mL) and (iii) commercial Cu-NFs (0.5 mg/mL); Figure S2: Interaction of *C. gloeosporioides* with Cu-NFs analysed using FTIR Spectra of fungal mycelium. FTIR analyses were conducted as described earlier; Figure S3: Size distribution of (a) aggregated copper and (b) copper oxide nanoforms analysed by DLS technique using a Zetasizer Nano ZS90 (Malvern Panalytical, Japan). All measurements were conducted in water at room temperature (25 °C) and t approximately 200 kcps count rate.

Author Contributions: Conceptualization, M.C.V.; Methodology, M.C.V., Y.N., L.T.T.T., X.Z., M.N., A.F.O.-A. and T.N.; Formal analysis, M.C.V., N.S. and X.Z.; Investigation, Y.N., L.T.T.T., X.Z., A.F.O.-A. and T.N.; Resources, M.C.V. and M.N.; Writing—original draft, M.C.V. and N.S.; Writing—review & editing, M.C.V., N.S. and M.N.; Supervision, M.C.V. All authors have read and agreed to the published version of the manuscript.

Funding: This study was financially supported by JST strategic International Collaborative Research Program (SCIORP), Grant Number JPMJSC22A2, Japan.

Data Availability Statement: Data are contained within the article and Supplementary Materials.

Acknowledgments: Some of this research was supported by the Japan Science and Technology Agency (JST) grant awarded to M.C.V. *Colletotrichum gloeosporioides* was kindly provided by the Plant Pathology Laboratory, Faculty of Agriculture, Kagoshima University, Japan; the authors thank the Gene Research Centre, Kagoshima University, for the instrumentation support; Instrument Analysis Facility, Research Support Centre, Kagoshima University.

Conflicts of Interest: The authors declare no conflict of interest.

References

1. Lemire, J.A.; Harrison, J.J.; Turner, R.J. Antimicrobial Activity of Metals: Mechanisms, Molecular Targets and Applications. *Nat. Rev. Microbiol.* **2013**, *11*, 371–384. [CrossRef]
2. Morones, J.R.; Elechiguerra, J.L.; Camacho, A.; Holt, K.; Kouri, J.B.; Ramírez, J.T.; Yacaman, M.J. The Bactericidal Effect of Silver Nanoparticles. *Nanotechnology* **2005**, *16*, 2346–2353. [CrossRef]
3. Wang, Y.; Wang, X.; Wang, G.; Wang, X. Effect of Ionic Strength on Freeze–Thaw Stability of Glycosylated Soy Protein Emulsion. *J. Am. Oil Chem. Soc.* **2020**, *98*, 891–901. [CrossRef]
4. Abbaszadegan, A.; Ghahramani, Y.; Gholami, A.; Hemmateenejad, B.; Dorostkar, S.; Nabavizadeh, M.; Sharghi, H. The Effect of Charge at the Surface of Silver Nanoparticles on Antimicrobial Activity against Gram-Positive and Gram-Negative Bacteria: A Preliminary Study. *J. Nanomater.* **2015**, *16*, 53. [CrossRef]
5. Samanta, T.; Cheeni, V.; Das, S.; Roy, A.B.; Ghosh, B.C.; Mitra, A. Assessing Biochemical Changes during Standardization of Fermentation Time and Temperature for Manufacturing Quality Black Tea. *J. Food Sci. Technol.* **2015**, *52*, 2387–2393. [CrossRef]
6. El-Batal, A.I.; Al-Hazmi, N.E.; Mosallam, F.M.; El-Sayyad, G.S. Biogenic Synthesis of Copper Nanoparticles by Natural Polysaccharides and *Pleurotus ostreatus* Fermented Fenugreek Using Gamma Rays with Antioxidant and Antimicrobial Potential towards Some Wound Pathogens. *Microb. Pathog.* **2018**, *118*, 159–169. [CrossRef]
7. Sánchez-López, E.; Gomes, D.; Esteruelas, G.; Bonilla, L.; Lopez-Machado, A.L.; Galindo, R.; Cano, A.; Espina, M.; Ettcheto, M.; Camins, A.; et al. Metal-Based Nanoparticles as Antimicrobial Agents: An Overview. *Nanomaterials* **2020**, *10*, 292. [CrossRef]
8. Ramteke, L.; Gawali, P.; Jadhav, B.L.; Chopade, B.A. Comparative Study on Antibacterial Activity of Metal Ions, Monometallic and Alloy Noble Metal Nanoparticles Against Nosocomial Pathogens. *Bionanoscience* **2020**, *10*, 1018–1036. [CrossRef]
9. Broglie, J.J.; Alston, B.; Yang, C.; Ma, L.; Adcock, A.F.; Chen, W.; Yang, L. Antiviral Activity of Gold/Copper Sulfide Core/Shell Nanoparticles against Human Norovirus Virus-Like Particles. *PLoS ONE* **2015**, *10*, e0141050. [CrossRef]
10. Rani, R.; Kumar, H.; Salar, R.K.; Purewal, S.S. Antibacterial activity of copper oxide nanoparticles against gram-negative bacterial strain synthesized by reverse micelle technique. *Int. J. Pharm. Res. Dev.* **2014**, *6*, 72–78.
11. Pazos-Ortiz, E.; Roque-Ruiz, J.H.; Hinojos-Márquez, E.A.; López-Esparza, J.; Donohué-Cornejo, A.; Cuevas-González, J.C.; Espinosa-Cristóbal, L.F.; Reyes-López, S.Y. Dose-Dependent Antimicrobial Activity of Silver Nanoparticles on Polycaprolactone Fibers against Gram-Positive and Gram-Negative Bacteria. *J. Nanomater.* **2017**, *2017*, 4752314. [CrossRef]
12. Yang, N.; Guo, H.; Cao, C.; Wang, X.; Song, X.; Wang, W.; Yang, D.; Xi, L.; Mou, X.; Dong, X. Infection Microenvironment-Activated Nanoparticles for NIR-II Photoacoustic Imaging-Guided Photothermal/Chemodynamic Synergistic Anti-Infective Therapy. *Biomaterials* **2021**, *275*, 120918. [CrossRef]
13. Yuan, P.; Ding, X.; Yang, Y.Y.; Xu, Q.-H. Metal Nanoparticles for Diagnosis and Therapy of Bacterial Infection. *Adv. Healthc. Mater.* **2018**, *7*, e1701392. [CrossRef]

14. Karlsson, H.L.; Cronholm, P.; Gustafsson, J.; Möller, L. Copper Oxide Nanoparticles Are Highly Toxic: A Comparison between Metal Oxide Nanoparticles and Carbon Nanotubes. *Chem. Res. Toxicol.* **2008**, *21*, 1726–1732. [CrossRef]
15. Wang, Z.; Li, N.; Zhao, J.; White, J.C.; Qu, P.; Xing, B. CuO Nanoparticle Interaction with Human Epithelial Cells: Cellular Uptake, Location, Export, and Genotoxicity. *Chem. Res. Toxicol.* **2012**, *25*, 1512–1521. [CrossRef]
16. Vanwinkle, B.A.; de Mesy Bentley, K.L.; Malecki, J.M.; Gunter, K.K.; Evans, I.M.; Elder, A.; Finkelstein, J.N.; Oberdörster, G.; Gunter, T.E. Nanoparticle (NP) Uptake by Type I Alveolar Epithelial Cells and Their Oxidant Stress Response. *Nanotoxicology* **2009**, *3*, 307–318. [CrossRef]
17. Cronholm, P.; Midander, K.; Karlsson, H.L.; Elihn, K.; Wallinder, I.O.; Möller, L. Effect of Sonication and Serum Proteins on Copper Release from Copper Nanoparticles and the Toxicity towards Lung Epithelial Cells. *Nanotoxicology* **2011**, *5*, 269–281. [CrossRef]
18. Applerot, G.; Lellouche, J.; Lipovsky, A.; Nitzan, Y.; Lubart, R.; Gedanken, A.; Banin, E. Understanding the Antibacterial Mechanism of CuO Nanoparticles: Revealing the Route of Induced Oxidative Stress. *Small* **2012**, *8*, 3326–3337. [CrossRef]
19. Laha, D.; Pramanik, A.; Laskar, A.; Jana, M.; Pramanik, P.; Karmakar, P. Shape-Dependent Bactericidal Activity of Copper Oxide Nanoparticle Mediated by DNA and Membrane Damage. *Mater. Res. Bull.* **2014**, *59*, 185–191. [CrossRef]
20. Dong, H.; Yang, K.; Zhang, Y.; Li, Q.; Xiu, W.; Ding, M.; Shan, J.; Mou, Y. Photocatalytic Cu_2WS_4 Nanocrystals for Efficient Bacterial Killing and Biofilm Disruption. *Int. J. Nanomed.* **2022**, *17*, 2735–2750. [CrossRef]
21. Oussou-Azo, A.F.; Nakama, T.; Nakamura, M.; Futagami, T.; Vestergaard, M.C.M. Antifungal Potential of Nanostructured Crystalline Copper and Its Oxide Forms. *Nanomaterials* **2020**, *10*, 1003. [CrossRef]
22. Sharma, N.; Phan, H.T.T.; Yoda, T.; Shimokawa, N.; Vestergaard, M.C.; Takagi, M. Effects of Capsaicin on Biomimetic Membranes. *Biomimetics* **2019**, *4*, 17. [CrossRef]
23. Vestergaard, M.C.; Morita, M.; Hamada, T.; Takagi, M. Membrane Fusion and Vesicular Transformation Induced by Alzheimer's Amyloid Beta. *Biochim. Biophys. Acta* **2013**, *1828*, 1314–1321. [CrossRef]
24. Vestergaard, M.; Hamada, T.; Takagi, M. Using Model Membranes for the Study of Amyloid Beta: Lipid Interactions and Neurotoxicity. *Biotechnol. Bioeng.* **2008**, *99*, 753–763. [CrossRef]
25. Hasan, M.; Karal, M.A.S.; Levadnyy, V.; Yamazaki, M. Mechanism of Initial Stage of Pore Formation Induced by Antimicrobial Peptide Magainin 2. *Langmuir* **2018**, *34*, 3349–3362. [CrossRef]
26. Sakamoto, K.; Morishita, T.; Aburai, K.; Ito, D.; Imura, T.; Sakai, K.; Abe, M.; Nakase, I.; Futaki, S.; Sakai, H. Direct Entry of Cell-Penetrating Peptide Can Be Controlled by Maneuvering the Membrane Curvature. *Sci. Rep.* **2021**, *11*, 31. [CrossRef]
27. Bhat, A.; Huan, K.; Cooks, T.; Boukari, H.; Lu, Q. Probing Interactions between AuNPs/AgNPs and Giant Unilamellar Vesicles (GUVs) Using Hyperspectral Dark-Field Microscopy. *Int. J. Mol. Sci.* **2018**, *19*, 1014. [CrossRef]
28. Walbrück, K.; Kuellmer, F.; Witzleben, S.; Guenther, K. Synthesis and Characterization of PVP-Stabilized Palladium Nanoparticles by XRD, SAXS, SP-ICP-MS, and SEM. *J. Nanomater.* **2019**, *2019*, 4758108. [CrossRef]
29. Lamprell, H.; Mazerolles, G.; Kodjo, A.; Chamba, J.F.; Noël, Y.; Beuvier, E. Discrimination of Staphylococcus Aureus Strains from Different Species of Staphylococcus Using Fourier Transform Infrared (FTIR) Spectroscopy. *Int. J. Food Microbiol.* **2006**, *108*, 125–129. [CrossRef]
30. Lefier, D.; Hirst, D.; Holt, C.; Williams, A.G. Effect of Sampling Procedure and Strain Variation in Listeria Monocytogenes on the Discrimination of Species in the Genus Listeria by Fourier Transform Infrared Spectroscopy and Canonical Variates Analysis. *FEMS Microbiol. Lett.* **1997**, *147*, 45–50. [CrossRef]
31. Ishii, K.; Hamada, T.; Hatakeyama, M.; Sugimoto, R.; Nagasaki, T.; Takagi, M. Reversible Control of Exo- and Endo-Budding Transitions in a Photosensitive Lipid Membrane. *ChemBioChem* **2009**, *10*, 251–256. [CrossRef]
32. Fernandes, B.; Matamá, T.; Guimarães, D.; Gomes, A.; Cavaco-Paulo, A. Fluorescent Quantification of Melanin. *Pigment Cell Melanoma Res.* **2016**, *29*, 707–712. [CrossRef]
33. Saad, N.A.; Dar, M.H.; Ramya, E.; Naraharisetty, S.R.G.; Narayana Rao, D. Saturable and Reverse Saturable Absorption of a Cu_2O–Ag Nanoheterostructure. *J. Mater. Sci.* **2019**, *54*, 188–199. [CrossRef]
34. Yang, Z.; Cingarapu, S.; Klabunde, K.J. Synthesis of Magnesium Oxychloride Nanorods with Controllable Morphology and Their Transformation to Magnesium Hydroxide Nanorods via Treatment with Sodium Hydroxide. *J. Sol-Gel Sci. Technol.* **2010**, *53*, 359–365. [CrossRef]
35. Prabhu, B.M.; Ali, S.F.; Murdock, R.C.; Hussain, S.M.; Srivatsan, M. Copper Nanoparticles Exert Size and Concentration Dependent Toxicity on Somatosensory Neurons of Rat. *Nanotoxicology* **2010**, *4*, 150–160. [CrossRef]
36. Menichetti, A.; Mavridi-Printezi, A.; Mordini, D.; Montalti, M. Effect of Size, Shape and Surface Functionalization on the Antibacterial Activity of Silver Nanoparticles. *J. Funct. Biomater.* **2023**, *14*, 244. [CrossRef]
37. Zapotoczny, S.; Jurkiewicz, A.; Tylko, G.; Anielska, T.; Turnau, K. Accumulation of Copper by Acremonium Pinkertoniae, a Fungus Isolated from Industrial Wastes. *Microbiol. Res.* **2007**, *162*, 219–228. [CrossRef]
38. Salman, A.; Tsror, L.; Pomerantz, A.; Moreh, R.; Mordechai, S.; Huleihel, M. FTIR Spectroscopy for Detection and Identification of Fungal Phytopathogenes. *Spectroscopy* **2010**, *24*, 261–267. [CrossRef]
39. Girometta, C.; Dondi, D.; Baiguera, R.M.; Bracco, F.; Branciforti, D.S.; Buratti, S.; Lazzaroni, S.; Savino, E. Characterization of Mycelia from Wood-Decay Species by TGA and IR Spectroscopy. *Cellulose* **2020**, *27*, 6133–6148. [CrossRef]
40. Aguilar-Méndez, M.A.; San Martín-Martínez, E.; Ortega-Arroyo, L.; Cobián-Portillo, G.; Sánchez-Espíndola, E. Synthesis and Characterization of Silver Nanoparticles: Effect on Phytopathogen Colletotrichum Gloesporioides. *J. Nanopart. Res.* **2011**, *13*, 2525–2532. [CrossRef]

41. Thakker, J.N.; Dalwadi, P.; Dhandhukia, P.C. Biosynthesis of Gold Nanoparticles Using *Fusarium oxysporum* f. sp. *cubense* JT1, a Plant Pathogenic Fungus. *ISRN Biotechnol.* **2013**, *2013*, 515091. [CrossRef]
42. Yoon, K.-Y.; Hoon Byeon, J.; Park, J.-H.; Hwang, J. Susceptibility Constants of *Escherichia coli* and *Bacillus subtilis* to Silver and Copper Nanoparticles. *Sci. Total Environ.* **2007**, *373*, 572–575. [CrossRef]
43. Israelachvili, J.N. *Intermolecules and Surface Forces*, 2nd ed.; Academic Press: New York, NY, USA, 1992.
44. Sakhtianchi, R.; Minchin, R.F.; Lee, K.-B.; Alkilany, A.M.; Serpooshan, V.; Mahmoudi, M. Exocytosis of Nanoparticles from Cells: Role in Cellular Retention and Toxicity. *Adv. Colloid Interface Sci.* **2013**, *201–202*, 18–29. [CrossRef]
45. Dullah, S.; Hazarika, D.J.; Goswami, G.; Borgohain, T.; Ghosh, A.; Barooah, M.; Bhattacharyya, A.; Boro, R.C. Melanin Production and Laccase Mediated Oxidative Stress Alleviation during Fungal-Fungal Interaction among Basidiomycete Fungi. *IMA Fungus* **2021**, *12*, 33. [CrossRef]

nanomaterials

Article

Liquid-Phase Partial Hydrogenation of Phenylacetylene at Ambient Conditions Catalyzed by Pd-Fe-O Nanoparticles Supported on Silica

Anastasiya A. Shesterkina [1,2], Olga A. Kirichenko [3], Olga P. Tkachenko [3], Alexander L. Kustov [1,2] and Leonid M. Kustov [1,2,3,*]

1 Chemistry Department, Lomonosov Moscow State University, Leninskie Gory 1/3, 119234 Moscow, Russia; anastasiia.strelkova@mail.ru (A.A.S.); kyst@list.ru (A.L.K.)
2 Laboratory of Nanochemistry and Ecology, Institute of Ecotechnologies, National University of Science and Technology MISIS, Leninsky Prospect 4, 119049 Moscow, Russia
3 Laboratory of Development and Research of Polyfunctional Catalysts, Zelinsky Institute of Organic Chemistry, Russian Academy of Sciences, Leninsky Prospekt 47, 119991 Moscow, Russia; okiriche@ioc.ac.ru (O.A.K.); ot113@mail.ru (O.P.T.)
* Correspondence: lmkustov@mail.ru

Abstract: Catalysts with no hazardous or toxic components are required for the selective hydrogenation of acetylenic bonds in the synthesis of pharmaceuticals, vitamins, nutraceuticals, and fragrances. The present work demonstrates that a high selectivity to alkene can be reached over a Pd-Fe-O/SiO_2 system prepared by the co-impregnation of a silica support with a solution of the metal precursors $(NH_4)_3[Fe(C_2O_4)_3]$ and $[Pd(NH_3)_4]Cl_2$ followed by thermal treatment in hydrogen or in air at 400 °C. A DRIFT spectroscopic study of CO adsorption revealed large shifts in the position of the Pd^{n+}-CO bands for this system, indicating the strong effect of Fe^{n+} on the Pd electronic state, resulting in a decreased rate of double C=C bond hydrogenation and an increased selectivity of alkyne hydrogenation to alkene. The prepared catalysts consisted of mono- and bimetallic nanoparticles on an SiO_2 carrier and exhibited a selectivity as high as that of the commonly used Lindlar catalyst (which contains such hazardous components as lead and barium), while the activity of the Fe-Pd-O/SiO_2 catalyst was an order of magnitude higher. The hydrogenation of a triple bond over the proposed Pd-Fe catalyst opens the way to selective hydrogenation over nontoxic catalysts with a high yield and productivity. Taking into account a simple procedure of catalyst preparation, this direction provides a rationale for the large-scale implementation of these catalysts.

Keywords: alkyne selective hydrogenation; Pd-Fe catalysts; DRIFT spectroscopy; CO adsorption

Citation: Shesterkina, A.A.; Kirichenko, O.A.; Tkachenko, O.P.; Kustov, A.L.; Kustov, L.M. Liquid-Phase Partial Hydrogenation of Phenylacetylene at Ambient Conditions Catalyzed by Pd-Fe-O Nanoparticles Supported on Silica. *Nanomaterials* **2023**, *13*, 2247. https://doi.org/10.3390/nano13152247

Academic Editors: Detlef W. Bahnemann and Chengbin Liu

Received: 30 June 2023
Revised: 30 July 2023
Accepted: 31 July 2023
Published: 3 August 2023

1. Introduction

Supported palladium catalysts are widely used for the selective hydrogenation of alkynes both in industry and in laboratory practice [1–3]. One of the most selective commercial catalysts of the liquid-phase hydrogenation to alkenes is a Lindlar catalyst that comprises a Pd catalyst partially poisoned with lead (5 wt.% palladium supported on calcium carbonate or barium sulfate and treated with various lead compounds [4]). The hydrogenation of alkynes, for the last decade, has been increasingly used in the synthesis of pharmaceuticals, vitamins, nutraceuticals, fragrances, and catalysts with no hazardous or toxic components, which are in strong demand. The economic reasons associated with the high palladium price and recent environmental regulations stimulate searching for nontoxic alternatives to the Lindlar catalyst as well. Several strategies have been applied [1], and most recent studies focus on the preparation of less toxic catalysts based on the supported modified Pd nanoparticles (NPs) with a high catalytic activity and selectivity at mild conditions in a three-phase liquid–solid–gas system [5–17]. Among the modifying additives in the catalyst developed, Ag, Au, Bi, C, Cu, Ga, Ni, Si, Sn, Ti, W, and Zn have been used [1,5,12–17]. Iron and its oxides could be the least toxic modifying additives, yet their effect was weak [15] or even negative [17] as compared with that of other elements. Different inorganic, hybrid, and organic materials have been studied as Pd catalyst supports, and SiO_2 is one of the most appropriate carriers [10–12]. The catalytic activity of supported Pd nanoparticles in the selective liquid-phase hydrogenation depends both on the support nature and the conditions of catalyst preparation that affect the Pd particle size and electronic state [18]. The outstanding selectivity of the highly dispersed Pd mono and bimetallic nanoparticles is supposed to be due to suppressing the β-PdH phase formation [1,18]. The complete elimination of β-PdH formation was revealed in Pd-Fe/SiO_2 systems of different compositions [19–21], yet they were not studied in alkyne hydrogenation. Also, the promoting effect of iron on palladium was observed in the hydrogenation of nitro compounds [22].

The selective hydrogenation of phenylacetylene (PhA) is considered as a convenient model reaction for the evaluation of catalysts for the selective hydrogenation of alkynes to alkenes under mild conditions [10,12,14,16]. On the other hand, phenylacetylene removal from styrene (St) feedstocks by selective liquid-phase hydrogenation is a process of great industrial importance [23,24].

One of the major problems in the hydrogenation of alkynes to alkenes over Pd catalysts is the drop of selectivity, even to zero, when approaching the high PhA conversion in the hydrogenation process. This makes it difficult to control the process, especially at high alkyne concentrations and in a large-scale production [8,12,13,16]. This phenomenon is especially pronounced for the Pd/SiO_2 catalysts prepared on commercial SiO_2 supports by simple, commercially scalable methods of ion exchange or impregnation followed by calcination [12]. In our previous publications [20,21], a new method of the synthesis of bimetallic Pd-Fe-O/SiO_2 materials was presented. The catalysts consist of Pd^0, $Pd_{1-x}Fe_x$, and FeO_x nanoparticles supported on silica. Some results on the liquid-phase hydrogenation of phenylacetylene to styrene with hydrogen at ambient conditions over the prepared materials have also been reported [25,26].

The goals of the present study are (i) to evaluate in detail the variation in the selectivity to styrene near the complete PhA hydrogenation over recently developed Pd-Fe-O/SiO_2 catalysts prepared using a commercial support; (ii) to compare the catalytic properties of developed catalysts with a commercial Lindlar catalyst; (ii) to reveal the factors regulating the selectivity and the rate of St hydrogenation.

2. Materials and Methods

2.1. Preparation of the Materials

The supported nanoparticles were prepared by the deposition of the metal precursors on granulated (diameter 4–6 mm) commercial silica carriers with a high specific surface area S_{BET} = 220 m^2g^{-1} (carrier HS in [20,21]). Commercial chemicals, $(NH_4)_3Fe(C_2O_4)_3{\cdot}3H_2O$ (98%, Acrus Organics) and $[Pd(NH_3)_4]Cl_2{\cdot}H_2O$ (41.42% Pd; Aurat, Russia), were used as metal precursors. The precursor deposition was performed by the method of incipient wetness impregnation, similar to the previously described procedure [21], at 60 °C (sample PdFe-1) or at a higher temperature of 80 °C (sample PdFe-2) instead of room temperature in order to avoid the formation of large crystallites on the surface of the support grain, which further resulted in the production of XRD detectable Pd^0 nanoparticles [27,28]. Both metal precursors were introduced simultaneously in the same solution. Before impregnation, the carrier sample was dried in a rotor evaporator at 60 °C and 40 mbar. After impregnation for 16 h, the samples were dried in a rotor evaporator at 60 °C. The Pd and Fe loadings were the same in all samples (3 and 7.6 mass. % calculated for reduced samples). The prepared materials were denoted as PdFe-X-CA-RB, where "X" is the sample number, "CA" is the temperature of calcination, if any, and "RB" denotes the sample reduction ("R"—reduced with hydrogen at 400 °C or at "B" °C). The sample of the same composition described in our previous publications was studied in this work as well (PdFe-3). The commercial Lindlar catalyst (Acros Organics, Geel, Belgium) was used as a reference sample.

2.2. Characterization of the Materials

X-ray diffraction patterns were recorded using a DRON-2 diffractometer with Ni-filtered Cu Ka radiation, as described previously [20,21]. Diffuse reflectance infrared Fourier transform (DRIFT) spectra were recorded at room temperature (RT) using a NICOLET "Protege" 460 spectrometer (Madison, WI, USA) equipped with a diffuse reflectance attachment in the interval of 6000–400 cm^{-1} at a resolution of 4 cm^{-1} (500 scans). CaF_2 powder was used as a reference. The fractioned sample (0.10–0.25 mm) was placed inside a special ampoule with a CaF_2 window. Before the spectroscopic measurements, a sample was evacuated at 350 °C (or at 250 °C if the sample calcined at 250 °C was studied) for 120 min. CO was used as a probe molecule, and it was adsorbed at RT and an equilibrium pressure of 1000–1500 Pa. The spectra were recorded at RT. The calcined samples were evacuated again at a high temperature and treated at RT with hydrogen (6000 Pa) for 15 min. Then, hydrogen was outgassed at room temperature, and spectra of adsorbed CO were recorded. The assignment of bands was performed using published reviews [27,28].

X-ray photoelectron spectra were recorded using an ES-2403 spectrometer equipped with a PHOIBOS 100 MCD analyzer (Thermo Fisher Scientific, Waltham, MA, USA).

The microstructure and morphological characteristics of the samples were studied by scanning electron microscopy using a Hitachi SU8000 electron microscope (Tokyo, Japan). The samples were examined by X-ray microanalysis (SEM-EDX) using an Oxford Instruments X-max 80 energy dispersive X-ray spectrometer (Abingdon, UK) at an accelerating voltage of 20 kV and an operating distance of 15 mm.

The microstructure of the samples was studied by transmission electron microscopy (TEM) with a Hitachi HT7700 electron microscope (Tokyo, Japan). The images were taken in the light field mode at an accelerating voltage of 100 kV. The average size of the supported nanoparticles was calculated based on the analysis of 250–350 nanoparticles for each sample of the catalyst.

The characterization of the catalytic activity was performed in the model reaction of liquid-phase phenylacetylene (PhA) hydrogenation at room temperature and atmospheric hydrogen pressure, as described in our previous publications [25,26]. Initial specific reaction rates (r_0) were calculated as moles of the substrate per mole of Pd per second at a conversion below 30%.

3. Results

3.1. Catalytic Activity of the Samples

The prepared monometallic palladium catalyst Pd-R exhibits a high catalytic activity in the reaction of PhA hydrogenation at the chosen reaction conditions, the selectivity to styrene being quite high but significantly lower than that of the Lindlar catalyst (Figure 1). Moreover, the rate of St hydrogenation after reaching the complete PhA conversion is rather high (Table 1) with complete hydrogenation to EtB, as has already been pointed out [25]. At the same reaction conditions, there is no PhA adsorption or conversion over the monometallic iron oxide samples calcined at 250–400 °C or further reduced at 400 °C.

The bimetallic catalysts prepared by the reduction of the dried supported precursors are more selective but less active than the monometallic Pd catalyst (Figure 1, Table 1). Their selectivity dependences on the PhA conversion approach that of the Lindlar catalyst (Figure 1b), yet the prepared catalysts are considerably more active than the Lindlar catalyst. This is confirmed both by the higher initial rate of hydrogenation and the shorter time required for the complete PhA conversion (Table 1, Figure 1a). The good reproducibility of the results regarding the catalytic activity of the samples should be mentioned (points for PdFe-1-R and PdFe-1/2-R in Figure 1b), as well as the agreement between the values of the reaction rates of PhA uptake and hydrogen consumption (Table 1). During the reproduction of the samples, care should be taken regarding the conditions used for the support impregnation and further thermal treatment. Slight variation results in a strong difference in the PhA hydrogenation rate (Table 1, samples PdFe-1-R and PdFe-3-R).

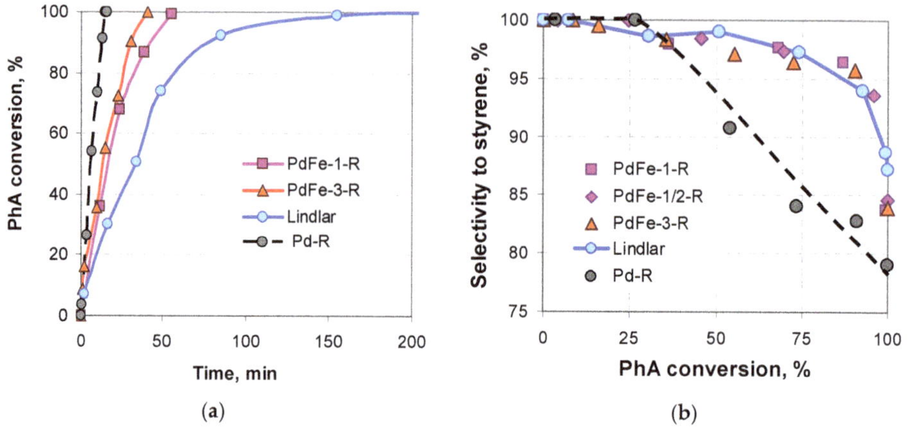

Figure 1. Time dependence of the phenylacetylene conversion (**a**) and variation in the selectivity to styrene vs. phenylacetylene conversion (**b**) over the samples prepared by the reduction of the dried supported precursors and over the Lindlar catalyst.

Table 1. The catalytic activity of the samples reduced in a hydrogen flow at 400 °C.

Sample	r_{St} [a], s^{-1}	η_{St} [b], $g_{St} \cdot g_{cat}^{-1} \cdot h^{-1}$
PdFe-1-R	0.02	5.7
PdFe-1/2-R [c]	0.05	4.5
PdFe-2-C250-R	0.20	17
PdFe-2-C250-R430 [d]	0.0025	34
PdFe-2-C250	0.46	15
PdFe-2-C350	0.62	48
PdFe-2-C400	1.0	31
Pd-R	0.51	23
Pd-R [e]	0.53	23
PdFe-3-R	0.02	14
Lindlar catalyst	<0.000	2.0

Reaction conditions: 0.13 M PhA, molar ratio PhA: Pd = 300 ± 20; [a] selectivity to styrene at complete PhA conversion; [b] r_0—the initial rates of PhA uptake and H_2 consumption; [c] St mass produced over 1 g of catalyst per hour; [d] r_{St}—the rate of St hydrogenation after the complete PhA conversion; [e]—repeated test.

The preparation procedure is supposed to be more essential for the catalytic properties than the structure of the silica support. It has been revealed that it is possible to prepare catalysts that exhibit high values of the selectivity to styrene at complete PhA conversion and a high catalytic activity in PhA hydrogenation to St combined with a low initial rate of further St hydrogenation by using another low-surface (LS) silica support (the surface area as low as 30 m^2g^{-1}; no micropores—LS in [20,21]) (Table 1, samples PdFe-4-R and PdFe-4-R-550Ar). When Pd-Fe/SiO_2 samples are prepared by known procedures [19,22], the selectivity and activity are good enough (Table 1, sample PdFe-5-C450-R440), yet the rate of St conversion is high, and significant Pd and Fe leaching is observed.

The bimetallic samples synthesized by the thermal decomposition of dried supported precursors in air at 250 °C (the minimal temperature of their decomposition found in our previous studies [20,21]) exhibit a higher activity as compared with the monometallic Pd catalyst (Figure 2a), yet their selectivity is low (Figure 2b). The reduction of these samples in a hydrogen flow at 400 °C results in a slight decrease in the activity, yet the selectivity to styrene increases to the values typical for the Lindlar catalyst (Figure 2). The reduced catalyst PdFe-2-C250-R is more selective than the monometallic Pd catalyst (Figure 2b, Table 1) and considerably more active than the Lindlar catalyst. The values of the St mass produced over 1 g of the catalyst per hour is an order of magnitude larger for the prepared reduced catalysts as compared with the Lindlar catalyst (Table 1) and may be increased by 60–100% while varying the reduction temperature, PhA:Pd ratio, and PhA concentration. For example, at PhA:Pd = 950 and C = 0.167 M, the sample PdFe-2-C250-R productivity is 34 $g_{St} \cdot g_{cat}{}^{-1} \cdot h^{-1}$ at a selectivity of 86%.

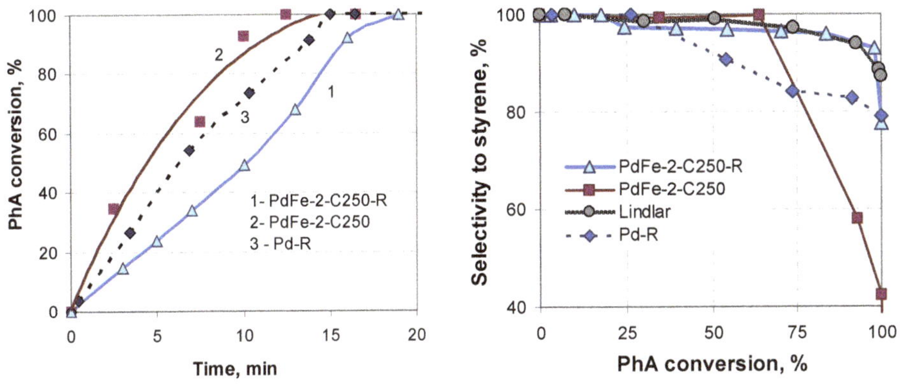

(**a**) (**b**)

Figure 2. Time dependence of the phenylacetylene conversion (**a**) and variation in the selectivity to styrene vs. phenylacetylene conversion (**b**) over the sample reduced after calcination and the reference catalysts.

The advantage of bimetallic Pd-Fe catalysts prepared by the reduction of the dried supported precursors is the lower rate of St hydrogenation at complete PhA hydrogenation (Table 1). The slow St hydrogenation extends the period with a high selectivity (Figure 3) and could make it possible to achieve and maintain a high selectivity when the process is to be scaled up.

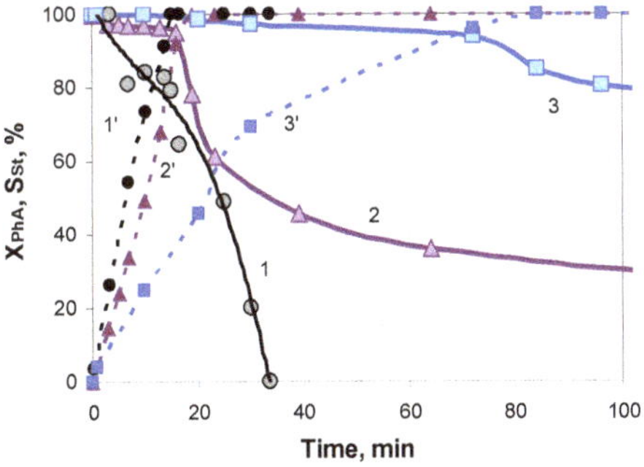

Figure 3. Time dependence of the selectivity to styrene and phenylacetylene conversion (dashed lines) over the catalysts Pd-R (1,1′), PdFe-2-C250-R (2,2′), and PdFe-1-R (3,3′).

In our recent studies [25,26], we revealed that the Pd-Fe samples calcined in an air flow catalyze both PhA and St hydrogenation, even with no preliminary reduction of the sample. The catalytic activity is comparable to or exceeds the activity of the reduced monometallic

Pd catalyst with the same Pd loading. The catalytic properties depend on the temperature of calcination. In the present study, it is shown that, among the bimetallic catalysts with a Pd loading of 3%, the most active is the sample calcined at 350 °C, yet its selectivity is not high enough at the conversion values approaching the complete PhA conversion in hydrogenation (Figure 4). The selectivity to St increases with the calcination temperature being up to 400 °C and becomes comparable with the selectivity of the Lindlar catalyst (Figure 4). However, in contradiction to the preliminary reduced samples, the calcined samples exhibit a high rate of styrene hydrogenation to ethylbenzene, increasing with the calcination temperature from 0.46 s^{-1} to 1.0 s^{-1}. On the other hand, the value of the St mass produced over 1 g of the catalyst per hour is 31, i.e., 1.5–2 times larger than that for other selective Pd-Fe catalysts (Table 1) and 15 times larger than that for the Lindlar catalyst.

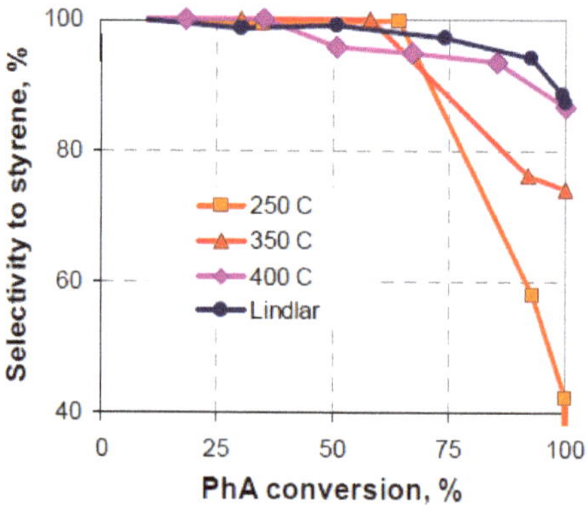

Figure 4. Variation in the selectivity to styrene vs. phenylacetylene conversion over the sample PdFe-2 calcined at different temperatures and over the Lindlar catalyst.

As can be seen from Figure 5, the sample PdFe-2-C400 has a high stability and does not deactivate during four reaction cycles without first washing it from the reaction medium. This confirms the fact that there is no leaching of the active phase and deactivation of the catalyst by reaction products.

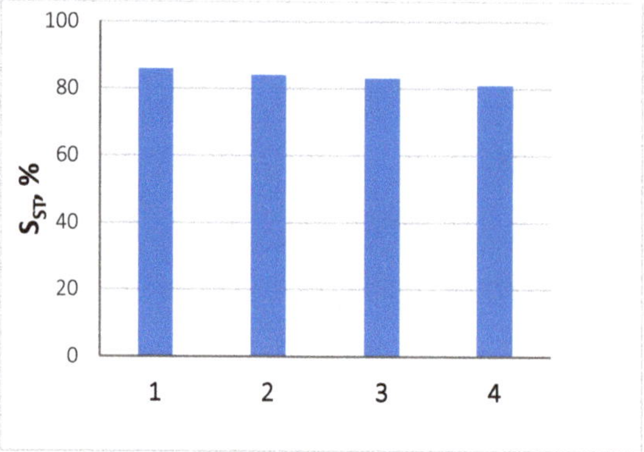

Figure 5. The stability of the PdFe-2-C400 sample, expressed in terms of the selectivity of styrene formation at full conversion.

Considering a significant difference in the catalytic behavior of the Pd-Fe/SiO_2 samples with the same composition and starting steps of preparation, it seems important to gain an understanding of the Fe influence on the Pd active surface sites and the effect of preparation variables on it.

3.2. Characterization of the Samples

The reduced monometallic Pd/SiO_2 sample contains Pd^0 crystallites of an average particle size of 25 nm. The supported nanoparticles in the bimetallic samples prepared in the present work are X-ray amorphous, except for the sample PdFe-2-C250-R, reduced after calcination. The Pd-Fe samples prepared previously are characterized by a number of physico-chemical methods in our publications [20,21,25,26], and the results are summarized in Table 2. The feature of these Pd-Fe samples is the presence of nano crystallites of both Pd^0 and $Pd_{1-x}Fe_x$ or Fe^0 and $Pd_{1-x}Fe_x$ phases. In the present work, to study the state of metals on the surface, DRIFT spectroscopy of chemisorbed CO is applied, and the obtained spectra are depicted in Figures 6–8.

Table 2. The surface sites in supported nanoparticles revealed with DRIFT spectroscopy and the average particle size and phase based on the XRD method [21,22].

Sample	Possible Surface Sites of Pd and Fe	Phase and Average Particle Size, nm
PdFe-2-C250-R	Pd^0 (top > facet cites), Pd^+	XRD: Pd^0-10 nm, Fe_3O_4-17 nm TEM: isolated nanoparticles (<5 nm)
PdFe-2-C250	Pd^{2+} reducible to Pd^+ with CO or to Pd^0 with H_2 at RT; Fe^{2+}, Fe^{3+}	XRD: Amorphous TEM: isolated nanoparticles (<5 nm)
PdFe-2-C350	Pd^{2+} reducible to Pd^+ with CO at RT; Fe^{2+}, Fe^{3+}	XRD: Amorphous
PdFe-2-C400	Pd^{2+} reducible to Pd^0 with CO or both to Pd^+ and Pd^0 with H_2 at RT; Fe^{2+}, Fe^{3+}	XRD: Amorphous
PdFe-3-R	Pd^0 (I: facet > top > X1 cites), Pd^+, Fe^{2+}	XRD: Pd^0 (15 nm)
Pd-R	Pd^0	XRD: Pd^0 (25 nm)

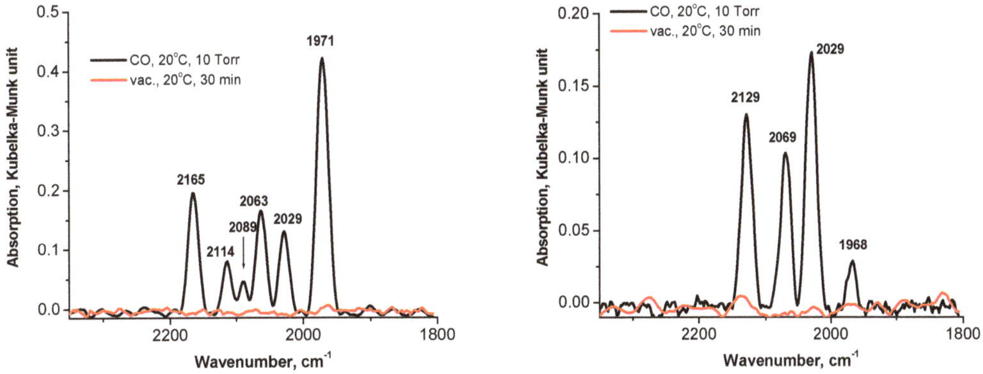

(a) (b)

Figure 6. DRIFT spectra recorded after CO adsorption on the PdFe-3-R (**a**) and PdFe-2-C250-R (**b**) samples (preliminarily reduced in hydrogen at 400 °C).

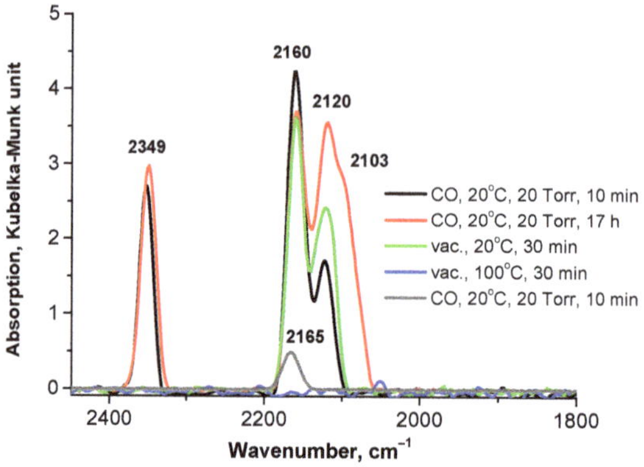

Figure 7. DRIFT spectra recorded after CO adsorption on the calcined PdFe-2-C350 sample.

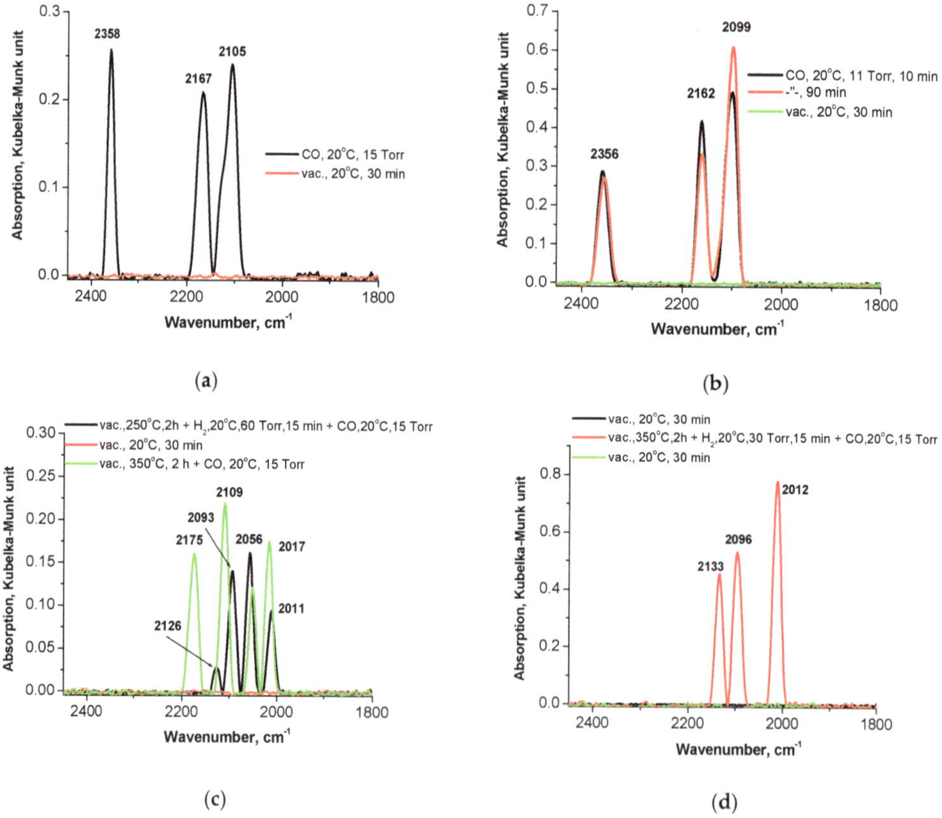

Figure 8. DRIFT spectra recorded after CO adsorption on the calcined PdFe-2-C250 (**a**,**c**) and PdFe-2-C400 (**b**,**d**) samples before (**a**,**b**) and after in situ treatment with H_2 at room temperature (**c**,**d**).

3.2.1. Preliminarily Reduced Samples

There are several bands in the spectral ranges of surface Pd carbonyls in DRIFT spectra recorded after CO adsorption on the samples reduced preliminarily in a hydrogen flow at 400 °C, yet the spectra differ depending on whether a dried or calcined sample is reduced

(Figure 6). The bands of linear complexes are observed in both spectra at 2114–2129 cm^{-1} (Pd^+—CO complexes) and at 2063–2089 cm^{-1} (Pd^0—CO complexes), exhibiting the high dispersion of Pd and its strong interaction with Fe species. The appearance of the Pd^+—CO band in the samples reduced in hydrogen at such a high temperature may be due to the formation of Pd^{n+}-Fe^{2+} species that further crystallize into alloy $Pd_{1-x}Fe_x$ after annealing in Ar at 550 °C [20]. The bands of Pd^0—CO complexes at 2084 cm^{-1} were observed by another research group for the silica-supported Pd-Fe samples after a reduction in hydrogen at 300 °C, the red shift from 2084 to 2064 being revealed with an increase in the Fe/Pd ratio [22]. The bands are assigned to CO adsorbed on on-top sites exposed at (111) microfacets.

The intensive band at 1971 cm^{-1} of bridging carbonyls Pd^+—CO—Pd^+ assigned to CO adsorbed on the (100) sites [22] indicates the presence of a considerable number of large-enough Pd nanoparticles in the sample PdFe-3-R prepared by the reduction of dried supported precursors, which agrees with the results of EXAFS and XRD studies (Table 2). The intensity of this band and its ratio to the intensity of the bands of linear complexes are an order of magnitude lower in the spectra of the sample PdFe-2-C250-R, supposing the higher Pd dispersion in the sample reduced after calcination, which looks contradictory to the results presented in a recent paper [21]. The only reason for this difference may be the difference in the preparation procedure, namely, the higher temperature of the support impregnation and drying used for the preparation of PdFe-2. The conclusion follows from this observation: the formation of large Pd^0 nanoparticles occurs via the fast sintering of metallic species generated under the reductive decomposition of supported precursors in hydrogen. Decomposition in hydrogen at low temperatures of 250 °C supposes a smaller nuclei size than the reduction of oxide to metal and, therefore, higher rates of sintering and particle growth. The absence of the band at 1971 cm^{-1} in the spectra of calcined samples (Figures 7 and 8) is commonly considered to be proof of the high Pd^0 dispersion, or it can be an indication that iron atoms or ions are preferentially located on the (100) sites of Pd^0 nanoparticles [22]. It is known that the higher alkene selectivity of the Pd nanocubes is attributed to the large adsorption energy of the carbon–carbon triple bond on the {100} facets [29].

The presence of Fe^{2+} in the sample PdFe-3-R is confirmed by the presence of the band at 2165 cm^{-1}, although it is absent in the spectra of the PdFe-2-C250-R sample. The latter fact may be explained both by the higher extent of Fe^{n+} reduction enhanced by the presence of Pd^0 nanoparticles formed upon calcination and by the enrichment of the surface of bimetallic nanoparticles and Fe_3O_4 crystallites with Pd^0 atoms (generation of core-shell structures). This band is not observed in the spectra of Pd-Fe/SiO_2 bimetallic catalysts prepared by other procedures [22].

3.2.2. Preliminarily Calcined Samples

The adsorption of CO on the oxidized Pd-containing samples in a CO atmosphere is almost always accompanied by a reduction of Pd^{n+} to Pd^0, which is indicated by the appearance of bands at wavenumbers below 2100 cm^{-1} in the IR spectra of adsorbed CO. This phenomenon is observed for the calcined samples (Figures 7 and 8a,b), especially at the prolonged treatment of the PdFe-2-C350 sample with CO (Figure 7). The appearance of the band of adsorbed CO_2 at 2349 cm^{-1} [28] is an indication of the oxidation of CO adsorbed on Fe^{3+} ions with lattice O^{2-} ions. The band at 2160 cm^{-1} may be attributed to the linear form of CO adsorption on Fe^{2+}, but the intensity of this band at 2165 cm^{-1} in the sample containing no Pd is considerably lower (Figure 7, grey line). It should be mentioned that its intensity decreases slightly in the spectra of bimetallic samples upon exposition to CO (Figures 7 and 8b). Therefore, it seems preferable to attribute the bands at 2160 cm^{-1} and 2120 cm^{-1} to the symmetric and antisymmetric stretching vibrations of $Pd^{2+}(CO)_2$ complexes, respectively. The disappearance of the band at 2162–2167 cm^{-1} after the sample treatment with hydrogen at room temperature (Figure 8c,d) confirms such an assignment of the bands. In the case of the samples calcined at 250 and 400 °C that strongly differ in the selectivity to styrene (Figure 4), DRIFT spectroscopic studies with adsorbed CO reveal at least two types of Pd^{2+} surface sites that can be easily reduced with hydrogen even at room temperature, resulting in the formation of Pd^{+} (the bands at 2125 cm^{-1} and 2133 cm^{-1}) and two states of Pd^0 (the bands at 2093–2096 cm^{-1} and 2012 cm^{-1}). The bands at 2063–2089 cm^{-1} can be attributed to Pd^0—CO complexes exhibiting the high dispersion of Pd and its strong interaction with Fe species. The band at 2012 cm^{-1} was not observed previously for any Pd-Fe/SiO_2 systems [22,27]. This band may be assigned to CO adsorbed on the (100) Pd sites affected by interaction with the Fe_2O_3 surface. As follows from X-ray photoelectron spectroscopy and theoretical calculation results [29–31], the interaction between Pd and the Fe_2O_3 surface occurs through the exchange of electrons with the surface Fe and O atoms. This bonding between Pd and surface oxide elements causes Pd to partially donate electrons to the oxide surface. This may be the reason for the blue shift of the band of bridging carbonyls Pd^{+}—CO—Pd^{+} from 1971 cm^{-1} to 2011–2017 cm^{-1}. One more state of Pd^0 (the band at 2056 cm^{-1}) is revealed for the sample calcined at 250 °C. Thus, formed Pd^0 sites should be present on the surface before the introduction of PhA, and therefore, they participate in the reaction process. The Pd^0 sites (the band at 2011–2017 cm^{-1}) formed in the calcined samples during reduction with H_2 at RT seem to be responsible for St hydrogenation. The most selective sample, PdFe-2-C400, contains neither Fe^{2+} nor Pd^0 sites, which are specific for the band at 2056 cm^{-1} after treatment with H_2. This fact makes us suppose their responsibility for the direct PhA hydrogenation to ethylbenzene and the low selectivity to styrene. After reduction in hydrogen at RT, the Pd^0 sites seem to be partly blocked with H_2O molecules, whereas Fe^{2+} are completely blocked. The increase in the intensity of the Pd^0 bands and the appearance of the Fe^{2+} band at 2176 cm^{-1} after evacuation at 350 °C (Figure 8c) prove this assumption. Such a blocking effect can modify the catalytic properties as well.

3.2.3. XPS Investigation

The study of the catalysts PdFe-2-C250 and PdFe-2-C250-R by the XPS method showed that, depending on the conditions of the heat treatment, both the electronic state of the supported metal nanoparticles and the distribution of metals in the surface layer of catalysts changed. Figure 9 shows the photoelectron spectra in the Fe 2p region. The energy position of Fe $2p_{3/2}$ at 710.5 eV and the shape of the Fe 2p doublet lines of photoelectrons in the spectra of both samples differ. In the spectrum of PdFe-2-C250-R, a shoulder is observed from the side of lower binding energies at 709.2 eV, which indicates the presence of mainly the FeO oxide in the reduced sample, while the oxidation degree of Fe^{3+} is characteristic of the calcined sample of PdFe-2-C250 [32]. The increase in the Fe/Si atomic ratio during the reduction in the catalyst by almost 1.5 times may be due to the segregation of iron in the surface layers of this sample (Table 3). An analysis of the X-ray spectra in the Pd 3d region of photoelectrons (Figure 9) at 336.7 and 335.2 eV allows us to conclude that palladium in the surface layers of the sample PdFe-2-C250 exists in two oxidation states, Pd^{1+} and Pd^0, respectively [33,34]. According to the XPS data for the sample PdFe-2-C250-R, all of the palladium is present in a metallic state [34]. The energy position of Pd $3d_{5/2}$ electrons and the atomic ratio Pd/Si are presented in Table 3.

Table 3. Binding energies and atomic ratios according to XPS data.

Sample	Binding Energy, eV		Surface Atomic Ratio			Atomic Ratio
	Fe $2p_{3/2}$	Pd $3d_{5/2}$	Fe/Si	Pd/Si	Fe/Pd	Fe/Pd
PdFe-2-C250	710.5	336.7 335.2	0.0094	0.0091	1.03	2.67
PdFe-2-C250-R	710.3 709.2	334.7	0.0135	0.0013	1.03	2.67

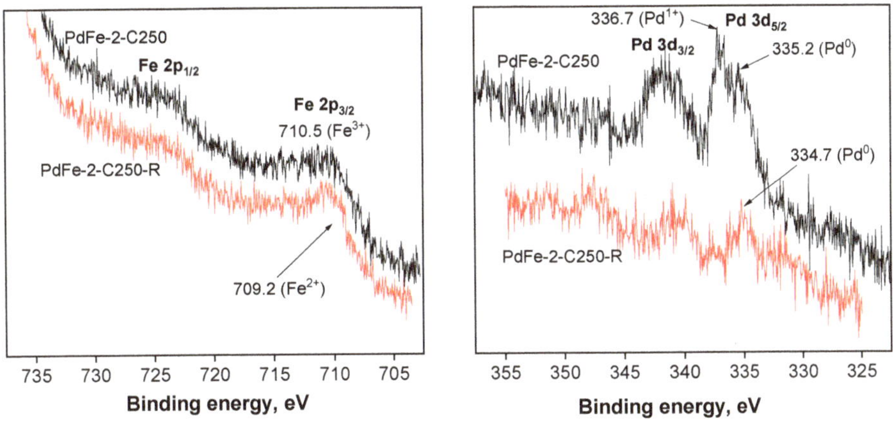

Figure 9. XPS data for PdFe-2 catalysts.

3.2.4. SEM and TEM

The morphological characteristics of bimetallic PdFe-2 catalysts that have undergone various heat treatments have been studied by TEM and SEM methods. The TEM micrograph (Figure 10a) of the sample PdFe-2-C250 shows spherical particles of an average diameter of 5 nm, evenly distributed over the surface of the carrier. On the SEM micrograph (Figure 10b) of this sample, the crystalline phase of iron oxides can be noted.

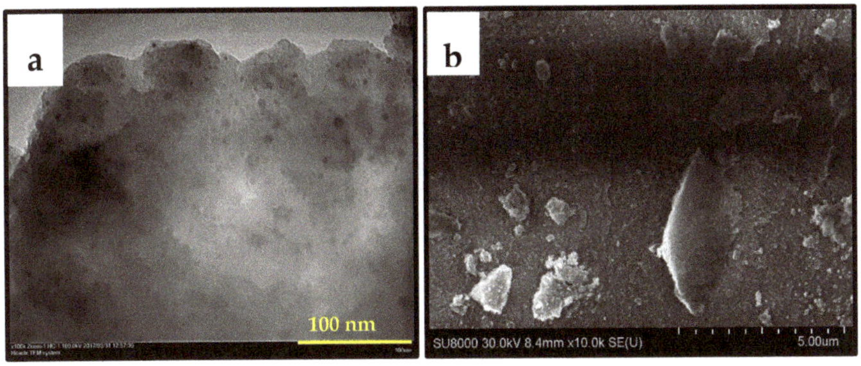

Figure 10. Micrographs of TEM (**a**) and SEM (**b**) of the PdFe-2-C250 catalyst.

Micrographs were also obtained for the PdFe-2-C250-R sample with its subsequent reduction in an H_2 flow at 400 °C. The SEM image (Figure 11a) clearly shows crystals of an ordered structure. SEM-EDX data show that these crystal structures consist mainly of iron atoms. Palladium is evenly distributed over the entire surface of the carrier in the form of much smaller particles, which also include iron. It should be noted that the particles in the reduced sample PdFe-2-C250-R are evenly distributed over the entire surface of the carrier, and the average size of the nanoparticles is 5 nm (Figure 11b).

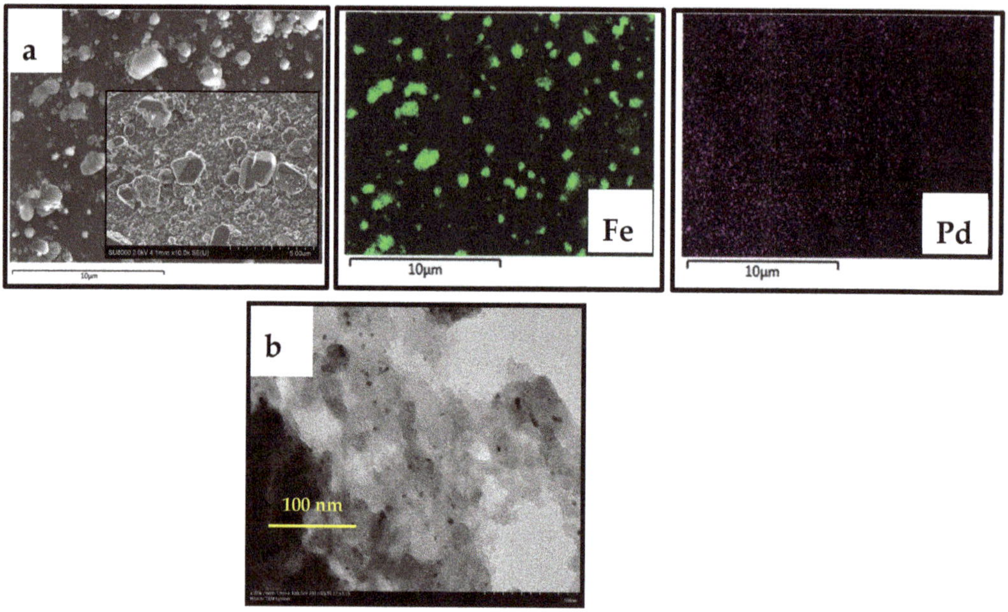

Figure 11. Micrographs of SEM-EDX (**a**) and TEM (**b**) of the PdFe-2-C250-R catalyst.

4. Conclusions

Thus, based on our investigation of the electronic state of Pd and Fe metals in calcined bimetallic PdFe-2 catalysts by CO chemisorption (DRIFTS-CO), the existence of a strong contact interaction due to the electronic effect between Pd (Pd^+, Pd^0) and Fe (Fe^{2+}, Fe^{3+}) has been established. By the TEM method, it was found that the addition of iron to palladium nanoparticles in the PdFe-2 catalysts leads to a significant decrease in the aggregation of crystallites and an increase in the dispersion of the active metal relative to the monometallic palladium Pd-R catalyst [25]. In conclusion, the high dispersed phase, the strong contact interaction of the metals Pd-Fe, and the absence of palladium hydride formation in bimetallic catalysts lead to the competitive adsorption of the C≡C bond on the surface of bimetallic PdFe-2 catalysts, which is the reason for the high catalytic activity in the selective hydrogenation of phenylacetylene to styrene with a high selectivity of 84–90% at the complete conversion of phenylacetylene. It has been shown that the formation of nanoparticles of the solid solution $Pd_{1-x}Fe_x$ is essential for the selective hydrogenation in order to considerably decrease the rate of styrene hydrogenation to ethylbenzene.

The bimetallic Pd-Fe/SiO_2 catalysts have wide potential practical implications. At least three relatively simple preparation procedures can be used for their preparation: (i) the direct reduction of silica-supported metal precursors, $(NH_4)_3[Fe(C_2O_4)_3]$ and $[Pd(NH_3)_4]Cl_2$, with a hydrogen flow at 400 °C; (ii) the thermal decomposition of the silica-supported metal precursors in air at 250 °C followed by a reduction in the hydrogen flow at 400–430 °C; (iii) the thermal decomposition of silica-supported metal precursors in air at 400 °C. Considering the strong sensitivity of the selectivity and reaction rate to the preparation variables, care should be taken when scaling up the catalyst production. The choice of the preparation procedure will depend on environmental regulations, technical and economical possibilities, as well as the principal direction of implementation.

Summarizing the results obtained, the prepared bimetallic Pd-Fe/SiO_2 catalysts can be proposed as a new non-toxic alternative to the Lindlar catalyst for the selective hydrogenation of triple C≡C bonds in the liquid phase at room temperature and atmospheric hydrogen pressure. These novel catalytic materials are as selective in the hydrogenation of the triple C≡C bond as the commercial Lindlar catalyst, and their productivity values are up to an order of magnitude higher.

Author Contributions: Conceptualization, A.A.S. and O.A.K.; methodology, A.A.S.; software, A.A.S.; validation, A.A.S., O.A.K. and A.L.K.; formal analysis, O.P.T.; investigation, A.A.S. and O.P.T.; resources, A.L.K.; data curation, L.M.K.; writing—original draft preparation, O.A.K. and A.A.S.; writing—review and editing, A.L.K. and L.M.K.; visualization, A.L.K.; supervision, L.M.K.; project administration, A.L.K.; funding acquisition, L.M.K. All authors have read and agreed to the published version of the manuscript.

Funding: The study of catalysts by physicochemical methods was funded by the Ministry of Science and Higher Education of the Russian Federation, project number 075-15-2021-591, and the preparation and catalytic tests were carried out with financial support from the Russian Science Foundation, grant No. 23-73-30007. L. M. Kustov thanks the «Priority-2030» academic leadership selectivity program, project number K7-2022-062.

Data Availability Statement: Data are available from the authors upon request.

Conflicts of Interest: The authors declare no conflict of interest.

References

1. Vilé, G.; Albani, D.; Almora-Barrios, N.; López, N.; Pérez-Ramírez, J. Advances in the design of nanostructured catalysts for selective hydrogenation. *ChemCatChem J.* **2016**, *8*, 21–33. [CrossRef]
2. Rehm, T.H.; Berguerand, C.; Ek, S.; Zapf, R.; Löb, P.; Nikoshvili, L.; Kiwi-Minsker, L. Continuously operated falling film microreactor for selective hydrogenation of carbon–carbon triple bonds. *Chem. Eng. J.* **2016**, *293*, 345–354. [CrossRef]
3. Vilé, G.; Almora-Barrios, N.; Mitchell, S.; López, N.; Pérez-Ramírez, J. From the Lindlar Catalyst to Supported Ligand-Modified Palladium Nanoparticles: Selectivity Patterns and Accessibility Constraints in the Continuous-Flow Three-Phase Hydrogenation of Acetylenic Compounds. *Chem. Eur. J.* **2014**, *20*, 5926–5937. [CrossRef] [PubMed]
4. Lindlar, H.; Dubuis, R. Palladium catalyst for partial reduction of acetylenes. *Org. Synth.* **1973**, *5*, 880.
5. Petrucci, G.; Oberhauser, W.; Bartoli, M.; Giachi, G.; Frediani, M.; Passaglia, E.; Capozzoli, L.; Rosi, L. Pd-nanoparticles supported onto functionalized poly(lactic acid)-based stereocomplexes for partial alkyne hydrogenation. *Appl. Catal. A Gen.* **2014**, *469*, 132–138. [CrossRef]
6. Maccarrone, M.J.; Lederhosa, C.R.; Torresa, G.; Betti, C.; Coloma-Pascual, F.; Quiroga, M.E.; Yori, J.C. Partial hydrogenation of 3-hexyne over low-loaded palladium mono and bimetallic catalysts. *Appl. Catal. A Gen.* **2012**, *441–442*, 90–98. [CrossRef]
7. Papp, A.; Molnar, A.; Mastalir, A. Catalytic Investigation of Pd Particles Supported on MCM-41 for the Selective Hydrogenations of Terminal and Internal Alkynes. *Appl. Catal. A Gen.* **2005**, *289*, 256–266. [CrossRef]
8. Hu, J.; Zhou, Z.; Zhang, R.; Li, L.; Cheng, Z. Selective hydrogenation of phenylacetylene over a nano-Pd/α-Al_2O_3 catalyst. *J. Mol. Catal. A Chem* **2014**, *381*, 61–69. [CrossRef]
9. Dominguez-Dominguez, S.; Berenguer-Murcia, A.; Pradhan, B.K.; Cazorla-Amoros, D.; Linares-Solano, A. Semihydrogenation of phenylacetylene catalyzed by palladium nanoparticles supported on carbon materials. *J. Phys. Chem. C* **2008**, *112*, 3827–3834. [CrossRef]
10. Dominguez-Dominguez, S.; Berenguer-Murcia, A.; Linares-Solano, A.; Cazorla-Amoros, D. Inorganic materials as supports for palladium nanoparticles: Application in the semi-hydrogenation of phenylacetylene. *J. Catal.* **2008**, *257*, 87–95. [CrossRef]
11. Marin-Astorga, N.; Pecchi, G.; Fierro, J.L.G.; Reyes, P. Alkynes hydrogenation over Pd-supported catalysts. *Catal. Lett.* **2003**, *91*, 115–121. [CrossRef]
12. Panpranot, J.; Phandinthong, K.; Sirikajorn, T.; Arai, M.; Praserthdam, P. Impact of palladium silicide formation on the catalytic properties of Pd/SiO_2 catalysts in liquid-phase semihydrogenation of phenylacetylene. *J. Mol. Catal. A Chem.* **2007**, *261*, 29–35. [CrossRef]
13. Yoshida, H.; Zama, T.; Fujita, S.; Panpranot, J.; Arai, M. Liquid phase hydrogenation of phenylacetylene over Pd and PdZn catalysts in toluene: Effects of alloying and CO_2 pressurization. *RSC Adv.* **2014**, *4*, 24922–24928. [CrossRef]
14. Lopez, N.; Vargas-Fuentes, C. Promoters in the hydrogenation of alkynes in mixtures: Insights from density functional theory. *Chem. Commun.* **2012**, *48*, 1379–1391. [CrossRef] [PubMed]
15. Shesterkina, A.A.; Kozlova, L.M.; Mishin, I.V.; Tkachenko, O.P.; Kapustin, G.I.; Zakharov, V.P.; Vlaskin, M.S.; Zhuk, A.Z.; Kirichenko, O.A.; Kustov, L.M. Novel Fe-Pd/γ-Al2O3 catalysts for the selective hydrogenation of C≡C bonds under mild conditions. *Mendeleev Commun.* **2019**, *259*, 339–342. [CrossRef]
16. Weerachawanasak, P.; Mekasuwandumrong, O.; Arai, M.; Fujita, S.-I.; Praserthdam, P.; Panpranot, J. Effect of strong metal–support interaction on the catalytic performance of Pd/TiO_2 in the liquid-phase semihydrogenation of phenylacetylene. *J. Catal.* **2009**, *262*, 199–205. [CrossRef]
17. Domínguez-Domínguez, S.; Berenguer-Murcia, Á.; Cazorla-Amorós, D.; Linares-Solano, Á. Semihydrogenation of phenylacetylene catalyzed by metallic nanoparticles containing noble metals. *J. Catal.* **2006**, *243*, 74–81. [CrossRef]

18. Stakheev, A.Y.; Mashkovskii, I.S.; Baeva, G.N.; Telegina, N.S. Specific features of the catalytic behavior of supported palladium nanoparticles in heterogeneous catalytic reactions. *J. Gen. Chem.* **2010**, *80*, 618–629. [CrossRef]
19. Juszczyk, W.; Pielaszek, J.; Karpiński, Z.; Pinna, F. Reaction of 2,2-dimethylpropane with dihydrogen over silica-supported PdFe catalysts. *Appl. Catal. A Gen.* **1996**, *144*, 281–291. [CrossRef]
20. Kirichenko, O.A.; Kapustin, G.I.; Nissenbaum, V.D.; Mishin, I.V.; Kustov, L.M. Evaluation of Stability of Silica-Supported Fe–Pd and Fe–Pt Nanoparticles in Aerobic Conditions Using Thermal Analysis. *J. Therm. Anal. Calorim.* **2014**, *118*, 749–758. [CrossRef]
21. Kustov, L.M.; Al-Abed, S.R.; Virkutyte, J.; Kirichenko, O.A.; Shuvalova, E.V.; Kapustin, G.I.; Mishin, I.V.; Nissenbaum, V.D.; Tkachenko, O.P.; Finashina, E.D. Novel Fe-Pd/SiO_2 catalytic materials for degradation of chlorinated organic compounds in water. *Pure Appl. Chem.* **2014**, *86*, 1141–1158. [CrossRef]
22. Boccuzzi, F.; Guglielminotti, E.; Pinna, F.; Signoretto, M. Surface composition of Pd–Fe catalysts supported on silica. *J. Chem. Soc. Faraday Trans.* **1995**, *91*, 3237–3244. [CrossRef]
23. Vergunst, T.; Kapteijn, F.; Moulijn, J.A. Optimization of geometric properties of a monolithic catalyst for the. selective hydrogenation of phenylacetylene. *Ind. Eng. Chem. Res.* **2001**, *40*, 2801–2809. [CrossRef]
24. Huang, X.; Wilhite, B.; McCready, M.J.; Varma, A. Phenylacetylene Hydrogenation in a Three-Phase Catalytic Packed-Bed Reactor: Experiments and Model. *Chem. Eng. Sci.* **2003**, *58*, 3465–3471. [CrossRef]
25. Shesterkina, A.A.; Kozlova, L.M.; Kirichenko, O.A.; Kapustin, G.I.; Mishin, I.V.; Kustov, L.M. Influence of the thermal treatment conditions and composition of bimetallic catalysts Fe—Pd/SiO_2 on the catalytic properties in phenylacetylene hydrogenation. *Russ. Chem. Bull., Int. Ed.* **2016**, *2*, 432–439. [CrossRef]
26. Shesterkina, A.A.; Kirichenko, O.A.; Kozlova, L.M.; Kapustin, G.I.; Mishin, I.V.; Strelkova, A.A.; Kustov, L.M. Liquid-phase hydrogenation of phenylacetylene to styrene on silica - supported Pd-Fe nanoparticles. *Mendeleev Commun.* **2016**, *26*, 228–230. [CrossRef]
27. Hadjivanov, K.J.; Vasilev, G.N. Characterization of oxide surfaces and zeolites by carbon monoxide as an IR probe molecule. *Adv. Catal.* **2002**, *47*, 307–511. [CrossRef]
28. Davydov, A. *Molecular Spectroscopy of Oxide Catalyst Surfaces*; Willey: London, UK, 2003.
29. Chung, J.; Kim, C.; Jeong, H.; Yu, T.; Binh, D.H.; Jang, J.; Lee, J.; Kim, B.M.; Lim, B. Selective semihydrogenation of alkynes on shape -controlled palladium nanocrystals. *Chem. Asian J.* **2013**, *8*, 919–925. [CrossRef]
30. Hensley, A.J.R.; Hong, Y.; Zhang, R.; Zhang, H.; Sun, J.; Wang, Y.; McEwen, J.-S. Enhanced Fe_2O_3 reducibility via surface modification with Pd: Characterizing the synergy within Pd/Fe catalysts for hydrodeoxygenation reactions. *ACS Catal.* **2014**, *4*, 3381–3392. [CrossRef]
31. Pan, T.J.; Li, Y.S.; Yang, Q.; Korotin, D.M.; Kurmaev, E.Z.; Sanchez-Pasten, M.; Zhidkov, I.S.; Cholakh, S.O. Diamond deposition on Fe-Cr-Al alloy substrates: Effect of native oxidation by XPS and XAS investigation. *J. Alloys Compd.* **2018**, *740*, 887–894. [CrossRef]
32. Voog, E.H.; Mens, A.J.M.; Gijzeman, O.L.J.; Geu, J.W. XPS analysis of palladium oxide layers and particles. *Surf. Sci.* **1996**, *350*, 21–31. [CrossRef]
33. Beketov, G.; Heinrichs, B.; Pirard, J.-P.; Chenakin, S.; Kruse, N. Xps structural characterization of Pd/SiO_2 catalysts prepared by cogelation. *Appl. Surf. Sci.* **2013**, *287*, 293–298. [CrossRef]
34. Hoflund, G.B.; Hagelin, H.A.E.; Weaver, J.F.; Salaita, G.N. ELS and XPS study of Pd/PdO methane oxidation catalysts. *Appl. Surf. Sci.* **2003**, *205*, 102–112. [CrossRef]

nanomaterials

MDPI

Article

Hard–Soft Core–Shell Architecture Formation from Cubic Cobalt Ferrite Nanoparticles

Marco Sanna Angotzi [1,2], Valentina Mameli [1,2,*], Dominika Zákutná [3], Fausto Secci [1], Huolin L. Xin [4] and Carla Cannas [1,2,*]

1 Department of Chemical and Geological Sciences, University of Cagliari, Cittadella Universitaria S.S. 554 Bivio per Sestu, 09042 Monserrato, Italy; marcosanna@unica.it (M.S.A.); fausto.secci@unica.it (F.S.)
2 Consorzio Interuniversitario Nazionale per la Scienza e Tecnologia dei Materiali (INSTM), Via Giuseppe Giusti 9, 50121 Florence, Italy
3 Department of Inorganic Chemistry, Charles University, Hlavova 2030, 128 40 Prague 2, Czech Republic; dominika.zakutna@natur.cuni.cz
4 Department of Physics and Astronomy, University of California, Irvine, CA 92617, USA; huolinx@uci.edu
* Correspondence: valentina.mameli@unica.it (V.M.); ccannas@unica.it (C.C.)

Abstract: Cubic bi-magnetic hard–soft core–shell nanoarchitectures were prepared starting from cobalt ferrite nanoparticles, prevalently with cubic shape, as seeds to grow a manganese ferrite shell. The combined use of direct (nanoscale chemical mapping via STEM-EDX) and indirect (DC magnetometry) tools was adopted to verify the formation of the heterostructures at the nanoscale and bulk level, respectively. The results showed the obtainment of core–shell NPs ($CoFe_2O_4$@$MnFe_2O_4$) with a thin shell (heterogenous nucleation). In addition, manganese ferrite was found to homogeneously nucleate to form a secondary nanoparticle population (homogenous nucleation). This study shed light on the competitive formation mechanism of homogenous and heterogenous nucleation, suggesting the existence of a critical size, beyond which, phase separation occurs and seeds are no longer available in the reaction medium for heterogenous nucleation. These findings may allow one to tailor the synthesis process in order to achieve better control of the materials' features affecting the magnetic behaviour, and consequently, the performances as heat mediators or components for data storage devices.

Keywords: cobalt ferrite; core–shell; heterostructures; cubic shape; STEM-EDX

Citation: Sanna Angotzi, M.; Mameli, V.; Zákutná, D.; Secci, F.; Xin, H.L.; Cannas, C. Hard–Soft Core–Shell Architecture Formation from Cubic Cobalt Ferrite Nanoparticles. *Nanomaterials* **2023**, *13*, 1679. https://doi.org/10.3390/nano13101679

Academic Editors: Jordi Sort, Lyudmila M. Bronstein, Andrei Honciuc and Mirela Teodorescu

Received: 3 April 2023
Revised: 8 May 2023
Accepted: 17 May 2023
Published: 19 May 2023

1. Introduction

In the field of nanotechnology, the combination of different materials for the obtainment of heterostructured multifunctional nanoparticles (NPs) [1–3] or nanocomposites [4–8] has driven the efforts of researchers with the aim of exploiting the physical (magnetic, optical, electrical, etc.) and chemical (thermal stability, reactivity, bonding ability, dispersibility, solubility, etc.) properties of single components in a combined or synergistic manner for selected applications. When two (or more) different phases are put in contact at the nanoscale, new phenomena may arise from their coupling, with a crucial role played by the interface extent. One of the most prominent examples is core–shell nanoarchitectures, consisting of an inner core covered by an external layer (shell) of a different material. The nature of the contact between the core and the shell is fundamental for these structures to show their peculiar properties; particularly, a net change in the chemical composition with no chemically mixed layers between the phases is desired. Diverse materials such as ferrites, perovskites, silica, titania, noble metals, elemental carbon, alloys, and polymers have been coupled to form inorganic/organic or inorganic/inorganic core–shell systems [9]. The choice of the materials to be coupled, the core–shell volume ratio, and the synthesis method are elements to play with. Core–shell heterostructures featuring a heterojunction between two different materials have an extensive range of applications, such as photocatalysis [10,11], gas sensing [12,13], water purification [14], and hydrogen production [15,16].

In these materials, the electron phenomena taking place at the interface between the two phases play a crucial role [17,18].

In the case of magnetic materials, since the discovery of exchange bias phenomena in a ferromagnet (FM)/antiferromagnet (AFM) system (Co/CoO core–shell NPs) by Meiklejohn and Bean [19], and by Kneller and Hawig [20], other interfaces have been built, such as AFM/ferrimagnet (FiM), or hard–soft FM or FiM [21–25]. The development of the core–shell heterostructure is favoured by the choice of isostructural crystalline phases, which should grow epitaxially one upon the other. In this context, spinel ferrite ($M^{II}Fe_2O_4$, $M^{II} = Fe^{2+}$, Co^{2+}, Mn^{2+}, Ni^{2+}, Zn^{2+}, etc.) bi-magnetic core–shell NPs have been studied and proposed in different application fields, such as magnetic recording, permanent magnets, spintronics, microwave absorption, biomedicine, magnetic heat generation for catalysis [26–29], or magnetic fluid hyperthermia (MFH) [30–45]. In this framework, the adoption of core–shell heterostructures was demonstrated to be efficient to improve the heating abilities of spinel ferrites or their energy product as permanent magnets [21,25,32,39,46]. The spinel ferrite core–shell NPs offer a tuneable range of hard and soft magnetic behaviours by properly changing the chemical composition. They can be successfully prepared with good crystallinity and good control over the composition, the particle's morphology, and heterostructure interface by seed-mediated approaches applied to thermal decomposition and solvothermal methods [32,39,47–61].

Seed-mediated synthesis approaches consist of introducing pre-formed nanoparticles (seeds) into the medium containing the precursors of the shell material. The nucleation of the secondary material onto the surface of the seeds (heterogeneous nucleation) has a lower energy barrier than the formation of new particles of the secondary material (homogeneous nucleation), thus giving rise to the formation of the core–shell structures [62].

In particular, the magnetic behaviour of the core–shell NPs can also be modified according to the shape, in addition to the size, for instance, due to surface anisotropy effects. Regarding this effect, a remarkable influence of the NPs' shape on their magnetic properties was observed by several authors [9,63–69]. Concentric spherical core–shell NPs are the most common systems [9].

Recently, we set up a versatile solvothermal synthesis method able to produce single-phase spherical particles [70,71], chemically mixed spinel phases [72,73], homogeneous spherical bi-magnetic core–shell NPs [46,70,74,75], and flower-like Ag-spinel ferrite nanoarchitectures [1] with high crystallinity, low size dispersity, and precise control of the shell growth.

Herein, we studied the formation of cubic cobalt ferrite nanoparticles and their bi-magnetic heterostructure counterparts by coating cubic cobalt ferrite nanoseeds with manganese ferrite to form $CoFe_2O_4@MnFe_2O_4$ with the same oleate-based solvothermal method adopted in our previous studies. High-resolution transmission electron microscopy (HRTEM) and chemical mapping at the nanoscale, performed through STEM-EDX, together with the Rietveld refinement of X-ray diffraction (XRD) patterns and DC magnetometry, were employed to study morphological, structural, and magnetic properties.

The study aimed to provide precious details on the formation mechanism of core–shell heterostructures by starting from larger (15 nm versus 6–8 nm) and anisotropic (cubic versus spherical) cobalt ferrite nanoparticles compared with our previous studies. To understand how the different morphologies of the seeds affect the final product, a combination of direct microscopic and indirect macroscopic characterisation techniques was adopted. The chemical mapping at the nanoscale via STEM-EDX allowed us to probe the local chemical composition of a few nanoparticles with a good resolution, while DC magnetometry provided a bulk-level view of the average physical behaviour, beyond the representativeness limits of electron microscopy.

2. Materials and Methods

2.1. Chemicals

Oleic acid (>99.99%), 1-pentanol (99.89%), hexane (84.67%), and toluene (99.26%) were purchased from Lach-Ner, Neratovice, Czech Republic; 1-octanol (>99.99%) and

$Mn(NO_3)_2 \cdot 4H_2O$ (>97.0%) from Sigma-Aldrich, St. Louis, MO, USA; absolute ethanol and $Co(NO_3)_2 \cdot 6H_2O$ (99.0%) from Penta, Prague 10, Czech Republic; NaOH (>98.0%) from Fluka, Muskegon, MI, USA; $Fe(NO_3)_3 \cdot 9H_2O$ (98.0%) from Lachema, Brno, Czech Republic.

2.2. Methods

Cobalt ferrite NPs and core–shells were prepared as described in previous work [46,70], starting from metal oleates. The synthesis conditions are summarised in Table S1.

2.3. Characterisation

Fourier Transform Infrared (FT-IR) spectra were recorded in the region from 400 to 4000 cm^{-1} by using a FT/IR-4X Spectrometer from Jasco, Easton, MD, USA. Samples were measured in a KBr pellet.

Thermogravimetric analysis (TGA) curves were obtained by using a PerkinElmer (Waltham, MA, USA) STA 6000, in the 25–850 °C range, with a heating rate of 10 °C min^{-1} under 40 mL min^{-1} O_2 flow.

The samples were characterised via X-ray diffraction (XRD), using a PANalytical X'Pert PRO (Malvern PANalytical, Malvern, UK) with Cu Kα radiation (1.5418 Å), a secondary monochromator, and a PIXcel position-sensitive detector. The peak position and instrumental width were calibrated using powder LaB_6 from NIST. The refinement of the structural parameters was carried out using the Rietveld method in the FullProf software [76] using a pseudo-Voigt profile function. For mean crystallite shapes, the spherical harmonics function was used [77]:

$$T(\boldsymbol{h}) = T(\vartheta, \varphi) = \sum_{l=0,2,4\ldots}^{n} \sum_{m=-l}^{l} a_{l,m} Y_{l,m}(\vartheta, \varphi) \tag{1}$$

where ϑ and φ are polar and azimuthal angles describing the direction of the normal to the family of the lattice plane in a Cartesian coordinate system, a is the lattice parameter, and Y is the Lorentzian isotropic size broadening.

TEM micrographs were acquired using a JEOL 200CX electron microscope (Jeol Ltd., Tokyo, Japan) operating at 160 kV. The size of more than 1000 particles was measured using Pebbles software [78], selecting a cubic or spherical shape in order to determine the particle size distribution. The average particle diameter was calculated together with the percentage ratio between the standard deviation and the mean value to provide the size dispersity. Additional TEM micrographs and EDX data for the Co:Mn:Fe molar ratios were obtained by using a JEOL JEM 1400 Plus (Jeol Ltd., Tokyo, Japan) operating at 120 kV.

HRTEM micrographs and STEM-EDX measurements were carried out using an FEI Talos F200X (Thermo Fisher Scientific, Waltham, MA, USA) equipped with a field-emission gun operating at 200 kV and a four-quadrant 0.9 sr energy-dispersive X-ray spectrometer.

The Quantum Design PPMS DynaCool system with a maximum magnetic field of 9 T and the VSM module was used to investigate the DC magnetic properties of powders. The magnetisation values were normalised based on thermogravimetric analyses to account for the inorganic phase. Various magnetic measurements were conducted, including studying the field dependence of magnetisation at 10 K and 300 K within a range of 7 to −7 T. The saturation magnetisation values (Ms^{300K} and Ms^{10K}) were evaluated using the following equation:

$$M = M_s\left(1 - \frac{a}{H} - \frac{b}{H^2}\right) \tag{2}$$

For H tending to ∞, the magnetisation curve was fit from 4 to 7 T [79]. The anisotropy field was calculated as a 3% difference between the magnetisation and demagnetisation curves at 10 K. The temperature dependence of magnetisation was analysed using the zero-field-cooled (ZFC) and field-cooled (FC) protocols. The sample was cooled to 5 K without any magnetic field, and then, the signals were recorded under a static magnetic field of 100 Oe. During the warm-up from 5 to 380 K, M_{ZFC} was measured, and M_{FC} was recorded during the cooling process.

3. Results and Discussion

Cobalt ferrite NPs (labelled Co) with cubic shapes were obtained by following an adapted one-pot solvothermal procedure previously set up for spherical NPs [46,70,74,80–82], decreasing the concentration of the mixed Co-Fe oleate precursor to produce larger NPs (Figure 1, Table S1). This sample was used as seed material for the growth of a manganese ferrite shell through a second solvothermal treatment in the presence of mixed Fe^{III}-Mn^{II} oleate to synthesise $CoFe_2O_4$@$MnFe_2O_4$ NPs (labelled Co@Mn). Indeed, in previous works [46,70,74], the seed-mediated growth approach in solvothermal conditions permitted the preparation of manganese-ferrite-coated cobalt ferrite NPs with different core sizes and shell thicknesses, starting from spherical particles.

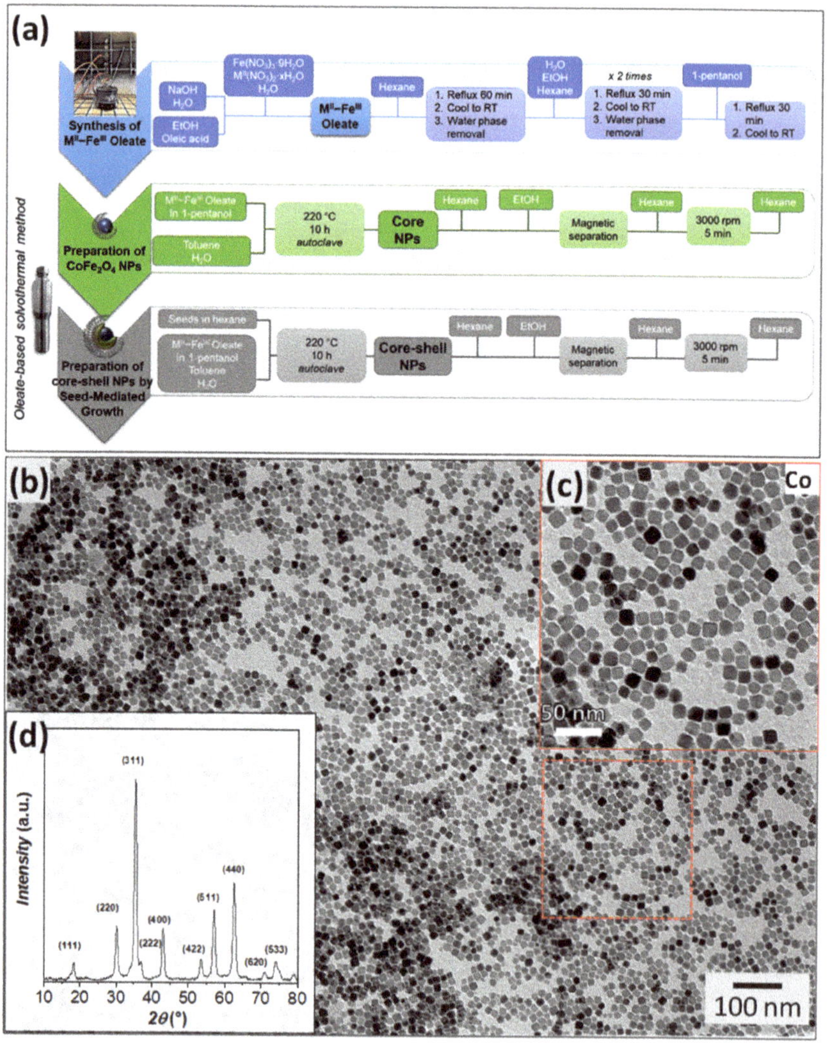

Figure 1. *Cont.*

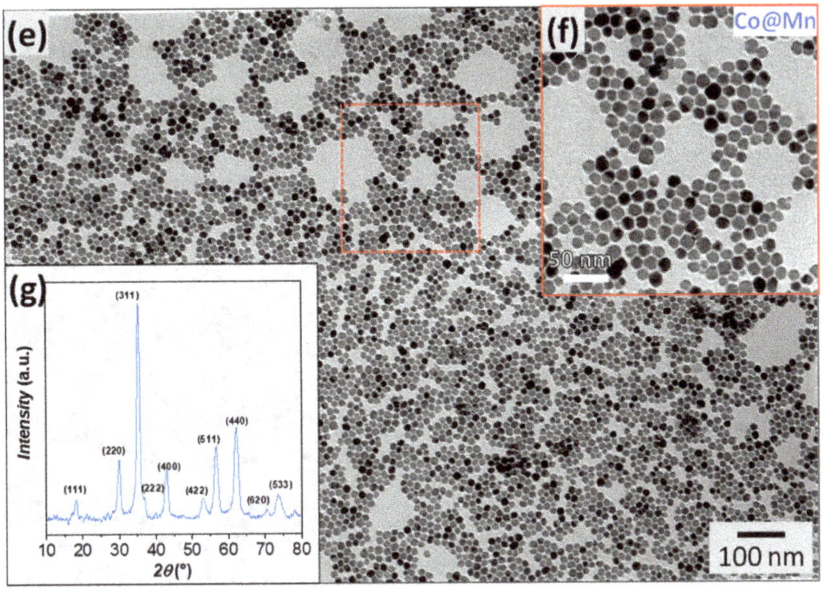

Figure 1. Scheme of the synthesis method: (**a**) XRD patterns and TEM images with size distributions of the Co (**b–d**) and Co@Mn (**e–g**) samples; the zones inside red boxes have been magnified as insets.

The samples were characterised via FTIR and TGA to ascertain the presence of oleate molecules capping at the NPs' surfaces (Figure S1). The FTIR spectra (Figure S1a) displayed vibrational modes of carboxylates ($\nu_{as}(COO^-)$, $\nu_s(COO^-)$) at around 1550 and 1430 cm^{-1}, respectively, as well as hydrocarbon chain modes in the 3000 cm^{-1} region. Thermogravimetric analyses (TGA, Figure S1b) demonstrated a comparable decrease in weight for the two samples in the temperature range of 200–350 °C, attribuTable to the decomposition of about 10%wt. of oleate molecules (Table 1). The slight shift in the temperature of the TGA curves and dTGA curves (Figure S1c) towards lower values for the Co@Mn sample (minima in the dTGA shifted by 15 °C) might suggest weaker bonds between the oleate molecules and the manganese cations [83], different coordinations [84], or differences in the morphological properties of the two samples (e.g., different curvature) having effects in stabilising the bonds between the carboxylate and the cations (e.g., through the proximity of the hydrocarbon chains) [85].

Table 1. Lattice parameter (*a*), crystallite size (D_{XRD}), volumetric particle size (D_{TEM_V}), size distribution (σ), organic content calculated by TGA, metal atomic percentages, and molar ratio Co:Mn:Fe (with Fe moles fixed to 2) calculated via TEM-EDX of the samples. *a* and D_{XRD} values are computed via Rietveld refinement using two spinel ferrite phases for the Co@Mn sample: the first row of the Table refers to $CoFe_2O_4$ and the second one to $MnFe_2O_4$.

Sample	*a* (Å)	D_{XRD} (nm)	D_{TEM_V} (nm)	σ_{TEM} (%)	Org. Phase (%wt.)	Mn (%)	Co (%)	Fe (%)	Mn:Co:Fe
Co	8.3914(2)	10.1(3)	14.2 (64% Cube) 14.9 (36% Sphere)	11 15	10	-	32.6(7)	67.4(7)	0:0.96:2
Co@Mn	8.4030(9) 8.4684(7)	10.7(9) 8.3(4)	15.3	12	11	25.0(2)	9.6(3)	65.4(4)	0.76:0.30:2

TEM micrographs of $CoFe_2O_4$ present well-defined cubic and spheroidal particles, quantified as about 64% and 36%, and having similar particle sizes of about 14 nm and 15 nm, respectively (Figures 1b,c and S2, Table 1) with low size dispersity (σ_{TEM} = 11%, 15%). On the contrary, the TEM micrographs of the core–shell system (Figures 1e,f and S3) reveal the obtainment of NPs with various shapes from spheroidal to faceted of about 15 nm (Table 1) and only a few well-defined cubic particles. EDX analysis was adopted to determine the molar ratio between the cations in the two samples. Almost stoichiometric cobalt ferrite (Co/Fe = 0.48) was obtained, in agreement with previous results on other spherical cobalt ferrite nanoparticles prepared using the same synthesis method [70]. For the core–shell sample, assuming that the cobalt content is associated with a cobalt ferrite with the molar ratio obtained for the Co sample, we can hypothesise a Mn/Fe molar ratio equal to 0.55, also revealing an almost stoichiometric manganese ferrite.

The XRD patterns of both samples show the formation of a spinel ferrite structure (Figure 1d,g). The crystallite size of Co, obtained from Rietveld refinement (Figure S4a, Table S2), is smaller (10.1(3) nm) than the mean physical size of the NPs obtained via TEM (14–15 nm, Table 1). This discrepancy might be due to the presence of structural disorder at the nanoparticle surface or to the limits of the Rietveld method in describing samples with populations of differently shaped NPs (cubic and spherical). In addition, the refinement of the Co@Mn sample using only one structural spinel phase leads to a coherent domain size smaller than the original core size (7.2(4) nm and 10.1(3) nm, respectively, Table S2). Therefore, an attempt to refine the XRD data with two spinel structures (Figure S4b, Table S3), one corresponding to $CoFe_2O_4$ and another to $MnFe_2O_4$, resulted in a better description of the XRD data. The crystallite size of the Co@Mn sample was, in this case, equal to 10.7(9) nm for the cobalt ferrite phase and 8.3(4) nm for the manganese ferrite counterpart. Moreover, the core–shell sample features a larger cell parameter (8.4030(9) Å for cobalt ferrite phase and 8.4684(7) Å for manganese ferrite) than the original core (8.3914(2) Å), in agreement with the inclusion of manganese in the spinel structure (Table 1). The adoption of the spherical harmonics function in the Rietveld refinement allowed us to visualise an average shape of the crystallites, revealing a cubic shape with rounded corners for the core sample and more faceted cuboidal particles for the core–shell one (Figure S4). The two contributions for the interpretation of the XRD pattern of the Co@Mn sample with almost constant or smaller crystallite sizes compared to the core and a slightly higher lattice parameter suggests the homogeneous nucleation of manganese ferrite, along with the formation of core–shell heterostructures with thin manganese ferrite shells, and also the occurrence of structural disorder phenomena. These findings suggest the role of the shape and the chemical nature of the metal cations involved in determining the success of the obtainment of core–shell heterostructures. To further investigate these features, HRTEM and STEM-EDX chemical mapping at the nanoscale were adopted.

STEM-EDX chemical mapping and HRTEM images of the cobalt ferrite sample are reported in Figures 2a–c and S5. The nanoscale chemical mapping (Figures 2a–c and S5) reveals the spread of cobalt and iron throughout the particles, which appear differently oriented towards the electron beam, with hexagonal projections besides cubic ones being visible. Interestingly, the presence of the organic capping seems to be visible in the micrograph reported in Figure S5b with a size of the layer coherent with the hydrocarbon chain length [86]. HRTEM images (Figure 2d–f) show highly crystalline cubic-shaped particles with rounded corners, revealing the typical inter-lattice distances of spinel ferrite crystals and with the planes regularly aligned through the particle to the surface. The high crystallinity of the cobalt ferrite nanoparticles proved via HRTEM further suggests that the observed discrepancy between the particle and crystallite sizes is more related to the difficulties in describing two populations of differently shaped NPs via the Rietveld method.

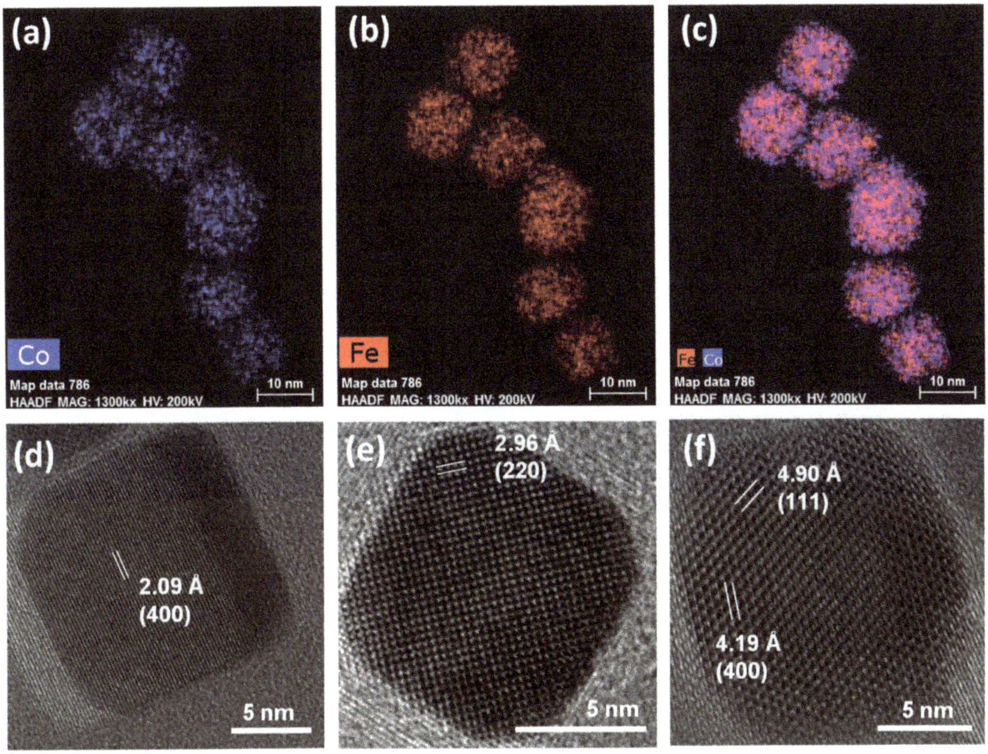

Figure 2. STEM-EDX chemical mapping (**a–c**) and HRTEM images with inter-lattice distances and Miller's indexes of some cubic nanoparticles of the sample Co (**d–f**).

STEM-EDX images of the core–shell sample are shown in Figure 3, where the obtainment of core–shell NPs with very thin shells (Figure 3a,b) accompanied by the formation of manganese ferrite NPs with similar sizes and shapes (Figure 3c,d) is proved. Single-phase cobalt ferrite NPs were not detected. These results indicate that the cubic cobalt ferrite seeds of about 14 nm were homogenously coated and that the heterogeneous and homogeneous nucleations of $MnFe_2O_4$ are competitive phenomena in the selected synthetic conditions. On the contrary, the obtainment of highly crystalline NPs exclusively in a core–shell architecture was previously achieved on cobalt ferrite spherical seeds of 6 and 8 nm in size with both manganese ferrite and spinel iron oxide as the coating shell, as proven via STEM-EDX, STEM-EELS, and HRTEM [46,70]. Therefore, we can hypothesise that the bigger size as well as the faceted shape of the cobalt ferrite seeds might be the main reason for the different formation mechanism of the final products. Nevertheless, it is worth noting that the presence of uncoated seeds was never revealed in the final products obtained using this oleate-based solvothermal method, regardless of the shape and size of the seeds. As expected, at the beginning of the synthesis process, the heterogenous nucleation seems to be favoured in comparison with the homogenous nucleation. Moreover, it seems that the heterogeneous nucleation (i.e., formation of the core–shell heterostructures) goes on until a critical size is reached (about 15 nm), at which, the particles lose their colloidal stability in the liquid medium and settle at the bottom of the Teflon liner. Therefore, in the liquid medium, only Mn and Fe oleates remain, and manganese ferrite nanoparticles nucleate and grow. Figure 3a depicts cubic core–shell NPs with manganese ferrite shells of about 1.4 nm, while Figure 3b represents a core–shell nanoparticle where the manganese ferrite grew in a different direction, generating a staggered cube with respect to the cubic core.

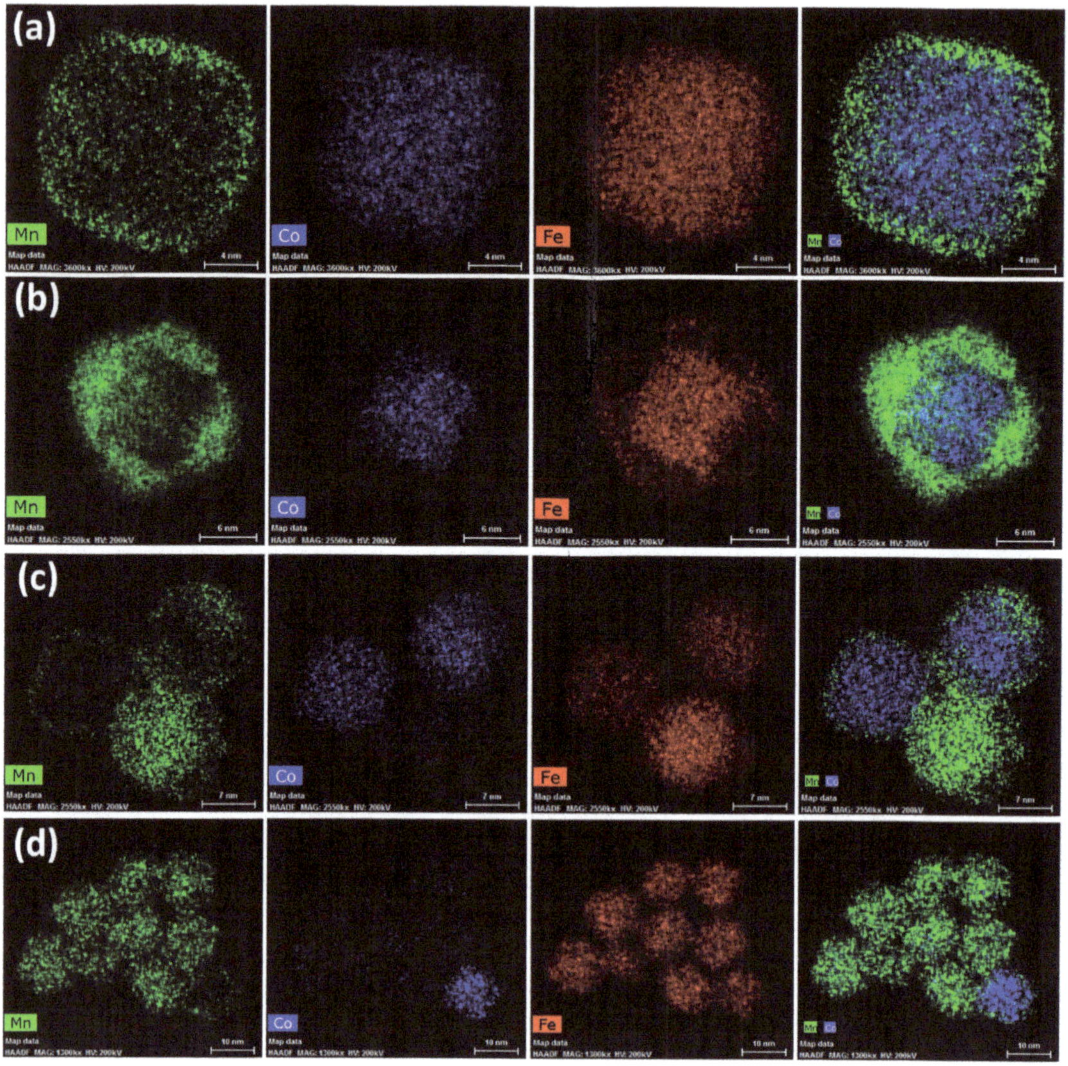

Figure 3. STEM-EDX chemical mapping for the sample Co@Mn, depicting core–shell heterostructures (**a**,**b**) and manganese ferrite NPs (**c**,**d**).

This behaviour can also be observed in the HRTEM image of the same nanoparticle, reported in Figure 4. The particle, pointed along the <110> zone axis, reveals twins between the {220} planes of the cubic core and the {111} planes of the shell, as better evidenced in the ellipsoidal spots visible in the FFT image (Figure 4b) and in the dashed-line white squares in the inversed masked FFT images (Figure 4e,f). Figures S6 and S7 report other structural defects found for the sample. The results obtained via HRTEM and STEM-EDX agree with the scenario hypothesised for this sample on the basis of the XRD and conventional TEM data, as previously discussed. In order to verify if these compositional and structural inhomogeneities at the nanoscale level strictly related to the faceted shape of the NPs affect the macroscopic behaviour of the sample, the magnetic properties were investigated through DC magnetometry (Figure 5).

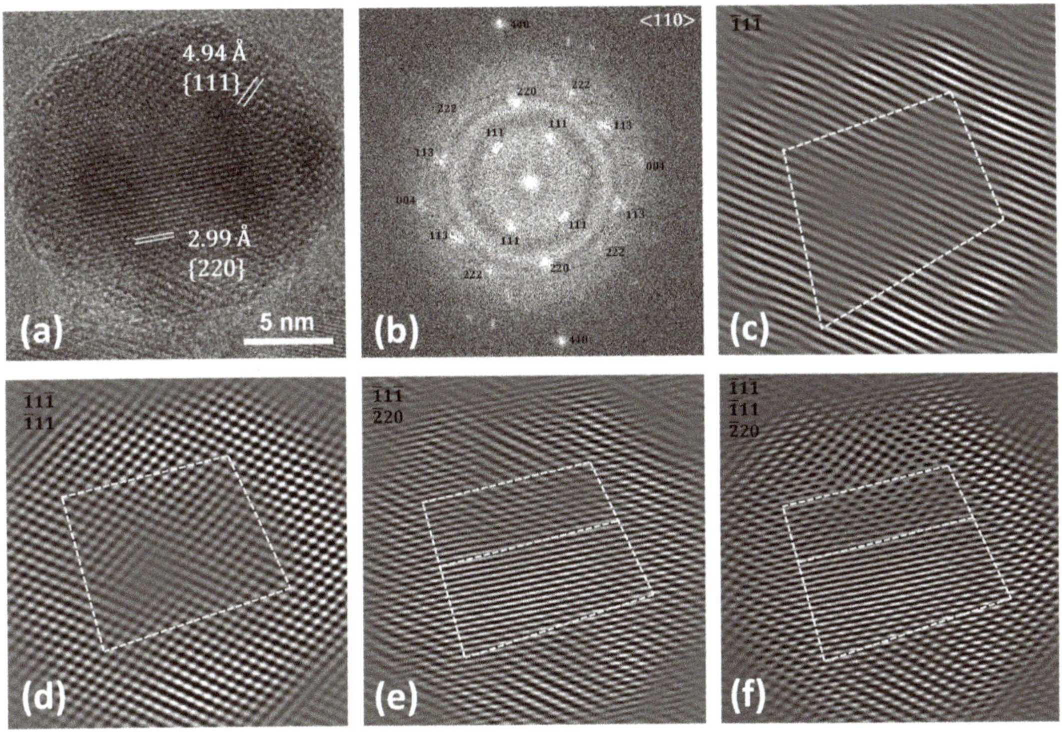

Figure 4. HRTEM image (**a**), FFT (**b**), and masked inverse FFT images (**c**–**f**) of Co@Mn.

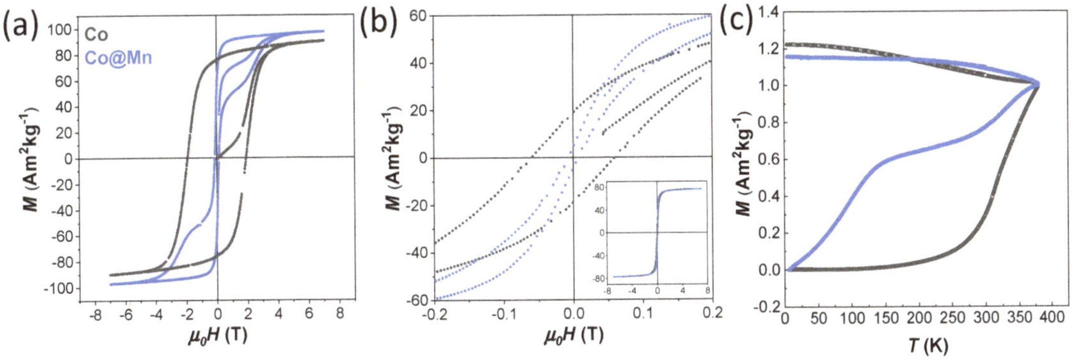

Figure 5. Field-dependent magnetisation curves recorded at 10 K (**a**) and 300 K (**b**), and zero-field cooled (ZFC) and field-cooled (FC) curves (**c**) of the samples Co (grey lines) and Co@Mn (blue lines).

The Co sample shows hysteresis at 10 K with a large coercive field of 1.91 T and high saturation magnetisation of about 90 Am^2kg^{-1}, while the Co@Mn sample coercivity is only 0.13 T with wasp-waisted-shaped hysteresis that arises from the superposition of the hard behaviour of the core–shell architectures and the soft behaviour of the manganese ferrite NPs (Figure 5a, Table 2). This two-stage hysteresis loop at 10 K resembles those of a physical mixture, as previously observed for an ad hoc reference sample prepared by mixing two single-phase NPs ($CoFe_2O_4$ and $MnFe_2O_4$) of similar sizes (Figure 4 of [46]). A single-stage hysteresis loop with lower coercivity compared with the cobalt ferrite seeds would be instead expected as a unique contribution for core–shell heterostructures with thin shells, proving the rigid coupling between the two spinel ferrite phases. Magnetisation isotherms at 300 K (Figure 5b) still show small hysteresis with coercivity of 0.06 T for the Co sample, indicating the presence of magnetically blocked particles. Furthermore, the slightly wasp-waisted shape is still present at 300 K for the Co@Mn sample, consistent with the expected behaviour of the physical mixture of the core–shell and manganese ferrite NPs.

Table 2. Coercive field (H_c) at 10 K and 300 K, anisotropy field (H_K) at 10 K, saturation magnetisation (M_s) at 10 K and 300 K, and remanent magnetisation (M_r) at 10 K of the samples.

Sample	H_c^{10K} (T)	H_K^{10K} (T)	M_s^{10K} (Am^2kg^{-1})	M_r^{10K} (Am^2kg^{-1})	M_r/M_s^{10K}	H_c^{300K} (T)	M_s^{300K} (Am^2kg^{-1})
Co	1.91(1)	3.7(1)	92(3)	75(2)	0.82(2)	0.06(1)	77(2)
Co@Mn	0.13(1)	3.8(1)	98(3)	51(1)	0.52(2)	0.01(1)	77(2)

The temperature-dependent magnetisation curves (ZFC-FC curves of Figure 5c) confirmed the presence of magnetically blocked NPs in the two samples at room temperature (with a temperature of the maximum above 350 K), but with the presence of a shallow maximum in the ZFC curve (at about 140 K)) for the Co@Mn sample was associated with the presence of the magnetically soft $MnFe_2O_4$ NPs.

4. Conclusions

A one-pot solvothermal approach was exploited with the aim of obtaining cubic cobalt ferrite NPs and further coupled with a seed-mediated growth strategy to build core–shell hard–soft heterostructures. Direct proof of the obtainment of highly crystalline, cubic and spherical, stoichiometric cobalt ferrite nanoparticles was obtained via TEM, STEM-EDX, and HRTEM. These techniques, and in particular the chemical mapping at the nanoscale via STEM-EDX, allowed us to verify the production of bi-magnetic hard–soft core–shell NPs ($CoFe_2O_4@MnFe_2O_4$) with a very thin shell in a physical mixture with manganese ferrite NPs. Additionally, the magnetic properties were diagnostic of the superposition of two contributions: the hard behaviour of the core–shell NPs and the soft one of the manganese ferrite NPs. Indeed, a double-stage hysteresis loop in the 10 K magnetisation isotherm and a maximum in the ZFC at about 140 K associated with a magnetically soft phase were observed. These findings allowed us to hypothesise a formation mechanism in which the heterogeneous nucleation (growth of a shell on the preformed seeds) and homogeneous nucleation (formation of new particles from the metal precursors) compete. From the comparison with previous achievements in which only core–shell NPs starting from small (6–8 nm) spherical NPs were achieved, it seems that starting from bigger and faceted spinel ferrite seeds might direct the formation process of the core–shell heterostructures towards systems with very thin shells, due to a critical size being reached that made these particles no longer sTable in the mother solution and made them settle down at the bottom of the synthesis reactor. Thus, the formation of new manganese ferrite nuclei occurred after this phase separation.

In this view, the obtainment of a homogenous sample of core–shell NPs might be achieved in these experimental conditions via the proper selection of the ratio between the number of preformed seeds and the metal oleate added in the seed-mediated growth process in order to avoid single-phase nucleation.

Supplementary Materials: The following supporting information can be downloaded at: https://www.mdpi.com/article/10.3390/nano13101679/s1, Table S1: Synthesis condition for the samples; Table S2: Refined structural parameters obtained via Rietveld refinement of the XRD patterns of Co and Co@Mn samples by using one spinel ferrite phase ($CoFe_2O_4$) for both samples; Table S3: Refined structural parameters obtained via Rietveld refinement of the XRD patterns of Co and Co@Mn samples by using two spinel ferrite phases ($CoFe_2O_4$, $MnFe_2O_4$) for the Co@Mn sample; Figure S1: FTIR spectra (a), TGA curves (b), and corresponding derivatives (c) of the Co and Co@Mn samples; Figure S2: TEM micrographs of the Co sample; Figure S3: TEM micrographs of the Co@Mn sample; Figure S4: Rietveld refinement via FullProf software of the XRD patterns of Co (pattern at the bottom) and Co@Mn (pattern in the upper part) samples by using (a) one spinel ferrite phase ($CoFe_2O_4$) for both samples and (b) two spinel ferrite phases ($CoFe_2O_4$, $MnFe_2O_4$); Figure S5: STEM-EDX chemical mapping (a–c) for the sample Co; Figure S6: HRTEM images (a,c,d) of the sample Co@Mn revealing structural defects; (b) represents the masked inversed FFT image of (a); Figure S7: HRTEM micrographs (a–c) of the sample Co@Mn revealing structural defects. Reference [87] is cited in the supplementary materials.

Author Contributions: Conceptualisation, M.S.A. and C.C.; investigation, M.S.A., V.M., D.Z. and F.S.; resources, H.L.X. and C.C.; data curation, M.S.A., V.M., D.Z. and F.S.; writing—original draft preparation, M.S.A., V.M., F.S. and C.C.; writing—review and editing, M.S.A., V.M., D.Z., F.S., H.L.X. and C.C.; supervision, C.C.; project administration, C.C.; funding acquisition, C.C., D.Z. and H.L.X. All authors have read and agreed to the published version of the manuscript.

Funding: CESA Project—RAS Piano Sulcis (CUP: E58C16000080003) is gratefully acknowledged for financing the post-doctoral fellowship of V. Mameli and M. Sanna Angotzi. MIUR—National Program PON Ricerca e Innovazione 2014–2020 (CUP J88D19001040001) is acknowledged for the Ph.D. grant of Fausto Secci. "Fondazione di Sardegna, Italy (FdS)" Project: CUP F72F20000240007(2019): "Surface-tailored Materials for Sustainable Environmental Applications" is acknowledged for the post-doctoral fellowship of Fausto Secci. This work was supported by Charles University Research Centre program No. UNCE/SCI/014. This research used resources of the Center for Functional Nanomaterials, which is a U.S. DOE Office of Science Facility, at Brookhaven National Laboratory under Contract No. DE-SC0012704.

Institutional Review Board Statement: Not applicable.

Informed Consent Statement: Not applicable.

Data Availability Statement: Data are contained within the article or Supplementary Materials.

Acknowledgments: Thanks are due to Andrea Ardu and to the "Centro Servizi di Ateneo per la Ricerca (CeSAR)" for the TEM measurements performed with JEOL JEM 1400 PLUS. Thanks are also due to Centro Servizi di Ateneo per la Ricerca (CeSAR) for the use of the PPMS DynaCool facility for the DC magnetometry measurements.

Conflicts of Interest: The authors declare no conflict of interest.

References

1. Sanna Angotzi, M.; Mameli, V.; Cara, C.; Grillo, V.; Enzo, S.; Musinu, A.; Cannas, C. Defect-assisted synthesis of magneto-plasmonic silver-spinel ferrite heterostructures in a flower-like architecture. *Sci. Rep.* **2020**, *10*, 17015. [CrossRef] [PubMed]
2. Sun, X.; Frey Huls, N.; Sigdel, A.; Sun, S. Tuning Exchange Bias in Core/Shell FeO/Fe_3O_4 Nanoparticles. *Nano Lett.* **2012**, *12*, 246–251. [CrossRef] [PubMed]
3. Salazar-Alvarez, G.; Lidbaum, H.; López-Ortega, A.; Estrader, M.; Leifer, K.; Sort, J.; Suriñach, S.; Baró, M.D.; Nogués, J. Two-, Three-, and four-component magnetic multilayer onion nanoparticles based on iron oxides and manganese oxides. *J. Am. Chem. Soc.* **2011**, *133*, 16738–16741. [CrossRef]
4. Liu, P.; Yao, Z.; Ng, V.M.H.; Zhou, J.; Kong, L.B.; Yue, K. Facile synthesis of ultrasmall Fe_3O_4 nanoparticles on MXenes for high microwave absorption performance. *Compos. Part A Appl. Sci. Manuf.* **2018**, *115*, 371–382. [CrossRef]
5. Liu, P.; Yao, Z.; Zhou, J.; Yang, Z.; Kong, L.B. Small magnetic Co-doped NiZn ferrite/graphene nanocomposites and their dual-region microwave absorption performance. *J. Mater. Chem. C* **2016**, *4*, 9738–9749. [CrossRef]
6. Vacca, M.A.; Cara, C.; Mameli, V.; Sanna Angotzi, M.; Scorciapino, M.A.; Cutrufello, M.G.; Musinu, A.; Tyrpekl, V.; Pala, L.; Cannas, C. Hexafluorosilicic Acid (FSA): From Hazardous Waste to Precious Resource in Obtaining High Value-Added Mesostructured Silica. *ACS Sustain. Chem. Eng.* **2020**, *8*, 14286–14300. [CrossRef]

7. Sanna Angotzi, M.; Mameli, V.; Cara, C.; Borchert, K.B.L.; Steinbach, C.; Boldt, R.; Schwarz, D.; Cannas, C. Meso- and macroporous silica-based arsenic adsorbents: Effect of pore size, nature of the active phase, and silicon release. *Nanoscale Adv.* **2021**, *3*, 6100–6113. [CrossRef]
8. Domingo, N.; Testa, A.M.; Fiorani, D.; Binns, C.; Baker, S.; Tejada, J. Exchange bias in Co nanoparticles embedded in an Mn matrix. *J. Magn. Magn. Mater.* **2007**, *316*, 155–158. [CrossRef]
9. Kurian, M.; Thankachan, S. Structural diversity and applications of spinel ferrite core—Shell nanostructures—A review. *Open Ceram.* **2021**, *8*, 100179. [CrossRef]
10. Zhang, Y.; Chen, D.; Li, N.; Xu, Q.; Li, H.; Lu, J. Fabricating 1D/2D Co_3O_4/$ZnIn_2S_4$ core–shell heterostructures with boosted charge transfer for photocatalytic hydrogen production. *Appl. Surf. Sci.* **2023**, *610*, 155272. [CrossRef]
11. Ma, B.; Zhang, C.; Jia, D.; Zhao, Q.; Yang, P. NiAl-LDH-Modified Core–Shell Rod-like ZnO@ZnS Heterostructures for Enhanced Photocatalytic Hydrogen Precipitation. *J. Phys. Chem. C* **2023**, *127*, 2908–2917. [CrossRef]
12. Wang, X.; Gao, Y.; Zhang, Q.; He, X.; Wang, X. Synthesis of MoO_3 (1D) @SnO_2 (2D) core-shell heterostructures for enhanced ethanol gas sensing performance. *Sens. Actuators B Chem.* **2023**, *382*, 133484. [CrossRef]
13. Chen, K.; Jiang, Y.; Tao, W.; Wang, T.; Liu, F.; Wang, C.; Yan, X.; Lu, G.; Sun, P. MOF Structure engineering to synthesize core-shell heterostructures with controllable shell layer thickness: Regulating gas selectivity and sensitivity. *Sens. Actuators B Chem.* **2023**, *378*, 133117. [CrossRef]
14. Xiao, Y.; Tao, Y.; Jiang, Y.; Wang, J.; Zhang, W.; Liu, Y.; Zhang, J.; Wu, X.; Liu, Z. Construction of core–shell CeO_2 nanorods/$SnIn_4S_8$ nanosheets heterojunction with rapid spatial electronic migration for effective wastewater purification and H_2O_2 production. *Sep. Purif. Technol.* **2023**, *304*, 122385. [CrossRef]
15. Yuan, W.; Jiang, T.; Fang, X.; Fan, Y.; Qian, S.; Gao, Y.; Cheng, N.; Xue, H.; Tian, J. Interface engineering of S-doped Co_2P@Ni_2P core–shell heterostructures for efficient and energy-saving water splitting. *Chem. Eng. J.* **2022**, *439*, 135743. [CrossRef]
16. Xiao, Y.; Wang, H.; Jiang, Y.; Zhang, W.; Zhang, J.; Wu, X.; Liu, Z.; Deng, W. Hierarchical Sb_2S_3/$ZnIn_2S_4$ core–shell heterostructure for highly efficient photocatalytic hydrogen production and pollutant degradation. *J. Colloid Interface Sci.* **2022**, *623*, 109–123. [CrossRef]
17. Shafiee, A.; Rabiee, N.; Ahmadi, S.; Baneshi, M.; Khatami, M.; Iravani, S.; Varma, R.S. Core–Shell Nanophotocatalysts: Review of Materials and Applications. *ACS Appl. Nano Mater.* **2022**, *5*, 55–86. [CrossRef]
18. Mulla, R.; Dunnill, C.W. Core–shell nanostructures for better thermoelectrics. *Mater. Adv.* **2022**, *3*, 125–141. [CrossRef]
19. Meiklejohn, W.H.; Bean, C.P. New Magnetic Anisotropy. *Phys. Rev.* **1957**, *105*, 904–913. [CrossRef]
20. Kneller, E.F.; Hawig, R. The exchange-spring magnet: A new material principle for permanent magnets. *IEEE Trans. Magn.* **1991**, *27*, 3588–3600. [CrossRef]
21. López-ortega, A.; Estrader, M.; Salazar-alvarez, G. Applications of exchange coupled bi-magnetic hard/soft and soft/hard magnetic core/shell nanoparticles. *Phys. Rep.* **2015**, *553*, 1–32. [CrossRef]
22. López-Ortega, A.; Tobia, D.; Winkler, E.; Golosovsky, I.V.; Salazar-Alvarez, G.; Estradé, S.; Estrader, M.; Sort, J.; González, M.A.; Suriñach, S.; et al. Size-Dependent Passivation Shell and Magnetic Properties in Antiferromagnetic/Ferrimagnetic Core/Shell MnO Nanoparticles. *J. Am. Chem. Soc.* **2010**, *132*, 9398–9407. [CrossRef]
23. Kavich, D.W.; Dickerson, J.H.; Mahajan, S.V.; Hasan, S.A.; Park, J.-H. Exchange bias of singly inverted FeO/Fe_3O_4 core-shell nanocrystals. *Phys. Rev. B* **2008**, *78*, 174414. [CrossRef]
24. Winkler, E.L.; Lima, E.; Tobia, D.; Saleta, M.E.; Troiani, H.E.; Agostinelli, E.; Fiorani, D.; Zysler, R.D. Origin of magnetic anisotropy in ZnO/CoF_e2O_4 and CoO/$CoFe_2O_4$ core/shell nanoparticle systems. *Appl. Phys. Lett.* **2012**, *101*, 252405. [CrossRef]
25. Lottini, E.; López-Ortega, A.; Bertoni, G.; Turner, S.; Meledina, M.; Van Tendeloo, G.; de Julián Fernández, C.; Sangregorio, C. Strongly Exchange Coupled Core | Shell Nanoparticles with High Magnetic Anisotropy: A Strategy toward Rare-Earth-Free Permanent Magnets. *Chem. Mater.* **2016**, *28*, 4214–4222. [CrossRef]
26. Kirschning, A.; Kupracz, L.; Hartwig, J. New Synthetic Opportunities in Miniaturized Flow Reactors with Inductive Heating. *Chem. Lett.* **2012**, *41*, 562–570. [CrossRef]
27. Baig, R.B.N.; Varma, R.S. Magnetically retrievable catalysts for organic synthesis. *Chem. Commun.* **2013**, *49*, 752–770. [CrossRef] [PubMed]
28. Polshettiwar, V.; Luque, R.; Fihri, A.; Zhu, H.; Bouhrara, M.; Basset, J.-M. Magnetically Recoverable Nanocatalysts. *Chem. Rev.* **2011**, *111*, 3036–3075. [CrossRef]
29. Zhang, D.; Zhou, C.; Sun, Z.; Wu, L.-Z.; Tung, C.-H.; Zhang, T. Magnetically recyclable nanocatalysts (MRNCs): A versatile integration of high catalytic activity and facile recovery. *Nanoscale* **2012**, *4*, 6244. [CrossRef]
30. Viñas, S.L.; Simeonidis, K.; Li, Z.-A.; Ma, Z.; Myrovali, E.; Makridis, A.; Sakellari, D.; Angelakeris, M.; Wiedwald, U.; Spasova, M.; et al. Tuning the magnetism of ferrite nanoparticles. *J. Magn. Magn. Mater.* **2016**, *415*, 20–23. [CrossRef]
31. Choi, H.; An, M.; Eom, W.; Lim, S.W.; Shim, I.-B.; Kim, C.S.; Kim, S.J. Crystallographic and magnetic properties of the hyperthermia material $CoFe_2O_4$@$AlFe_2O_4$. *J. Korean Phys. Soc.* **2017**, *70*, 173–176. [CrossRef]
32. Lee, J.-H.; Jang, J.; Choi, J.; Moon, S.H.; Noh, S.; Kim, J.; Kim, J.-G.; Kim, I.-S.; Park, K.I.; Cheon, J. Exchange-coupled magnetic nanoparticles for efficient heat induction. *Nat. Nanotechnol.* **2011**, *6*, 418–422. [CrossRef] [PubMed]
33. Liébana-Viñas, S.; Simeonidis, K.; Wiedwald, U.; Li, Z.-A.; Ma, Z.; Myrovali, E.; Makridis, A.; Sakellari, D.; Vourlias, G.; Spasova, M.; et al. Optimum nanoscale design in ferrite based nanoparticles for magnetic particle hyperthermia. *RSC Adv.* **2016**, *6*, 72918–72925. [CrossRef]

34. Zhang, Q.; Castellanos-Rubio, I.; Munshi, R.; Orue, I.; Pelaz, B.; Gries, K.I.; Parak, W.J.; del Pino, P.; Pralle, A. Model Driven Optimization of Magnetic Anisotropy of Exchange-Coupled Core–Shell Ferrite Nanoparticles for Maximal Hysteretic Loss. *Chem. Mater.* **2015**, *27*, 7380–7387. [CrossRef] [PubMed]
35. Robles, J.; Das, R.; Glassell, M.; Phan, M.H.; Srikanth, H. Exchange-coupled $Fe_3O_4/CoFe_2O_4$ nanoparticles for advanced magnetic hyperthermia. *AIP Adv.* **2018**, *8*, 056719. [CrossRef]
36. Hammad, M.; Nica, V.; Hempelmann, R. Synthesis and Characterization of Bi-Magnetic Core/Shell Nanoparticles for Hyperthermia Applications. *IEEE Trans. Magn.* **2017**, *53*, 1–6. [CrossRef]
37. Yelenich, O.V.; Solopan, S.O.; Greneche, J.M.; Belous, A.G. Synthesis and properties MFe_2O_4 (M = Fe, Co) nanoparticles and core–shell structures. *Solid State Sci.* **2015**, *46*, 19–26. [CrossRef]
38. Kim, M.; Kim, C.S.; Kim, H.J.; Yoo, K.-H.; Hahn, E. Effect hyperthermia in $CoFe_2O_4$@$MnFe_2O_4$ nanoparticles studied by using field-induced Mössbauer spectroscopy. *J. Korean Phys. Soc.* **2013**, *63*, 2175–2178. [CrossRef]
39. Noh, S.; Na, W.; Jang, J.; Lee, J.-H.; Lee, E.J.; Moon, S.H.; Lim, Y.; Shin, J.-S.; Cheon, J. Nanoscale Magnetism Control via Surface and Exchange Anisotropy for Optimized Ferrimagnetic Hysteresis. *Nano Lett.* **2012**, *12*, 3716–3721. [CrossRef]
40. Wang, J.; Zhou, Z.; Wang, L.; Wei, J.; Yang, H.; Yang, S.; Zhao, J. $CoFe_2O_4$ @$MnFe_2O_4$/polypyrrole nanocomposites for in vitro photothermal/magnetothermal combined therapy. *RSC Adv.* **2015**, *5*, 7349–7355. [CrossRef]
41. Pilati, V.; Cabreira Gomes, R.; Gomide, G.; Coppola, P.; Silva, F.G.; Paula, F.L.O.; Perzynski, R.; Goya, G.F.; Aquino, R.; Depeyrot, J. Core/Shell Nanoparticles of Non-Stoichiometric Zn–Mn and Zn–Co Ferrites as Thermosensitive Heat Sources for Magnetic Fluid Hyperthermia. *J. Phys. Chem. C* **2018**, *122*, 3028–3038. [CrossRef]
42. Wang, L.; Yan, Y.; Wang, M.; Yang, H.; Zhou, Z.; Peng, C.; Yang, S. An integrated nanoplatform for theranostics via multifunctional core–shell ferrite nanocubes. *J. Mater. Chem. B* **2016**, *4*, 1908–1914. [CrossRef] [PubMed]
43. Solopan, S.O.; Nedelko, N.; Lewińska, S.; Ślawska-Waniewska, A.; Zamorskyi, V.O.; Tovstolytkin, A.I.; Belous, A.G. Core/shell architecture as an efficient tool to tune DC magnetic parameters and AC losses in spinel ferrite nanoparticles. *J. Alloys Compd.* **2019**, *788*, 1203–1210. [CrossRef]
44. Fabris, F.; Lima, E.; De Biasi, E.; Troiani, H.E.; Vásquez Mansilla, M.; Torres, T.E.; Fernández Pacheco, R.; Ibarra, M.R.; Goya, G.F.; Zysler, R.D.; et al. Controlling the dominant magnetic relaxation mechanisms for magnetic hyperthermia in bimagnetic core–shell nanoparticles. *Nanoscale* **2019**, *11*, 3164–3172. [CrossRef] [PubMed]
45. Angelakeris, M.; Li, Z.A.; Hilgendorff, M.; Simeonidis, K.; Sakellari, D.; Filippousi, M.; Tian, H.; Van Tendeloo, G.; Spasova, M.; Acet, M.; et al. Enhanced biomedical heat-triggered carriers via nanomagnetism tuning in ferrite-based nanoparticles. *J. Magn. Magn. Mater.* **2015**, *381*, 179–187. [CrossRef]
46. Sanna Angotzi, M.; Mameli, V.; Cara, C.; Musinu, A.; Sangregorio, C.; Niznansky, D.; Xin, H.L.; Vejpravova, J.; Cannas, C. Coupled hard–soft spinel ferrite-based core–shell nanoarchitectures: Magnetic properties and heating abilities. *Nanoscale Adv.* **2020**, *2*, 3191–3201. [CrossRef]
47. Cannas, C.; Ardu, A.; Musinu, A.; Suber, L.; Ciasca, G.; Amenitsch, H.; Campi, G. Hierarchical Formation Mechanism of $CoFe_2O_4$ Mesoporous Assemblies. *ACS Nano* **2015**, *9*, 7277–7286. [CrossRef]
48. Zeng, H.; Li, J.; Wang, Z.L.; Liu, J.P.; Sun, S. Bimagnetic Core/Shell FePt/Fe_3O_4 Nanoparticles. *Nano Lett.* **2004**, *4*, 187–190. [CrossRef]
49. Masala, O.; Hoffman, D.; Sundaram, N.; Page, K.; Proffen, T.; Lawes, G.; Seshadri, R. Preparation of magnetic spinel ferrite core/shell nanoparticles: Soft ferrites on hard ferrites and vice versa. *Solid State Sci.* **2006**, *8*, 1015–1022. [CrossRef]
50. López-Ortega, A.; Estrader, M.; Salazar-Alvarez, G.; Estradé, S.; Golosovsky, I.V.; Dumas, R.K.; Keavney, D.J.; Vasilakaki, M.; Trohidou, K.N.; Sort, J.; et al. Strongly exchange coupled inverse ferrimagnetic soft/hard, $MnxFe_{3-x}O_4/FexMn_{3-x}O_4$, core/shell heterostructured nanoparticles. *Nanoscale* **2012**, *4*, 5138. [CrossRef]
51. Chen, J.; Ye, X.; Oh, S.J.; Kikkawa, J.M.; Kagan, C.R.; Murray, C.B. BisTable Magnetoresistance Switching in Exchange-Coupled CoFe 2 O 4 -Fe 3 O 4 Binary Nanocrystal Superlattices by Self-Assembly and Thermal Annealing. *ACS Nano* **2013**, *7*, 1478–1486. [CrossRef]
52. Estrader, M.; López-Ortega, A.; Estradé, S.; Golosovsky, I.V.; Salazar-Alvarez, G.; Vasilakaki, M.; Trohidou, K.N.; Varela, M.; Stanley, D.C.; Sinko, M.; et al. Robust antiferromagnetic coupling in hard-soft bi-magnetic core/shell nanoparticles. *Nat. Commun.* **2013**, *4*, 2960. [CrossRef] [PubMed]
53. Gavrilov-Isaac, V.; Neveu, S.; Dupuis, V.; Taverna, D.; Gloter, A.; Cabuil, V. Synthesis of Trimagnetic Multishell $MnFe_2O_4$ @$CoFe_2O_4$ @$NiFe_2O_4$ Nanoparticles. *Small* **2015**, *11*, 2614–2618. [CrossRef] [PubMed]
54. Song, Q.; Zhang, Z.J. Controlled synthesis and magnetic properties of bimagnetic spinel ferrite $CoFe_2O_4$ and $MnFe_2O_4$ nanocrystals with core-shell architecture. *J. Am. Chem. Soc.* **2012**, *134*, 10182–10190. [CrossRef]
55. Thanh, N.T.K.; Maclean, N.; Mahiddine, S. Mechanisms of nucleation and growth of nanoparticles in solution. *Chem. Rev.* **2014**, *114*, 7610–7630. [CrossRef] [PubMed]
56. Xia, Y.; Gilroy, K.D.; Peng, H.-C.; Xia, X. Seed-Mediated Growth of Colloidal Metal Nanocrystals. *Angew. Chem. Int. Ed.* **2017**, *56*, 60–95. [CrossRef] [PubMed]
57. Cannas, C.; Musinu, A.; Peddis, D.; Piccaluga, G. Synthesis and Characterization of $CoFe_2O_4$ Nanoparticles Dispersed in a Silica Matrix by a Sol−Gel Autocombustion Method. *Chem. Mater.* **2006**, *18*, 3835–3842. [CrossRef]
58. Sun, S.; Zeng, H.; Robinson, D.B.; Raoux, S.; Rice, P.M.; Wang, S.X.; Li, G. Monodisperse MFe_2O_4 (M = Fe, Co, Mn) Nanoparticles. *J. Am. Chem. Soc.* **2004**, *126*, 273–279. [CrossRef]

59. Niederberger, M.; Pinna, N. *Metal Oxide Nanoparticles in Organic Solvents. Synthesis, Formation, Assembly and Application*; Springer: Berlin/Heidelberg, Germany, 2009.
60. Cannas, C.; Musinu, A.; Ardu, A.; Orrù, F.; Peddis, D.; Casu, M.; Sanna, R.; Angius, F.; Diaz, G.; Piccaluga, G. $CoFe_2O_4$ and $CoFe_2O_4/SiO_2$ core/shell nanoparticles: Magnetic and spectroscopic study. *Chem. Mater.* **2010**, *22*, 3353–3361. [CrossRef]
61. Cannas, C.; Ardu, A.; Peddis, D.; Sangregorio, C.; Piccaluga, G.; Musinu, A. Surfactant-assisted route to fabricate $CoFe_2O_4$ individual nanoparticles and spherical assemblies. *J. Colloid Interface Sci.* **2010**, *343*, 415–422. [CrossRef]
62. Nobile, C.; Cozzoli, P.D. Synthetic Approaches to Colloidal Nanocrystal Heterostructures Based on Metal and Metal-Oxide Materials. *Nanomaterials* **2022**, *12*, 1729. [CrossRef] [PubMed]
63. Ghosh Chaudhuri, R.; Paria, S. Core/Shell Nanoparticles: Classes, Properties, Synthesis Mechanisms, Characterization, and Applications. *Chem. Rev.* **2012**, *112*, 2373–2433. [CrossRef]
64. Salazar-Alvarez, G.; Qin, J.; Šepelák, V.; Bergmann, I.; Vasilakaki, M.; Trohidou, K.N.; Ardisson, J.D.; Macedo, W.A.A.; Mikhaylova, M.; Muhammed, M.; et al. Cubic versus Spherical Magnetic Nanoparticles: The Role of Surface Anisotropy. *J. Am. Chem. Soc.* **2008**, *130*, 13234–13239. [CrossRef] [PubMed]
65. Gavilán, H.; Posth, O.; Bogart, L.K.; Steinhoff, U.; Gutiérrez, L.; Morales, M.P. How shape and internal structure affect the magnetic properties of anisometric magnetite nanoparticles. *Acta Mater.* **2017**, *125*, 416–424. [CrossRef]
66. Niraula, G.; Coaquira, J.A.H.; Zoppellaro, G.; Villar, B.M.G.; Garcia, F.; Bakuzis, A.F.; Longo, J.P.F.; Rodrigues, M.C.; Muraca, D.; Ayesh, A.I.; et al. Engineering Shape Anisotropy of Fe_3O_4-γ-Fe_2O_3 Hollow Nanoparticles for Magnetic Hyperthermia. *ACS Appl. Nano Mater.* **2021**, *4*, 3148–3158. [CrossRef]
67. Roca, A.G.; Gutiérrez, L.; Gavilán, H.; Fortes Brollo, M.E.; Veintemillas-Verdaguer, S.; Morales, M. del P. Design strategies for shape-controlled magnetic iron oxide nanoparticles. *Adv. Drug Deliv. Rev.* **2019**, *138*, 68–104. [CrossRef]
68. Ma, Z.; Mohapatra, J.; Wei, K.; Liu, J.P.; Sun, S. Magnetic Nanoparticles: Synthesis, Anisotropy, and Applications. *Chem. Rev.* **2021**, *123*, 3904–3943. [CrossRef]
69. Peddis, D.; Muscas, G.; Mathieu, R.; Kumar, P.A.; Varvaro, G.; Singh, G.; Orue, I.; Gil-Carton, D.; Marcano, L.; Muela, A.; et al. Studying nanoparticles' 3D shape by aspect maps: Determination of the morphology of bacterial magnetic nanoparticles. *Faraday Discuss.* **2016**, *191*, 177–188. [CrossRef]
70. Sanna Angotzi, M.; Musinu, A.; Mameli, V.; Ardu, A.; Cara, C.; Niznansky, D.; Xin, H.L.; Cannas, C. Spinel Ferrite Core-Shell Nanostructures by a Versatile Solvothermal Seed-Mediated Growth Approach and Study of Their Nanointerfaces. *ACS Nano* **2017**, *11*, 7889–7900. [CrossRef]
71. Sanna Angotzi, M.; Mameli, V.; Cara, C.; Ardu, A.; Nizňanský, D.; Musinu, A. Oleate-Based Solvothermal Approach for Size Control of $MIIFe_2IIIO_4$ (MII = MnII, FeII) Colloidal Nanoparticles. *J. Nanosci. Nanotechnol.* **2019**, *19*, 4954–4963. [CrossRef]
72. Sanna Angotzi, M.; Mameli, V.; Musinu, A.; Nizňanský, D. 57Fe Mössbauer Spectroscopy for the Study of Nanostructured Mixed Mn–Co Spinel Ferrites. *J. Nanosci. Nanotechnol.* **2019**, *19*, 5008–5013. [CrossRef] [PubMed]
73. Sanna Angotzi, M.; Mameli, V.; Zákutná, D.; Kubániová, D.; Cara, C.; Cannas, C. Evolution of the Magnetic and Structural Properties with the Chemical Composition in Oleate-Capped $Mn_xCo_{1-x}Fe_2O_4$ Nanoparticles. *J. Phys. Chem. C* **2021**, *125*, 20626–20638. [CrossRef]
74. Sanna Angotzi, M.; Mameli, V.; Cara, C.; Peddis, D.; Xin, H.L.; Sangregorio, C.; Mercuri, M.L.; Cannas, C. On the synthesis of bi-magnetic manganese ferrite-based core–shell nanoparticles. *Nanoscale Adv.* **2021**, *3*, 1612–1623. [CrossRef] [PubMed]
75. Khanal, S.; Sanna Angotzi, M.; Mameli, V.; Veverka, M.; Xin, H.L.; Cannas, C.; Vejpravová, J. Self-Limitations of Heat Release in Coupled Core-Shell Spinel Ferrite Nanoparticles: Frequency, Time, and Temperature Dependencies. *Nanomaterials* **2021**, *11*, 2848. [CrossRef]
76. Rodríguez-Carvajal, J. Recent advances in magnetic structure determination by neutron powder diffraction. *Phys. B Condens. Matter* **1993**, *192*, 55–69. [CrossRef]
77. Bergmann, J.; Monecke, T.; Kleeberg, R. Alternative algorithm for the correction of preferred orientation in Rietveld analysis. *J. Appl. Crystallogr.* **2001**, *34*, 16–19. [CrossRef]
78. Mondini, S.; Ferretti, A.M.; Puglisi, A.; Ponti, A. Pebbles and PebbleJuggler: Software for accurate, unbiased, and fast measurement and analysis of nanoparticle morphology from transmission electron microscopy (TEM) micrographs. *Nanoscale* **2012**, *4*, 5356. [CrossRef]
79. Morrish, A.H. *The Physical Principles of Magnetism*; Wiley-VCH: Weinheim, Germany, 1965; ISBN 9780780360297.
80. Repko, A.; Nižňanský, D.; Poltierová-Vejpravová, J. A study of oleic acid-based hydrothermal preparation of $CoFe_2O_4$ nanoparticles. *J. Nanoparticle Res.* **2011**, *13*, 5021–5031. [CrossRef]
81. Repko, A.; Nižňanský, D.; Matulková, I.; Kalbáč, M.; Vejpravová, J. Hydrothermal preparation of hydrophobic and hydrophilic nanoparticles of iron oxide and a modification with CM-dextran. *J. Nanopart. Res.* **2013**, *15*, 1767. [CrossRef]
82. Repko, A.; Vejpravová, J.; Vacková, T.; Zákutná, D.; Nižňanský, D. Oleate-based hydrothermal preparation of $CoFe_2O_4$ nanoparticles, and their magnetic properties with respect to particle size and surface coating. *J. Magn. Magn. Mater.* **2015**, *390*, 142–151. [CrossRef]
83. Chang, H.; Kim, B.H.; Lim, S.G.; Baek, H.; Park, J.; Hyeon, T. Role of the Precursor Composition in the Synthesis of Metal Ferrite Nanoparticles. *Inorg. Chem.* **2021**, *60*, 4261–4268. [CrossRef] [PubMed]
84. Bronstein, L.; Huang, X. Influence of iron oleate complex structure on iron oxide nanoparticle formation. *Chem. Mater.* **2007**, *19*, 3624–3632. [CrossRef]

85. Creutzburg, M.; Konuk, M.; Tober, S.; Chung, S.; Arndt, B.; Noei, H.; Meißner, R.H.; Stierle, A. Adsorption of oleic acid on magnetite facets. *Commun. Chem.* **2022**, *5*, 1–9. [CrossRef] [PubMed]
86. Mameli, V.; Musinu, A.; Ardu, A.; Ennas, G.; Peddis, D.; Niznansky, D.; Sangregorio, C.; Innocenti, C.; Thanh, N.T.K.; Cannas, C. Studying the effect of Zn-substitution on the magnetic and hyperthermic properties of cobalt ferrite nanoparticles. *Nanoscale* **2016**, *8*, 10124–10137. [CrossRef] [PubMed]
87. White, W.B.; De Angelis, B.A. Interpretation of the vibrational spectra of spinels. Spectrochim. *Acta Part A Mol. Spectrosc.* **1967**, *23*, 985–995. [CrossRef]

nanomaterials

Article

Solvent Influence on the Magnetization and Phase of Fe-Ni Alloy Nanoparticles Generated by Laser Ablation in Liquids

Inna Y. Khairani [1], Qiyuan Lin [2], Joachim Landers [3], Soma Salamon [3], Carlos Doñate-Buendía [1], Evguenia Karapetrova [4], Heiko Wende [3], Giovanni Zangari [2] and Bilal Gökce [1,*]

[1] Chair of Materials Science and Additive Manufacturing, School of Mechanical Engineering and Safety Engineering, University of Wuppertal, 42119 Wuppertal, Germany
[2] Department of Materials Science and Engineering, University of Virginia, Charlottesville, VA 22903, USA
[3] Faculty of Physics and Center for Nanointegration Duisburg-Essen (CENIDE), University of Duisburg-Essen, 47057 Duisburg, Germany
[4] Advanced Photon Source, Argonne National Laboratory, Argonne, IL 60439, USA
* Correspondence: goekce@uni-wuppertal.de

Abstract: The synthesis of bimetallic iron-nickel nanoparticles with control over the synthesized phases, particle size, surface chemistry, and oxidation level remains a challenge that limits the application of these nanoparticles. Pulsed laser ablation in liquid allows the properties tuning of the generated nanoparticles by changing the ablation solvent. Organic solvents such as acetone can minimize nanoparticle oxidation. Yet, economical laboratory and technical grade solvents that allow cost-effective production of FeNi nanoparticles contain water impurities, which are a potential source of oxidation. Here, we investigated the influence of water impurities in acetone on the properties of FeNi nanoparticles generated by pulsed laser ablation in liquids. To remove water impurities and produce "dried acetone", cost-effective and reusable molecular sieves (3 Å) are employed. The results show that the $Fe_{50}Ni_{50}$ nanoparticles' properties are influenced by the water content of the solvent. The metastable HCP FeNi phase is found in NPs prepared in acetone, while only the FCC phase is observed in NPs formed in water. Mössbauer spectroscopy revealed that the FeNi nanoparticles oxidation in dried acetone is reduced by 8% compared to acetone. The high-field magnetization of $Fe_{50}Ni_{50}$ nanoparticles in water is the highest, 68 Am^2/kg, followed by the nanoparticles obtained after ablation in acetone without water impurities, 59 Am^2/kg, and acetone, 52 Am^2/kg. The core-shell structures formed in these three liquids are also distinctive, demonstrating that a core-shell structure with an outer oxide layer is formed in water, while carbon external layers are obtained in acetone without water impurity. The results confirm that the size, structure, phase, and oxidation of FeNi nanoparticles produced by pulsed laser ablation in liquids can be modified by changing the solvent or just reducing the water impurities in the organic solvent.

Keywords: iron-nickel alloy; core-shell nanoalloys; nickel ferrite; hexagonal closed packed phase; carbon shell; laser synthesis of colloids

Citation: Khairani, I.Y.; Lin, Q.; Landers, J.; Salamon, S.; Doñate-Buendía, C.; Karapetrova, E.; Wende, H.; Zangari, G.; Gökce, B. Solvent Influence on the Magnetization and Phase of Fe-Ni Alloy Nanoparticles Generated by Laser Ablation in Liquids. *Nanomaterials* **2023**, *13*, 227. https://doi.org/10.3390/nano13020227

Academic Editor: Vincenzo Amendola

Received: 6 December 2022
Revised: 28 December 2022
Accepted: 28 December 2022
Published: 4 January 2023

1. Introduction

Iron nickel alloys are one of the most studied magnetic materials due to the abundance of their constituting elements on Earth [1,2] and owing to the interesting properties exhibited depending on their atomic ratio. For example, Invar ($Fe_{64}Ni_{36}$) exhibits very little thermal expansion (almost zero) over a wide temperature variation, while Permalloy ($Fe_{20}Ni_{80}$) offers a notably high magnetic permeability, low coercivity, and small magnetostriction [3–5]. Due to these interesting properties, iron-nickel alloys are employed in various key technologies such as transformers [6], magnetic actuators [7], magnetic sensors [8], electromagnetic shielding [9], spintronics [4,10], and catalysis [11–13]. The equiatomic iron-nickel alloy ($Fe_{50}Ni_{50}$), in particular, gained popularity as an electrocatalyst for the oxygen evolution reaction (OER) [12,14] and as a potential candidate for a permanent magnet after the discovery of the high-coercive tetrataenite mineral with $L1_0$ structure found in a meteorite [2,15]. In both cases, the high material abundance of the alloy constituent elements on Earth represents a fundamental advantage, envisioned to overcome the price and supply chain problems associated with the current rare-earth-based OER catalysts (RuO_2 and IrO_2) and permanent magnets (NdFeB) in strategic technologies, such as electric mobility and energy storage. Other than the atomic ratio, the iron-nickel alloy particle size is also a crucial parameter defining their properties and performance in certain applications, especially for the catalytic activity of $Fe_{50}Ni_{50}$ alloy in the OER process [13], where nanosized materials are desired. By reducing the size to the nanometer range, especially below 10 nm, the specific surface area of the $Fe_{50}Ni_{50}$ catalyst is significantly increased, thus exposing more of its active sites for the reactions to take place. In addition, the recent report on the non-cubic symmetry in $Fe_{50}Ni_{50}$ nanoparticles [16] sparks the possibility of employing nanosized $Fe_{50}Ni_{50}$ as a rare-earth-free alternative to permanent magnets. These findings highlight the relevance of understanding and controlling the formation of $Fe_{50}Ni_{50}$ nanoparticles and explore novel synthesis techniques that allow $Fe_{50}Ni_{50}$ nanoparticles' phase control.

Conventional fabrication methods of $Fe_{50}Ni_{50}$ nanoparticles (NPs) include chemical reduction and gas condensation routes. The chemical reduction of the iron and nickel salts with hydrazine in the presence of polyvinylpyrrolidone (PVP) resulted in face-centered cubic (FCC) $Fe_{50}Ni_{50}$ with an average diameter of 29 nm [17], and 96 nm without PVP [18]. Gas condensation of iron-nickel alloy in a helium atmosphere was sought, but oxidation took place on the surface of the particle after exposure to oxygen, resulting in the formation of core-shell NPs with FeNi γ-phase and oxides of γ-Fe_2O_3 or Fe_3O_4. These two methods, unfortunately, do not follow the green chemistry principles [19] due to the multi-step processes and the use of hazardous materials and inert gases to reduce oxidation. Meanwhile, pulsed laser ablation in liquid (PLAL) offers a one-step method to produce NPs directly in the desired liquid and avoids the generation of by-products, hence removing purification steps and the generation of extra chemical waste [20]. This technique does not require high vacuum or temperature conditions, making it easily implementable and transferable to industrial environments [21]. PLAL is based on the ablation of the bulk target in the desired liquid providing the versatility to tune the laser parameters and the ablation liquid to influence the temperature, pressure, and surrounding media [22]. By changing the liquid employed for PLAL, properties such as the composition and phase of the produced NPs can be modified [23].

Another NP property that is influenced by the liquid employed in the PLAL is the NP oxidation. Due to their small size and large surface area, NPs are prone to oxidation upon exposure to oxygen or oxidizing agents. In the PLAL, it is reported that the NP oxidation is influenced by the redox activity of the target material [24] and the choice of the ablation liquid [23]. For example, almost 100% of the surface of laser-generated Ti in water is oxidized, while less than 5% of the gold surface is oxidized [24]. For the ablation liquid, complete or partial oxidation of the NPs is observed in the ablation of Ti and Mn in water, which results in the generation of TiO_2 [25] and Mn_3O_4 [26] NPs, respectively. NP oxidation might be purposely performed in some contexts where oxide NPs are desired, such as the generation of TiO_2 by ablating Ti target immersed in water. However, in the cases where oxidation needs to be avoided, organic solvents such as acetone are known to reduce the oxidation of laser-generated NPs [27]. However, the oxidation itself is caused by the exposure of NPs inside the cavitation bubble to the oxygen species generated from the breakdown of liquid or gas in the nearby vicinity of the ablation spot; hence, all species with oxygen atoms might contribute to the oxidation of the NPs, including dissolved oxygen gas. To investigate this matter, Marzun et al. [28] ablated Cu in different ablation liquids, i.e., H_2O (H_2O_{air}), H_2O purged with Ar gas (H_2O_{ar}), and acetone. They reported the formation of Cu oxides in both H_2O_{air} and H_2O_{ar}, with H_2O_{air} having a higher oxidation level and amorphous phase in acetone. This indicates that not only the dissolved oxygen gases in the water contribute to the Cu oxidation, but also the water itself. Meanwhile, acetone with technical grade or laboratory grade, even with ACS reagent and HPLC grades, still contains water impurity to some extent (≤0.5%) [29,30], which contributes to the oxidation of the produced NPs.

In this study, the influence of water removal in acetone using molecular sieves on the oxidation level of the laser-generated $Fe_{50}Ni_{50}$ NPs is investigated. A molecular sieve is an adsorbent with three-dimensional frameworks of alumina-silicate, which is capable of reducing the water content down to 0.001% [31,32]. The molecular sieves provide an inexpensive and easy way to remove water from acetone, complying with the green chemistry principle due to its reusability. The removal of water in the organic solvents is not only intended to reduce the oxidation but also to directly encapsulate the NPs in a carbon shell during the PLAL synthesis that enhances their catalytic activity [33,34]. In addition to reducing the oxidation level and altering the shell formation, the generation of the non-cubic metastable hexagonal closed packed (HCP) in the organic solvents was investigated. It has been proposed that non-cubic phases might be used as a precursor to generating FeNi with $L1_0$ structure [35], but the suggested methods to fabricate the non-cubic FeNi involve the use of high-pressure and high-temperature conditions such as in the diamond anvil cell (DAC) [36] or high-strain process [35]. Here, we propose PLAL as a method to produce the non-cubic HCP FeNi phase at room conditions, taking advantage of the locally high-pressure and -temperature conditions achieved by the high-intensity laser interaction with the target surface and surrounding liquid.

2. Materials and Methods

2.1. $Fe_{50}Ni_{50}$ Colloidal Nanoparticles Production

A picosecond laser Nd:YAG with a wavelength of 1064 nm, a pulse duration of 10 ps, a power of 8 W, a repetition rate of 100 kHz, a raw beam diameter of 2 mm, and a pulse energy of 80 μJ was employed to produce nanoparticles by PLAL (Figure 1). The laser beam was focused on the immersed (6 mm liquid layer) equiatomic FeNi alloy target by a galvanometric scanner coupled with an f-theta lens (focal length of 100 mm) following an Archimedean spiral pattern (6 mm diameter) with a speed of 2 m/s. The beam radius and peak fluence at the processing plane were 65 μm and 1.2 J/cm^2, respectively. To avoid shielding of the laser beam by the produced nanoparticles, the liquid was pumped by a peristaltic pump at a flow rate of 150 mL/min (calibrated before the experiments of each liquid). The investigated liquids are distilled water, acetone, and "dried" acetone (obtained by immersing molecular sieves type 3 Å for 24 h to capture water molecules in acetone). The FeNi samples ablated in different liquids will be further referred to as FeNi in water, FeNi in acetone, and FeNi in dried acetone, respectively. All colloids have a similar absorbance value at the laser wavelength as shown in Figure S1. To dry the colloids and obtain nanopowders suitable for characterization, we performed magnetic separation using a permanent magnet (NdFeB) followed by liquid evaporation using an exhaust fan.

2.2. Analytical Methods

The generated FeNi NPs colloids were analyzed by transmission electron microscopy (TEM, JEOL JEM-2200FS, 200 kV, ZrO_2/W emitter) and energy dispersive X-ray spectroscopy (EDX) to determine the size distribution, morphology, elemental composition, and oxide formation of the NPs. The sample was drop-casted on a copper grid with lacey carbon coating and was measured within one week after production to ensure minimum particle growth and oxidation due to aging.

Synchrotron X-ray diffraction (SXRD) was used to analyze the phase of the produced NPs qualitatively (using the peak matching technique) and quantitively (using the Rietveld refinement technique). Measurements were carried out at the 33-BM-C beamline of the Advanced Photon Source (APS) at the Argonne National Laboratory, United States with a beam wavelength of 0.77 Å. Since a wavelength of 0.77 Å was employed, the 2θ value is shifted to a lower degree compared to the standard 1.54 Å wavelength. The measurements were performed using the transmission (i.e., Debye-Scherrer) geometry. The colloids were sealed in Special Glass 10 capillaries (Hampton Research Corp.) by Beeswax (Hampton Research Corp.). The NPs generated in water were transferred to dried acetone to mitigate post-synthesis particle aging (i.e., oxidation and particle growth) before being loaded into capillaries for SXRD measurements.

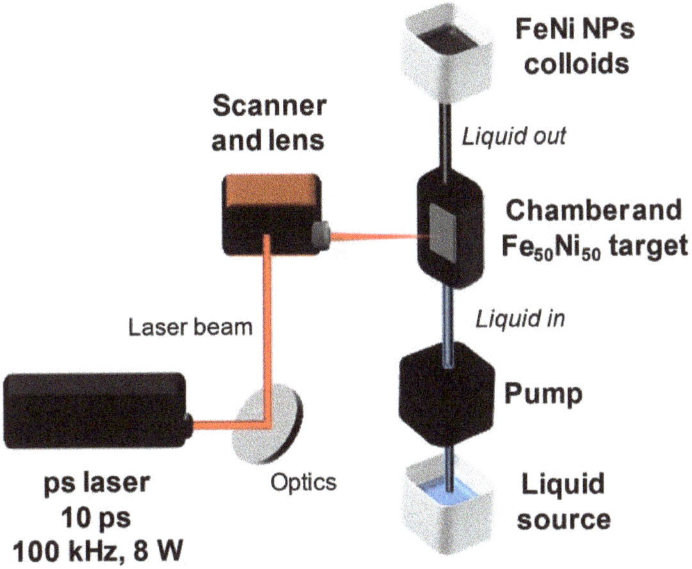

Figure 1. Schematic illustration of pulsed laser ablation in liquid using a flow chamber. The selected liquid flowed through the ablation chamber from the bottom to the top by a pump while the ablation took place. A $Fe_{50}Ni_{50}$ target was placed perpendicular to the incoming laser beam.

Mössbauer spectroscopy was employed to determine the oxidation level and magnetic structure of the FeNi samples in acetone and dried acetone. Spectra of both powder and colloid samples were recorded in transmission geometry, with the latter being placed in an airtight sample container of appropriate geometry. A ^{57}Co(Rh) radiation source was used, mounted on a constant-acceleration driving unit (WissEl GmbH), with low temperatures down to 4.3 K being achieved via a closed-cycle cryostat (Lake Shore Cryotronics). Spectra measured in external magnetic fields up to 8 T were recorded using a magnet cryostat (Oxford Instruments). Subspectra of magnetically ordered phases have been reproduced using hyperfine field distributions; isomer shifts are given relative to α-Fe at room temperature.

Nanoparticle magnetic behavior was studied using vibrating sample magnetometry (PPMS DynaCool, Quantum Design). Field-dependent magnetization loops M(H) were recorded at temperatures between 4.3–300 K and a magnetic field range of ±9 T.

3. Results and Discussion

3.1. Crystallographic Phases

To determine the influence of different liquids on the phase formation of FeNi NPs, XRD phase analysis was performed. The SXRD profiles of FeNi NPs in dried acetone, acetone, and water are presented in Figure 2, and the complete indexing can be found in the supplementary (Figure S2). The FeNi NPs generated in water (Figure 2) show the diffraction peaks of the face-centered cubic (FCC) FeNi and the spinel $NiFe_2O_4$ structure. Meanwhile, the FeNi NPs generated in acetone and dried acetone (Figure 2) consist of the hexagonal closed-packed (HCP) FeNi phase in addition to the FCC FeNi phases and the spinel $NiFe_2O_4$ phases. To quantify the weight fraction (wt%) of the HCP phase, Rietveld refinement was performed (Figure S3) and the results are presented in Table 1.

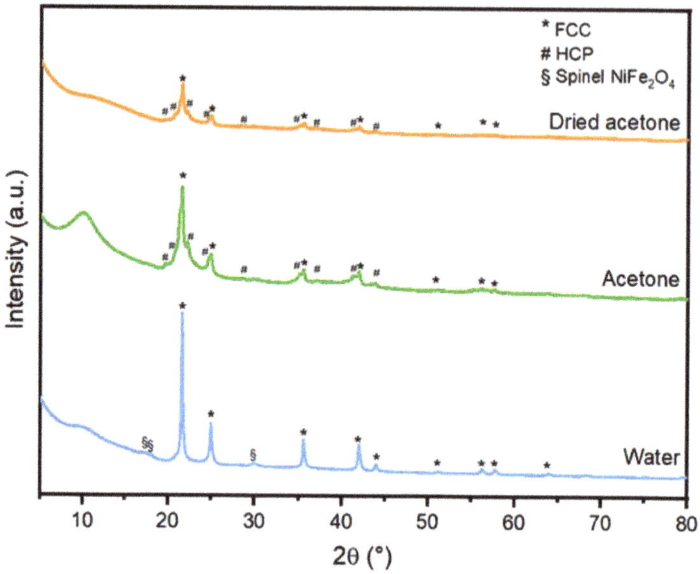

Figure 2. Synchrotron XRD profiles of the FeNi NPs ablated in different liquids. The ablation in acetone or dried acetone produced the FCC phase and the HCP phase, with a small volume of spinel $NiFe_2O_4$ phases (barely visible in this graph). Meanwhile, the ablation in water produced the FCC phase and $NiFe_2O_4$ phase. The complete indexing is presented in the supplementary information (Figure S2).

Table 1. The crystalline phase composition of FeNi NPs synthesized in different ablation liquids. The weight fraction of the HCP FeNi phase was extracted from the SXRD profile by Rietveld refinement.

Ablation liquid	Crystalline Phase Composition	HCP Content (wt%) *
Dried acetone	FCC FeNi, HCP FeNi, $NiFe_2O_4$	35.2 ± 1.0
Acetone	FCC FeNi, HCP FeNi, $NiFe_2O_4$	38.4 ± 0.2
Water	FCC FeNi, $NiFe_2O_4$	0

* The HCP content reported here is the weight fraction of the FeNi HCP phase with respect to the weight of the nanoparticle core. The Rietveld refinement was performed on a structure model of the nanoparticle core that consisted of the FCC FeNi phase(s) and the HCP FeNi phase if existing. The $NiFe_2O_4$ phase in the particle shell was not included in the structure model being refined.

All peaks corresponding to FCC and HCP phases were taken into account, while the $NiFe_2O_4$ peaks are excluded since the contribution of crystalline oxides in the XRD results is significantly low compared to the other phases. The NPs in dried acetone consist of 35.2 ± 1.0 wt% of the HCP phase, while the NPs in acetone account for 38.4 ± 0.2 wt% of the HCP phase, and the NPs in water contain no HCP phase (Table 1). It is interesting to observe that the ablation in acetone and dried acetone produces a mixture of the metastable HCP and stable FCC phases in the NP core, while only the FCC phase was formed in water. Meta- and stable phase mixtures in NPs produced by PLAL of different targets were previously reported. The formation of metastable zinc-blende and the stable diamond structures in silicon NPs [37], metastable hexagonal and stable cubic structures in diamond nanocrystals [38], the metastable γ-Fe and stable cubic FeO and α-Fe phases [39], and Ni NPs with stable FCC and metastable HCP phases [40,41]. Different arguments were postulated regarding the formation of stable and metastable phases during PLAL. (i) The specific heat capacity of the solvent, which influences the cooling rate of the ablation plasma plume generated during PLAL [40]. (ii) Shorter cavitation bubble lifetime compared to the theoretical lifetime according to the Rayleigh–Plesset theory [41], and (iii) the confinement of the cavitation bubble by the surrounding liquid, which induces the high temperature, high pressure, and high density (HTHPHD) state and shorter quenching time of the plasma plume in the liquid [37]. The above-mentioned hypotheses all pointed to the freezing of the metastable phase during the cooling (quenching), which preserves the metastable phases.

For FeNi alloy, the formation of a metastable FeNi HCP phase is usually associated with high-pressure and high-temperature conditions, such as in the Earth's core [42–46]. It has been produced synthetically using a diamond anvil cell (DAC) from the bulk FeNi with a face-centered cubic (FCC) phase [36,46,47], where the sample is placed in a tiny space (3–4 mm) between two diamonds, which are pressed to each other [48]. Laser ablation in liquid also provides a high-pressure and high-temperature state to the nuclei inside the cavitation bubble (CB) and its collapse [49]. The bubble pressure during laser ablation might provide a suitable environment for the formation of the HCP phase; however, this cannot be the sole reason since the ablation in water does not produce HCP phases. The cavitation bubble dynamic study from the Choi group also showed that the cavitation bubble size was larger for hexane and acetonitrile compared to water [41]. The larger cavitation bubble and longer lifetime indicate lower pressure inside the bubble, as formerly reported from the laser ablation of aluminum oxide in ethanol, water, and isopropanol [50]. Hence, the pressure difference due to the cavitation bubble geometry would favor the HCP formation in water; however, it is only observed in organic solvents. Consequently, the liquid composition seems to be a significant factor influencing the FeNi NP's phase.

Based on the results in Table 1, we have observed that the HCP phase does not scale with the fraction of water content in the organic solvent, consequently, this factor can be ruled out. However, the fact that the ablation in acetone (and dried acetone) produced an HCP phase, while the ablation in water only provided the FCC phase, suggests that the carbon-based solvent plays a significant role in the HCP phase formation. During PLAL, the presence of carbon species in the cavitation bubble generated from the interaction of the high-intensity laser with the organic solvent can influence the nucleation kinetics of the HCP and FCC phases. Hence, not only the FCC phase forms but also the HCP phase. When the cavitation bubble finally collapses, the fast temperature quenching freezes this metastable phase. Nevertheless, many factors related to the liquid and the laser ablation dynamics might form a complex system that contributes to the formation of the HCP phase in the FeNi NPs.

3.2. Oxide Formation

Oxidation of NPs, either partially or completely, changes the NP properties such as catalytic activity [51] and magnetization [52]. Controlling the oxidation level of laser-generated NPs is therefore important to produce NPs with the desired functionality. In this section, the influence of water impurity in acetone on the oxidation of laser-generated FeNi NPs is studied. Based on the XRD results (Figure 2), formations of minor amounts of spinel iron-nickel oxide $NiFe_2O_4$ (ICSD No. 241661) are observed in all studied samples, which shows that oxidation occurs even in dried acetone where most of the water molecules are captured by molecular sieves. Nevertheless, the amount of crystalline oxides in all of the samples is significantly low, approximately 0.7 wt% for FeNi in water, while for FeNi synthesized in acetone and dried acetone, the quantities are lower than the quantification error of the measurement/device, hence, the values are not of significance. Based on the study by Marzun et al., the ablation of a Cu target in water with an inert Ar atmosphere still resulted in oxidized species, due to the splitting of water molecules to reactive OH species. To avoid water impurities in acetone, we used molecular sieves. It was formerly reported that using the molecular sieve with the size of 4 Å for 21 h reduced the water content from 0.45% to 0.001% (*w*/*w*) [31]. Meanwhile, the water molecule has a diameter of 2.8 Å, hence, molecular sieves with a pore size of 3 Å were used to capture the water molecules in acetone and produce the "dried acetone". Nadarajah et al. have investigated the influence of 3 Å molecular sieves to capture water molecules in acetone and reduce the oxidation level of the laser-ablated FeRh NPs. They reported that the use of molecular sieves resulted in less nanoparticle oxidation compared to NPs produced in untreated acetone [53] and they suggest that the bound oxygen atoms in acetone contribute to NP oxidation. The dissolved oxygen gas in the liquid is also found to partially oxidize NPs due to aging [28], which means that the oxidation occurs also due to the possibly prolonged NPs storage in the liquid before the analysis. Hence, the surface oxidation of FeNi NPs into spinel $NiFe_2O_4$ was likely caused by the NPs' exposure to the oxygen species generated from the pyrolysis of the ablation liquid and later followed by the dissolved gas due to aging.

3.3. Morphology and Particle Size Distribution

The morphologies of the NPs ablated in dried acetone, water, and acetone are presented in Figure 3. Based on the bright field images of NPs in dried acetone (Figure 3a–c), core-shell structures with a core and two layers are formed, independently of the particle size. The thickness of the first layer (inner shell) ranges from 1.5 to 2.9 nm and has an average of 1.9 nm, whereas the average thickness of the second layer (outmost shell) was measured to be 2.4 nm, with a size range of 1.1–4.2 nm (Table 2). The core part shows a darker contrast in comparison to the shell, which can be explained as the change of electron scattering due to the electron density. The high electron density of the core part can be associated with the high material density, and in our case, it is $Fe_{50}Ni_{50}$ with a density of approximately 8.4 g/cm^3. For the inner shell, the formation of a carbide or oxide layer is likely, as the ablation was performed in a solvent with molecularly bound carbon and oxygen atoms. The density of iron and nickel carbide are approximately 4.93 and 7.99 g/cm^3, respectively, while iron, nickel, and iron-nickel oxide densities are between 5–7 g/cm^3, which explains the lower contrast of the inner shell compared to the core. The formation of iron and nickel carbides and oxides after the ablation of $Ni_{50}Fe_{50}$ in acetone was previously reported, but there were still unidentified peaks around 52°, 71°, and between 75–90° despite efforts from the authors [33], which are identified as FCC and HCP peaks of FeNi in this study (Figure 2). XRD results in Figure 2 and the lattice distance observed in Figure 3c confirm that the inner shell of this sample is constituted by spinel iron-nickel oxide ($NiFe_2O_4$). Meanwhile, the outmost layer with the brightest contrast can be attributed to a carbon layer, which was formed due to the pyrolysis of organic solvent by the high-intensity pulses [27]. The laser radiation pyrolyzes the organic solvent and yields carbon species [54], which then become the building block of the outer NP layer. A small part of graphitic carbon is observed in this sample (Figure S4), but most of the observed carbon layers are amorphous.

Table 2. FeNi NPs in dried acetone, acetone, and water shell thicknesses as obtained by TEM.

Ablation Liquid	Average Particle Size (x_c, nm)	Core Phase	Shell Phase	Shell Thickness (nm) Average (Mean)	Shell Thickness (nm) Range (Min to Max)
Dried acetone	10.2 ± 0.3	HCP/FCC FeNi	$NiFe_2O_4$	2.4	1.1–4.2
			Amorphous carbon	1.9	1.5–2.9
Acetone	12.0 ± 0.2	HCP/FCC FeNi	$NiFe_2O_4$	2.3	1.4–3.5
			Graphitic carbon	1.2	0.7–1.9
Water	17.7 ± 0.6	FCC FeNi	$NiFe_2O_4$	4.9	2.4–9.8

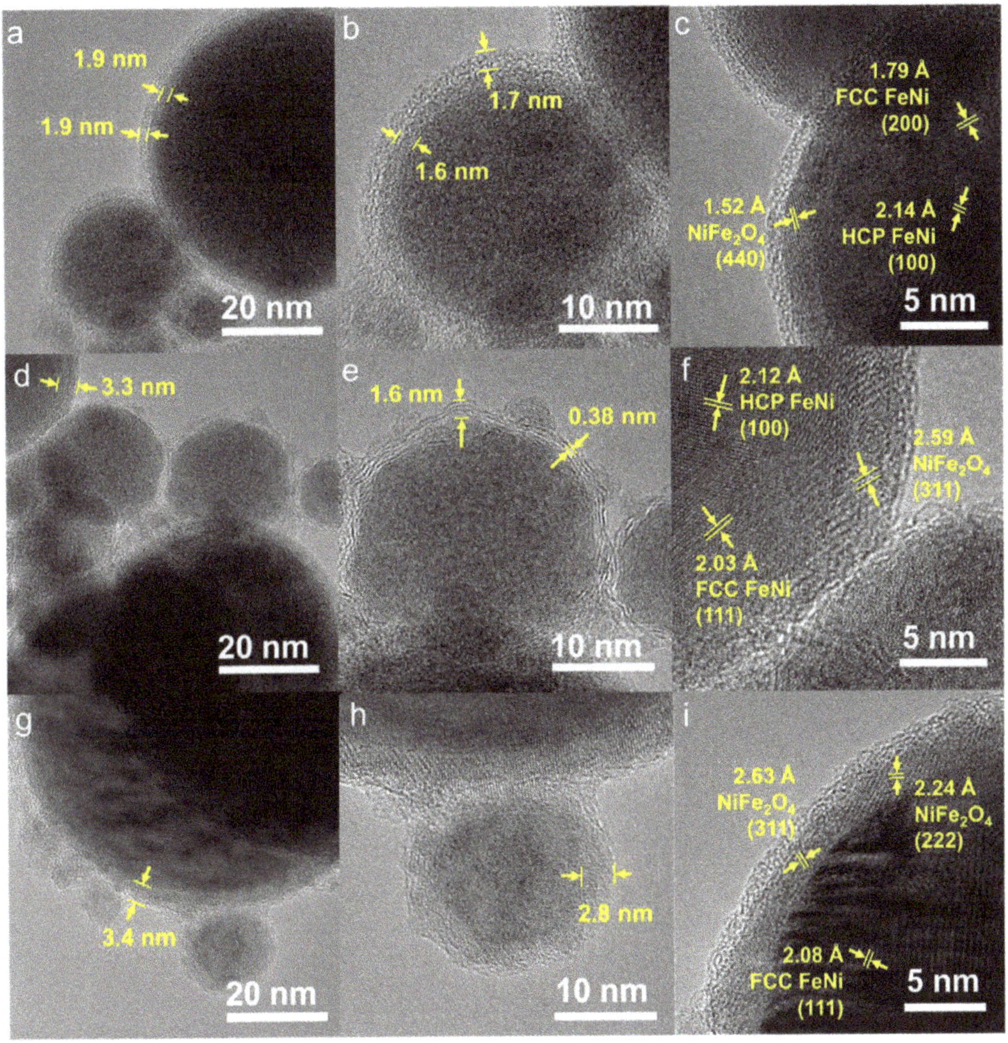

Figure 3. Morphology of FeNi NPs ablated in dried acetone (**a**–**c**), acetone (**d**–**f**), and water (**g**–**i**). (**a**,**d**,**g**) show the core-shell structure and the shell thickness of larger NPs, while (**b**,**e**,**h**) represent the smaller NPs. (**c**,**f**,**i**) confirm the phases observed in the XRD by measuring the lattice distance between the core and the shell.

Contrarily to the FeNi NPs in dried acetone, which exhibit the same core-shell structure for both small and large nanoparticle sizes, the sample in acetone (Figure 3) has two types of core-shell structures. Large NPs (d > 50 nm) form a core-shell structure, and the small NPs (d~20 nm) lean towards the formation of a core with outer graphitic carbon layers. The formation of a graphitic carbon layer was formerly reported after PLAL of metal targets in organic solvents, where the metal acts as a catalyst for the graphitization of the pyrolyzed carbon-based solvent [27,33]. Regarding the ablation of FeNi NPs in water, the formation of a core and a single shell structure for all NPs sizes was found. The formation of a single layer (without the carbon layer) is expected as water decomposes to H_2 and O_2 [54]. Based on the standard reduction potential, O_2 acts as an oxidizing agent in the reaction with Fe and Ni, hence, the shell is most likely to be composed of oxides as supported by the XRD data (Figure 2).

The particle size distribution of each sample was measured for at least 400 particles (Figure 4). All the histograms of the particle size distribution fit the log-normal distribution, which is common in PLAL-produced NPs. Meanwhile, the NPs produced through chemical synthesis methods, such as coprecipitation, hydrothermal, and sol-gel methods, usually have a Gaussian-type size distribution [18,55–57]. The average particle size of the sample is defined based on the center value of the log-normal fitting curve (x_c) and the polydispersity index (PDI) is calculated from the square of standard deviation divided by the square of the mean value (σ^2/μ^2). The PDI is used to define whether the NPs are monodisperse or polydisperse, where a value of less than 0.3 is considered monodisperse [58]. The NPs size in dried acetone shows the lowest average particle size (x_c) of 10.2 $\pm$ 0.3 nm, followed by NPs in acetone (12.0 $\pm$ 0.2 nm), and NPs in water (17.7 $\pm$ 0.6 nm). The PDI values of NPs in dried acetone, acetone and water are found to be 0.28, 0.28, and 0.91, respectively. Based on these results, the FeNi NPs in dried acetone and acetone can be considered monodisperse, while the FeNi NPs produced in water are polydisperse. The FeNi NPs in dried acetone and acetone are smaller than the FeNi in water due to the carbon coating on the NPs surface, which prevent the growth and coalescence of the NPs [59]. Nevertheless, it should be noted that further growth during storage cannot be completely ruled out even with carbon coating [60].

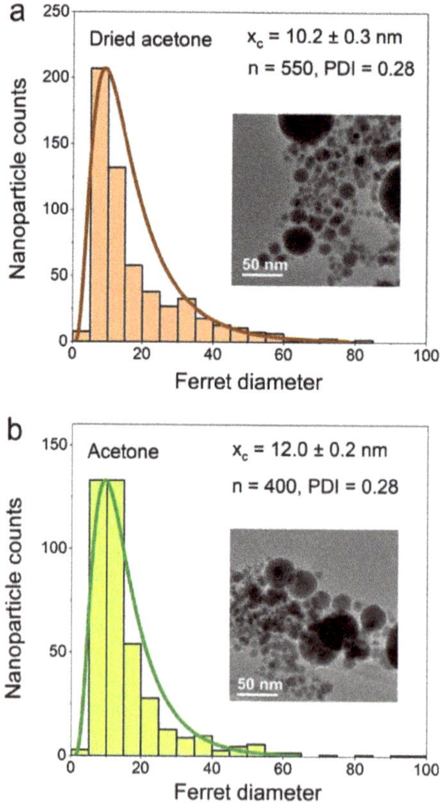

Figure 4. *Cont.*

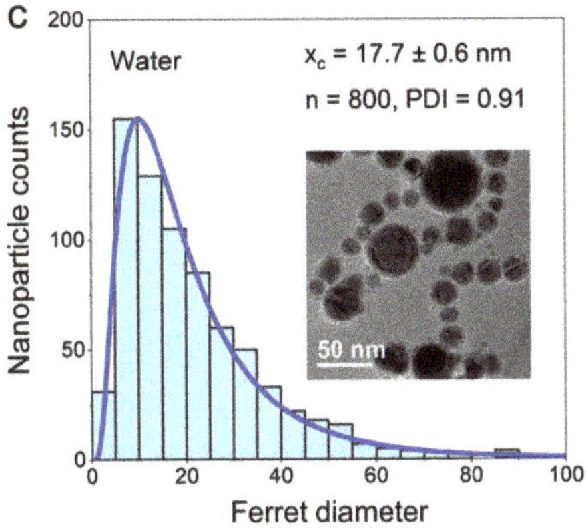

Figure 4. Number-weighted particle size distribution of FeNi NPs in (**a**) dried acetone, (**b**) acetone, and (**c**) water. FeNi NPs produced in dried acetone have the smallest median size and PDI, followed by FeNi produced in acetone, and FeNi produced in water. The number of counted particles (n) is denoted in the figures.

3.4. Elemental Composition

To determine the elemental composition of the NPs' core and shell, elemental scans using EDX-TEM were performed (Figure 5). A particle size of around 50 nm was selected as representative since the NPs generally have a distinct core-shell structure. Smaller NPs (10–20 nm) show distinct core-shell structures as well, but the oxygen signal from the environment sometimes provides a stronger contribution than the actual oxygen level on the NPs, hindering the oxidation analysis of the shell part (Figure S5). Hence, the discussion related to the EDX line scanning is limited to the larger NPs with a diameter of approximately 50 nm.

The EDX line scans (Figure 5) show that the Fe intensity is generally higher than the Ni intensity on the nanoparticle's surface. This signal difference between Ni and Fe represents the composition of the shell, where Fe is present in a higher percentage compared to Ni. The oxygen intensity in all samples increases from the start of the shell where Fe is detected, then the value is constant throughout the particle. This shows that oxidation only occurs on the surface of the particle, but not in the core, where the $Fe_{50}Ni_{50}$ composition of the initial target is preserved. By assuming that all O atomic % (at%) belongs to the shell with a composition of $NiFe_2O_4$, the approximate Fe and Ni at% in the core part were calculated, as shown in Table 3. Note that the drop-casting was not performed in a glovebox; hence, it is likely that some oxygen adsorbs to the grid during the sample preparation prior to the TEM analysis. In addition, the accuracy of the device is around 1 at%, which might influence the estimation of the core composition. The at% of Fe and Ni in the core part of dried acetone, acetone, and water samples show almost similar values with a difference of around 1–3 at%, which means that the bulk composition is maintained. Jakobi et al. argued that similar heat of evaporation and density of Pt and Ir during the ablation of $Pt_{91}Ir_9$ in acetone produced NPs with similar stoichiometry as the target material [61]. The heat of evaporation of Ni and Fe are 370.4 kJ/mol and 349.6 kJ/mol, while the densities are 8.9 g/cm^3 and 7.9 g/cm^3, respectively. These similar values of heat of evaporation and density (1.06 and 1.13-factor difference, respectively) induce the simultaneous evaporation and condensation of the FeNi NPs alloy, which preserves the target's elemental ratio. However, the oxidation level of the sample in dried acetone showed an unexpectedly high O at% value, which is even higher than water and twice the value of the sample in acetone. We have also measured the elemental composition using the EDX map scanning, which represents a larger area covering a larger number of NPs and also different NP sizes. As shown in Figure S6 and Table S1, the O at% of the dried acetone sample is the lowest, with a 15 at% difference compared to the acetone sample. There is also an anomalous trend where O at% of the water sample is slightly lower by almost 3 at% compared to the acetone sample. Therefore, we believe that the measurement of O at% from EDX-TEM fails to provide a complete representative value for the whole sample and includes the contribution of all the NP sizes, leading to a variation of the O at% values obtained for different NPs or analyzed areas. Thus, we sought another measurement, i.e., Mössbauer spectroscopy, to define the oxidation level of the whole sample with higher statistical confidence.

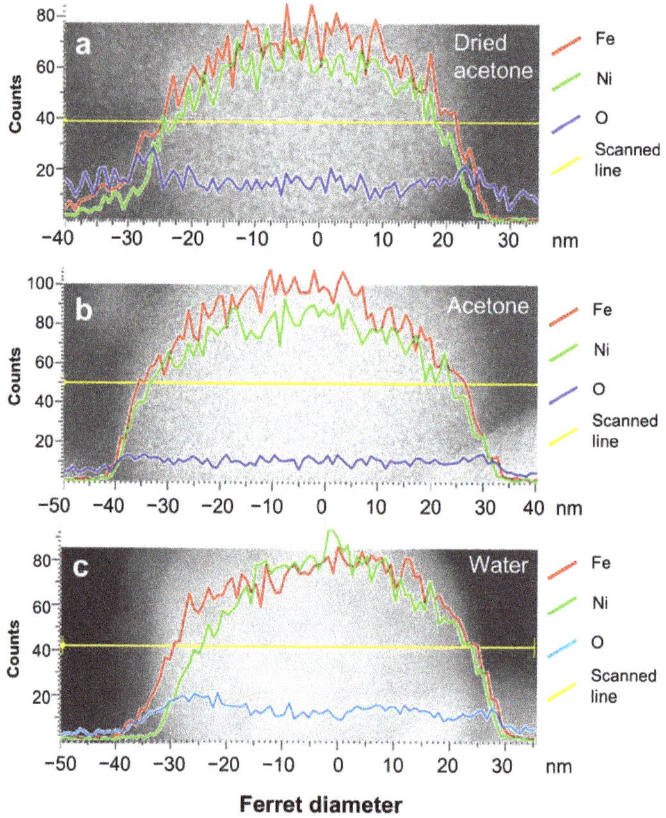

Figure 5. EDX-line scans of FeNi NPs ablated in (**a**) dried acetone, (**b**) acetone, and (**c**) water. The constant oxygen signals, which do not follow the Fe and Ni signals, indicate that oxidation only occurs on the NP surface.

Table 3. Elemental composition (in at%) of the NPs shown in Figure 5 and the estimation of Fe and Ni at% in the core part, assuming that all oxygen at% belongs to the $NiFe_2O_4$ shell.

Ablation Liquid	Whole Particle Composition			Shell Composition *			Core Composition **	
	Fe at%	Ni at%	O at%	Ni at%	Fe_2 at%	O_4 at%	Fe at%	Ni at%
Dried acetone	35.1	29.3	35.6	8.9	17.8	35.6	17.3	20.4
Acetone	43.6	38.1	18.2	4.6	9.1	18.2	34.5	33.6
Water	38.7	33.6	27.7	6.9	13.9	27.7	24.9	26.7

* with the assumption that all O at% of the particle comes from the $NiFe_2O_4$ shell, ** subtracting the whole particle composition with the shell composition.

Mössbauer spectroscopy was employed to quantify the total oxide fraction of the FeNi NPs and their aging behavior for longer oxidation times. The measurements were performed in transmission geometry, providing a measurement signal averaged over the total sample volume, thus, giving a comprehensive overview of the composition of Fe-bearing phases in the nanoparticles as well as their general magnetic structure. Due to different hyperfine interactions of the Fe nuclei with their surroundings, metallic and oxidic Fe-bearing phases result in distinctively different sub-spectra, as visible in Figure 6a. At ca. 4.3 K, two broadened sextet distributions can be identified for the aged, dried acetone sample: a larger one with moderate hyperfine magnetic fields B_{hf} and an average isomer shift of ca. 0.30 mm/s (green) assigned to metallic FeNi, and a second one with a larger B_{hf} and an isomer shift of ca. 0.47 mm/s. The latter is usually indicative of ferric oxides [62,63], whereby this distribution is assigned to iron atoms in the $NiFe_2O_4$ shell. Due to the very broad structure of the metallic FeNi subspectrum, a resolution of HCP- and FCC-components was not feasible.

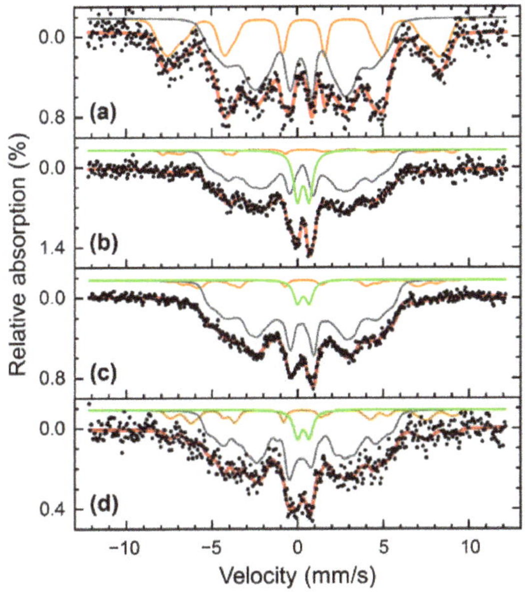

Figure 6. Mössbauer spectra of FeNi NP colloids: Prepared from dried acetone, aged for four months measured at 4.3 K (**a**) and 80 K (**b**), from dried acetone in the initial state at 80 K (**c**) and acetone in the initial state at 80 K (**d**). Spectra consist of an outer sextet distribution (orange) assigned to $NiFe_2O_4$, an inner sextet distribution corresponding to metallic FeNi (grey), and a doublet contribution (green) assigned to oxide material in the para- or superparamagnetic state.

Studying the spectrum at 80 K in Figure 6b in comparison, we observe only minor changes in the metallic FeNi subspectrum, while the oxidic sextet almost vanished, now manifesting mainly in a para- or superparamagnetic doublet state (green), both being mainly identical in spectral intensity and isomer shift. A more detailed analysis can be found in Figure S7, showing the dried acetone FeNi nanopowder spectra between 5 K and room temperature without an external magnetic field. This was done to study whether complete evaporation of the liquid to produce a powdered sample resulted in an oxidation increase due to the exposure to air, and the possibility of storing the colloids as a powder without influencing the oxidation level.

From the spectrum at 5 K, where the sub-spectra can be well resolved, 26% of the spectral area is assigned to the oxide fraction, which would suggest that further oxidation of this sample takes place following drying and storage before the measurement was completed. This proves our earlier point, that it is important to use freshly produced colloids without extended storage time, either in their original liquid or as dried NP powder. To further reduce the oxidation level, it is also possible to use an organic solvent with no molecularly bound oxygen, such as acetonitrile, or H_2 as a reducing gas. However, the reduced price and the reusability of the molecular sieves employed in this work, which could be re-activated by heat treatment at around 300 °C, offer a beneficial option for the oxidation control of PLAL-generated NPs and the cost-effective upscaling of the production as required for catalysis applications. At higher temperatures, it is found that the sextet to doublet transition of the $NiFe_2O_4$ component mainly takes place between 30 and 60 K. No considerable changes in the spectral structure are visible above ca. 100 K. Corresponding measurements up to room temperature were not attainable for the colloidal samples since Brownian nanoparticle motion leads to severe line broadening, hindering a detailed analysis [64,65].

For the dried acetone colloid 4 months after production shown in Figure 6a,b, the $NiFe_2O_4$ sub-spectra contains roughly 27% of the spectral area. Assuming the oxide shell consists of stoichiometric $NiFe_2O_4$ based on the previous XRD results (Figure 2) and expecting similar Debye-Waller factors for metallic FeNi and $NiFe_2O_4$ at cryogenic temperatures, relative spectral areas represent a simple approximation of the weight percentage (wt%) of the corresponding phase due to very similar atomic Fe fractions per mass. To evaluate the effect of reducing water content on the total oxide fraction as well as the stability of the prepared nanoparticles, the oxide spectral area in aged, and dried acetone colloid is compared to fresh dried acetone (14%) and fresh acetone colloid (22%) shown in Figure 6c,d. The results clearly show a lower oxide fraction after preparation in mole-sieved acetone and minor ongoing oxidation upon a longer aging time. It can be concluded that while the drying process is effective, the reduced oxidation of the sample is lost again after extended storage time, and results in a similar oxide fraction as the fresh colloid made from the commercial, untreated acetone. This also means that the carbon shell and the $NiFe_2O_4$ shell on the NPs did not completely stop further oxidation of the NPs during longer storage time. Oxidation might occur from the presence of molecularly bound O atoms in acetone or the dissolved O_2 gas. Therefore, it is important to use fresh colloids in the posterior intended catalysis or magnetic application to avoid further oxidation that can detriment the produced FeNi NPs performance.

3.5. Magnetic Properties

The magnetic field-dependent magnetization M(H) curves of FeNi NPs formed in different liquids are shown in Figure 7, recorded at 300 K up to a maximum magnetic field of 1 T. A similar saturation alignment for the three samples can be observed, with the overall character of the M(H) curves being comparable, reaching high magnetization values already at ca. 0.4 T and showing a gradual further increase in the high-field region. Based on Mössbauer spectroscopy in-field experiments as shown in Figure S8, this M(H) shape can be explained as follows: A distinctively reduced intensity of lines 2 and 5 of the FeNi subspectrum can be observed already at a magnetic field of 1 T visible in Figure S8b, revealing a state of almost complete magnetic alignment for the metallic core of the nanoparticles [66]. The $NiFe_2O_4$ shell, on the other hand, displays high intensities of lines 2 and 5 even up to 8 T (Figure S8c), proving that magnetic moments here are still relatively random in their orientation, resulting in a limited oxide contribution to magnetization, slowly increasing when going to higher fields. The incomplete magnetic alignment of the oxide shell is also clearly evident by the only partial resolution of the contributions from A- and B-spinel lattice positions at 8 T.

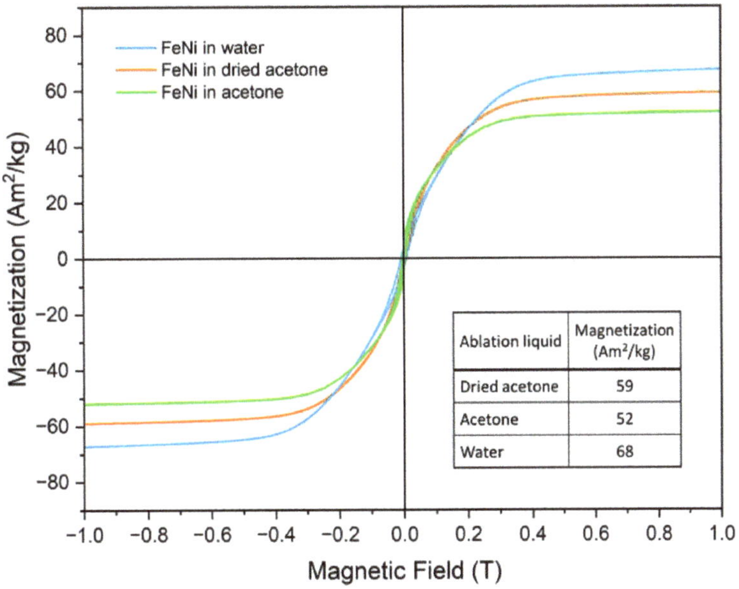

Ablation liquid	Magnetization (Am²/kg)
Dried acetone	59
Acetone	52
Water	68

Figure 7. M(H) curves of FeNi nanoparticle powders from colloids prepared in different liquids at room temperature (300 K). FeNi in acetone data reproduced from [67].

Clear differences are apparent when regarding the 1 T magnetization values. When comparing the acetone and dried acetone samples, the effect of the drying process becomes clear, as the mole-sieved sample has a higher magnetization of ca. 59 Am^2/kg compared to the 52 Am^2/kg of the non-sieved sample, which can presumably be attributed to the lower oxidation of the former. However, the sample produced in water shows an even higher magnetization at 68 Am^2/kg. It would be prudent to assume that this difference stems from the fact that the particle size of the water-based sample is significantly higher than that of the two acetone-based ones, which would lead to a lower surface-to-volume ratio and thus a reduced amount of surface spin canting. To discern this difference, additional magnetometry measurements were performed, up to higher fields of 9 T and in a wide range of temperatures from 4.3 K up to 300 K, as shown in Figure S9 for the dried acetone and water-based samples. Here, two aspects can be discussed: on the one hand, the low temperature, and high field measurements show that the water-based sample still retains a slightly higher magnetization value at 9 T of 82 Am^2/kg compared to 76 Am^2/kg for the dried acetone sample. Interestingly, the slightly more pronounced shape of the M(H) curves for the water-based sample indicates that full saturation has not yet been reached at 9 T, which would suggest that the higher magnetization value compared to the acetone-based samples is not solely due to a reduced occurrence of spin canting due to the larger average particle size. An explanation can be provided by the paramagnetic HCP phase being present in the acetone-based samples, but not in the water-based one, leading to a reduction of the overall measured magnetization. Additionally, the previously mentioned carbon shell formation can also contribute to this effect. However, despite this slight decrease relative to the water-based sample, the difference in magnetization visible between the sample formed in dried and regular acetone clearly shows the viability of the method presented here to prevent undesired oxidation of the FeNi NPs.

4. Conclusions

Reducing water impurities in acetone for the generation of $Fe_{50}Ni_{50}$ nanoalloys by PLAL influences the phases, core-shell structure, oxidation, and magnetic property of the produced NPs. FeNi NPs in dried acetone with reduced water impurity exhibit FCC and HCP phases in the core, an inner $NiFe_2O_4$ shell, and an outer amorphous carbon shell (FCC/HCP FeNi@$NiFe_2O_4$@amorphous carbon). The NPs in commercial, untreated acetone (water impurity of 0.3–0.5%) produced a mixture of FCC and HCP phases in the core with either $NiFe_2O_4$ shell or graphitic carbon (FCC/HCP FeNi@$NiFe_2O_4$ and FCC/HCP FeNi@graphitic carbon). Meanwhile, ablating FeNi alloy in water produced FCC core and $NiFe_2O_4$ shell (FCC FeNi@ $NiFe_2O_4$) NPs without any traces of the HCP phase or carbon shell. Reducing water impurity in acetone was found to lower the oxidation level by 8% (total oxide fraction, as measured by Mössbauer spectroscopy) compared to the NPs in untreated acetone. The magnetization of the dried acetone sample (59 Am^2/kg) was higher than the acetone sample (52 Am^2/kg) due to the lower degree of oxidation. The NPs produced in water exhibit a higher magnetization of 68 Am^2/kg. The higher magnetization in the water sample is due to the larger average NPs size (17 nm), compared to the NPs in dried acetone (10 nm) and acetone (12 nm). The smaller average size of NPs in acetone-based liquid is related to the carbon layer formed in the ablation plume, which constrains the particle growth. The diverse core-shell structure and the modified FeNi NPs properties observed in this study show that FeNi NPs with different phase and shell structures can be generated just by reducing the amount of water impurity in the organic solvent or modifying the solvent employed in PLAL. This opens up a straightforward synthesis approach of different core-shell FeNi NPs that can be adapted to the broad fields where FeNi NPs are applied, such as sensors and actuators development, catalysis, magnetism, or biomedicine.

Supplementary Materials: The following supporting information can be downloaded at: https://www.mdpi.com/article/10.3390/nano13020227/s1, Figure S1: The normalized absorption spectra of Fe50Ni50 nanoparticles generated in dried acetone (black), acetone (red), and water (blue) within the wavelength of 400–1100 nm; Figure S2: Phase identification of FeNi NPs in dried acetone, acetone, and water based on the Synchrotron XRD results; Figure S3: Rietveld refinement of the synchrotron XRD results without the contribution of the oxide phase; Figure S4: HR-TEM image of NP in dried acetone which shows the formation of graphitic carbon; Figure S5: Line scanning EDX-TEM of small NP (d = 16 nm) of FeNi ablated in acetone (top) and in dried acetone (bottom); Figure S6: EDX map scanning of the FeNi NPs in different ablation liquids. Figure S7: Mössbauer spectra of the dried acetone colloid powder sample recorded between 5 K and room temperature. Figure S8: Mössbauer spectra of an aged dried acetone powder sample recorded at ca. 4.3 K in external magnetic fields of (a) 0 T, (b) 1 T, and (c) 8 T parallel to γ-ray incidence direction. Figure S9: Field-dependent magnetization of FeNi nanoparticle powder from the (a) dried acetone and (b) the water-based sample recorded at 4.3 K to 300 K in magnetic fields up to 9 T. Table S1: The elemental composition of the whole area (in at%) obtained by map scanning as shown in Figure S5 and the estimation of Fe and Ni at% in the core part, assuming that all O at% belongs to the $NiFe_2O_4$ shell. Reference [68] is cited in the supplementary materials.

Author Contributions: Conceptualization, I.Y.K., C.D.-B., and B.G.; methodology, I.Y.K., C.D.-B., and B.G.; software, I.Y.K., C.D.-B., Q.L., J.L., S.S., E.K.; validation, C.D.-B., B.G., G.Z..; formal analysis, I.Y.K., C.D.-B., Q.L., J.L., S.S., E.K.; investigation, I.Y.K., C.D.-B., Q.L., J.L., S.S., E.K.; resources, B.G., H.W., G.Z., E.K.; data curation, I.Y.K., Q.L., J.L., S.S.; writing—original draft preparation, I.Y.K.; writing—review and editing, I.Y.K., C.D.-B., B.G., Q.L., J.L., S.S.; visualization, I.Y.K., Q.L., J.L., S.S., C.D.-B.; supervision, C.D.-B., B.G.; project administration, B.G., G.Z., H.W.; funding acquisition, B.G., G.Z., H.W. All authors have read and agreed to the published version of the manuscript.

Funding: Financial funding from the European Union's Horizon 2020 research and innovation program under the grant agreement No 952068 (project LESGO) and funding from the DFG via the CRC/TRR 270 (Project-ID 405553726, projects B08 and B05), the CRC/TRR 247 (Project-ID 388390466, project B02) and GO 2566/10-1-(Project-ID 445127149) are gratefully acknowledged. The synchrotron X-ray diffraction experiments were performed at beamline 33-BM-C at the Advanced Photon Source (APS) at Argonne National Laboratory (ANL). This research used resources of the Advanced Photon Source, a U.S. Department of Energy (DOE) Office of Science user facility operated for the DOE Office of Science by Argonne National Laboratory under Contract No. DE-AC02-06CH11357. Q. Lin and G.

Zangari would like to thank the financial support from the Department of Energy (DOE), Office of Basic Energy Sciences (BES), United States (Grant number: DE-SC0019191).

Data Availability Statement: Besides the data published in this article, no new data were created.

Acknowledgments: The authors acknowledge support from the Open Access Publication Fund of the University of Wuppertal. We would like to thank Markus Heidelmann from ICAN, University of Duisburg-Essen, for the TEM measurements, and Karsten Albe, TU Darmstadt, for the valuable discussion.

Conflicts of Interest: The authors declare no conflict of interest.

References

1. Frey, P.A.; Reed, G.H. The ubiquity of iron. *ACS Chem. Biol.* **2012**, *7*, 1477–1481. [CrossRef]
2. Cui, J.; Kramer, M.; Zhou, L.; Liu, F.; Gabay, A.; Hadjipanayis, G.; Balasubramanian, B.; Sellmyer, D. Current progress and future challenges in rare-earth-free permanent magnets. *Acta Mater.* **2018**, *158*, 118–137. [CrossRef]
3. van Schilfgaarde, M.; Abrikosov, I.A.; Johansson, B. Origin of the Invar effect in iron–nickel alloys. *Nature* **1999**, *400*, 46–49. [CrossRef]
4. Shinjo, T.; Okuno, T.; Hassdorf, R.; Shigeto, K.; Ono, T. Magnetic vortex core observation in circular dots of permalloy. *Science* **2000**, *289*, 930–932. [CrossRef] [PubMed]
5. Clarke, R.S.; Scott, E.R.D. Tetrataenite-ordered FeNi, a new mineral in meteorites. *Am. Mineral.* **1980**, *65*, 624–630.
6. Spooner, T. Current Transformers with Nickel-Iron Cores. *Trans. Am. Inst. Electr. Eng.* **1926**, *45*, 701–707. [CrossRef]
7. Ahn, C.H.; Allen, M.G. Micromachined planar inductors on silicon wafers for MEMS applications. *IEEE Trans. Ind. Electron.* **1998**, *45*, 866–876. [CrossRef]
8. Ahn, C.H.; Kim, Y.J.; Allen, M.G. A fully integrated planar toroidal inductor with a micromachined nickel-iron magnetic bar. *IEEE Trans. Compon. Packag. Manuf. Technol. Part A* **1994**, *17*, 463–469. [CrossRef]
9. Dijith, K.S.; Aiswarya, R.; Praveen, M.; Pillai, S.; Surendran, K.P. Polyol derived Ni and NiFe alloys for effective shielding of electromagnetic interference. *Mater. Chem. Front.* **2018**, *2*, 1829–1841. [CrossRef]
10. Giordano, M.C.; Escobar Steinvall, S.; Watanabe, S.; Fontcuberta i Morral, A.; Grundler, D. Ni 80 Fe 20 nanotubes with optimized spintronic functionalities prepared by atomic layer deposition. *Nanoscale* **2021**, *13*, 13451–13462. [CrossRef]
11. Bokare, A.D.; Chikate, R.C.; Rode, C.V.; Paknikar, K.M. Iron-nickel bimetallic nanoparticles for reductive degradation of azo dye Orange G in aqueous solution. *Appl. Catal. B Environ.* **2008**, *79*, 270–278. [CrossRef]
12. Gong, M.; Dai, H. A mini review of NiFe-based materials as highly active oxygen evolution reaction electrocatalysts. *Nano Res.* **2014**, *8*, 23–39. [CrossRef]
13. Zhang, X.; Xu, H.; Li, X.; Li, Y.; Yang, T.; Liang, Y. Facile Synthesis of Nickel-Iron/Nanocarbon Hybrids as Advanced Electrocatalysts for Efficient Water Splitting. *ACS Catal.* **2016**, *6*, 580–588. [CrossRef]
14. Potvin, E.; Brossard, L. Electrocatalytic activity of Ni-Fe anodes for alkaline water electrolysis. *Mater. Chem. Phys.* **1992**, *31*, 311–318. [CrossRef]
15. Lewis, L.H.; Mubarok, A.; Poirier, E.; Bordeaux, N.; Manchanda, P.; Kashyap, A.; Skomski, R.; Goldstein, J.; Pinkerton, F.E.; Mishra, R.K.; et al. Inspired by nature: Investigating tetrataenite for permanent magnet applications. *J. Phys. Condens. Matter* **2014**, *26*, 10. [CrossRef]
16. Lin, Q.; Nadarajah, R.; Hoglund, E.; Semisalova, A.; Howe, J.M.; Gökce, B.; Zangari, G. Towards synthetic L10-FeNi: Detecting the absence of cubic symmetry in Laser-Ablated Fe-Ni nanoparticles. *Appl. Surf. Sci.* **2021**, *567*, 150664. [CrossRef]
17. Chokprasombat, K.; Pinitsoontorn, S.; Maensiri, S. Effects of Ni content on nanocrystalline Fe–Co–Ni ternary alloys synthesized by a chemical reduction method. *J. Magn. Magn. Mater.* **2016**, *405*, 174–180. [CrossRef]
18. Lima, E.; Drago, V.; Bolsoni, R.; Fichtner, P.F.P. Nanostructured Fe50Ni50 alloy formed by chemical reduction. *Solid State Commun.* **2003**, *125*, 265–270. [CrossRef]
19. Anastas, P.; Eghbali, N. Green Chemistry: Principles and Practice. *Chem. Soc. Rev.* **2010**, *39*, 301–312. [CrossRef]
20. Gökce, B.; Amendola, V.; Barcikowski, S. Opportunities and Challenges for Laser Synthesis of Colloids. *ChemPhysChem* **2017**, *18*, 983–985. [CrossRef]
21. Amendola, V.; Amans, D.; Ishikawa, Y.; Koshizaki, N.; Scirè, S.; Compagnini, G.; Reichenberger, S.; Barcikowski, S. Room-Temperature Laser Synthesis in Liquid of Oxide, Metal-Oxide Core-Shells, and Doped Oxide Nanoparticles. *Chem. Eur. J.* **2020**, *26*, 9206–9242. [CrossRef] [PubMed]
22. Zhang, D.; Gökce, B.; Barcikowski, S. Laser Synthesis and Processing of Colloids: Fundamentals and Applications. *Chem. Rev.* **2017**, *117*, 3990–4103. [CrossRef] [PubMed]
23. Zhang, D.; Li, Z.; Sugioka, K. Laser ablation in liquids for nanomaterial synthesis: Diversities of targets and liquids. *J. Phys. Photonics* **2021**, *3*, 042002. [CrossRef]
24. Kalus, M.R.; Lanyumba, R.; Lorenzo-Parodi, N.; Jochmann, M.A.; Kerpen, K.; Hagemann, U.; Schmidt, T.C.; Barcikowski, S.; Gökce, B. Determining the role of redox-active materials during laser-induced water decomposition. *Phys. Chem. Chem. Phys.* **2019**, *21*, 18636–18651. [CrossRef]

25. Chaturvedi, A.; Joshi, M.P.; Mondal, P.; Sinha, A.K.; Srivastava, A.K. Growth of anatase and rutile phase TiO_2 nanoparticles using pulsed laser ablation in liquid: Influence of surfactant addition and ablation time variation. *Appl. Surf. Sci.* **2017**, *396*, 303–309. [CrossRef]
26. Zhang, D.; Lu, S.; Gökce, B.; Ma, Z.; Spasova, M.; Yelsukova, A.E.; Farle, M.; Wiedwald, U.; Zhang, D.; Ma, Z.; et al. Formation Mechanism of Laser Synthesized Iron-Manganese Alloy Nanoparticles, Manganese Oxide Nanosheets and Nanofibers. *Part. Part. Syst. Charact.* **2017**, *34*, 1600225. [CrossRef]
27. Zhang, D.; Zhang, C.; Liu, J.; Chen, Q.; Zhu, X.; Liang, C. Carbon-Encapsulated Metal/Metal Carbide/Metal Oxide Core-Shell Nanostructures Generated by Laser Ablation of Metals in Organic Solvents. *ACS Appl. Nano Mater.* **2019**, *2*, 28–39. [CrossRef]
28. Marzun, G.; Bönnemann, H.; Lehmann, C.; Spliethoff, B.; Weidenthaler, C.; Barcikowski, S. Role of Dissolved and Molecular Oxygen on Cu and PtCu Alloy Particle Structure during Laser Ablation Synthesis in Liquids. *ChemPhysChem* **2017**, *18*, 1175–1184. [CrossRef]
29. Sigma Aldrich. Acetone, ACS Reagent, ≥99.5%. Available online: https://www.sigmaaldrich.com/DE/de/product/sigald/179124 (accessed on 7 September 2022).
30. Sigma Aldrich. Acetone, Suitable for HPLC, ≥99.9%. Available online: https://www.sigmaaldrich.com/DE/de/product/sigald/270725 (accessed on 7 September 2022).
31. Meeker, R.L.; Critchfield, F.E.; Bishop, E.T. Water Determination by Near Infrared Spectrophotometry. *Anal. Chem.* **1962**, *34*, 1510–1511. [CrossRef]
32. Lin, R.; Ladshaw, A.; Nan, Y.; Liu, J.; Yiacoumi, S.; Tsouris, C.; DePaoli, D.W.; Tavlarides, L.L. Isotherms for Water Adsorption on Molecular Sieve 3A: Influence of Cation Composition. *Ind. Eng. Chem. Res.* **2015**, *54*, 10442–10448. [CrossRef]
33. Davodi, F.; Mühlhausen, E.; Settipani, D.; Rautama, E.L.; Honkanen, A.P.; Huotari, S.; Marzun, G.; Taskinen, P.; Kallio, T. Comprehensive study to design advanced metal-carbide@garaphene and metal-carbide@iron oxide nanoparticles with tunable structure by the laser ablation in liquid. *J. Colloid Interface Sci.* **2019**, *556*, 180–192. [CrossRef] [PubMed]
34. Li, G.; Yang, B.; Xu, X.; Cao, S.; Shi, Y.; Yan, Y.; Song, X.; Hao, C. FeNi Alloy Nanoparticles Encapsulated in Carbon Shells Supported on N-Doped Graphene-Like Carbon as Efficient and Stable Bifunctional Oxygen Electrocatalysts. *Chem. Eur. J.* **2020**, *26*, 2890–2896. [CrossRef] [PubMed]
35. Montes-Arango, A.M.; Marshall, L.G.; Fortes, A.D.; Bordeaux, N.C.; Langridge, S.; Barmak, K.; Lewis, L.H. Discovery of process-induced tetragonality in equiatomic ferromagnetic FeNi. *Acta Mater.* **2016**, *116*, 263–269. [CrossRef]
36. Komabayashi, T.; Hirose, K.; Ohishi, Y. In situ X-ray diffraction measurements of the fcc-hcp phase transition boundary of an Fe-Ni alloy in an internally heated diamond anvil cell. *Phys. Chem. Miner.* **2012**, *39*, 329–338. [CrossRef]
37. Liu, P.; Cao, Y.L.; Cui, H.; Chen, X.Y.; Yang, G.W. Micro- and nanocubes of silicon with zinc-blende structure. *Chem. Mater.* **2008**, *20*, 494–502. [CrossRef]
38. Wang, J.B.; Zhang, C.Y.; Zhong, X.L.; Yang, G.W. Cubic and hexagonal structures of diamond nanocrystals formed upon pulsed laser induced liquid-solid interfacial reaction. *Chem. Phys. Lett.* **2002**, *361*, 86–90. [CrossRef]
39. Patil, P.P.; Phase, D.M.; Kulkarni, S.A.; Ghaisas, S.V.; Kulkarni, S.K.; Kanetkar, S.M.; Ogale, S.B.; Bhide, V.G. Pulsed-laser—Induced reactive quenching at liquid-solid interface: Aqueous oxidation of iron. *Phys. Rev. Lett.* **1987**, *58*, 238–241. [CrossRef] [PubMed]
40. Jung, H.J.; Choi, M.Y. Specific solvent produces specific phase Ni nanoparticles: A pulsed laser ablation in solvents. *J. Phys. Chem. C* **2014**, *118*, 14647–14654. [CrossRef]
41. Lee, S.J.; Theerthagiri, J.; Choi, M.Y. Time-resolved dynamics of laser-induced cavitation bubbles during production of Ni nanoparticles via pulsed laser ablation in different solvents and their electrocatalytic activity for determination of toxic nitroaromatics. *Chem. Eng. J.* **2022**, *427*, 130970. [CrossRef]
42. Tateno, S.; Hirose, K.; Komabayashi, T.; Ozawa, H.; Ohishi, Y. The structure of Fe-Ni alloy in Earth's inner core. *Geophys. Res. Lett.* **2012**, *39*. [CrossRef]
43. Shen, G.; Mao, H.; Hemley, R.J.; Duffy, T.S.; Rivers, M.L. Melting and crystal structure of iron at high pressures and temperatures. *Geophys. Res. Lett.* **1998**, *25*, 373–376. [CrossRef]
44. Torchio, R.; Boccato, S.; Miozzi, F.; Rosa, A.D.; Ishimatsu, N.; Kantor, I.; Sévelin-Radiguet, N.; Briggs, R.; Meneghini, C.; Irifune, T.; et al. Melting Curve and Phase Relations of Fe-Ni Alloys: Implications for the Earth's Core Composition. *Geophys. Res. Lett.* **2020**, *47*, e2020GL088169. [CrossRef]
45. Lin, J.-F.; Heinz, D.L.; Campbell, A.J.; Devine, J.M.; Mao, W.L.; Shen, G. Iron-Nickel alloy in the Earth's core. *Geophys. Res. Lett.* **2002**, *29*, 109-1–109-3. [CrossRef]
46. Huang, E.; Bassett, W.A.; Weathers, M.S. Phase relationships in Fe-Ni alloys at high pressures and temperatures. *J. Geophys. Res.* **1988**, *93*, 7741. [CrossRef]
47. Kuwayama, Y.; Hirose, K.; Sata, N.; Ohishi, Y. Phase relations of iron and iron-nickel alloys up to 300 GPa: Implications for composition and structure of the Earth's inner core. *Earth Planet. Sci. Lett.* **2008**, *273*, 379–385. [CrossRef]
48. Boehler, R. Diamond cells and new materials. *Mater. Today* **2005**, *8*, 34–42. [CrossRef]
49. Soliman, W.; Nakano, T.; Takada, N.; Sasaki, K. Modification of Rayleigh-Plesset theory for reproducing dynamics of cavitation bubbles in liquid-phase laser ablation. *Jpn. J. Appl. Phys.* **2010**, *49*, 116202. [CrossRef]
50. Lam, J.; Lombard, J.; Dujardin, C.; Ledoux, G.; Merabia, S.; Amans, D. Dynamical study of bubble expansion following laser ablation in liquids. *Appl. Phys. Lett.* **2016**, *108*, 074104. [CrossRef]

51. Cuenya, B.R. Synthesis and catalytic properties of metal nanoparticles: Size, shape, support, composition, and oxidation state effects. *Thin Solid Films* **2010**, *518*, 3127–3150. [CrossRef]
52. Rebodos, R.L.; Vikesland, P.J. Effects of oxidation on the magnetization of nanoparticulate magnetite. *Langmuir* **2010**, *26*, 16745–16753. [CrossRef]
53. Nadarajah, R.; Tahir, S.; Landers, J.; Koch, D.; Semisalova, A.S.; Wiemeler, J.; El-Zoka, A.; Kim, S.H.; Utzat, D.; Möller, R.; et al. Controlling the oxidation of magnetic and electrically conductive solid-solution iron-rhodium nanoparticles synthesized by laser ablation in liquids. *Nanomaterials* **2020**, *10*, 2362. [CrossRef] [PubMed]
54. Kalus, M.R.; Bärsch, N.; Streubel, R.; Gökce, E.; Barcikowski, S.; Gökce, B. How persistent microbubbles shield nanoparticle productivity in laser synthesis of colloids—Quantification of their volume, dwell dynamics, and gas composition. *Phys. Chem. Chem. Phys.* **2017**, *19*, 7112–7123. [CrossRef] [PubMed]
55. Khairani, I.Y.; Arifiadi, A.N.; Lee, J.-H.; Bhoi, B.; Patel, S.K.S.; Kim, S. Fabrication, Structure, and Magnetic Properties of Pure-Phase $BiFeO_3$ and $MnFe_2O_4$ Nanoparticles and their Nanocomposites. *J. Magn.* **2020**, *25*, 140–149. [CrossRef]
56. Arifiadi, A.N.; Kim, K.-T.; Khairani, I.Y.; Park, C.B.; Kim, K.H.; Kim, S.-K. Synthesis and multiferroic properties of high-purity CoFe2O4–BiFeO3 nanocomposites. *J. Alloys Compd.* **2021**, *867*, 159008. [CrossRef]
57. Tao, K.; Dou, H.; Sun, K. Interfacial coprecipitation to prepare magnetite nanoparticles: Concentration and temperature dependence. *Colloids Surf. A Physicochem. Eng. Asp.* **2008**, *320*, 115–122. [CrossRef]
58. Barcikowski, S.; Amendola, V.; Lau, M.; Marzun, G.; Rehbock, C.; Reichenberger, S.; Zhang, D.; Gökce, B. *Handbook of Laser Synthesis & Processing of Colloids*, 2nd ed.; Duisburg-Essen Publication Online: Essen, Germany, 2019.
59. Amendola, V.; Riello, P.; Meneghetti, M. Magnetic nanoparticles of iron carbide, iron oxide, iron@iron oxide, and metal iron synthesized by laser ablation in organic solvents. *J. Phys. Chem. C* **2011**, *115*, 5140–5146. [CrossRef]
60. Zhang, D.; Choi, W.; Jakobi, J.; Kalus, M.R.; Barcikowski, S.; Cho, S.H.; Sugioka, K. Spontaneous shape alteration and size separation of surfactant-free silver particles synthesized by laser ablation in acetone during long-period storage. *Nanomaterials* **2018**, *8*, 529. [CrossRef]
61. Jakobi, J.; Menéndez-Manjón, A.; Chakravadhanula, V.S.K.; Kienle, L.; Wagener, P.; Barcikowski, S. Stoichiometry of alloy nanoparticles from laser ablation of PtIr in acetone and their electrophoretic deposition on PtIr electrodes. *Nanotechnology* **2011**, *22*, 145601. [CrossRef]
62. Cornell, R.M.; Schwertmann, U. *The Iron Oxides: Structure, Properties, Reactions, Occurrences and Uses*; Wiley Online Library, Wiley: Weinheim, Germany, 2003; ISBN 9783527302741.
63. Wareppam, B.; Kuzmann, E.; Garg, V.K.; Singh, L.H. Mössbauer spectroscopic investigations on iron oxides and modified nanostructures: A review. *J. Mater. Res.* **2022**. [CrossRef]
64. Keller, H.; Kündig, W. Mössbauer studies of Brownian motion. *Solid State Commun.* **1975**, *16*, 253–256. [CrossRef]
65. Landers, J.; Salamon, S.; Remmer, H.; Ludwig, F.; Wende, H. Simultaneous Study of Brownian and Néel Relaxation Phenomena in Ferrofluids by Mössbauer Spectroscopy. *Nano Lett.* **2016**, *16*, 1150–1155. [CrossRef] [PubMed]
66. Ammar, S.; Jouini, N.; Fiévet, F.; Beji, Z.; Smiri, L.; Moliné, P.; Danot, M.; Grenèche, J.M. Magnetic properties of zinc ferrite nanoparticles synthesized by hydrolysis in a polyol medium. *J. Phys. Condens. Matter* **2006**, *18*, 9055–9069. [CrossRef]
67. Nadarajah, R.; Tasdemir, L.; Thiel, C.; Salamon, S.; Semisalova, A.S.; Wende, H.; Farle, M.; Barcikowski, S.; Erni, D.; Gökce, B. Article formation of fe-ni nanoparticle strands in macroscopic polymer composites: Experiment and simulation. *Nanomaterials* **2021**, *11*, 2095. [CrossRef] [PubMed]
68. Toby, B.H.; Von Dreele, R.B. GSAS-II: The genesis of a modern open-source all purpose crystallography software package. *J. Appl. Crystallogr.* **2013**, *46*, 544–549. [CrossRef]

nanomaterials

MDPI

Article

A Versatile Route for Synthesis of Metal Nanoalloys by Discharges at the Interface of Two Immiscible Liquids

Ahmad Hamdan * and **Luc Stafford**

Département de Physique, Université de Montréal, 1375 Avenue Thérèse-Lavoie-Roux, Montreal, QC H2V 0B3, Canada

* Correspondence: ahmad.hamdan@umontreal.ca

Abstract: Discharge in liquid is a promising technique to produce nanomaterials by electrode erosion. Although its feasibility was demonstrated in many conditions, the production of nanoalloys by in-liquid discharges remains a challenge. Here, we show that spark discharge in liquid cyclohexane that is in contact with conductive solution, made of a combination of Ni-nitrate and/or Fe-nitrate and/or Co-nitrate, is suitable to produce nanoalloys (<10 nm) of Ni-Fe, Ni-Co, Co-Fe, and Ni-Co-Fe. The nanoparticles are synthesized by the reduction of metal ions during discharge, and they are individually embedded in C-matrix; this latter originates from the decomposition of cyclohexane. The results open novel ways to produce a wide spectrum of nanoalloys; they are needed for many applications, such as in catalysis, plasmonic, and energy conversion.

Keywords: nanosecond discharge; discharge in liquid; plasma-liquid interface; nanoalloys

check for updates

Citation: Hamdan, A.; Stafford, L. A Versatile Route for Synthesis of Metal Nanoalloys by Discharges at the Interface of Two Immiscible Liquids. *Nanomaterials* **2022**, *12*, 3603. https://doi.org/10.3390/nano12203603

Academic Editors: Andrei Honciuc and Mirela Teodorescu

Received: 19 September 2022
Accepted: 11 October 2022
Published: 14 October 2022

1. Introduction

Reducing the dimension of a material to the nanoscale often reveals properties unattainable at the macroscopic level. Such discovery has emerged novel field of research: synthesis of nanomaterials. Be it 0D (e.g., nanoparticles), 1D (e.g., nanowires), or 2D (e.g., nanosheets), the properties of nanomaterials can be linked to enhanced chemical reactivity, remarkable electron and thermal transport, quantum behavior, etc. [1–3]. In addition to the dimension, chemical composition and spatial distribution of the elements play a crucial role on the functionalities of nano-objects [4–6]. In the case of nanoparticles, for example, core–shell and bi- or tri-metallic nanoalloys offer promising properties for many applications, including electronic, photonic, catalysis, and plasmonic [7,8].

Over the last two decades, several physical, chemical, or biological methods were developed to produce nanoalloys [9,10]. However, simple, versatile, and environmentally friendly techniques providing non-agglomerated and well-controlled nanostructures remain scarce. Gas-phase plasmas are often used in production of nanomaterials, including thermal as well as non-thermal processes [11,12]. For example, plasma torch is a well-known atmospheric pressure technique to produce various types of nanomaterials by processing solid, liquid, or suspension precursors [13–15]. Laser ablation is also used to generate nanomaterials through ablation of solid targets under different gas pressure conditions [16,17]. Very recently, plasma-liquid systems were proposed as novel method to efficiently produce nanomaterials [18,19]. The plasma can be either generated directly in liquid or in gas in contact with liquid [20]. The nanomaterials produced using the former type of discharges are strongly affected by both plasma-electrode and plasma-liquid interactions. In-liquid plasmas exhibit non-conventional properties of temperature (several thousands of Kelvin), pressure (several tens of bars), and density of charged species (10^{17}–10^{19} cm^{-3}) over very short time scales (rise and decay time less than 1 μs) [21–24]. Moreover, the flexibility of in-liquid plasmas allows discharge ignition in different liquids (e.g., water, hydrocarbons, cryogenics, etc.) and between electrodes of varying chemical nature (e.g., Al,

Cu, Pt, Co, Ni, etc.) [25–28]. This has led to the production of a wide range of nanomaterials, including nanocomposites [27,29] and materials with novel crystallographic phases [30,31]. Although nanoalloys are searched for many applications, their synthesis is, however, not straightforward as it requires either multiple stage experiments [32] or the use of alloy electrodes [33,34].

Coupling a gas-phase plasma with a liquid that contain ions (supplied either by precursors or by metal electrode dissolution) was also proposed as efficient technique to produce nanomaterials through the reduction of the dissolved ions by the reactive species at the plasma-liquid interface. Using this approach, Velusamy et al. [35] synthesized CuO nanoparticles with a tailored energy-band diagram, Patel et al. [36] produced surfactant-free electrostatically stabilized gold nanoparticles, and Richmonds and Sankaran [37] synthesized Ag and Au nanoparticles. Recently, the feasibility of adding two kinds of ions (Eu and Ce) in solution was demonstrated to produce Eu doped Ce oxide nanoparticles by processing the mixture with the jet of an atmospheric pressure Ar plasma [38].

In a previous study, we introduced a novel plasma-liquid system, that combines the advantages of both in-liquid and in-contact with liquid systems. Indeed, discharges were sustained in liquid heptane that is in contact with a conductive solution (water + Ag-nitrate) [39]. This system successfully produced ultrasmall Ag nanoparticles embedded in a hydrocarbon network. The presence of carbon network is essential to prohibit the agglomeration between the nanoparticles and, therefore, to maintain their property as individual nanoparticles. Here, we examine the possibility to extent this technique for nanoalloys production. This is done by using solution with various kinds of metal salts (combination of Co-nitrate, Fe-nitrate, and Ni-nitrate), yielding non-agglomerated binary and ternary nanoalloys.

2. Materials and Methods

The experimental setup is schematically shown in Figure 1a. The discharge was ignited using a nanosecond positive pulsed generator (NSP 120-20-P-500-TG-H, Eagle Harbor Technologies, Seattle, WA, USA). The amplitude and width of the applied voltage were 22 kV and 500 ns, respectively. The discharge repetition rate was set to 10 Hz, and the duration of the experiment was 30 min. This low frequency has been chosen not to have correlation between two successive discharges through interaction between the bubble (induced by previous discharge) and the following discharge. Although this parameter has not been investigated here, we believe that it can be increased up to a few of kHz, which may significantly increase the yield of synthesis. The upper electrode, a carbon rod (99.99% pure; Goodfellow) that is mechanically polished to a curvature radius of $\sim$10 μm, was immersed in liquid cyclohexane, and the distance between its tip and the interface was kept at $\sim$1 mm. Meanwhile, the lower electrode, a carbon rod that is polished to a flat surface, was placed in the conductive solution at 3 mm below the interface. Note, the electrode erosion was insignificant. Current-voltage characteristics were recorded using a high-voltage probe (P6015A, Tektronix) for the applied voltage and a current coil (6585, Pearson) for the total current (discharge and displacement). Both probes were connected to an oscilloscope (DPO5420B, Tektronix) to record the corresponding voltage and current waveforms. Typical current-voltage waveforms for condition (i) (mixture of Co-nitrate and Ni-nitrate) are shown in Figure 1b. A drop in the voltage (from ~22 to ~5 kV) and a current peak (~35–40 A) can be seen; the initial current peak (~10 A in the period 0–50 ns) is due to the displacement current. Notably, rather identical characteristics were observed for all conductive solutions (not shown in Figure 1b). Usually, in-liquid spark discharge is characterized by a drop in the voltage to zero [40,41]. Here, the fact that the voltage drops to ~5 kV (and not to 0) indicates the discharge mode in not a conventional spark, but rather a spark-like. We believe that the presence of solution in electrical circuit adds a "new component" that should be considered in the analysis of the electrical characteristics. This can be conducted by, e.g., equivalent electrical circuit model, which is beyond the scope the study.

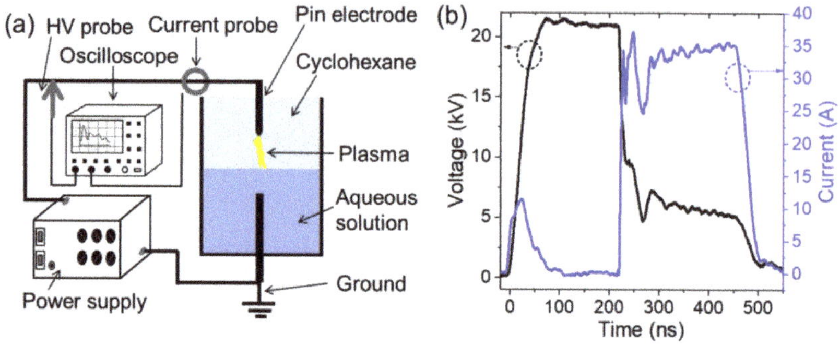

Figure 1. (**a**) Scheme of the experimental setup used to generate discharges in liquid cyclohexane that is in contact with a conductive solution. (**b**) Electrical characteristics (voltage and current) of a typical discharge in cyclohexane in contact with conductive solution (water + Co-nitrate + Ni-nitrate) generated at 22 kV voltage amplitude and 500 ns pulse width.

The synthesized nanostructures were characterized using a Transmission Electron Microscope (TEM, JEOL JEM-2100F) operated at 200 kV. For this purpose, the liquid samples collected after discharge treatment were sonicated for 5 min and then drop-casted on Cu TEM grids endowed with a lacey C-film (Electron Microscopy Science). The nanostructures were subsequently analyzed by bright field TEM imaging and Electron Dispersive Spectroscopy (EDS). After 30 min of discharge treatment at 10 Hz, both liquids change color, which indicates that they contain nanoparticles. In the previous study [39] where a solution of Ag-nitrate was used, Ag particles were synthesized in both liquids. Most particles collected from heptane were Ag nanoparticles (<10 nm) embedded in hydrocarbon network; meanwhile, the material collected from the silver nitrate solution was Ag nanoparticles (10–150 nm of diameter). Here, only the particles collected from cyclohexane side were characterized.

3. Results

Figure 2 shows some morphological characteristics of the particles synthesized in condition (i), i.e., in a mixture of Co-nitrate and Ni-nitrate. Figure 2a is a low-resolution TEM image of the collected particles, and it clearly shows that they are not agglomerated. The inset is an electron diffraction pattern performed on the imaged zone; the rings indicate the crystalline nature of the nanoparticles. The imaged zone was also analyzed by EDS, and the obtained spectrum is shown in Figure 2b. The detected elements in the analyzed sample are C, O, Co, Ni, and Cu. Carbon is due to both, the lacey C-film on the TEM grid as well as to the matrix synthesized by the discharge due to its interaction with cyclohexane. Cu is also due to TEM grid. As for Co and Ni, they are due to the nanoparticles synthesized by the discharge. Finally, O is probably due to the oxidation of the nanoparticles. The oxidation may happen during synthesis (by the oxidative species during discharge such as OH and O) [39] or after being exposed to ambient air [42]. Figure 2c shows the particle size distribution deduced by measuring the diameter of many particles imaged by TEM. This distribution indicates that the majority of the particles have diameter between 1 and 5 nm.

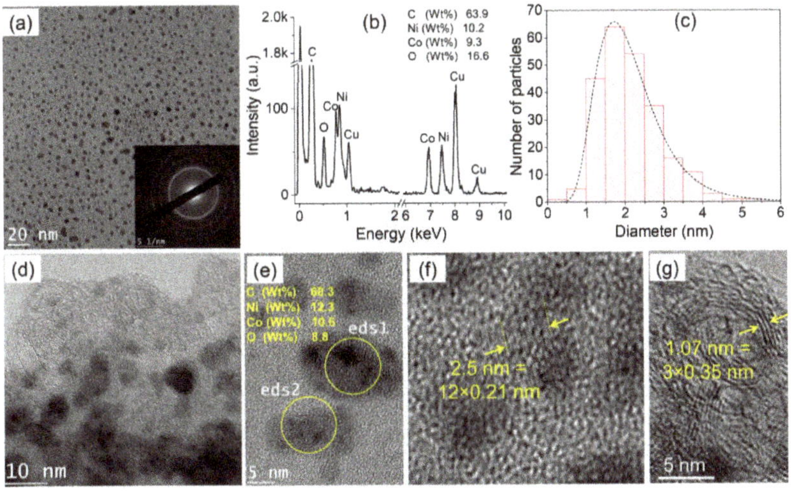

Figure 2. TEM characterization of the particles generated by discharges in cyclohexane in contact with water + Ni-nitrate + Co-nitrate collected from cyclohexane: (**a**) Low-resolution TEM image (inset: electron diffraction conducted on the imaged zone), (**b**) EDS spectrum acquired on the imaged zone in a), (**c**) size distribution of the synthesized nanoparticles, (**d**) intermediate-resolution TEM image, (**e**) high-resolution TEM image showing two typical nanoparticles as well as their composition deduced from EDS analysis, and (**f**,**g**) high-resolution TEM images showing the arranged atoms in the particles and in the matrix, respectively.

Figure 2d is an intermediate TEM image showing that the produced nanoparticles are embedded in a film-like matrix. High-resolution TEM images are shown in Figure 2e–g. Local EDS analysis performed on two selected particles show very similar composition of 68.3 Wt% of C, 12.3 Wt% of Ni, 10.6 Wt% of Co, and 8.8 Wt% of O (Figure 2e). The high-resolution TEM image in Figure 2f highlights the arranged atoms in one particle; the measured interplanar distance (averaged on 12 layers) is ~0.21 nm, which could correspond to either Ni (111) or Co (111) interplanar distance (very close values 0.20 and 0.21 nm) [26]. Therefore, it is not possible to use such a measurement as criteria to differentiate Ni and Co distribution in a nanoparticle (this is also true when Fe is added). However, the local EDS analysis is more reliable, and it reveals the co-existence of Ni and Co in individual nanoparticles. Another high-resolution TEM image (Figure 2g) performed on the edge of the sample (as in Figure 2d) shows arranged structure of 0.35 nm interplanar distance, which corresponds to the graphitic carbon [43]. At this stage, one concludes that discharges in cyclohexane in-contact with a solution of Co-nitrate and Ni-nitrate produce individual (non-agglomerated) nanoalloy of Co-Ni embedded in carbon network.

Figure 3 shows the characteristics of the particles produced in condition (ii), i.e., in a mixture of Co-nitrate and Fe-nitrate. Figure 3a shows the low-resolution TEM image of the produced particles. EDS analyses were performed on a large region (eds1) as well as on selected nanoparticles (eds2 and eds3). The three spectra are shown in Figure 3b, and all of them show the presence of Fe and Co, in addition to C, O, and Cu. The particles size distribution was performed on the imaged particles (not shown here) and shows that the majority of the particles are <10 nm. Figure 3c is high resolution TEM image that shows one nanoparticle as well as a film-like structure in which the particle is embedded. The average interplanar distance measured in the nanoparticle is ~0.22 nm, while the interplanar distance of the film around the particle is around 0.42 nm. Local EDS analysis performed on the particle shows the presence of Fe and Co, in addition to the other species (Figure 3d). Here, also, one concludes that discharges in cyclohexane in-contact with a solution of Co-nitrate and Fe-nitrate produce individual nanoalloys of Co-Fe embedded in carbon matrix.

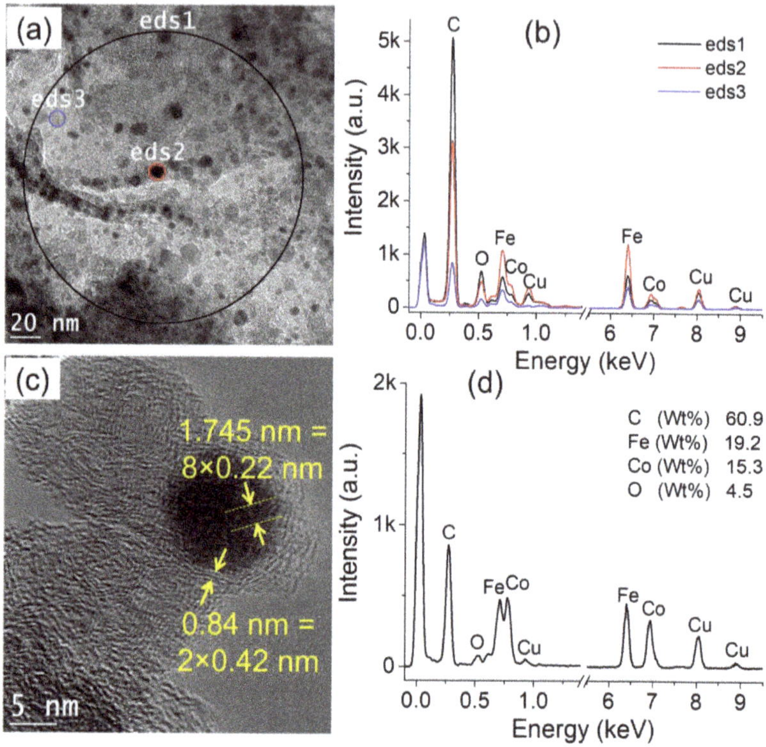

Figure 3. TEM characterization of the particles generated by discharges in cyclohexane in contact with water + Fe-nitrate + Co-nitrate collected from cyclohexane: (**a**) low-resolution TEM image, (**b**) EDS spectra acquired on the encircled zones in a), (**c**) high-resolution TEM image showing one nanoparticle and the matrix around it, and (**d**) EDS analysis performed on the imaged zone in (**c**).

Figure 4 summarizes the characteristics of the particles produced in condition (iii), i.e., in a mixture of Ni-nitrate and Fe-nitrate. The low-resolution TEM image (Figure 4a) shows that the majority of the particles are ultrasmall and are embedded in a film-like structure. Global and local EDS analysis were performed on the imaged zone, and the obtained spectra are shown in Figure 4b. Both spectra clearly show the presence of Fe and Ni, in addition to the other elements. A high resolution TEM image performed on the synthesized nanoparticles is depicted in Figure 4c. The measurement of the interplanar distance (average value) is ~0.22 nm (inset of Figure 4c). Local EDS analyses (Figure 4d) performed on a ~20 nm-diameter particle (eds1) as well as on ultrasmall particles (eds2) further indicate the presence of Fe and Ni in individual particles. As stated above, one concludes that discharges in cyclohexane in-contact with a solution of Ni-nitrate and Fe-nitrate produce isolated nanoalloys of Ni-Fe embedded in carbon network.

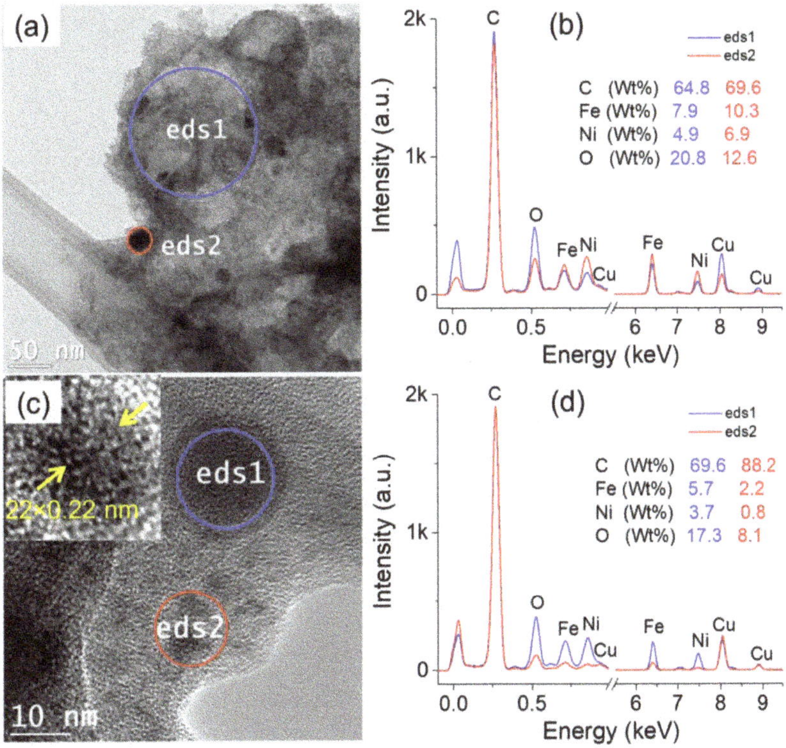

Figure 4. TEM characterization of the particles generated by discharges in cyclohexane in contact with water + Fe-nitrate + Ni-nitrate collected from cyclohexane: (**a**) low-resolution TEM image, (**b**) EDS spectra acquired on the encircled zones in a), (**c**) high-resolution TEM image showing individual nanoparticles and the matrix around it, and (**d**) EDS analysis performed on the encircled zones in (**c**).

Finally, Figure 5 summarizes the observations performed on the particles synthesized in condition (iv), i.e., in a mixture of Ni-nitrate, Co-nitrate, and Fe-nitrate. TEM image (Figure 5a) shows that the majority of the particles are ultrasmall (<10 nm); it was also possible to find few large particles (50–100 nm), and a typical one is depicted in Figure 5b. EDS analysis performed on both particles' population, the small as well as the large particles, shows the presence of Fe, Co, and Ni elements (Figure 5c), in addition to the other elements. Figure 5d,e show high-resolution TEM images of typical particles. The average interplanar distance measured on the particles is ~0.21 nm, and the measurement of the interplanar distance of the matrix is ~0.44–0.45 nm. These findings indicate, once again, that discharges in cyclohexane in-contact with a solution of Ni-nitrate, Co-nitrate, and Fe-nitrate produce nanoalloys of Ni-Fe-Co embedded in carbon matrix.

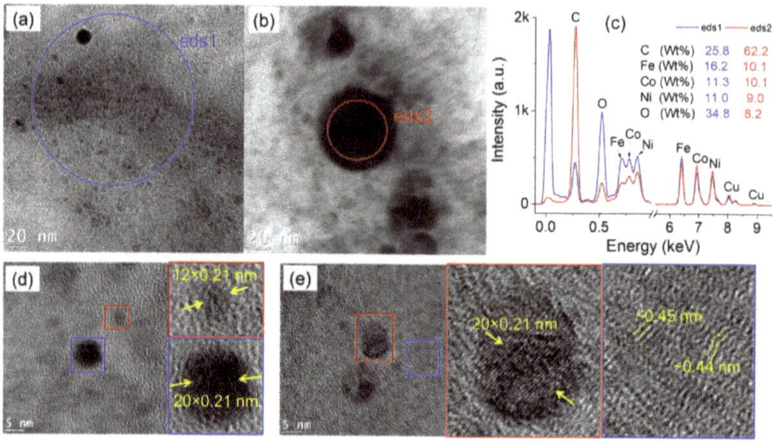

Figure 5. TEM characterization of the particles generated by discharges in cyclohexane in contact with water + Fe-nitrate + Ni-nitrate + Co-nitrate collected from cyclohexane: (**a**,**b**) low-resolution TEM images, (**c**) EDS spectra acquired on the encircled zones in (**a**,**b**), and (**d**,**e**) high-resolution TEM images showing individual nanoparticles and the matrix.

4. Discussion

The results presented above demonstrate that discharges in dielectric liquid (cyclohexane) that is in contact with conductive solution (water + metals nitrate) is a promising technology to efficiently produce non-agglomerated and well-controlled metal nanoalloys. All the four conditions tested here have led to various nanoalloys, namely Ni-Co, Ni-Fe, Fe-Co, and Ni-Fe-Co. The synthesis mechanisms are expected to be similar to those often highlighted in processes of plasma in-contact with solution [20]. Indeed, the reactive species in a typical gas phase plasma (e.g., electrons, radicals, etc.) reduce the ions in solution to form atoms that nucleate to form nanoparticles. Two major reactions are usually utilized to describe the growth [44]:

$$A^{n+} + ne \rightarrow A^0 \quad (1)$$

$$A^{n+} + nH \rightarrow A^0 + nH^+ \quad (2)$$

where A^{n+} is an ion in solution (n = 3 for Fe-nitrate and 2 for Ni- and Co-nitrate), e is the electron, A^0 is the reduced species, and H and H^+ are the hydrogen atom and ion, respectively.

The simultaneous and homogeneous presence of multiple ions in solution induces a instantaneous reduction of the different ions, which then leads to their nucleation as nanoparticles. On the other hand, the decomposition of cyclohexane by plasma produces many carbonaceous species (e.g., C_2, C_xH_y, etc.) that contribute to the formation of the carbon matrix. Because of the variety of carbon structures observed here, it is not straightforward to advance its mechanisms of formation. However, in some specific cases (e.g., Figures 2g and 3c), it is possible to identify carbon structures grown around the particle, which could be related to catalysis effect. Indeed, as it is well known in chemical vapor deposition (CVD) techniques, the growth of carbon nanostructures (e.g., nanotubes) requires the presence of three ingredients: carbon precursor, high temperature, and catalysis [45,46]. Over the range of experimental conditions investigated, all the three ingredients are present, such that it can be proposed that the specific carbon nanostructures (Figures 2g and 3c) are related to a CVD-like growth. Furthermore, it is worthy to note that the production of the matrix, simultaneously with the nanoparticles, reduces the particle-particle interactions, which prohibits their agglomeration. Such characteristic is extensively searched for, especially in the field of deposition of multifunctional nanocomposite thin films [47,48]. Note that the production of nanoalloys using the present process may be used in other processes to obtain devices with specific properties (e.g., catalytic properties). In the cases when the presence of carbon matrix is undesired, we believe that it can be removed by an appropriate method before or during processing. Finally, this study is considered as proof-of-concept to produce metal nanoalloys by a novel discharge-based technique. The production yield may be enhanced by increasing, e.g., the discharge frequency. Furthermore, instead of conducting discharges in stationary liquids, it is also feasible to conduct discharges in microfluidic devices. Although this has not been tested so far, we believe that it can be one of the methods to produce nanoalloys at large-scale.

5. Conclusions

In summary, compared with conventional synthesis techniques, in-liquid discharges offer some advantages and deserve further development. First, in-liquid discharges produce transient plasmas with high temperature (several thousands of Kelvin), high pressure (several tens of bar), and high density of reactive species (e.g., electron density of ~10^{17}–10^{19} cm^{-3}), which significantly increases the yield of synthesis. Second, the upper liquid (here cyclohexane) may be used as additional precursor, and it contributes to the final product. Therefore, its composition could be adjusted to fit a targeted material. We are convinced that such discharge conditions open the way and can be used as building blocks to synthesize 'novel' nanomaterial, probably with novel properties. In terms of perspective, additional systematic work should be conducted with the aim to address the influence of the relative concentration of different ions in water on the final product; this parameter may be used to finely control the composition of the nanoalloys.

Author Contributions: Conceptualization, A.H.; methodology, A.H.; formal analysis, A.H.; investigation, A.H.; resources, A.H.; data curation, A.H.; writing—original draft preparation, A.H. and L.S.; writing—review and editing, A.H. and L.S.; visualization, A.H.; supervision, A.H.; project administration, A.H.; funding acquisition, A.H. All authors have read and agreed to the published version of the manuscript.

Funding: This research was funded by the Natural Sciences and Engineering Research Council of Canada (NSERC), under award number RGPIN-2018-04869. The authors thank the Fonds de Recherche du Québec—Nature et Technologie (FRQ-NT) and the Canada Foundation for Innovation (CFI) for funding the research infrastructure. A Hamdan and L Stafford acknowledge the Fondation Courtois for the financial support.

Institutional Review Board Statement: Not applicable.

Informed Consent Statement: Not applicable.

Data Availability Statement: The data that supports the findings of this study are available within the article.

Acknowledgments: A. Hamdan thanks K. Mohammadi for the statistical analysis and the determination of the particles size distribution.

Conflicts of Interest: The authors declare no conflict of interest.

References

1. Li, D.; Gao, J.; Cheng, P.; He, J.; Yin, Y.; Hu, Y.; Chen, L.; Cheng, Y.; Zhao, J. 2D Boron Sheets: Structure, Growth, and Electronic and Thermal Transport Properties. *Adv. Funct. Mater.* **2020**, *30*, 1904349. [CrossRef]
2. Lou, Y.; Xu, J.; Zhang, Y.; Pan, C.; Dong, Y.; Zhu, Y. Metal-support interaction for heterogeneous catalysis: From nanoparticles to single atoms. *Mater. Today Nano* **2020**, *12*, 100093. [CrossRef]
3. Stewart, S.; Wei, Q.; Sun, Y. Surface chemistry of quantum-sized metal nanoparticles under light illumination. *Chem. Sci.* **2021**, *12*, 1227–1239. [CrossRef]
4. Tiwari, K.; Devi, M.M.; Biswas, K.; Chattopadhyay, K. Phase transformation behavior in nanoalloys. *Prog. Mater. Sci.* **2021**, *121*, 100794. [CrossRef]
5. Gonzalez-Reyna, M.; Rodriguez-Lopez, A.; Pérez-Robles, J.F. One-step synthesis of carbon nanospheres with an encapsulated iron-nickel nanoalloy and its potential use as an electrocatalyst. *Nanotechnology* **2020**, *32*, 095706. [CrossRef] [PubMed]

6. Ferrando, R. Determining the equilibrium structures of nanoalloys by computational methods. *J. Nanopart. Res.* **2018**, *20*, 179. [CrossRef]
7. Calvo, F. (Ed.) *Nanoalloys: From Fundamentals to Emergent Applications*; Elsevier: Amsterdam, The Netherlands, 2020.
8. Pandey, P.; Kunwar, S.; Sui, M.; Bastola, S.; Lee, J. Modulation of Morphology and Optical Property of Multi-Metallic PdAuAg and PdAg Alloy Nanostructures. *Nanoscale Res. Lett.* **2018**, *13*, 151. [CrossRef] [PubMed]
9. Iravani, S.; Korbekandi, H.; Mirmohammadi, S.V.; Zolfaghari, B. Synthesis of silver nanoparticles: Chemical, physical and biological methods. *Res. Pharm. Sci.* **2014**, *9*, 385–406.
10. Rane, A.V.; Kanny, K.; Abitha, V.K.; Thomas, S. Methods for Synthesis of Nanoparticles and Fabrication of Nanocomposites. In *Synthesis of Inorganic Nanomaterials*; Woodhead Publishing: Cambridge, UK, 2018; pp. 121–139. [CrossRef]
11. Das, D.; Roy, A. Synthesis of diameter controlled multiwall carbon nanotubes by microwave plasma-CVD on low-temperature and chemically processed Fe nanoparticle catalysts. *Appl. Surf. Sci.* **2020**, *515*, 146043. [CrossRef]
12. Woodard, A.; Xu, L.; Barragan, A.A.; Nava, G.; Wong, B.M.; Mangolini, L. On the non-thermal plasma synthesis of nickel nanoparticles. *Plasma Process. Polym.* **2018**, *15*, 1700104. [CrossRef]
13. Klébert, S.; Keszler, A.M.; Sajó, I.; Drotár, E.; Bertóti, I.; Bódis, E.; Fazekas, P.; Károly, Z.; Szépvölgyi, J. Effect of the solid precursors on the formation of nanosized TiBx powders in RF thermal plasma. *Ceram. Int.* **2014**, *40*, 3925–3931. [CrossRef]
14. Mohai, I.; Gál, L.; Szépvölgyi, J.; Gubicza, J.; Farkas, Z. Synthesis of nanosized zinc ferrites from liquid precursors in RF thermal plasma reactor. *J. Eur. Ceram. Soc.* **2007**, *27*, 941–945. [CrossRef]
15. Mavier, F.; Rat, V.; Bienia, M.; Lejeune, M.; Coudert, J.-F. Suspension and precursor solution plasma spraying by means of synchronous injection in a pulsed arc plasma. *Surf. Coat. Technol.* **2017**, *318*, 18–27. [CrossRef]
16. Schuffenhauer, C.; Parkinson, B.A.; Jin-Phillipp, N.Y.; Joly-Pottuz, L.; Martin, J.-M.; Popovitz-Biro, R.; Tenne, R. Synthesis of Fullerene-Like Tantalum Disulfide Nanoparticles by a Gas-Phase Reaction and Laser Ablation. *Small* **2005**, *1*, 1100–1109. [CrossRef]
17. Hartanto, A.; Ning, X.; Nakata, Y.; Okada, T. Growth mechanism of ZnO nanorods from nanoparticles formed in a laser ablation plume. *Appl. Phys. A* **2004**, *78*, 299–301. [CrossRef]
18. Chen, Q.; Li, J.; Li, Y. A review of plasma–liquid interactions for nanomaterial synthesis. *J. Phys. D Appl. Phys.* **2015**, *48*, 424005. [CrossRef]
19. Belmonte, T.; Hamdan, A.; Kosior, F.; Noël, C.; Henrion, G. Interaction of discharges with electrode surfaces in dielectric liquids: Application to nanoparticle synthesis. *J. Phys. D Appl. Phys.* **2014**, *47*, 224016. [CrossRef]
20. Mariotti, D.; Patel, J.; Švrček, V.; Maguire, P. Plasma-Liquid Interactions at Atmospheric Pressure for Nanomaterials Synthesis and Surface Engineering. *Plasma Process. Polym.* **2012**, *9*, 1074–1085. [CrossRef]
21. Merciris, T.; Valensi, F.; Hamdan, A. Determination of the Electrical Circuit Equivalent to a Pulsed Discharge in Water: Assessment of the Temporal Evolution of Electron Density and Temperature. *IEEE Trans. Plasma Sci.* **2020**, *48*, 3193–3202. [CrossRef]
22. Taylor, N.D.; Fridman, G.; Fridman, A.; Dobrynin, D. Non-equilibrium microsecond pulsed spark discharge in liquid as a source of pressure waves. *Int. J. Heat Mass Transf.* **2018**, *126*, 1104–1110. [CrossRef]
23. Bian, D.C.; Yan, D.; Zhao, J.C.; Niu, S.Q. Experimental Study of Pulsed Discharge Underwater Shock-Related Properties in Pressurized Liquid Water. *Adv. Mater. Sci. Eng.* **2018**, *2018*, 8025708. [CrossRef]
24. Descoeudres, A.; Hollenstein, C.; Wälder, G.; Demellayer, R.; Perez, R. Time- and spatially-resolved characterization of electrical discharge machining plasma. *Plasma Sources Sci. Technol.* **2008**, *17*, 024008. [CrossRef]
25. Saito, G.; Akiyama, T. Nanomaterial Synthesis Using Plasma Generation in Liquid. *J. Nanomater.* **2015**, *2015*, 1–21. [CrossRef]
26. Merciris, T.; Valensi, F.; Hamdan, A. Synthesis of nickel and cobalt oxide nanoparticles by pulsed underwater spark discharges. *J. Appl. Phys.* **2021**, *129*, 063303. [CrossRef]
27. Hamdan, A.; Noël, C.; Ghanbaja, J.; Migot-Choux, S.; Belmonte, T. Synthesis of platinum embedded in amorphous carbon by micro-gap discharge in heptane. *Mater. Chem. Phys.* **2013**, *142*, 199–206. [CrossRef]
28. Tabrizi, N.S.; Ullmann, M.; Vons, V.A.; Lafont, U.; Schmidt-Ott, A. Generation of nanoparticles by spark discharge. *J. Nanopart. Res.* **2009**, *11*, 315–332. [CrossRef]
29. Glad, X.; Gorry, J.; Cha, M.S.; Hamdan, A. Synthesis of core–shell copper–graphite submicronic particles and carbon nano-onions by spark discharges in liquid hydrocarbons. *Sci. Rep.* **2021**, *11*, 7516. [CrossRef] [PubMed]
30. Hamdan, A.; Kabbara, H.; Noël, C.; Ghanbaja, J.; Redjaimia, A.; Belmonte, T. Synthesis of two-dimensional lead sheets by spark discharge in liquid nitrogen. *Particuology* **2018**, *40*, 152–159. [CrossRef]
31. Kabbara, H.; Ghanbaja, J.; Redjaïmia, A.; Belmonte, T. Crystal structure, morphology and formation mechanism of a novel polymorph of lead dioxide, γ-PbO_2. *J. Appl. Crystallogr.* **2019**, *52*, 304–311. [CrossRef]
32. Trad, M.; Nominé, A.; Noël, C.; Ghanbaja, J.; Tabbal, M.; Belmonte, T. Evidence of alloy formation in CoNi nanoparticles synthesized by nanosecond-pulsed discharges in liquid nitrogen. *Plasma Process. Polym.* **2020**, *17*, 1900255. [CrossRef]
33. Saito, G.; Nakasugi, Y.; Yamashita, T.; Akiyama, T. Solution plasma synthesis of bimetallic nanoparticles. *Nanotechnology* **2014**, *25*, 135603. [CrossRef] [PubMed]
34. Yatsu, S.; Takahashi, H.; Sasaki, H.; Sakaguchi, N.; Ohkubo, K.; Muramoto, T.; Watanabe, S. Fabrication of Nanoparticles by Electric Discharge Plasma in Liquid. *Arch. Met. Mater.* **2013**, *58*, 425–429. [CrossRef]

35. Velusamy, T.; Liguori, A.; Macias-Montero, M.; Padmanaban, D.B.; Carolan, D.; Gherardi, M.; Colombo, V.; Maguire, P.; Svrcek, V.; Mariotti, D. Ultra-small CuO nanoparticles with tailored energy-band diagram synthesized by a hybrid plasma-liquid process. *Plasma Process. Polym.* **2017**, *14*, 1600224. [CrossRef]
36. Patel, J.; Němcová, L.; Maguire, P.; Graham, W.G.; Mariotti, D. Synthesis of surfactant-free electrostatically stabilized gold nanoparticles by plasma-induced liquid chemistry. *Nanotechnology* **2013**, *24*, 245604. [CrossRef] [PubMed]
37. Richmonds, C.; Sankaran, R.M. Plasma-liquid electrochemistry: Rapid synthesis of colloidal metal nanoparticles by microplasma reduction of aqueous cations. *Appl. Phys. Lett.* **2008**, *93*, 131501. [CrossRef]
38. Lin, L.; Ma, X.; Li, S.; Wouters, M.; Hessel, V. Plasma-electrochemical synthesis of europium doped cerium oxide nanoparticles. *Front. Chem. Sci. Eng.* **2019**, *13*, 501–510. [CrossRef]
39. Mohammadi, K.; Hamdan, A. Spark discharges in liquid heptane in contact with silver nitrate solution: Investigation of the synthesized particles. *Plasma Process. Polym.* **2021**, *18*, e2100083. [CrossRef]
40. Li, Y.; Wen, J.-Y.; Huang, Y.-F.; Zhang, G.-J. Streamer-to-spark transitions in deionized water: Unsymmetrical structure and two-stage model. *Plasma Sources Sci. Technol.* **2022**, *31*, 07LT02. [CrossRef]
41. Belmonte, T.; Kabbara, H.; Noel, C.; Pflieger, R. Analysis of Zn I emission lines observed during a spark discharge in liquid nitrogen for zinc nanosheet synthesis. *Plasma Sources Sci. Technol.* **2018**, *27*, 074004. [CrossRef]
42. Hamdan, A.; Noël, C.; Ghanbaja, J.; Belmonte, T. Comparison of Aluminium Nanostructures Created by Discharges in Various Dielectric Liquids. *Plasma Chem. Plasma Process.* **2014**, *34*, 1101–1114. [CrossRef]
43. Lin, Z.; Shao, G.; Liu, W.; Wang, Y.; Wang, H.; Wang, H.; Fan, B.; Lu, H.; Xu, H.; Zhang, R. In-situ TEM observations of the structural stability in carbon nanotubes, nanodiamonds and carbon nano-onions under electron irradiation. *Carbon* **2022**, *192*, 356–365. [CrossRef]
44. Kondeti, V.S.S.K.; Gangal, U.; Yatom, S.; Bruggeman, P.J. Ag^+ reduction and silver nanoparticle synthesis at the plasma–liquid interface by an RF driven atmospheric pressure plasma jet: Mechanisms and the effect of surfactant. *J. Vac. Sci. Technol. Vac. Surf. Films* **2017**, *35*, 061302. [CrossRef]
45. Li, M.; Li, Z.; Lin, Q.; Cao, J.; Liu, F.; Kawi, S. Synthesis strategies of carbon nanotube supported and confined catalysts for thermal catalysis. *Chem. Eng. J.* **2022**, *431*, 133970. [CrossRef]
46. Shoukat, R.; Khan, M.I. Carbon nanotubes/nanofibers (CNTs/CNFs): A review on state of the art synthesis methods. *Microsyst. Technol.* **2022**, *28*, 885–901. [CrossRef]
47. Mitronika, M.; Profili, J.; Goullet, A.; Gautier, N.; Stephant, N.; Stafford, L.; Granier, A.; Richard-Plouet, M. TiO_2–SiO_2 nanocomposite thin films deposited by direct liquid injection of colloidal solution in an O2/HMDSO low-pressure plasma. *J. Phys. D Appl. Phys.* **2020**, *54*, 085206. [CrossRef]
48. Wang, Z.; Zhao, X.; Guo, Z.; Miao, P.; Gong, X. Carbon dots based nanocomposite thin film for highly efficient luminescent solar concentrators. *Org. Electron.* **2018**, *62*, 284–289. [CrossRef]

nanomaterials

MDPI

Article

Electrochemical Synthesis of Nb-Doped $BaTiO_3$ Nanoparticles with Titanium-Niobium Alloy as Electrode

Qi Yuan [1], Wencai Hu [2], Tao Wang [3], Sen Wang [1,*], Gaobin Liu [1], Xueyan Han [1], Feixiang Guo [1] and Yongheng Fan [1]

1 School of Materials and Metallurgy, University of Science and Technology Liaoning, Anshan 114051, China
2 Fujian Huaqing Electronic Material Technology Co., Ltd., Quanzhou 362000, China
3 Jiangsu Can Qin Technology Co., Ltd., Suzhou 215633, China
* Correspondence: wsenl@yeah.net

Abstract: In this paper, Nb-doped $BaTiO_3$ nanoparticles ($BaNb_{0.47}Ti_{0.53}O_3$) were prepared using an electrochemical method in an alkaline solution, with titanium-niobium alloy as the electrode. The results indicated that under relatively mild conditions (normal temperature and pressure, V < 60 V, I < 5 A), cubic perovskite phase Nb-doped $BaTiO_3$ nanoparticles with high crystallinity and uniform distribution can be synthesized. With this increase in alkalinity, the crystallinity of the sample increases, the crystal grain size decreases, and the particles become more equally dispersed. Furthermore, in our study, the average grain size of the nanoparticles was 5–20 nm, and the particles with good crystallinity were obtained at a concentration of 3 mol/L of NaOH. This provides a new idea and method for introducing foreign ions under high alkalinity conditions.

Keywords: Nb-$BaTiO_3$; electrochemical synthesis; microstructure; nanoparticles

check for updates

Citation: Yuan, Q.; Hu, W.; Wang, T.; Wang, S.; Liu, G.; Han, X.; Guo, F.; Fan, Y. Electrochemical Synthesis of Nb-Doped $BaTiO_3$ Nanoparticles with Titanium-Niobium Alloy as Electrode. *Nanomaterials* **2023**, *13*, 252. https://doi.org/10.3390/nano13020252

Academic Editors: Andrei Honciuc, Mirela Teodorescu and Kenji Kaneko

Received: 23 November 2022
Revised: 23 December 2022
Accepted: 3 January 2023
Published: 6 January 2023

1. Introduction

$BaTiO_3$ is a significant component of electronic functional ceramics. It is frequently utilized in electronic devices because of its excellent dielectric, piezoelectric and ferroelectric characteristics [1,2]. With the rapid development of electronic information and fifth-generation industry, electronic components are developing in the direction of high integration, multi-functioning, and miniaturization; higher requirements are put forward to prepare $BaTiO_3$ nanomaterials that meet the performance requirements [3,4].

At present, the most common preparation methods of $BaTiO_3$ materials mainly include the solid-state method [5] and a range of chemical synthesis techniques for preparing ultrafine $BaTiO_3$ powders, such as the polymer precursor [6–8], hydrothermal, precipitation [9], sol-gel [10–12], and other methods. What is worthy of recognition is that chemical synthesis technology can fundamentally optimize the performance of electronic materials, especially in terms of the grain-size effect mechanism [13–16]. Despite these advantages, there are also many defects, such as long reaction times, demanding equipment requirements, and high costs. In contrast, the electrochemical synthesis reaction device is simple, and grain size control is achieved by varying the current density and the applied potential. Tao et al. [1] reported an electrochemical method in which $BaTiO_3$ nanoparticles were synthesized in a H_2O/EtOH (ethyl alcohol) solution containing KOH and $Ba(OH)_2$, using two titanium plates as electrodes. The results indicated that the composition of the electrolyte affects the size of the $BaTiO_3$ nanoparticles. Subsequently, Santos et al. [17] investigated the influence of solvent type and different electrolyte compositions on the synthesis of $BaTiO_3$ nanoparticles. The results showed that highly crystalline nanoparticles could be synthesized using ethanol and methanol in a high alkalinity environment, while changing the alkalinity of the electrolyte might viably control the size of the synthesized nanoparticles.

In order to meet the performance requirements of materials in the industry and practical applications, doped $BaTiO_3$-based ceramics are mostly used [18]. For instance, the anti-reduction performance of MLCC dielectric materials can be improved by doping

Mn^{2+}, Ca^{2+}, and Mg^{2+} [19–21]. The modification of Curie points by introducing foreign ions into $BaTiO_3$ ceramics is also an important means by which to study the preparation of widely temperature-stable MLCC. This group of common dopants includes La, Nb, Ce, and Zr ions [22–28]. The traditional solid-state method and a series of wet chemical synthesis methods to synthesize doped barium titanate powder have a long preparation cycle and time-consuming process, whereas the electrochemical method can provide a quick, simple, and one-time way to synthesize doped $BaTiO_3$ powder.

So far as we know, the electrochemical synthesis of doped $BaTiO_3$ nanopowders with alloy material as the electrode has not previously been reported. Hence, we report for the first time the electrochemical synthesis of Nb-doped $BaTiO_3$, using titanium–niobium alloy plates as the electrode. The influence of the alkalinity of electrolyte on the grain size, phase composition, and morphology of the Nb-doped $BaTiO_3$ is investigated. The results provide a novel method for the electrochemical synthesis of doped $BaTiO_3$ powder.

2. Materials and Methods

The schematic diagram of the electrochemical synthesis of Nb-doped $BaTiO_3$ nanoparticles is shown in Figure 1. The titanium–niobium alloy plate (purity > 99.0%, Xinbiao Metal Materials Co., Jiangsu, Wuxi, China) with a thickness of 1 mm was cut into a rectangle of 50 mm × 20 mm, to be used as an electrode for the electrolytic reaction. The surface of the electrode was mechanically polished with 800 Cw and 1200 Cw sandpaper. After cleaning with ultrapure water and EtOH, the electrode was washed with an ultrasonic wave in acetone to remove the remaining oil on the surface of the electrode. A heat-resistant glass electrolytic cell with a capacity of 250 mL is used for the electrolysis reaction. The electrolyte is composed of 50 mL EtOH, 150 mL ultrapure water (pre-boiled for 30 min to remove as much CO_2 as possible), 0.05 $mol \cdot L^{-1}$ $Ba(OH)_2$, and different concentrations of NaOH (0.5 $mol \cdot L^{-1}$, 1 $mol \cdot L^{-1}$, 1.5 $mol \cdot L^{-1}$, 2 $mol \cdot L^{-1}$, 2.5 $mol \cdot L^{-1}$, 3 $mol \cdot L^{-1}$) composition. The titanium-niobium alloy plate was fixed to the electrode holder, then two-thirds of its length was immersed in the electrolyte at a distance of 20 mm. The synthesis process uses a DC power supply (RXN-650D) and magnetic stirring. The initial voltage is 60 V, and the current is kept constant at 3.25 A.

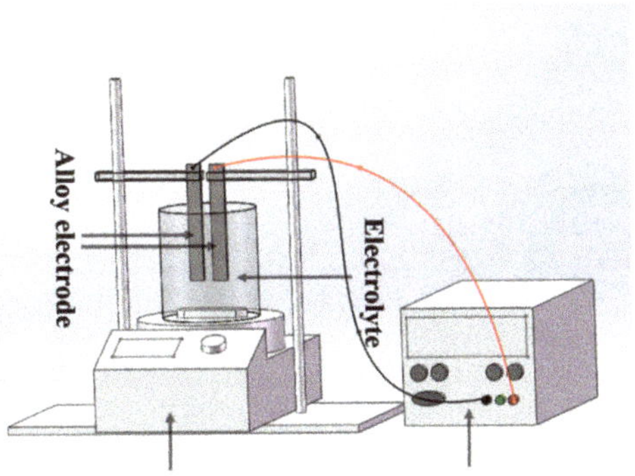

Figure 1. Schematic diagram of the electrochemical electrolytic cell.

The obtained powder sample was washed with 0.1 $mol \cdot L^{-1}$ of hot dilute nitric acid to remove the impurity $BaCO_3$ and was then repeatedly filtered and washed with ultrapure water and ethanol. The sample was placed in an electrothermal constant temperature blast-drying box and dried at 100 °C for 2 h. Nano $BaTiO_3$ (purity 99.9%, size < 100 nm, Aladdin, Shanghai, China) was selected as the reference material.

The crystal phase of the synthesized powder was analyzed by X-ray diffraction (Panalytical X'Pert-MPD type, Cu Kα radiation source, 40 kV, 40 mA). All scans were carried out at 5°/min for 2θ values between 10° and 90°. Small-angle X-ray scattering (SAXS) was used to calculate the grain size, based on the X-ray scattering intensity from the sample, measured at a scattering angle from 0.1° to 5°. The chemical composition of the samples was analyzed using a monochromatic Al Kα X-ray source and X-ray photoelectron spectroscopy (XPS, Thermo Scientific K-alpha, MA, USA). All XPS spectra were calibrated in terms of C1s at 284.8 eV. An ultra-high-resolution field emission scanning electron microscope (SEM, Zeiss, Ober-Kochen, Germany), energy spectrometer (EDS), and spherical aberration-corrected transmission electron microscope (TEM, HF5000, 200 kV) were used to observe the morphology, grain size, and microstructure of the powder.

3. Results and Discussions

Figure 2 shows the XRD diffraction pattern of the prepared $BaNb_{0.47}Ti_{0.53}O_3$ (hereafter referred to as BNTO) powder and pure $BaTiO_3$. The obtained powders all have a cubic perovskite ABO_3 structure (a = b = c, α = γ = β = 90°) and are consistent with the inorganic crystal structure database ICSD standard card (PDF-98-008-3902). Figure 2b shows a partially enlarged view of the XRD patterns of the two samples in the range of 30°~50°. The diffraction peak position of the BNTO powder has changed to a low angle compared to that of $BaTiO_3$. The phenomenon can be explained by the Bragg equation:

$$2d\sin\theta = n\lambda \tag{1}$$

where d is the interplanar spacing, n is an integer, called the reflection order, and θ is the grazing angle (the ionic radii of Nb^{4+} and Ti^{4+} are 0.74 Å and 0.61 Å, respectively [29,30]). The Nb^{4+} ion radius is larger; the interplanar spacing likely becomes larger after replacing part of Ti^{4+} at the B site, which causes the diffraction peaks to shift to a low-angle direction (Figure 2b).

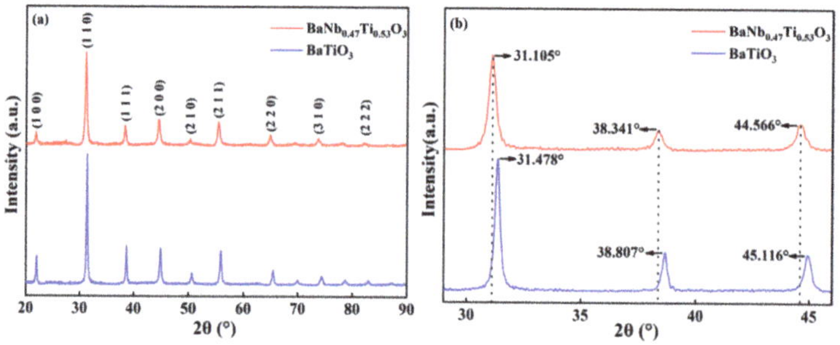

Figure 2. XRD pattern of electrochemically synthesized powder (**a**) and its partially enlarged schematic diagram (**b**).

To prove the above content, XPS measurement was performed on the synthesized BNTO nanoparticles. Figure 3a shows the XPS spectra of Ti2p, Nb3d, Ba3d, and O1s, indicating that the sample contains barium, titanium, and niobium. Figure 3b,c shows the fine spectra of Ti2p and Nb3d energy levels, respectively. It can be seen from the spectrum that the binding energies of the Ti element in $Ti2p_{1/2}$ and $Ti2p_{3/2}$ are 460.4 eV and 454.7 eV, which correspond to the two characteristic peaks of Ti^{4+} [31]. The binding energies of $3d_{3/2}$ and $3d_{5/2}$ are the 205.8 eV and 203.1 eV Nb3d doublet, respectively, as observed in the fine spectrum of the Nb3d energy level in Figure 3c. After querying the standard reference database, NIST, we found two characteristic peaks corresponding to Nb^{4+} [32] that were consistent with the conclusions drawn regarding the reason for the shift of XRD diffraction peaks mentioned above. The Nb/Ti ratio of 4.84/5.36 was obtained, using Avantage to fit the fine XPS spectra of Nb and Ti. XPS calculation is a semi-quantitative method, but the Nb/Ti ratio is very close to $BaNb_{0.47}Ti_{0.53}O_3$.

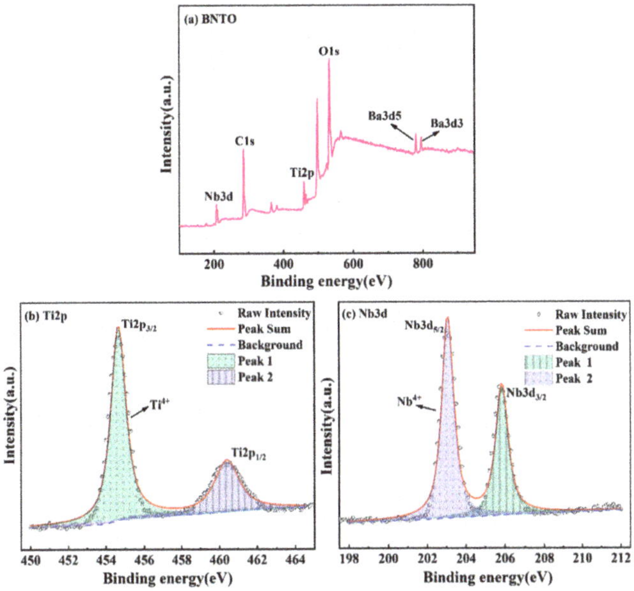

Figure 3. XPS spectra of electrochemically synthesized BNTO nanoparticles: (**a**) the XPS full spectrum of the sample, (**b**) the Ti2p fine spectrum, and (**c**) the Nb3d fine spectrum.

The relationship between the breakdown potential, the size of the crystal grain of the synthesized powder, and the NaOH concentration ($mol \cdot L^{-1}$) during the reaction process of electrochemically synthesized BNTO powder is shown in Figure 4 and Table 1. The crystallite size of the sample was obtained via the Guiner curve-fitting of SAXS data (See Supplementary Materials Figures S1–S5 for SAXS data map). When the NaOH concentration increases from 1 $mol \cdot L^{-1}$ to 3 $mol \cdot L^{-1}$, the breakdown potential gradually decreases; the increase in the concentration of NaOH in the electrolyte means that the conductive anions and cations increase, and the current conduction becomes easier, meaning that the electrochemical synthesis reaction can be completed with a lower applied potential. From Figure 4 and Table 1, it can be observed that the grain size and breakdown potential have the same trend concerning the change in NaOH concentration. During the reaction, the appearance of anode sparks is very important for forming BNTO nanoparticles [1]. As NaOH concentration in the electrolyte increases, the reaction becomes gentler, and anode sparks can be more evenly distributed on the electrode surface, meaning that the Ti and Nb are better separated and are distributed from the immersion area of the electrode.

Figure 5 shows the time-varying curves of the potential and current under different NaOH concentrations during the synthesis of BNTO nanoparticles. When the NaOH concentration is 1 $mol \cdot L^{-1}$, the required applied potential is higher, and the potential decreases significantly with the increase in concentration. Under the conditions of a single concentration, the potential was initially high. However, the potential decreased with the extension of the reaction time, and it was more stable after the reaction had proceeded for 40 min. Since the total surface area of the alloy, the electrode is large and the reaction is violent at the beginning of the reaction, the required potential is higher. After a while, the alloy electrode gradually dissolves, which leads first to a decrease in total surface area and then the decline of potential. This phenomenon can indicate that the high alkalinity environment of the electrolyte can ensure the current conduction during the whole reaction process, thereby promoting the effective separation of Ti and Nb from the electrode (as shown in Figure 6). The anodic reaction can be described as follows (2) and (3) [33,34]:

$$0.47Ti + 0.53Nb + 6OH^- \xrightarrow{spark} [Nb_{0.47}Ti_{0.53}O_3]^{2-} + 3H_2O + 4e' \quad (2)$$

$$Ba^{2+} + [Nb_{0.47}Ti_{0.53}O_3]^{2-} \longrightarrow BaNb_{0.47}Ti_{0.53}O_3(s). \quad (3)$$

With the increase in NaOH concentration, the grain size of the BNTO nanoparticles gradually decreases, which may be due to the more stable current conduction, milder reaction, and more uniform distribution of the anode spark on the electrode in the environment of a higher concentration of NaOH. The change makes the better separation and distribution of Ti and Nb at the interface between the electrode and electrolyte, promoting grain-size reduction.

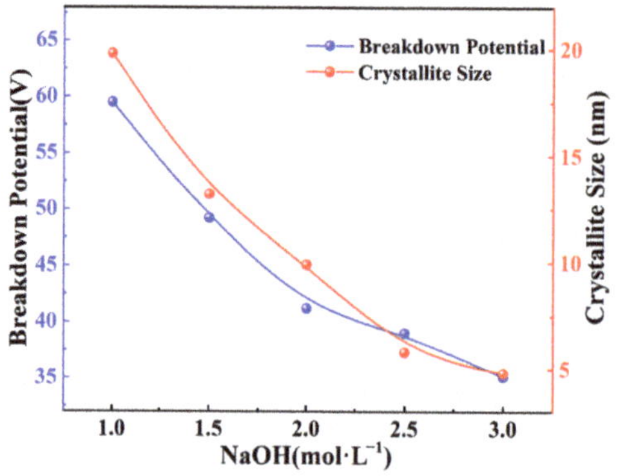

Figure 4. Relationship between breakdown potential, grain size, and NaOH concentration.

Table 1. Effects of different NaOH concentrations on breakdown potential and the grain size of the synthetic powder.

Breakdown Potential (V)	NaOH (mol·L^{-1})	Initial Current (I)	Crystallite Size (nm)
59.5	1	3.25	19.8
49.2	1.5	3.25	13.2
41.2	2	3.25	9.9
39.0	2.5	3.25	5.8
35.1	3	3.25	4.8

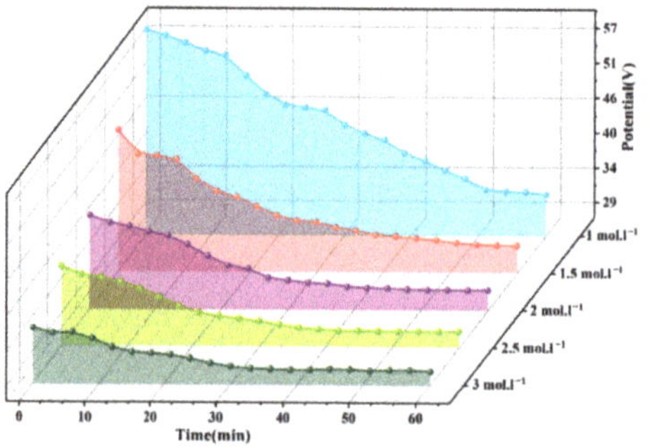

Figure 5. Relationship curve between potential and reaction time in the process of synthesizing BNTO nanoparticles under different NaOH concentrations.

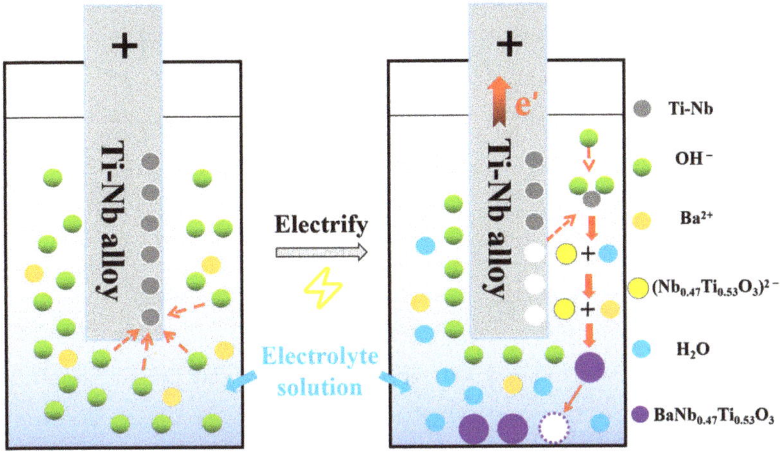

Figure 6. Anodic chemical reaction diagram.

Figure 7 shows the XRD patterns of samples prepared with different concentrations of NaOH (1/2/3 mol·L^{-1}). It can be seen that when the concentration of NaOH is low (1 mol·L^{-1}), obvious diffraction peaks of $BaCO_3$ will appear at 2θ = 24.27° and 34.06°. The intensity of the diffraction peak gradually weakens or even disappears. This means that the high-alkalinity environment will inhibit the generation of $BaCO_3$ impurities. The ultrapure water in the electrolyte has been boiled before being used, to remove as much CO_2 as possible. Although CO_2 in the air may still be dissolved in the electrolyte during the reaction (the solubility of CO_2 in water at room temperature and pressure at 0.033 mol·L^{-1}), the temperature of the electrolyte will increase and the solubility of CO_2 will decrease during the reaction. In addition, a small amount of $CO_3{}^{2-}$ will preferentially neutralize with OH^- in an alkaline environment, thereby inhibiting the formation of $BaCO_3$. The lattice parameter, cell volume, and average crystallite size were obtained by fitting the XRD data (a full spectrum fitting using the JADE (MDI. Jade. 6.0) software). The calculation results are shown in Table 2. With the increase in NaOH concentration, the lattice constant of the sample increases and the crystallite size decreases; the crystallite size is very close to the SAXS calculation result.

Figure 8a–c shows the SEM images of BNTO powder when prepared under different NaOH concentrations. It can be observed that a small number of amorphous particles appear when the NaOH concentration is low, indicating that the nucleation and crystal growth are insufficient and immature under these conditions, resulting in the obvious heterogeneity and certain agglomeration of the sample. With the increase in NaOH concentration, the particle shape becomes clearer. When the NaOH concentration is at 3.0 mol·L^{-1}, the particle shape is the most complete. Figure 8d and Table 3 show the EDS spectrum and quantitative composition of point A in Figure 8c. In Table 3, L/K represents the line system of the characteristic X-rays. The actual atomic ratio of Ti:Nb is almost equal to the stoichiometric ratio of 1:1.13 in $BaNb_{0.47}Ti_{0.53}O_3$, which indicates that Nb-doped $BaTiO_3$ powder was synthesized via the electrochemical method, using the titanium-niobium alloy material as the electrode. EDS data show that Ba:Ti:Nb = 1:0.69:0.96, and the proportion of Nb and Ti elements is higher than expected. By observing Figure 8d, it is clear that Ba and Ti peaks appear as "pathological overlaps" near the positions of 4.5 keV and 4.8 keV. This kind of "pathological overlap" may cause identification errors when detected by the EDS equipment, which makes the proportion of the Ti element higher than expected and the proportion of the Ba element lower than expected, resulting in an inconsistent proportion of elements when analyzed by EDS.

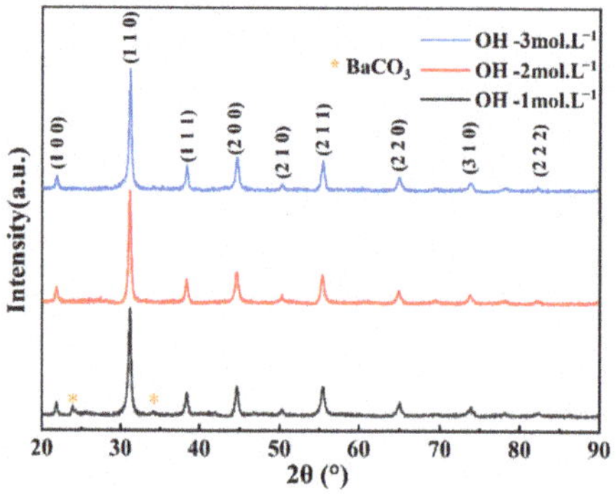

Figure 7. XRD patterns of samples prepared under different NaOH concentrations.

Table 2. The lattice parameter, cell volume, and crystallite size of the samples, prepared at different NaOH concentrations.

NaOH ($mol \cdot L^{-1}$)	a (Å)	V = a^3 ($Å^3$)	Crystallite Size (nm)
1	4.043 ± 0.010	66.1 ± 0.5	20.3 ± 1.6
2	4.059 ± 0.005	66.9 ± 0.2	10.5 ± 1.5
3	4.063 ± 0.006	67.1 ± 0.3	5.4 ± 1.0

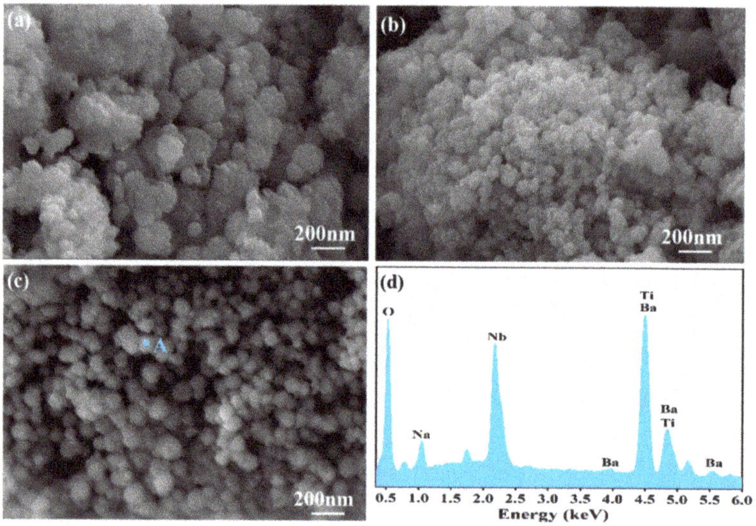

Figure 8. SEM image of BNTO powder, synthesized in solvents containing different concentrations of NaOH: (**a**) 1.0 $mol \cdot L^{-1}$ (**b**) 2.0 $mol \cdot L^{-1}$ (**c**) 3.0 $mol \cdot L^{-1}$; (**d**) EDS spectrum in point A.

Table 3. The EDS spectrum analysis results of point A in Figure 8c.

Elements	Weight (%)	Atomic (%)
O K	30.11	71.04
Na K	1.96	3.05
Ti K	12.59	9.37
Nb L	17.6	6.75
Ba L	37.74	9.79
Totals	100	100

To further study the morphology and microstructure of the BNTO powder, TEM observation was carried out. According to Figure 9a (TEM image), the particle size of the powder was roughly estimated to be about 5–10 nm, which was in good agreement with the particle size as measured by small-angle X-ray scattering (SAXS). It was observed that the powder has a certain degree of agglomeration, which may be due to the small grain size, high surface energy, and surface tension, which causes agglomeration and the formation of larger aggregates [35]. Figure 9b (TEM local magnification image) shows the measured values of lattice fringes at 0.291 and 0.239 nm, which correspond to the spacing of the (110) and (111) crystal planes of BNTO, respectively. The XRD diffraction peaks (110) and (111) (2θ angles of 28°~33° and 37°~40°) were fitted and calculated by applying origin software; the results were consistent with the interfacial distance values in the HRTEM images (NaOH concentration at 3.0 mol·L^{-1}). The clear lattice fringes in the TEM image also illustrate the synthesis of the BNTO nanoparticles, which have high crystallinity.

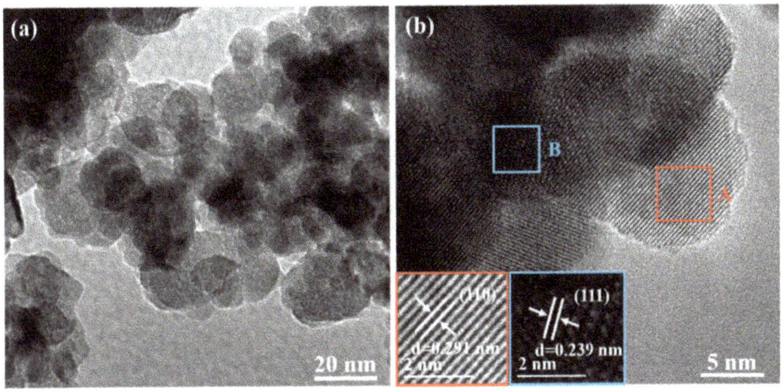

Figure 9. TEM image (**a**), local magnification image (**b**), crystal fringe image A and B of the electrochemically synthesized BNTO nanoparticles (NaOH concentration is 3.0 mol·L^{-1}).

4. Conclusions

Nb-doped $BaTiO_3$ nanoparticles were synthesized successfully by the electrochemical method using titanium-niobium alloy as the electrode, which overcomes the disadvantage that most metal ions are difficult to dissolve under high alkalinity and inhibit the electrochemical synthesis reaction. Compared with other Nb-doped $BaTiO_3$ synthesis methods, the experimental device of this scheme is simple, and well-crystallized $BaNb_{0.47}Ti_{0.53}O_3$ nanoparticles can be obtained in a short time under a normal temperature and pressure. Among them, the concentration of NaOH has an important effect on the potential change during the entire synthesis reaction, and the crystallinity and grain size of the sample. When the concentration of NaOH increases from 1 mol·L^{-1} to 3 mol·L^{-1}, the average grain size of the sample changes between 20 nm and 5 nm. Finally, the electrochemical synthesis of Nb-doped $BaTiO_3$ nanoparticles using alloy materials as electrodes is expected to provide a simple method for the synthesis of other metal element-doped titanate nanoparticles.

Supplementary Materials: The following supporting information can be downloaded at: https://www.mdpi.com/article/10.3390/nano13020252/s1, Figure S1. SAXS data (NaOH 1.0 mol·L^{-1}); Figure S2. SAXS data (NaOH 1.5 mol·L^{-1}); Figure S3. SAXS data (NaOH 2.0 mol·L^{-1}); Figure S4. SAXS data (NaOH 2.5 mol·L^{-1}); Figure S5. SAXS data (NaOH 3.0 mol·L^{-1}).

Author Contributions: Q.Y. and W.H.: Conceptualization, synthesis, formal analysis concerning XRD, XPS, SEM-EDX, TEM, etc., writing—original draft preparation. T.W., X.H., F.G. and Y.F.: Validation. S.W. and G.L.: supervision, project administration, funding acquisition, validation, visualization, writing—review and editing. All authors have read and agreed to the published version of the manuscript.

Funding: This work has been supported by the Liaoning Education Department Program under fellowship number 2019FWDF01.

Conflicts of Interest: The authors declare no conflict of interest.

References

1. Tao, J.; Ma, J.; Wang, Y.; Zhu, X.; Liu, J.; Jiang, X.; Lin, B.; Ren, Y. Synthesis of barium titanate nanoparticles via a novel electrochemical route. *Mater. Res. Bull.* **2008**, *43*, 639–644. [CrossRef]
2. Xia, B.; Lenggoro, I.W.; Okuyama, K. Novel Route to Nanoparticle Synthesis by Salt-Assisted Aerosol Decomposition. *Adv. Mater.* **2001**, *13*, 1579–1582. [CrossRef]
3. Boccardi, F.; Heath, R.W., Jr.; Lozano, A.; Marzetta, T.L.; Popovski, P. Five disruptive technology directions for 5G. *IEEE Commun. Mag.* **2014**, *52*, 74–80. [CrossRef]
4. Wong, P.C.; Luo, S. Flip the Chip. *Science* **2000**, *290*, 2269–2270. [CrossRef] [PubMed]
5. Buscaglia, M.T.; Bassoli, M.; Buscaglia, V.; Alessio, R. Solid-State Synthesis of Ultrafine $BaTiO_3$ Powders from Nanocrystalline $BaCO_3$ and TiO_2. *J. Am. Ceram. Soc.* **2005**, *88*, 2374–2379. [CrossRef]
6. Vijatovic, M.M.; Petrovic, J.D.; Bobic, T. Electrical properties of lanthanum doped barium titanate ceramics. *Mater. Charact.* **2011**, *62*, 1000–1006.
7. Durán, P.; Capel, F.; Tartaj, J.; Gutierrez, D.; Moure, C. Heating-rate effect on the $BaTiO_3$ formation by thermal decomposition of metal citrate polymeric precursors. *Solid State Ion.* **2001**, *141*, 529–539. [CrossRef]
8. Ischenko, V.; Pippel, E.; Köferstein, R.; Abicht, H.P.; Woltersdorf, J. Barium titanate via thermal decomposition of Ba,Ti-precursor complexes: The nature of the intermediate phases. *Solid State Sci.* **2007**, *9*, 21–26. [CrossRef]
9. Chen, J.F.; Shen, Z.G.; Liu, F.T. Preparation and properties of barium titanate nanopowder by conventional and high-gravity reactive precipitation methods. *Scr. Mater.* **2003**, *49*, 509–514. [CrossRef]
10. Zanfir, A.V.; Voicu, G.; Jinga, S.I.; Vasile, E.; Ionita, V. Low-temperature synthesis of $BaTiO_3$ nanopowders. *Ceram. Int.* **2016**, *42*, 1672–1678. [CrossRef]
11. Wang, W.; Cao, L.; Liu, W.; Su, G.; Zhang, W. Low-temperature synthesis of $BaTiO_3$ powders by the sol-gel-hydrothermal method. *Ceram. Int.* **2013**, *39*, 7127–7134. [CrossRef]
12. Zheng, H.; Zhu, K.; Wu, Q.; Liu, J.; Qiu, J. Preparation and characterization of monodispersed $BaTiO_3$ nanocrystals by sol–hydrothemal method. *J. Cryst. Growth* **2013**, *363*, 300–307. [CrossRef]
13. Kinoshita, K.; Yamaji, A. Grain-size effects on dielectric properties in barium titanate ceramics. *J. Appl. Phys.* **1976**, *47*, 371–373. [CrossRef]
14. Ding, S.; Song, T.; Yang, X.; Luo, G. Effect of Grain Size of $BaTiO_3$ Ceramics on Dielectric Properties. *Ferroelectrics* **2010**, *402*, 55–59. [CrossRef]
15. Hoshina, T.; Takizawa, K.; Li, J.; Kasama, T.; Kakemoto, H.; Tsurumi, T. Domain Size Effect on Dielectric Properties of Barium Titanate Ceramics. *Jap. J. Appl. Phys.* **2008**, *47*, 7607–7611. [CrossRef]
16. Zhao, Z.; Buscaglia, V.; Viviani, M.; Buscaglia, M.T.; Mitoseriu, L.; Testino, A.; Nygren, M.; Johnsson, M.; Nanni, P. Grain-size effects on the ferroelectric behavior of dense nanocrystalline $BaTiO_3$ ceramics. *Phys. Rev. B* **2004**, *70*, 024107. [CrossRef]
17. Santos, G.O.S.; Silva, R.S.; Costa, L.P.; Cellet, T.S.P.; Rubira, A.F.; Eguiluz, K.I.B.; Salazar-Banda, G.R. Influence of synthesis conditions on the properties of electrochemically synthesized $BaTiO_3$ nanoparticles. *Ceram. Int.* **2014**, *40*, 3603–3609. [CrossRef]
18. Ihlefeld, J.F.; Borland, W.J.; Maria, J.P. Enhanced dielectric tunability in barium titanate thin films with boron additions. *Scr. Mater.* **2008**, *58*, 549–552. [CrossRef]
19. Bernard, J.; Houivet, D.; Fallah, J. $MgTiO_3$ for Cu base metal multilayer ceramic capacitors. *J. Eur. Ceram. Soc.* **2004**, *24*, 1877–1881. [CrossRef]
20. Frey, M.H.; Xu, Z.; Han, P.; Payne, D.A. The role of interfaces on an apparent grain size effect on the dielectric properties for ferroelectric barium titanate ceramics. *Ferroelectrics* **1998**, *206*, 337–353. [CrossRef]
21. Hennings, D.F.K.; Schreinemacher, H. Ca-acceptors in dielectric ceramics sintered in reducive atmospheres. *J. Eur. Ceram. Soc.* **1995**, *15*, 795–800. [CrossRef]
22. Panwar, N.S.; Semwal, B.S. Dielectric properties of $BaTiO_3$ and La doped $BaTiO_3$ ceramics. *Pramana* **1991**, *36*, 163–166. [CrossRef]
23. Ren, P.; He, J.; Wang, X. Colossal permittivity in niobium doped $BaTiO_3$ ceramics annealed in N_2. *Scr. Mater.* **2018**, *146*, 110–114. [CrossRef]
24. Paunovic, V.; Mitic, V.V.; Djordjevic, M.; Prijic, Z. Niobium doping effect on $BaTiO_3$ structure and dielectric properties. *Ceram. Int.* **2020**, *46*, 8154–8164. [CrossRef]
25. Dechakupt, T.; Tangsritrakul, J.; Ketsuwan, P.; Yimnirun, R. Microstructure and Electrical Properties of Niobium Doped Barium Titanate Ceramics. *Ferroelectrics* **2011**, *415*, 141–148. [CrossRef]
26. Urek, S.; Drofenik, M. PTCR behaviour of highly donor doped $BaTiO_3$. *J. Eur. Ceram. Soc.* **1999**, *19*, 913–916. [CrossRef]
27. Chen, A.; Zhi, Y.; Zhi, J. Synthesis and characterization of $Ba(Ti_{1-x}Ce_x)O_3$ ceramics. *J. Eur. Ceram. Soc.* **1997**, *17*, 1217–1221. [CrossRef]
28. Hofer, C.; Meyer, R.; Böttger, U.; Waser, R. Characterization of $Ba(Ti, Zr)O_3$ ceramics sintered under reducing conditions. *J. Eur. Ceram. Soc.* **2004**, *24*, 1473–1477. [CrossRef]

29. Zhang, Z.H.; Wu, S.Y.; Xu, P. Studies on the local structures and spin Hamiltonian parameters for the rhombic Nb^{4+} centers in MO_2(M= Sn, Ti and Ge) crystals. *Eur. Phys. J. Appls.* **2011**, *53*, 20902. [CrossRef]
30. Liu, Y.; Zhu, D.; Feng, G. Effects of Ni^{2+}, Zr^{4+}, or Ti^{4+} doping on the electromagnetic and microwave absorption properties of $CaMnO_3$ particles. *Ceram. Int.* **2021**, *47*, 19995–20002. [CrossRef]
31. Chen, W.; Liang, H.; Shao, L.; Shu, J.; Wang, Z. Observation of the structural changes of sol-gel formed $Li_2MnTi_3O_8$ during electrochemical reaction by in-situ and ex-situ studies. *Electrochim. Acta* **2015**, *152*, 187–194. [CrossRef]
32. Yan, L.; Chen, G.; Sarker, S.; Richins, S.; Wang, H.; Xu, W.; Rui, X.; Luo, H. Ultrafine Nb_2O_5 Nanocrystal Coating on Reduced Graphene Oxide as Anode Material for High Performance Sodium Ion Battery. *ACS Appl. Mater. Inter.* **2016**, *8*, 22213–22219. [CrossRef]
33. Zeng, J.L.; Teng, H.P.; Lu, F.H. Electrochemical deposition of barium titanate thin films on TiN/Si substrates. *Surf. Coat. Tech.* **2013**, *231*, 297–300. [CrossRef]
34. Macova, Z.; Bouzek, K.; Híveš, J. Research progress in the electrochemical synthesis of ferrate (VI). *Electrochim. Acta* **2009**, *54*, 2673–2683. [CrossRef]
35. Schaafsma, S.H.; Vonk, P.; Segers, P.; Kossen, N.W. Description of agglomerate growth. *Powder Technol.* **1998**, *97*, 183–190. [CrossRef]

nanomaterials

MDPI

Article

Hydrothermal Synthesis and Properties of Yb^{3+}/Tm^{3+} Doped Sr_2LaF_7 Upconversion Nanoparticles

Bojana Milićević [1,*], Jovana Periša [1], Zoran Ristić [1], Katarina Milenković [1], Željka Antić [1], Krisjanis Smits [2], Meldra Kemere [2], Kaspars Vitols [2], Anatolijs Sarakovskis [2] and Miroslav D. Dramićanin [1,*]

1 Centre of Excellence for Photoconversion, Vinča Insitute of Nuclear Sciences—National Institute of the Republic of Serbia, University of Belgrade, P.O. Box 522, 11001 Belgrade, Serbia
2 Institute of Solid State Physics, University of Latvia, Kengaraga Street 8, LV-1063 Riga, Latvia
* Correspondence: bojana.milicevic85@gmail.com (B.M.); dramican@vinca.rs (M.D.D.)

Abstract: We report the procedure for hydrothermal synthesis of ultrasmall Yb^{3+}/Tm^{3+} co-doped Sr_2LaF_7 (SLF) upconversion phosphors. These phosphors were synthesized by varying the concentrations of Yb^{3+} (x = 10, 15, 20, and 25 mol%) and Tm^{3+} (y = 0.75, 1, 2, and 3 mol%) with the aim to analyze their emissions in the near IR spectral range. According to the detailed structural analysis, Yb^{3+} and Tm^{3+} occupy the La^{3+} sites in the SLF host. The addition of Yb^{3+}/Tm^{3+} ions has a huge impact on the lattice constant, particle size, and PL emission properties of the synthesized SLF nanophosphor. The results show that the optimal dopant concentrations for upconversion luminescence of Yb^{3+}/Tm^{3+} co-doped SLF are 20 mol% Yb^{3+} and 1 mol% Tm^{3+} with EDTA as the chelating agent. Under 980 nm light excitation, a strong upconversion emission of Tm^{3+} ions around 800 nm was achieved. In addition, the experimental photoluminescence lifetime of Tm^{3+} emission in the SLF host is reported. This study discovered that efficient near IR emission from ultrasmall Yb^{3+}/Tm^{3+} co-doped SLF phosphors may have potential applications in the fields of fluorescent labels in bioimaging and security applications.

Keywords: nanophosphor; fluoride; morphology; upconversion; Tm^{3+} emission; NIR emission

Citation: Milićević, B.; Periša, J.; Ristić, Z.; Milenković, K.; Antić, Ž.; Smits, K.; Kemere, M.; Vitols, K.; Sarakovskis, A.; Dramićanin, M.D. Hydrothermal Synthesis and Properties of Yb^{3+}/Tm^{3+} Doped Sr_2LaF_7 Upconversion Nanoparticles. *Nanomaterials* **2023**, *13*, 30. https://doi.org/10.3390/nano13010030

Academic Editor: Michele Back

Received: 1 December 2022
Revised: 14 December 2022
Accepted: 19 December 2022
Published: 21 December 2022

1. Introduction

Upconversion (UC) phosphor materials doped or co-doped with trivalent lanthanide (Ln^{3+}) ions have drawn considerable attention, since four-electron configurations of Ln^{3+} ions should split by electron–electron repulsion and spin–orbit coupling, resulting in a rich energy-level pattern that can be easily populated by the near infrared (NIR) laser source [1,2]. Under a NIR laser source, UC phosphors effectively convert energy into shorter wavelength emissions (NIR, visible, or ultraviolet) via multiphoton absorption or efficient energy transfer [3]. These unique features of Ln^{3+}-doped/co-doped UC phosphors enable a wide range of applications, such as LED devices [4], solar energy conversion [5], temperature sensors [6–8], latent fingerprint detection [9,10], biomedical imaging [11,12], food safety detection [13], etc. The NIR-to-NIR UC luminescence mechanism is gaining popularity due to its efficient way of producing NIR emission at nearly 800 nm when excited by a commercially available laser source (~980 nm) [9,14–21]. It is well known that the Yb^{3+} ion, with its simple energy level structure (ground state $^2F_{7/2}$ and excited state $^2F_{5/2}$), strong absorption band in the wavelength range of 860–1060 nm, and relatively long luminescence lifetime (1–2 ms), is an excellent sensitizer for energy transfer to RE ions [21–23]. Ln^{3+} ions such as Pr^{3+}, Er^{3+}, Tm^{3+}, and Ho^{3+} have been recognized as co-activators in Yb^{3+}/Ln^{3+} UC phosphors due to the position of their energy levels and the possibility of efficient radiative transitions under NIR light sources [9,14–24]. Among them, Yb^{3+}/Tm^{3+} UC phosphors have been intensively studied because NIR-to-NIR UC emission is crucial for biomedical imaging [17–20], security printing applications [15,16] and latent fingerprint detection [9].

The UC luminescence mechanism has been explored in a variety of host materials, including chlorides, fluorides, oxides, vanadates, and others. The selection of appropriate host materials with low phonon energy frequencies to prevent non-radiative relaxation processes and thus improve emission efficiency is essential for UC luminescence. Chlorides have low phonon frequencies ($\leq$300 cm^{-1}) and poor chemical stability, which limits their application possibilities, whereas oxide host materials have relatively high phonon frequencies (>500 cm^{-1}) and excellent chemical stability [25]. Fluoride materials are thus ideal hosts for UC luminescence due to their low phonon frequency (from 300 to 500 cm^{-1}), good chemical stability, and simplicity of dispersion in colloidal form with water or various nonpolar solvents [25,26].

Even though alkaline-Ln^{3+} tetrafluoride phosphors ($ALnF_4$, A = Na, K, Li) are the most preferred for efficient UC luminescence, their agglomeration limits applications that require nanoparticles, such as in biomedical imaging [27–30]. According to several recent studies, alkali–earth–Ln^{3+} nanophosphors (M_2LnF_7, M = Ca, Sr, Ba; Ln^{3+} = Y, La, Gd, Lu) are small enough for biomedical imaging applications while exhibiting extremely high UC luminescence [31–34]. Under 980 nm laser stimulation, Xie et al. observed efficient visible emission of Sr_2LaF_7:Yb^{3+}, Er^{3+} UC nanophosphors with average particles of around 25 nm [34]. Guo et al. reported that Sr_2GdF_7:Er^{3+}, Yb^{3+} nanocrystals incorporated into electrospun fibers promote energy transfer processes from Yb^{3+} to Er^{3+}, which is crucial for potential applications in the field of noncontact biomedical temperature sensors [8]. In this work, a set of $Sr_2La_{1\text{-x-y}}F_7$ phosphors with different concentrations of Yb^{3+} (x = 10, 15, 20, and 25 mol%) and Tm^{3+} (y = 0.75, 1, 2, and 3 mol%) ions with respect to La^{3+} ions are prepared hydrothermally at 180 °C. The room temperature photoluminescence spectra of SLF:Yb, Tm under 980 nm excitation clearly show intense NIR emissions in the wavelength range from 750 to 850 nm, with the highest intensity around 800 nm.

Herein, we propose a procedure for the hydrothermal synthesis of small Yb/Tm activated SFL nanoparticles. Further, we documented their NIR-to-NIR UC. This UC process has been given much less attention in Yb^{3+} and Tm^{3+} co-doped phosphors than blue and deep-red UC emissions, although it can considerably expand the fields of application of UC nanophosphors, especially as a suitable fluorescent marker in the development of latent fingerprints.

2. Materials and Methods

2.1. Chemicals

Strontium nitrate ($Sr(NO_3)_2$, Alfa Aesar Karlsruhe, Germany, 99%), lanthanum (III) nitrate hexahydrate ($La(NO_3)_3 \cdot 6H_2O$, Alfa Aesar, 99.99%), ytterbium (III) nitrate hexahydrate ($Yb(NO_3)_3 \cdot 5H_2O$, Alfa Aesar, 99.9%), thulium (III) nitrate hexahydrate ($Tm(NO_3)_3 \cdot 6H_2O$, Alfa Aesar, 99.9%), disodium ethylendiaminetetraacetate dihydrate (EDTA-2Na, $C_{10}H_{14}N_2O_8Na_2 \cdot 2H_2O$, Kemika, Zagreb, Croatia, 99%), ammonium fluoride (NH_4F, Alfa Aesar, 98%), 25% ammonium solution (NH_4OH, Fisher, Loughborough, Leicestershire, United Kingdom) and de-ionized water were used as starting materials without further purification.

2.2. Synthesis of SLF:Yb,Tm

$Sr_2La_{1\text{-x-y}}F_7$:xYb,yTm were synthesized hydrothermally using $Sr(NO_3)_2$, Ln^{3+} nitrates (Ln = La,Yb,Tm), and NH_4F as precursors and EDTA-2Na as a stabilizing agent (see Figure 1). Typically, for the synthesis of 1 g of the representative sample Sr_2LaF_7 co-doped with 20mol% Yb^{3+} and 1 mol% Tm^{3+}, all nitrates were weighed according to the stoichiometric ratio (precisely, 0.4762g Sr^{3+}-nitrate, 0.3849 g La^{3+}-nitrate, 0.1010 g Yb^{3+}-nitrate and 0.0050 g Tm^{3+}-nitrate) and then dissolved in 12.5 mL deionized water while stirring at room temperature. The above solution was then mixed for 30 min with a transparent solution of 0.4188 g EDTA-2Na in 12.5 mL in water (molar ratio EDTA-2Na:La = 1:1). Following that, a 10 mL aqueous solution containing 0.5001 g of NH_4F (molar ratio NH_4F:La = 12:1) was added and vigorously stirred for 1 h, yielding a white complex. Using 400 μL of

NH4OH, the pH of the mixture was adjusted to around 6. This mixture was placed in a 100-mL Teflon-lined autoclave and heated in the oven at 180 °C for 20 h. After natural cooling, the final products were centrifuged and washed twice with water, then once with an ethanol:water = 1:1 mixture to remove any possible remnants before drying in an air atmosphere at 80 °C for 4 h. Undoped SLF and SLF phosphors with varying concentrations of Yb^{3+} (x = 10, 15, 20, and 25 mol%) and Tm^{3+} (y = 0.75, 1, 2, and 3 mol%) ions with respect to La^{3+} ions were prepared using the described procedure.

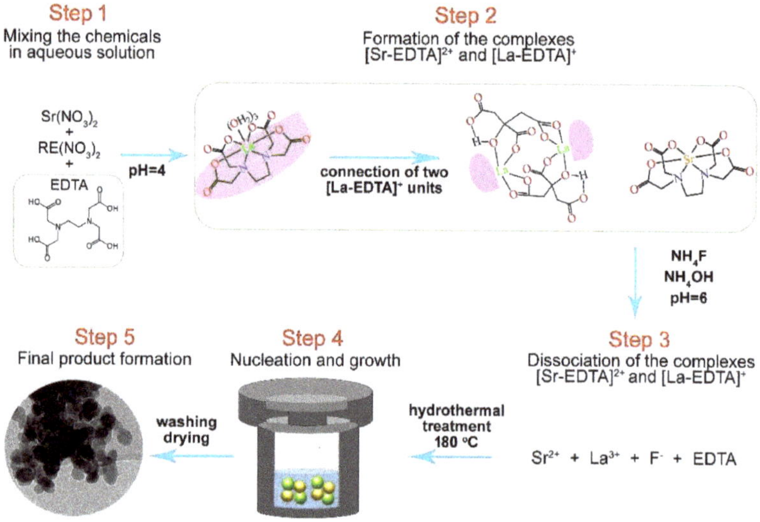

Figure 1. Synthesis of SLF:Yb,Tm nanophosphors using a simple EDTA-assisted hydrothermal method.

2.3. Measurement

X-ray diffraction (XRD) measurements were performed on a Rigaku SmartLab system operating with Cu Kα radiation (30 mA, 40 kV) in the 2θ range from 10° to 90°. Diffraction data were recorded with a step size of 0.02° and a counting time of 1°/min over the investigated 2θ. Results of the structural analysis (unit cell parameters, crystal coherence size, microstrain values, and data fit parameters) were obtained using the built-in PDXL2 software. The microstructure of the samples was characterized by a transmission electron microscope (TEM) Tecnai GF20 operated at 200 kV. The average particle size was calculated using ImageJ software. Diffuse reflectance measurements were performed with the Shimadzu UV-2600 (Shimadzu Corporation, Tokyo, Japan) spectrophotometer equipped with an integrated sphere (ISR-2600), using $BaSO_4$ as the standard reference. Luminescence characterization was done using a 980 nm high power (3W) solid state IR laser as an excitation source. Luminescence emissions were recorded using a FHR1000 monochromator (Horiba Jobin Yvon) and an ICCD camera (Horiba Jobin Yvon 3771). All the measurements were performed at room temperature.

3. Results and Discussion

3.1. XRD Analysis

The M_xLnF_{2x+3} fluorides crystallize in a cubic structure with $Fm\bar{3}m$ space group [31]. XRD patterns of SLF:xYb^{3+},1 mol%Tm^{3+} and SLF:20 mol%Yb^{3+},yTm^{3+} nanophosphors are shown in Figure 2a,b, respectively. Despite the addition of Yb^{3+} and Tm^{3+} ions, the main diffraction peaks observed around 2θ = 26.4, 30.7, 43.8, 51.9, 54.4, 63.7, 70.2, and 80.5°, correspond to the main reflections from 111, 200, 220, 311, 222, 400, 331, and 422 crystal planes, respectively, and are well-aligned with the standard data of ICDD No. 00–053–0774 for pure SLF. Diffraction peaks corresponding to other phases and/or impurities were not noticed. The sharp diffraction peaks indicate a good crystallinity of SLF nanophosphors. Tables 1 and 2 show the results of the structural analysis using whole pattern-fitting (WPF) refinement: crystallite coherence size (CS), microstrain values, unit cell parameters, unit cell volume (CV), and data fit parameters (R_{wp}, R_p, R_e and GOF) of SLF:xYb^{3+},1 mol%Tm^{3+} and SLF:20 mol%Yb^{3+},yTm^{3+} nanophosphors. The *CS* of pure SLF is estimated to be 27.1 nm, and the lattice constant *a* is 5.8451 Å (CV = 199.70 Å^3). The influence of Yb^{3+} doping in SLF lattice causes the linear host lattice shrinkage up to a = 5.8045 Å, CV = 195.57 Å^3 for the sample SLF:25 mol%Yb^{3+},1 mol%Tm^{3+}. This shrinkage could be ascribed to the fact that dopants with smaller ionic radii Yb^{3+} (0.868 Å) and Tm^{3+} (0.880 Å) replace the La^{3+} with larger ionic radii (1.032 Å) in SLF [35]. Tm^{3+} doping in SLF lattice also produces host lattice shrinkage up to a = 5.8107 Å, CV = 196.2 Å^3 for the sample SLF:20 mol%Yb^{3+}, 2 mol%Tm^{3+}. When the concentration of Tm^{3+} ions is increased further, the other doping strategy occurs due to the Tm^{3+} ability to occupy the interstitial sites, leading to crystal lattice expansion (a = 5.8288 Å, CV = 198.03 Å^3) [36]. To identify the strategy of Tm^{3+} and Yb^{3+} doping in SLF, the magnified (111) diffraction peak of the samples are shown in Figure 2c,d. With the addition of Yb^{3+} ions, the position of the (111) diffraction peak consistently shifts to higher degree values, and the shift becomes more notable with increasing dopant concentration (Figure 2c). Tm^{3+} doping in SLF has the same behavior at concentrations equal to or less than 2 mol%, while further addition of Tm^{3+} shifts the diffraction peak to lower degree values (Figure 2d). These findings confirm that both Yb^{3+} and Tm^{3+} (equal or less than 2 mol%) were successfully inserted into SLF by occupying La^{3+} sites.

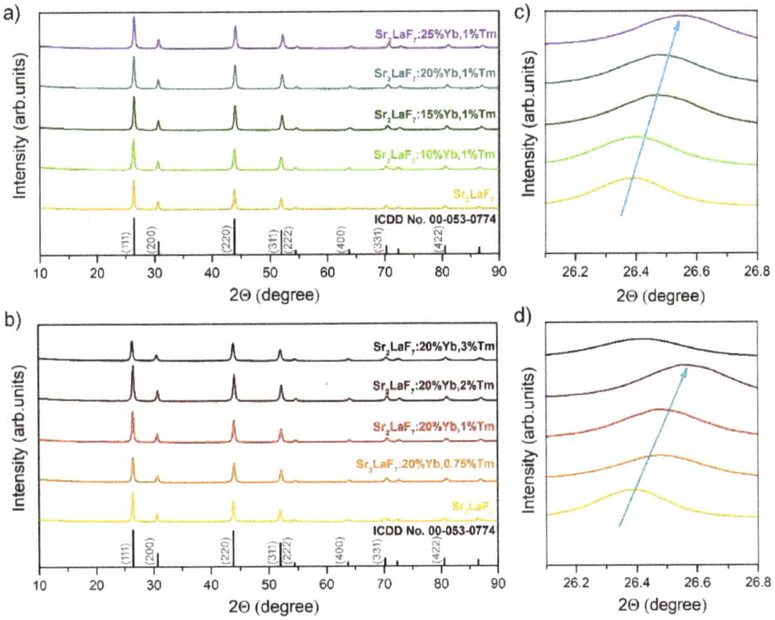

Figure 2. XRD patterns of (**a**) SLF:xYb^{3+},1 mol%Tm^{3+} and (**b**) SLF:20 mol%Yb^{3+},yTm^{3+} nanophosphors. (**c**) The evolution of the (111) diffraction peak magnified from (**a**). (**d**) The evolution of the (111) diffraction peak magnified from (**b**). Arrows indicate changes of the peak position maximum.

Table 1. Results of the structural analysis by using WPF refinement of SLF:xYb^{3+}, 1 mol%Tm^{3+} nanophosphors.

	SLF	SLF:10Yb,1Tm	SLF:15Yb,1Tm	SLF:20Yb,1 Tm	SLF:25Yb,1Tm
CS (nm)	27.1	23.5	25	25.4	24.4
Strain	0.144	0.254	0.254	0.272	0.258
* R_{wp}	6.00	4.74	4.58	4.33	4.09
** R_p	4.62	3.79	3.51	3.35	3.28
*** R_e	3.42	3.48	3.37	3.37	3.28
GOF	1.7569	1.3597	1.3597	1.2843	1.2484
$a = b = c$ (Å)	5.8451	5.8372	5.8190	5.8115	5.8045
CV (Å^3)	199.70	198.89	197.04	196.27	195.57

* R_{wp}—the weighted profile factor; ** R_p—the profile factor; *** R_e—the expected weighted profile factor; GOF—the goodness of fit.

Table 2. Results of the structural analysis by using WPF refinement of SLF:20 mol%Yb^{3+},yTm^{3+} nanophosphors.

	SLF	SLF:20Yb,0.75Tm	SLF:20Yb,1Tm	SLF:20Yb,2Tm	SLF:20Yb,3Tm
CS (nm)	27.1	22.6	25.4	21.9	17.8
Strain	0.144	0.211	0.272	0.185	0.02
* R_{wp}	6.00	4.12	4.33	4.91	5.97
** R_p	4.62	3.14	3.35	3.90	4.41
*** R_e	3.42	3.56	3.37	3.33	3.39
GOF	1.7569	1.1589	1.2843	1.4747	1.7607
$a = b = c$ (Å)	5.8451	5.8129	5.8115	5.8107	5.8288
CV (Å^3)	199.70	196.42	196.27	196.20	198.03

* R_{wp}—the weighted profile factor; ** R_p—the profile factor; *** R_e–the expected weighted profile factor; GOF—the goodness of fit.

3.2. Morphology Analysis

An assisted EDTA hydrothermal method was used to create SLF:Yb, Tm nanophosphors. EDTA is an efficient complexing agent for Ln^{3+} ions, with chelation constants (logK_1) of 19.51 and 15.50 for Yb^{3+} and La^{3+} ions, respectively [31]. Due to the ability to improve crystalline seed dispersibility by forming $[Sr\text{-}EDTA]^{2+}$ and $[La\text{-}EDTA]^{+}$ complexes after mixing all of the chemicals, EDTA prevented SLF particle aggregation during the subsequent hydrothermal treatment. On the other hand, $[La\text{-}EDTA]^{+}$ cations could be absorbed on the surfaces of SLF particles, limiting their further growth into large particles, and also increasing their stability [36,37].

TEM images of undoped SLF with different magnifications together with particle size distribution histogram are shown in Figure 3I. Nanoparticles show a similar, quasi-spherical shape as well as a high degree of crystallinity. HRTEM image of SLF phosphors (Figure 3Ie) shows that the measured *d*-spacing is around 3.4 Å, that corresponds to the (111) lattice plane of SLF, which agrees to the previous XRD data. The half-displayed particles were not considered when calculating the average particle size, and the histogram was fitted with a log-normal distribution. The average crystalline size of nanoparticles, considering around 120 particles, was estimated to be 38 ± 4 nm (see Figure 3If). The influence of Yb^{3+} and Tm^{3+} co-doping on the morphology of SLF samples can be observed by comparing features in Figure 3I, 3II and 3III, respectively. The average particle size of SLF nanophosphor doped with 10 mol% Yb^{3+} (Figure 3IIa–f) and SLF doped with 25 mol% Yb^{3+} (Figure 3IIIa–f) ions and a fixed concentration level of 1 mol% Tm^{3+} was calculated to be 25 ± 3 nm and 26 ± 2 nm, respectively. Therefore, the average particle size of SLF was reduced by doping from 38 to around 25 nm, which is well-aligned with the previous XRD analysis. HRTEM images of both SLF:10Yb,1Tm (Figure 3IIe) and SLF:25Yb,1Tm phosphors (Figure 3IIIe) show that the measured d-spacing is around 3.5 Å, which also corresponds to the (111) lattice plane of SLF. As previously explained, the addition of Yb^{3+}/Tm^{3+} ions has a slight impact on the lattice constant when compared to undoped

SLF because dopants with smaller ionic radii Yb^{3+} and Tm^{3+} replace the La^{3+} with larger ionic radii. The average particle size of SLF nanophosphor, on the other hand, is strongly influenced by the concentrations of Yb^{3+} and Tm^{3+}.

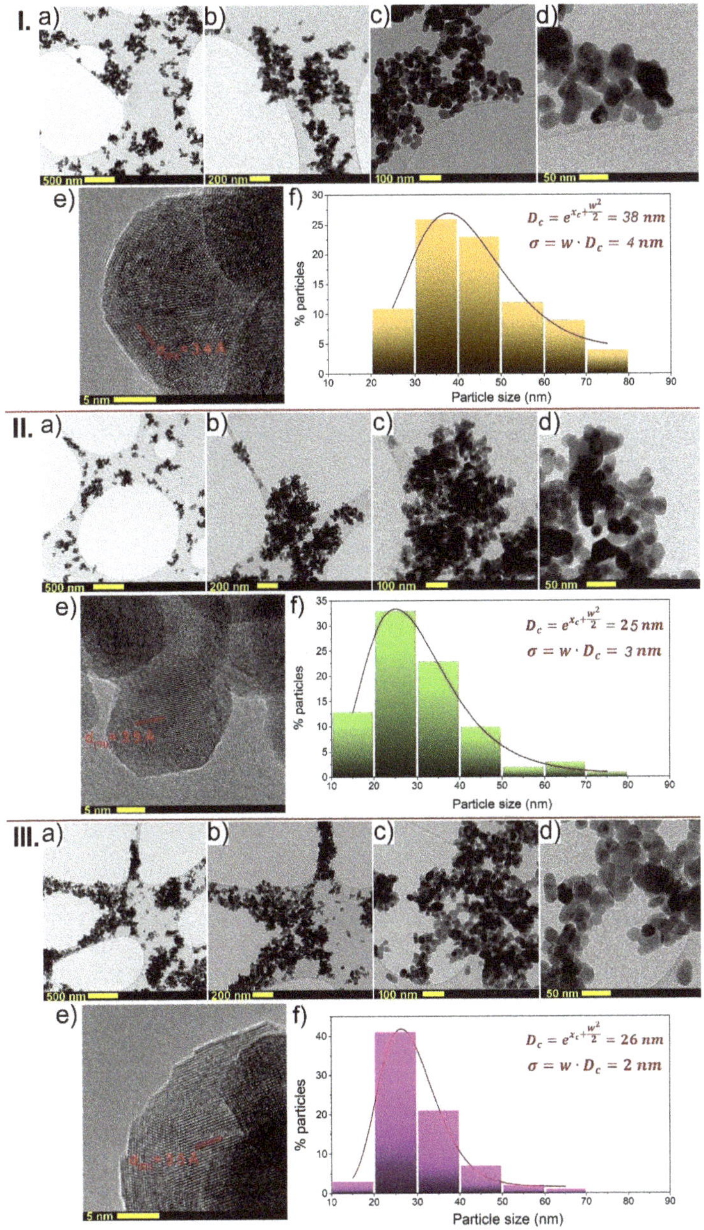

Figure 3. I. Undoped SLF: (**a–d**) TEM images of hydrothermally synthesized SLF, (**e**) HRTEM image of SLF, (**f**) particle size distribution of SLF. **II. SLF:10Yb,1Tm:** (**a–d**) TEM images (**e**) HRTEM image, (**f**) particle size distribution. **III. SLF:25Yb,1Tm:** (**a–d**) TEM images, (**e**) HRTEM image, (**f**) particle size distribution.

3.3. Spectroscopic Properties

Figure 4a shows the room temperature diffuse reflectance spectra of a representative SLF:20Yb,1Tm sample in the 400–1300 nm wavelength range with typical optical features of Yb^{3+} and Tm^{3+} ions [38]. The absorption peaks of Yb^{3+} ions appear in the 885–1060 nm wavelength range due to electronic transitions from $^2F_{7/2} \rightarrow {}^2F_{5/2}$, with the highest intensity around 980 nm. In the case of Tm^{3+} ions, three major electronic transitions are involved: $^3H_6 \rightarrow {}^3F_{2,3}$, $^3H_6 \rightarrow {}^3H_4$, and $^3H_6 \rightarrow {}^3H_5$, which correspond to absorption peaks at 677 nm, 770 nm, and 1206 nm, respectively.

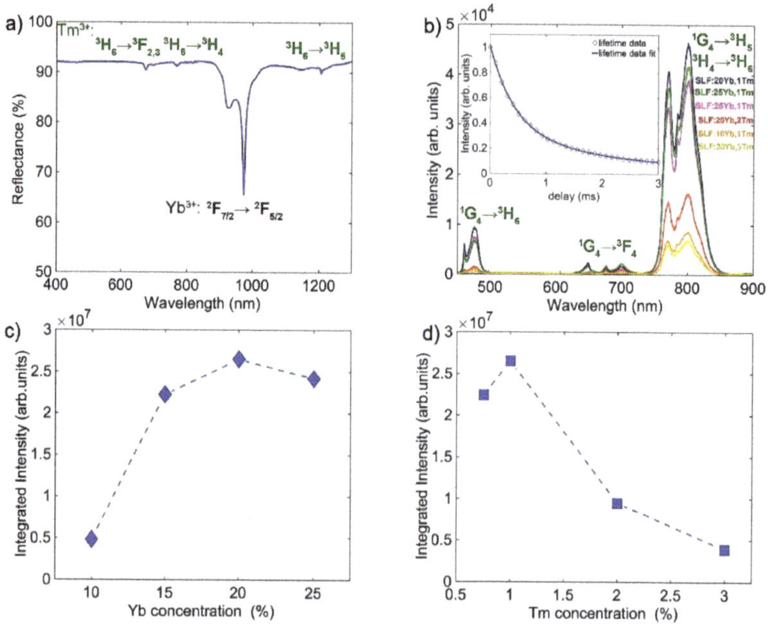

Figure 4. (**a**) Diffuse reflectance spectra of representative SLF:20Yb,1Tm nanophosphor. (**b**) PL spectra of SLF: xYb, yTm nanophosphors measured upon 980 nm excitation (Inset: UC lifetime profiles of Tm^{3+} ($^3H_4 \rightarrow {}^3H_6$ transition) of representative SLF:20Yb,1Tm). (**c**) Yb^{3+} concentration dependence of integrated UC emission intensity of SLF nanophosphor. (**d**) Tm^{3+} concentration dependence of integrated UC emission intensity of SLF nanophosphor.

The room temperature emission spectra of a representative SLF:20Yb,1Tm nanophosphor in the 450–900 nm wavelength range are shown in Figure 4b. In a typical multiphoton UC process, Yb^{3+} absorbs NIR radiation at around 980 nm which causes electron excitation from $^2F_{7/2}$ to $^2F_{5/2}$ energy level. Then, the Tm^{3+} is excited via cross-relaxation and energy transfer from excited Yb^{3+}. The deexcitation from multiple Tm^{3+} excited levels provide emissions that cover UV–VIS–NIR spectra. The observed emission peaks, which occur at wavelengths ranging from 455 to 500 nm, 625 to 720 nm, and 750 to 850 nm, are attributed to the transitions from the $^1G_4 \rightarrow {}^3H_6$, $^1G_4 \rightarrow {}^3F_4$, and $^1G_4 \rightarrow {}^3H_5$ / $^3H_4 \rightarrow {}^3H_6$ of excited Tm^{3+} ions, respectively. The PL decay curve at room temperature is shown in the inset of Figure 4b. The average emission time (τ_{av}), calculated based on the double exponential model, was used as a measurement of PL lifetime (τ). Through the fit of our experimental data to the double exponential model, the two values of τ_1 and τ_2 are obtained:

$$I(t) = A_1 e^{-\frac{t}{\tau_1}} + A_2 e^{-\frac{t}{\tau_2}} + bg. \quad (1)$$

where, A_1 and A_2 are arbitrary constants (magnitudes of short and long decay components), and bg is a background correction. Because the measured signal (I (t)) at delayed time t_d is proportional to the number of excited states at the moment t_d, the simple weighted average formula is used to calculate τ_{av}:

$$\tau_{av} = \frac{A_1\tau_1 + A_2\tau_2}{A_1 + A_2}. \quad (2)$$

The PL lifetime of Tm^{3+} ($^3H_4 \rightarrow {}^3H_6$ transition) in a representative SLF:20Yb,1Tm nanophosphor was estimated to be 1.05 ms. Table 3 summarizes the arbitrary constants, background correction, two values of PL lifetime (τ_1 and τ_2), and average PL lifetime of the representative SLF:20Yb,1Tm nanophosphor. The deviation of the $^3H_4 \rightarrow {}^3H_6$ emission decay from the single exponentiality indicates that energy back-transfer to Yb^{3+} occurs from $^3H_4$ level. This can be further investigated by measuring the variation in lifetimes of Yb^{3+} $^2F_{5/2}$ emission for different Tm^{3+} and Yb^{3+} concentrations.

Table 3. Summary of the different parameters used to calculate PL lifetime of the representative SLF:20Yb,1Tm nanophosphor.

A_1	τ_1 (ms)	A_2	τ_2 (ms)	bg	τ (ms)
0.6887	0.4503	0.3163	2.3607	0.0030	1.0516

The UC emission intensity relates to both Yb^{3+} and Tm^{3+} concentrations. Figure 4c presents the dependence of the integrated UC emission intensity of SLF with different concentrations of Yb^{3+} (x = 10, 15, 20, and 25 mol%) and a fixed Tm^{3+} concentration (1 mol%). With increasing Yb^{3+} concentration, the NIR emission intensity band increases, reaching a maximum value at 20 mol% of Yb^{3+}. Similarly, when Yb^{3+} concentration is fixed at 20 mol%, the NIR emission of SLF monitored at different concentrations of Tm^{3+} (x = 0.75, 1, 2, and 3 mol%) has the highest intensity for 1 mol% Tm^{3+}, as shown in Figure 4d. When the Tm^{3+} doping concentration is equal to or greater than 2, the emission intensity decreases gradually due to the concentration quenching. In contrast to the emission intensity, the shape and characteristic peak position of the UC emission spectra have not changed. For the Yb^{3+} and Tm^{3+} co-doped phosphors, blue ($^1G_4 \rightarrow {}^3H_6$) and deep-red ($^1G_4 \rightarrow {}^3F_4$) emissions have been widely investigated [27,38–40], while the efficient NIR-to-NIR Yb^{3+}/Tm^{3+} UC emission in ultrasmall SLF nanophosphor has received far less attention. Therefore, ultrasmall SLF nanoparticles with intense emission around 800 nm are promising candidates as fluorescent labels in bioimaging and security applications.

4. Conclusions

In conclusion, ultrasmall SLF:Yb^{3+}/Tm^{3+} nanoparticles were produced using a straightforward hydrothermal process at a variety of doping doses. With Yb^{3+} and Tm^{3+} ions present, the lattice constant and average particle size of SLF are both decreased from 38 nm to roughly 25 nm. At room temperature, Yb^{3+} and Tm^{3+} concentrations have a significant impact on the PL emission properties. When excited with a 980 nm high power (3W) solid state IR laser, these ultrasmall nanoparticles show simultaneous three-color (blue-green, deep-red, and NIR) UC emissions in the 450–900 nm wavelength range. The blue-green and deep-red emission bands are weak, while the NIR emission band is strong, which is beneficial for imaging biological tissues. Furthermore, the PL lifetime of Tm^{3+} ($^3H_4 \rightarrow {}^3H_6$ transition) in a representative SLF:20Yb,1Tm nanophosphor was estimated to be 1.05 ms. These findings also suggest that SLF:Yb,Tm could be a useful fluorescent marker in the development of latent fingerprints. Our future work will concentrate on the dual-mode fluorescent development of latent fingerprints using both NIR-to-VIS and NIR-to-NIR processes to achieve double fluorescent images in dark and bright fields, as well as additional contrast and sensitivity analysis of fingerprints or fingerprint residuals deposited on a variety of substrates. Furthermore, the proposed NIR-to-NIR UC mechanism of ultrasmall Yb^{3+}/Tm^{3+} co-doped SLF nanophosphors could be a useful tool in security applications.

Author Contributions: Conceptualization, A.S. and M.D.D.; methodology, Z.R., K.S., A.S. and M.D.D.; formal analysis, B.M., J.P., Z.R., K.S., M.K., K.V., A.S. and M.D.D.; investigation, B.M., J.P., K.M., M.K., Ž.A. and K.V.; data curation, B.M. and J.P.; writing—original draft preparation, B.M. and Z.R.; writing—review and editing, A.S. and M.D.D. All authors have read and agreed to the published version of the manuscript.

Funding: This research was funded by the Ministry of Science, Technological Development and Innovation of the Republic of Serbia and by the European Union's Horizon 2020 Framework Programme H2020-WIDESPREAD-01-2016-2017-TeamingPhase2 under grant agreement No. 739508, project CAMART2.

Data Availability Statement: The data presented in this study are available on request from the corresponding author.

Acknowledgments: The authors from Serbia acknowledge funding from the Ministry of Science, Technological Development and Innovation of the Republic of Serbia. The authors from Latvia acknowledge funding from the European Union's Horizon 2020 Framework Programme H2020-WIDESPREAD-01-2016-2017-TeamingPhase2 under grant agreement No. 739508, project CAMART2.

Conflicts of Interest: The authors declare no conflict of interest.

References

1. Peijzel, P.S.; Meijerink, A.; Wegh, R.T.; Reid, M.F.; Burdick, G.W. A complete $4f^n$ energy level diagram for all trivalent lanthanide ions. *J. Solid State Chem.* **2005**, *178*, 448–453. [CrossRef]
2. Zhou, B.; Shi, B.; Jin, D.; Liu, X. Controlling upconversion nanocrystals for emerging applications. *Nat. Nanotechnol.* **2015**, *10*, 924–936. [CrossRef] [PubMed]
3. Zou, W.; Visser, C.; Maduro, J.A.; Pshenichnikov, M.S.; Hummelen, J.C. Broadband dye-sensitized upconversion of near-infrared light. *Nat. Photonics* **2012**, *6*, 560–564. [CrossRef]
4. Erol, E.; Vahedigharehchopogh, N.; Kıbrıslı, O.; Ersundu, M.Ç.; Ersundu, A.E. Recent progress in lanthanide-doped luminescent glasses for solid-state lighting applications-a review. *J. Phys. Condens. Matter* **2021**, *33*, 483001. [CrossRef]
5. Chen, E.Y.; Milleville, C.; Zide, J.M.O.; Doty, M.F.; Zhang, J. Upconversion of low-energy photons in semiconductor nanostructures for solar energy harvesting. *MRS Energy Sustain.* **2018**, *5*, e16. [CrossRef]
6. Skwierczyńska, M.; Stopikowska, N.; Kulpiński, P.; Kłonowska, M.; Lis, S.; Runowski, M. Ratiometric Upconversion Temperature Sensor Based on Cellulose Fibers Modified with Yttrium Fluoride Nanoparticles. *Nanomaterials* **2022**, *12*, 1926. [CrossRef]
7. Ryszczyńska, S.; Trejgis, K.; Marciniak, Ł.; Grzyb, T. Upconverting $SrF_2:Er^{3+}$ Nanoparticles for Optical Temperature Sensors. *ACS Appl. Nano Mater.* **2021**, *4*, 10438–10448. [CrossRef]
8. Guo, M.Y.; Shen, L.F.; Pun, E.Y.B.; Lin, H. Sr_2GdF_7: Er^{3+}/Yb^{3+} nanocrystal-inlaid pliable fibers for synergistic feedback temperature monitoring. *J. Lumin.* **2022**, *252*, 119394. [CrossRef]
9. Baride, A.; Sigdel, G.; Cross, W.M.; Kellar, J.J.; May, P.S. Near Infrared-to-Near Infrared Upconversion Nanocrystals for Latent Fingerprint Development. *ACS Appl. Nano Mater.* **2019**, *2*, 4518–4527. [CrossRef]
10. Wang, M.; Li, M.; Yang, M.; Zhang, X.; Yu, A.; Zhu, Y.; Qiu, P.; Mao, C. NIR-induced highly sensitive detection of latent fingermarks by $NaYF_4$:Yb,Er upconversion nanoparticles in a dry powder state. *Nano Res.* **2015**, *8*, 1800–1810. [CrossRef]
11. Kavand, A.; Serra, C.A.; Blanck, C.; Lenertz, M.; Anton, N.; Vandamme, T.F.; Yves Mély, Y.; Przybilla, F.; Chan-Seng, D. Controlled Synthesis of $NaYF_4$:Yb,Er Upconversion Nanocrystals as Potential Probe for Bioimaging: A Focus on Heat Treatment. *ACS Appl. Nano Mater.* **2021**, *4*, 5319–5532. [CrossRef]
12. Chen, G.; Qiu, H.; Prasad, P.N.; Chen, X. Upconversion Nanoparticles: Design, Nanochemistry, and Applications in Theranostics. *Chem. Rev.* **2014**, *114*, 5161–5214. [CrossRef] [PubMed]
13. Ji, G.; Wang, Y.; Qin, Y.; Peng, Y.; Li, S.; Han, D.; Ren, S.; Qin, K.; Li, S.; Gao, Z.; et al. Latest developments in the upconversion nanotechnology for rapid detection of food safety: A review. *Nanotechnol. Rev.* **2022**, *11*, 2110–2122. [CrossRef]
14. May, P.S.; Baride, A.; Hossan, M.Y.; Berry, M. Measuring the Internal Quantum Yield of Upconversion Luminescence for Ytterbium-Sensitized Upconversion Phosphors Using the Ytterbium-(III) Emission as an Internal Standard. *Nanoscale* **2018**, *10*, 17212–17226. [CrossRef] [PubMed]
15. Gao, G.; Busko, D.; Joseph, R.; Howard, I.A.; Turshatov, A.; Richards, B.S. Highly Efficient $La_2O_3:Yb^{3+},Tm^{3+}$ Single-Band NIR-to-NIR Upconverting Microcrystals for Anti-Counterfeiting Applications. *ACS Appl. Mater. Interfaces* **2018**, *10*, 39851–39859. [CrossRef]
16. Baride, A.; Meruga, J.M.; Douma, C.; Langerman, D.; Crawford, G.; Kellar, J.J.; Cross, W.M.; May, P.S. A NIR-to-NIR Upconversion Luminescence System for Security Printing Applications. *RSC Adv.* **2015**, *5*, 101338–101346. [CrossRef]
17. Kim, J.; Kwon, J.H.; Jang, J.; Lee, H.; Kim, S.; Hahn, Y.K.; Kim, S.K.; Lee, K.H.; Lee, S.; Pyo, H.; et al. Rapid and Background-free Detection of Avian Influenza Virus in Opaque Sample using NIR-to-NIR Upconversion Nanoparticle-based Lateral Flow Immunoassay Platform. *Biosens. Bioelectron.* **2018**, *112*, 209–215. [CrossRef]

18. Levy, E.S.; Tajon, C.A.; Bischof, T.S.; Iafrati, J.; Fernandez-Bravo, A.; Garfield, D.J.; Chamanzar, M.; Maharbiz, M.M.; Sohal, V.S.; Schuck, P.J. Energy-Looping Nanoparticles: Harnessing Excited-State Absorption for Deep-Tissue Imaging. *ACS Nano* **2016**, *10*, 8423–8433. [CrossRef]
19. Xia, A.; Chen, M.; Gao, Y.; Wu, D.; Feng, W.; Li, F. Gd^{3+} complex-modified $NaLuF_4$-based upconversion nanophosphors for trimodality imaging of NIR-to-NIR upconversion luminescence, X-Ray computed tomography and magnetic resonance. *Biomaterials* **2012**, *33*, 5394–5405. [CrossRef]
20. Ortgies, D.H.; Tan, M.; Ximendes, E.C.; Del Rosal, B.; Hu, J.; Xu, L.; Wang, X.; Martín Rodríguez, E.; Jacinto, C.; Fernandez, N.; et al. Lifetime-encoded infrared-emitting nanoparticles for in vivo multiplexed imaging. *ACS Nano* **2018**, *12*, 4362–4368. [CrossRef]
21. Ragin, T.; Baranowska, A.; Kochanowicz, M.; Zmojda, J.; Miluski, P.; Dorosz, D. Study of Mid-Infrared Emission and Structural Properties of Heavy Metal Oxide Glass and Optical Fibre Co-Doped with Ho^{3+}/Yb^{3+} Ions. *Materials* **2019**, *12*, 1238. [CrossRef] [PubMed]
22. Cao, W.; Feifei Huang, F.; Ye, R.; Cai, M.; Lei, R.; Zhang, J.; Shiqing Xu, S.; Zhang, X.H. Structural and fluorescence properties of Ho^{3+}/Yb^{3+} doped germanosilicate glasses tailored by Lu_2O_3. *J. Alloys Compd.* **2018**, *746*, 540–548. [CrossRef]
23. Kowalska, K.; Kuwik, M.; Pisarska, J.; Pisarski, W.A. Near-IR Luminescence of Rare-Earth Ions (Er^{3+}, Pr^{3+}, Ho^{3+}, Tm^{3+}) in Titanate–Germanate Glasses under Excitation of Yb^{3+}. *Materials* **2022**, *15*, 3660. [CrossRef] [PubMed]
24. Yang, T.; Qin, J.; Zhang, J.; Guo, L.; Yang, M.; Wu, X.; You, M.; Peng, H. Recent Progresses in NIR-II Luminescent Bio/Chemo Sensors Based on Lanthanide Nanocrystals. *Chemosensors* **2022**, *10*, 206. [CrossRef]
25. Ledoux, G.; Joubert, M.F.; Mishra, S. Upconversion Phenomena in Nanofluorides. In *Photonic & Electronic Properties of Fluoride Materials*, 1st ed.; Tressaud, A., Poeppelmeir, K.R., Eds.; Elsevier: Paris, France, 2016; pp. 35–63.
26. Tiwari, S.P.; Maurya, S.K.; Yadav, R.S.; Kumar, A.; Kumar, V.; Joubert, M.-F.; Swart, H.C. Future prospects of fluoride based upconversion nanoparticles for emerging applications in biomedical and energy harvesting. *J. Vac. Sci. Technol. B* **2018**, *36*, 060801. [CrossRef]
27. Krämer, K.W.; Biner, D.; Frei, G.; Güdel, H.U.; Hehlen, M.P.; Lüthi, S.R. Hexagonal Sodium Yttrium Fluoride Based Green and Blue Emitting Upconversion Phosphors. *Chem. Mater.* **2004**, *16*, 1244–1251. [CrossRef]
28. Runowski, M.; Bartkowiak, A.; Majewska, M.; Martín, I.R.; Lis, S. Upconverting Lanthanide Doped Fluoride $NaLuF_4$:Yb^{3+}-Er^{3+}-Ho^{3+}- Optical Sensor for Multi-Range Fluorescence Intensity Ratio (FIR) Thermometry in Visible and NIR Regions. *J. Lumin.* **2018**, *201*, 104–109. [CrossRef]
29. Gonçalves, I.M.; Pessoa, A.R.; Hazra, C.; Correales, Y.S.; Ribeiro, S.J.L.; Menezes, L.d.S. Phonon-assisted NIR-To-Visible Upconversion in Single β-$NaYF_4$ Microcrystals Codoped with Er^{3+} and Yb^{3+} for Microthermometry Applications: Experiment and Theory. *J. Lumin.* **2021**, *231*, 117801. [CrossRef]
30. Li, J.; Wang, Y.; Zhang, X.; Li, L.; Hao, H. Up-Converting Luminescence and Temperature Sensing of $Er^{3+}/Tm^{3+}/Yb^{3+}$ Co-Doped $NaYF_4$ Phosphors Operating in Visible and the First Biological Window Range. *Nanomaterials* **2021**, *11*, 2660. [CrossRef]
31. Grzyb, T.; Przybylska, D. Formation Mechanism, Structural, and Upconversion Properties of Alkaline Rare-Earth Fluoride Nanocrystals Doped with Yb^{3+}/Er^{3+} Ions. *Inorg. Chem.* **2018**, *57*, 6410–6420. [CrossRef]
32. Xia, Z.; Du, P.; Liao, L. Facile hydrothermal synthesis and upconversion luminescence of tetragonal Sr_2LnF_7:Yb^{3+}/Er^{3+} (Ln=Y, Gd) nanocrystals. *Phys. Status Solidi A* **2013**, *210*, 1734–1737. [CrossRef]
33. Mao, Y.; Ma, M.; Gong, L.; Xu, C.; Ren, G.; Yang, Q. Controllable synthesis and upconversion emission of ultrasmall near-monodisperse lanthanide-doped Sr_2LaF_7 nanocrystals. *J. Alloys Compd.* **2014**, *609*, 262–267. [CrossRef]
34. Xie, J.; Bin, J.; Guan, M.; Liu, H.; Yang, D.; Xue, J.; Liao, L.; Mei, L. Hydrothermal Synthesis and Upconversion Luminescent Properties of Sr_2LaF_7 Doped with Yb^{3+} and Er^{3+} Nanophosphors. *J. Lumin.* **2018**, *200*, 133–140. [CrossRef]
35. Shannon, R.D. Revised Effective Ionic Radii and Systematic Studies of Interatomie Distances in Halides and Chaleogenides. *Acta Crystallogr.* **1976**, *A32*, 751–767. [CrossRef]
36. Wang, Z.; Li, Y.; Jiang, Q.; Zeng, H.; Ci, Z.; Sun, L. Pure near-infrared to near-infrared upconversion of multifunctional Tm^{3+} and Yb^{3+} co-doped $NaGd(WO_4)_2$ nanoparticles. *J. Mater. Chem. C* **2014**, *2*, 4495–4501. [CrossRef]
37. Alkahtani, M.; Alfahd, A.; Alsofyani, N.; Almuqhim, A.A.; Qassem, H.; Alshehri, A.A.; Almughem, F.A.; Hemmer, P. Photostable and Small YVO_4:Yb,Er Upconversion Nanoparticles in Water. *Nanomaterials* **2021**, *11*, 1535. [CrossRef]
38. Li, L.; Pan, Y.; Chang, W.; Feng, Z.; Chen, P.; Li, C.; Zeng, Z.; Zhou, X. Near-infrared downconversion luminescence of $SrMoO_4$:Tm^{3+},Yb^{3+} phosphors. *Mater. Res. Bull.* **2017**, *93*, 144–149. [CrossRef]
39. Simpson, D.A.; Gibbs, W.E.K.; Collins, S.F.; Blanc, W.; Dussardier, B.; Monnom, G.; Peterka, P.; Baxter, G.W. Visible and near infra-red up-conversion in Tm^{3+}/Yb^{3+} co-doped silica fibers under 980 nm excitation. *Opt. Express* **2008**, *16*, 13781–13799. [CrossRef]
40. Güell, F.; Solé, R.; Gavaldà, J.; Aguiló, M.; Galán, M.; Díaz, F.; Massons, J. Upconversion luminescence of Tm^{3+} sensitized by Yb^{3+} ions in monoclinic $KGd(WO_4)_2$ single crystals. *Opt. Mater.* **2007**, *30*, 222–226. [CrossRef]

nanomaterials

Article

Monitoring the Surface Energy Change of Nanoparticles in Functionalization Reactions with the NanoTraPPED Method

Andrei Honciuc * and **Oana-Iuliana Negru**

Petru Poni Institute of Macromolecular Chemistry, Aleea Gr. Ghica Voda 41A, 700487 Iasi, Romania
* Correspondence: honciuc.andrei@icmpp.ro

Abstract: Performing chemical functionalization on the surface of nanoparticles underlies their use in applications. Probing that a physicochemical transformation has indeed occurred on a nanoparticles' surface is rather difficult. For this reason, we propose that a macroscopic parameter, namely the surface energy γ, can monitor the physicochemical transformations taking place at the surface of nanoparticles. Determining the surface energy of macroscopic surfaces is trivial, but it is very challenging for nanoparticles. In this work we demonstrate that the Nanoparticles Trapped on Polymerized Pickering Emulsion Droplet (NanoTraPPED) method can be successfully deployed to monitor the evolution of surface energies γ, with its γ^p polar and γ^d dispersive components of the silica nanoparticles at each stage of two surface reactions: (i) amination by siloxane chemistry, coupling reaction of a 2,4-dihydroxy benzaldehyde and formation of a Schiff base ligand, followed by coordination of metal ions and (ii) epoxide ring opening and formation of azide. The change in surface energy and its components are discussed and analyzed for each step of the two reactions. It is observed that large variations in surface energy are observed with the complexity of the molecular structure attaching to nanoparticle surface, while functional group replacement leads to only small changes in the surface energies.

Keywords: surface energy; contact angle; nanoparticles; interfaces; Pickering emulsions; interfacial energy; self-assembly of nanoparticles

Citation: Honciuc, A.; Negru, O.-I. Monitoring the Surface Energy Change of Nanoparticles in Functionalization Reactions with the NanoTraPPED Method. *Nanomaterials* **2023**, *13*, 1246. https://doi.org/10.3390/nano13071246

Academic Editor: Antonio Guerrero-Ruiz

Received: 16 March 2023
Revised: 30 March 2023
Accepted: 30 March 2023
Published: 31 March 2023

1. Introduction

The functional groups present onto the surface of nanoparticles (NPs) determine the way the NPs interact with the environment, controlling their behavior in bulk powders but also in liquid or gaseous media, especially their chemical reactivity. Performing surface chemical modification on the surface of nanoparticles to make them useful in a variety of applications has become ubiquitous and routine [1]. The chemical transformations taking place at the surface of nanoparticles following chemical treatment, transformation of chemical functional groups via oxidation, coupling, condensation, polymerization reactions, or even physical adsorption and coordination of metal ions, is difficult to monitor but can be done in principle by employing advanced surface specific spectroscopy methods. Although one can enumerate some surface sensitive spectroscopy methods capable of resolving chemical composition and surface functional groups, these are generally difficult to deploy on powders consisting of nanoparticles. However, even though the presence of a certain functional group, of several or combinations thereof on the surface of NPs could be resolved through various spectroscopy methods, this gives little information on the actual physicochemical behavior of the nanoparticle in bulk powder or its preferred way to interact with the environment. In an earlier work, we have alluded at the possibility that a macroscopic parameter, namely the surface energy, which describes the capacity of a surface to interact with its environment through various physicochemical forces [2], can serve as one of the essential parameters to describe the physicochemical state of the NPs, and could help establish a causal relationship with the NPs' behavior in bulk to help design powders with good flowing ability [3], improve dispersion of pigments in polymer matrices [4], understand powder cohesion [5], flotation of ore and minerals [6], emulsification and creation of Pickering emulsions [7], stability of colloids [2], etc. Even more, based on the governing principle of independent actions, the total surface energy γ, regardless of scale, can be seen as the sum of various components that reflect the capacity of a surface to interact with the environment through different kind of interaction forces, such as dipole–dipole γ^p, dispersive van der Waals interaction forces γ^d, hydrogen bonding γ^{H-H}, acid γ^A and base γ^B, and so on. This constitutes a further advantage in the holistic characterization of the physicochemical state of NPs as the relative magnitude of the surface energy components provide meaningful information of the preferred type of interaction of a particular surface. For a macroscopic surface, the surface energy and its polar and dispersive components γ, γ^p and γ^d can be determined from the contact angle of at least two liquid droplets, see the Owens–Wendt–Rabel–Kaelble (OWRK) model [8]. However, in case of nanoparticle surfaces, making droplets smaller than the nanoparticles to determine their contact angle is technically very challenging. Although some methods have been developed to specifically address the measurement of the contact angle of a single NP with a liquid [9–11], none could be deployed yet in measurement of the contact angles of NPs with multiple liquids, which is necessary to determine the surface energy and its components. To overcome this, we have implemented a new method called Nanoparticles Trapped on Polymerized Pickering Emulsion Droplet (NanoTraPPED), and demonstrated that this method is capable of measuring the contact angles of NPs at various oil/water interfaces. Additionally, from these values of interfacial energies, the surface energies with the polar and disperse components can be determined [2]. We have shown that water/NP interfacial energies obtained with NanoTraPPED are extremely useful at predicting the dispersibility and stability of various NPs in water, while the surface energies are useful in predicting their emulsification ability and monitoring the physicochemical state of NPs in air. In this work we demonstrate that the NanoTraPPED method can be successfully deployed to monitor the evolution and change of interfacial NP/water energies $\gamma_{NP/water}$, $\gamma^p_{NP/water}$ and $\gamma^d_{NP/water}$ and surface energies γ_{NP}, γ^p_{NP} and γ^d_{NP} of fresh silica NPs with a hydroxylated surface (NP-OH). They undergo surface physicochemical transformations, amination by siloxane chemistry (NP-NH_2), the coupling reaction of a 2,4-dyhydroxy benzaldehyde and the formation of a Schiff base ligand (NP-L), followed by the capturing and coordination of Cu(II) (NP-LCu) or Co(II) (NP-LCu) ions, according to the reactions presented in Scheme 1. Interestingly, the total surface energy and its components increases with the increase in the complexity of the molecular structure of the nanoparticle surface.

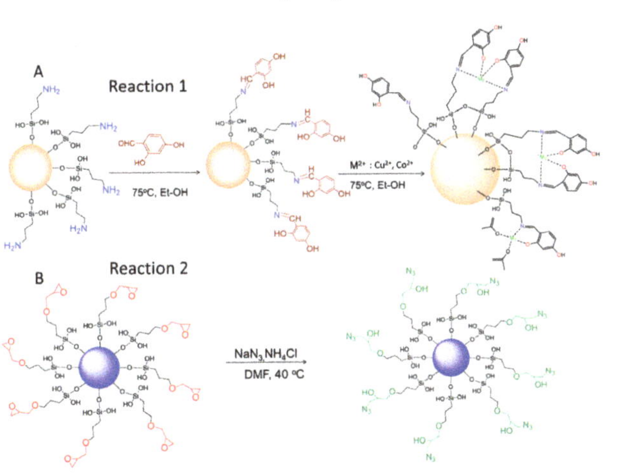

Scheme 1. (**A**) Formation of Schiff base interfacial ligand, NP-L, by condensation of the 2,4-dyhydroxy benzaldehyde with the amine surface functionalities, NP-NH_2, and formation of the complex by

coordination of a divalent metal ion Cu(II) or Co(II), giving rise to NP-LCu and NP-LCu. (**B**) Functional group replacement reaction of glycidyl with azide by ring opening and substitution reaction.

2. Materials and Methods

2.1. Materials

Tetraethylorthosilicate (TEOS) 99% and benzoin methyl ether (BME) 97% were purchased from ABCR; GmbH, 3-(trimethoxysilyl) propylamine 98% (APTES), 3,(3-glycidoxypropyl)trimethoxysilane (GLYMO), divinylbenzene (DVB) technical grade 80%, benzyl metacrylate (BM) 96% containing monomethyl ether hydroquinone as inhibitor, tert-butyl acrylate (tBA) 98% containing 10–20 ppm monomethyl ether hydroquinone as inhibitor, methyl methacrylate (MM) 99% stabilized for synthesis with monomethyl ether hydroquinone as inhibitor, 2-(Dimethylamino)ethyl methacrylate (DAEMA) containing 700–1000 ppm monomethyl ether hydroquinone as inhibitor 98%, ethyl methacrylate (EM) 98% stabilized with 200 ppm hydroquinone, 2,4-dyhydroxy benzaldehyde, ammonium chloride, sodium azide, aluminum oxide (active basic) Brockmann I and N,N-Dimethylformamide anhydrous 99.8% were purchased from Sigma-Aldrich; absolute ethanol (EtOH, 99.3%), hydrochloric acid (HCl), $Cu(CH_3COO)_2{\cdot}H_2O$ and $Co(CH_3COO)_2{\cdot}4H_2O$ were purchased from Chemical Company; ammonium hydroxide solution (28–30%) for analysis was purchased from EMSURE ACS. Reag. Ph Eur. Supelco.

2.2. Synthesis of Silica Nanoparticles

The preparation procedure for silica nanoparticles (NP) and silica nanoparticles functionalized with amine by reaction with APTES was reported previously [2]; the functionalization with epoxy by reaction with GLYMO was also reported [7].

2.3. Synthesis of NP-L, NP-LCu and NP-LCo

NP-Ligand (NP-L) was synthesized using the post-grafting method, as shown in Scheme 1. A total of 1.2 g of NP-NH2 was then refluxed with 50 mg of 2,4-dihydroxybenzaldehyde in ethanol for 24 h at 75 °C. The resulting NP-L dispersion was centrifuged, and the particles were then rinsed three times with ethanol and distilled water. Thereafter, we prepared a dispersion containing 0.5 g NP-L in of 20 mL absolute ethanol over which we added dropwise a solution of 45 mg $Cu(CH_3COO)_2{\cdot}H_2O$ dissolved in 10 mL absolute ethanol or 40 mg $Co(CH_3COO)_2{\cdot}4H_2O$ dissolved in 10 mL absolute ethanol. The reaction took place at 70 °C for 24 h. After the end of the reaction, the NPs were washed three times with ethanol and three times with distilled water and stored in water for subsequent use in emulsification.

2.4. Synthesis of NP-N_3

NP-N_3 was obtained following the nucleophilic substitution reaction at the epoxy ring in the presence of NaN_3 and NH_4Cl. In a typical procedure, we started with a suspension of 1 g NP-Gly in approx. 40 mL DMF to which we added 32.5 mg (5 mmol) of NaN_3 and 26.8 mg (5 mmol) of NH_4Cl. The reaction mixture was allowed to react at 40 °C for 24 h. After the end of the reaction, the nanoparticles were washed three times with ethanol and three times with water and stored in water for subsequent use in emulsification.

2.5. Pickering Emulsion Preparation and Polymerization

Water immiscible vinyl bearing monomers EM, BM, tBA, DAEMA and MM, having different polarities, whose chemical structures are given in Figure S3 in the Supplementary Material, and a crosslinking monomer DVB, were used for the Pickering emulsion preparation and polymerization. The Pickering emulsion was produced by first adding 20 mg of BME radical initiator to a 20 mL glass scintillator vial, followed by 1 mL of monomer and 0.1 mL of DVB crosslinker, followed by 5 min of waiting for the mixture to become homogeneous. Next, 5 mg of colloidal particles and 12 mL of water were added. The glass scintillator vial was then sonicated with a Branson 450 Sonifier equipped with a 7 mm diameter horn for 15 s at 30% amplitude. The Pickering emulsions were next exposed for 1 h to a UV lamp (wavelength = 365 nm, with 4 lamps, each with an intensity = 2.2 mW/cm^2). The synthesis procedures and recipes for each monomer are listed in Table S2. After the polymerization, the product was filtered and thoroughly washed with approximately 15 mL of ethanol to remove the unreacted monomer, and subsequently dried at room temperature.

3. Results

The starting silica NPs were synthesized according to a previously reported method [2], which is a modified version of the Stöber process, see Scheme S1 in the Supplementary Material.

3.1. Functionalization and Surface Reactions on Silica Nanoparticles

The pristine surface of the silica NPs has mostly hydroxyl surface functionalities, hence the hitherto notation NP-OH. The starting NP-OH have a diameter of $\approx$500 $\pm$ 15 nm, see Figure S1, and were employed in further functionalization such that the subsequent NPs have (i) amine surface functionalities, see Scheme S2, hence the hitherto notation NP-NH2, or (ii) glycidyl functionalities, hence the hitherto notation NP-Gly, see Scheme S2. The presence of the amine functional groups on the NP-NH_2 can be evidenced both from the ninhydrin reaction, which turns purple in the presence of primary amine, and from the presence of the peaks in the FTIR spectra, see Figure 1. The presence of glycidyl on the surface of the NPs was confirmed with FTIR. We also noted a change in the zeta ζ-potential from -53.7 ± 0.5 mV in NP-OH to lower absolute values -19.4 ± 0.8 mV in NP-NH_2 at neutral pH, see Table S1. This change in the ζ-potential value can be explained by partial protonation of the amine group even at neutral pH.

Further, the NP-NH_2 were used in a condensation reaction with 2,4-dyhydroxy benzaldehyde to generate a Schiff base interfacial ligand, NP-L, and subsequent capture and coordination of Cu(II) and Co(II) metal cations (Reaction 1), see Scheme 1A. Alternatively, the NP-Gly were employed in a substitution reaction of the glycidyl group with azide, NP-N_3, (Reaction 2), see Scheme 1B.

For Reaction 1, the successful functionalization and the formation of the interfacial Schiff base ligand can be confirmed using FTIR, according to the spectra present in Figure 1A. The stretching and bending vibrations of silanol groups (Si-OH) are represented by the bands at 3428 (broad) and 1630 cm^{-1} and are easily identified for the starting NP-OH and NP-NH_2, Figure 1A. However, with the coupling of the 2,4-dyhydroxy benzaldehyde and formation of a Schiff base ligand, the imine -C=N stretching vibration becomes somewhat visible between 1632 and1618 cm^{-1} Figure 1A. In addition, the formation of the Schiff base ligand is confirmed by the appearance of the several relevant new bands, in NP-L, NP-LCu and NP-LCo at 1461 cm^{-1}, which correspond to the aromatic stretching -C=C-. It is also confirmed by the one at 730 cm^{-1}, which is due to the deformation vibrations of C-H from the trisubstituted benzene nucleus, as well as the phenolic C-OH stretching vibrations at 1278 cm^{-1}. It is interesting to note that the participation of the -OH groups in the coordination of the metal ions is signaled by the shift in the -OH to the vibration of the phenolic group, υ (C-OH), from 1278 cm^{-1} in NP-L, to the lower value of 1253 cm^{-1} in NP-LCu and NP-LCo. The same shift is seen in the stretching frequency of the azomethine υ(C=N) group in the NP-L to lower wavenumbers from 1632 cm^{-1} in NP-L to 1618 cm^{-1} in NP-LCu and NP-LCo complexes due to the coordination of the nitrogen atom of the azomethine group to the metal ion. Peaks in the region of 2900–3200 cm^{-1}, corresponding to C-H stretching on NP-OH are mainly due to the incomplete hydrolysis of the precursors, or the capture of the organic hydrolysis fragments into the structure of the NPs; these are always present and disappear only after calcination. For this reason, no spectral interpretation was performed in the region.

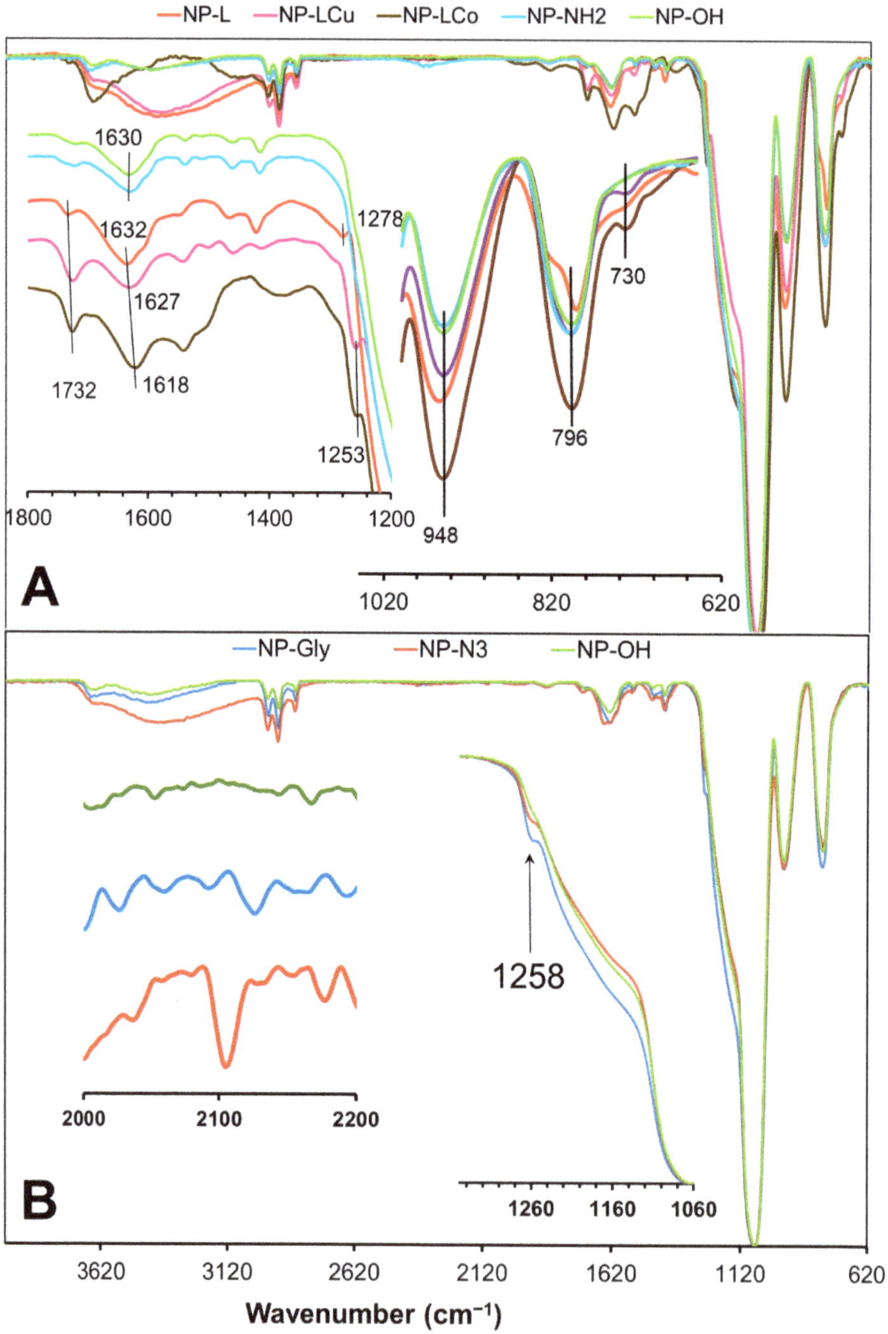

Figure 1. Normalized FTIR spectra of (**A**) Reaction 1: NP-OH (light green), $NP-NH_2$ (light blue) and modified NP-L (red), NP-LCu (violet) and NP-LCo (khaki); (**B**) Reaction 2: NP-OH (light green), NP-Gly (blue) and modified $NP-N_3$ (crimson). The insets show the offset of the normalized spectra in the regions of interest.

Furthermore, the formation of the interfacial Schiff base ligand, is also supported by the change in the zeta potential value, see Table S1, which decreases from -19.4 ± 0.8 mV in NP-NH_2 to -28.6 ± 0.6 mV in NP-L at normal pH. This evidences even further that the presence of primary amines has been reduced by their participation in the condensation reaction with the 2,4-dyhydroxy benzaldehyde. This is also confirmed by the surface zeta potential change, from -19.4 ± 0.8 mV of the former to a lower surface potential value of -28.6 ± 0.6 mV in the case of the latter. Surface zeta potential decreases to a lower value due to the disappearance of the -NH_2 and -NH_3^+ functionality after the condensation reaction with the aldehyde. Furthermore, the coordination of the Co(II) and Cu(II) ions can be evidenced by a decrease in the zeta potential value from -28.6 ± 0.6 mV in NP-L to -39.6 ± 0.3 mV in NP-LCu and -40.5 ± 0.2 mV in NP-LCo. This can be interpreted as either the coordination of the metal ions that have further diminished the primary amine functional groups on the surface of the NP-L, or that the imine functionalities are also diminished by the coordination of the metal ions. It is worth noting that mere physical adsorption of the divalent metal ions does not produce a dramatic change towards titration of surface charge. This is the case for the trivalent ions [1], as judged from the negative zeta potential values of NPSiCu and NPSiCo, which were NPs left in the presence of a 0.16×10^{-3} M solution of Co(II) and 0.25×10^{-3} M solution Cu(II) salts and then washed, see Table S1.

For Reaction 2, the successful functionalization and the formation of the NP-Gly and NP-N3 can be confirmed from the analysis of the FTIR spectra, Figure 1B. For example, the shoulder band at 1258 cm^{-1} appearing for the NP-Gly could be identified as the ring vibrations of epoxy, confirming that the surface modification was successful. The epoxides exhibit distinctive IR absorption bands related to ring breathing and asymmetrical ring stretching between 950 and 810 cm^{-1} and 1260 to 1240 cm^{-1}. This band at 1258 cm^{-1} is non-existent in NP-OH but can be seen in NP-N3 with a much weaker intensity, suggesting that majority of them had undergone ring opening. The nucleophilic reaction of the epoxy ring with sodium azide is confirmed by the appearance of a small signal in the region of 2200–2050 cm^{-1}, which is the characteristic stretching vibration of ν(–N-N≡N) in azide. In addition, changes are observed in the region of 1650–1300 cm^{-1}, such as the appearance of a new shoulder located at 1438 cm^{-1}. This corresponds to the deformation vibration of the CH_2 group, followed by the decrease in intensity of the band at 1258 cm^{-1}, signaling the epoxy ring opening in NP-N3.

3.2. *Pickering Emulsion Formation and Polymerization*

Pickering emulsions are emulsions stabilized by nanoparticles of different sizes and they were named after S. U. Pickering (1907). Particles stabilize the emulsion by adsorbing irreversibly at the oil/water interface and forming a shield-like monolayer film at the interface [12–14]. Some particles adsorb spontaneously at the oil/water interface, but others can be forced to adsorb by kinetic energy input, to overcome the interfacial adsorption barrier [13,15]. Once adsorbed they become trapped at the oil/water interface. It has been experimentally demonstrated that the emulsion droplet sizes and the emulsion phase are determined exclusively by the contact angle β of the nanoparticle at the water/oil interface, where the oil can be any organic liquid immiscible with water. In the current case, we convene that the contact angle β is the contact angle formed by the nanoparticle with the organic phase, see Figure S2. Our experiment is based on the principles laid forward by Finkle [16], as well as through the experimental observations of Binks and Clint [17] and Aveyard [18], who tried to predict the emulsion phase from the values of the contact angle. We have concluded, also through our experimental investigations [7], that the emulsion phase and oil droplet sizes generated by the spherical nanoparticles depend on the value of the contact angle β at the interface. Furthermore, if the oil phase is chosen among the vinyl bearing monomers, the Pickering emulsion can be easily polymerized either thermically or initiated by exposure to UV, such that the nanoparticles trapped at the oil/water interface remain frozen. This enables their observation with scanning electron microscopy (SEM).

This is the core principle of the NanoTraPPED method. In this work, we have chosen the following vinyl bearing monomers that are immiscible with water and can be polymerized: MM, EM, BM, DAEMA, tBA and DVB, whose chemical structures are given in Figure S3. For the Pickering emulsion stabilization we have chosen NP-OH of diameters of $\approx 500 \pm 15$ nm, shown in the SEM images, which we have subsequently modified, see Figure S1. Upon polymerization PMM, PEM, PBM, PDAEMA and PtBA microspheres covered with nanoparticles are obtained. The PEM, PBM and PtBA microspheres obtained from the Pickering emulsions of the corresponding monomers in water stabilized by the NP-L with the are presented in Figure 2. Those for NP-LCu and NP-LCo are given in Figures S4 and S5. From the inset of the SEM images, in Figure 2, the NPs self-assemble around the Pickering emulsion droplets of the corresponding monomers in water, stabilizing the emulsion, which after the polymerization remain trapped at the interface. Some of the NPs have already fallen, leaving on the surface of the polymer circular traces, whose diameter is direct evidence of their immersion at the interface as it will be discussed next.

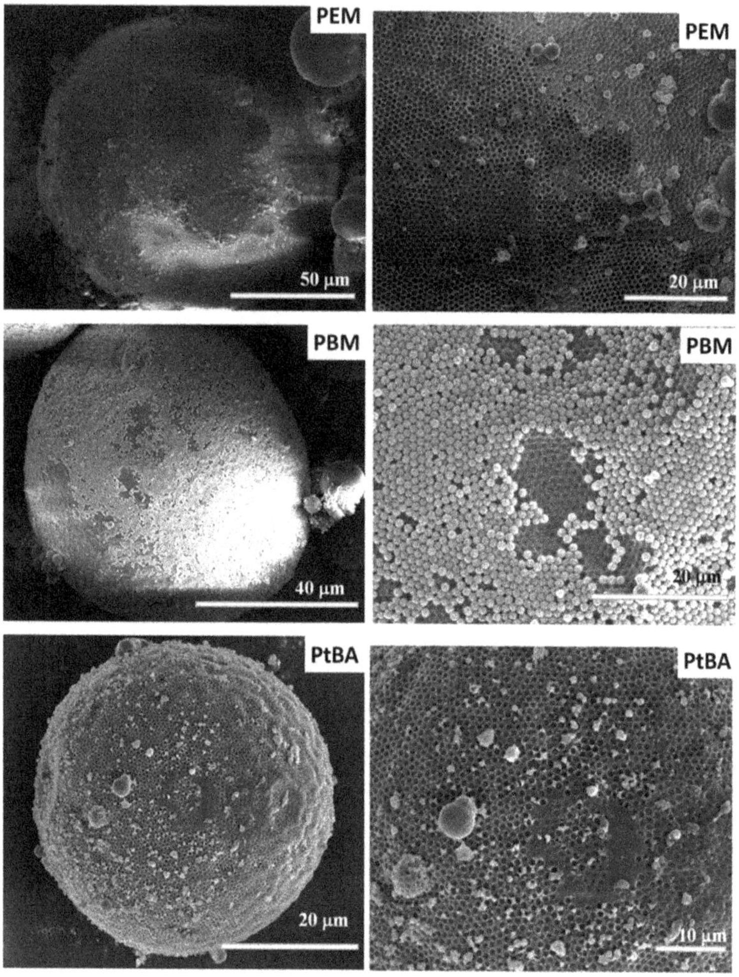

Figure 2. SEM images of the microspheres obtained after the polymerization of the Pickering emulsions of EM, BM and tBA in water, and the corresponding circular traces left by the stabilizing NP-L after Pickering emulsion polymerization.

Similarly, the PBM, PDAEMA, PMM, PEM, PtB and PBM microspheres resulting from the Pickering emulsion polymerization of the corresponding monomer emulsions stabilized with NP-Gly and NP-N3 are given in Figures S6 and S7, respectively. Generally, the microspheres obtained in this case are spherical, and their surface is sometimes slightly wavy or wrinkled, which is due to interfacial jamming of nanoparticles.

An overview of the composition of the Pickering emulsions recipes for each monomer as well as the polymerization conditions are given in the Table S2. The surface-trapped nanoparticles can be removed to observe the traces left by the nanoparticles, as we will cover next.

3.3. *Measuring the Contact Angles—The NanoTraPPED Method*

The principles of the NanoTraPPED method have been previously discussed [2]. In the current case, the diameter of the circular traces left on the surface of the microparticles obtained via Pickering emulsion polymerization were observed with scanning electron microscopy (SEM) after the removal of the silica nanoparticles by dissolving them with concentrated solution sodium hydroxide, or by simple ultrasonication. The diameter of the circular traces remaining on the surface of the microspheres depend on the immersion depth of the NPs at the interface between water and monomer before polymerization. As previously discussed [2,7], the immersion depth depends on the affinity of the surface of the NPs to the organic phase. The diameter of the circular traces can be geometrically related to the contact angle β of the nanoparticle, see Figure S2. $\theta = \sin^{-1}\frac{l}{r}$, $\beta = 180^\circ - \theta$, where l is the diameter of the circular trace imprinted by the nanoparticle on the surface of the polymer and r is the radius of the NPs. The contact angle β of the NP-L with PEM, PBM and PtBA, as well as the SEM images from which it was measured, are presented in Figure 3. The results for the NP-LCu and NP-LCo are given in Figures S8 and S9, respectively. It is important to note that the contact angle β measured with NanoTraPPED is the contact angle of the NPs at the three-phase line of NP–polymer (NP/P), NP–water (NP/W) and polymer–water (P/W), as indicated by the subscript to the Greek letter γ throughout this work. From Figure 3, it can be qualitatively seen that the β of NP-L at the interface increases with the decrease in the polarity of the polymer, where the decrease in polarity in the order PEM > PBM > PtBA is quantitatively ranked according to the ratio of the polarity to the disperse component of the surface energy $\sqrt{\frac{\gamma_P^p}{\gamma_P^d}}$. The γ_P^p and γ_P^d for each of the polymers used in this work were experimentally determined in a previous work [2].

Similarly, the circular traces left on the surface of the polymer microspheres by NP-Gly and NP-N_3 after surface reactions are given in Figure 4 and Figure S10, respectively. When comparing the contact angle β of the same polymers with different nanoparticles, for example NP-L in Figure 3 with NP-Gly in Figure 4, or NP-LCo in Figure S9 with NP-N_3 in Figure S10, these appear to be significantly different. Whereas, for the same nanoparticle but different polymers the contact angles vary much more subtly; compare the column of images contained within Figures 3, 4, S9 and S10. This proves that the method is extremely sensitive to a change in surface functional groups of NPs, and is less sensitive to the chemistry of the polymer. This can be explained by the fact that the nanoparticle–water, NP/w, interfacial tension unit vector, $\gamma_{NP/w}$, is larger than that of the nanoparticle–polymer, NP/P, $\gamma_{NP/P}$. The magnitude of these two interfacial tension unit vectors are the main parameters determining the value of the contact angle of the NPs at the oil/water interface [1]. Therefore, when a surface modification is performed on the NPs, it results in a much stronger response in the contact angle, due to a stronger change in $\gamma_{NP/w}$ than a change in $\gamma_{NP/P}$. This is valid for the o/w emulsions, for which NPs are more immersed in water phase than in the organic phase [7]. The opposite should be true for the inverse water-in-oil (w/o) Pickering emulsions.

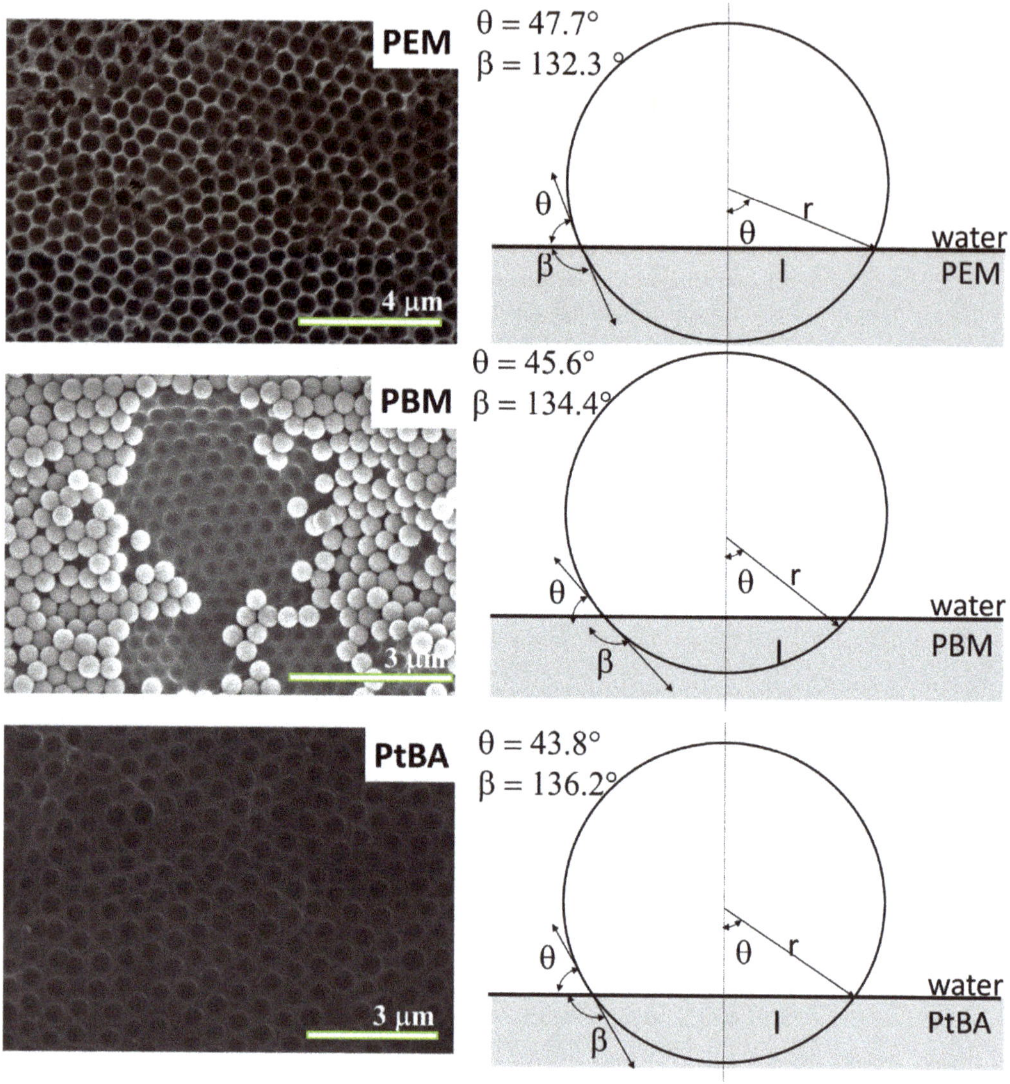

Figure 3. SEM images of the circular traces left by NP-L on the surface of PEM, PBM and PtBA microspheres. The cartoons on the right represent the immersion depth of the nanoparticles at the interface that can be fully described by the contact angle values θ with the water and β with the organic phase. In the image the l represents the diameter of the circular trace determined by SEM, and r represents theradius of the nanoparticle.

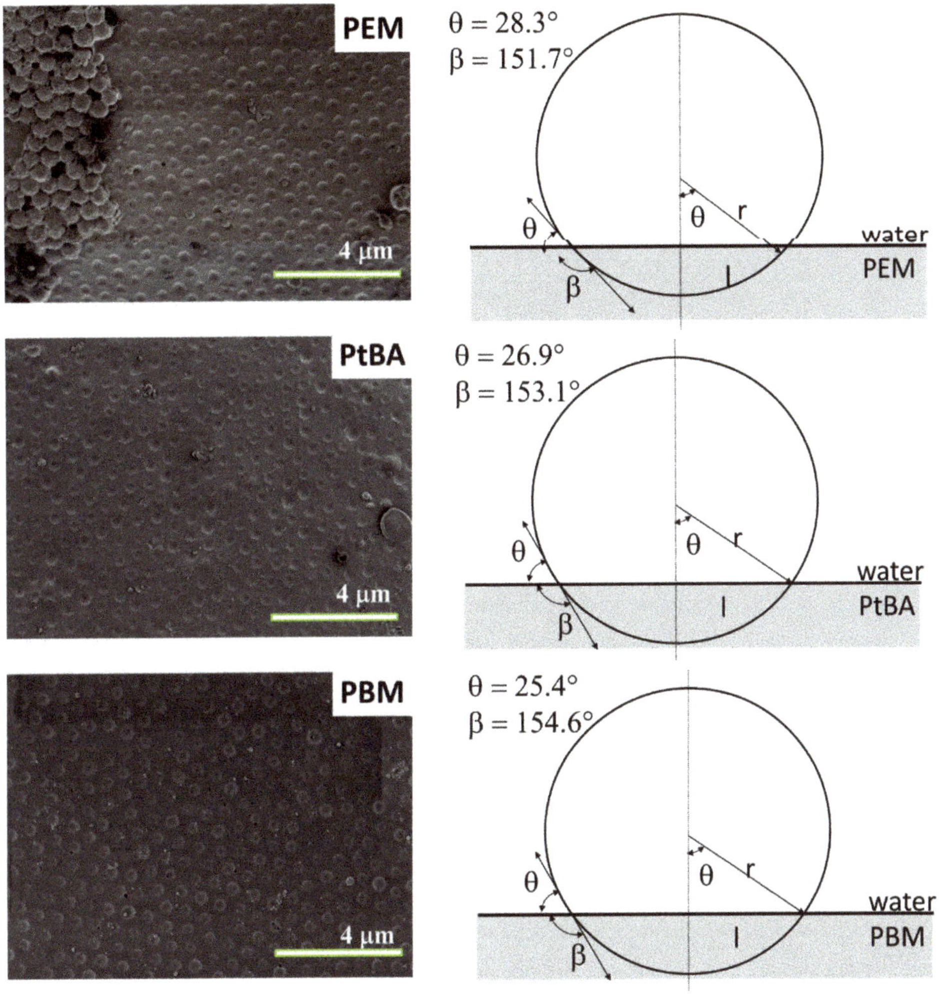

Figure 4. SEM images of the circular traces left by NP-Gly on the surface of PEM, PtBA and PBM microspheres. The cartoons on the right represent the immersion depth of the nanoparticles at the interface that can be fully described by the contact angle values θ with the water and β with the organic phase. In the image the l represents the diameter of the circular trace determined by SEM, and r represents the radius of the nanoparticle.

The experimental data of the circular hole diameters on the polymer colloidosomes left by the NPs and the calculated contact angles, are summarized in Table S3.

4. Discussion

4.1. Calculating the Surface Energy and Its Polar and Disperse Components

The surface energy of a surface and its polar and disperse components can be calculated if the contact angles with at least two liquids with known values for surface energy and the components are known. In the current case, the values of β—the contact angle of the NP at the interface with the polymer phases are used instead. Furthermore, we have used the Owens–Wendt–Rabel–Kaelble (OWRK) [8,19] model to determine the surface energy components from the contact angle β measured for the NPs with different polymers of known surface energy and components $\gamma^d_{P/W}$, $\gamma^p_{P/W}$ [2]:

$$\frac{\gamma_{P/W}(1+\cos\theta)}{2\sqrt{\gamma^d_{P/W}}} = \sqrt{\gamma^p_{NP/W}}\sqrt{\frac{\gamma^p_{P/W}}{\gamma^d_{P/W}}} + \sqrt{\gamma^d_{NP/W}} \quad (1)$$

Thus, by measuring the contact angle β of each nanoparticle with at least three different polymer/water interfaces, we calculated the total surface energy and its components $\gamma^p_{NP/W}$, $\gamma^d_{NP/W}$ in water by fitting the data to a linear equation where the $\sqrt{\gamma^p_{NP/W}}$ is the slope and the $\sqrt{\gamma^d_{NP/W}}$ is the intercept, as shown in the Figure 5 for Reaction 1 and Figure 6 for Reaction 2. Interestingly, for Reaction 1, from the data in Figure 5, one can see that the error bars for the NP-L and NP-LCo are overlapping throughout the range of the measurement, which suggests that these values are in fact indistinguishable from one another. On the other hand, the NP-LCu values differ for the latter portion of the curve, with clearly non-overlapping errors. Thus, we conclude that the NP-L and NP-LCo are indistinguishable, while the NP-LCu and NP-NH_2 could be clearly distinguished from the other groups of data, meaning that they are uniquely determined.

For Reaction 2, the slope of the NP-N_3 is markedly different from that of NP-Gly, suggesting that the polarity of the former is greater than that of the latter. For the latter portion of the curves in Figure 5 the error bars of the two curves do not overlap, suggesting that the group of data belonging to NP-N_3 is clearly distinguishable from that of NP-Gly.

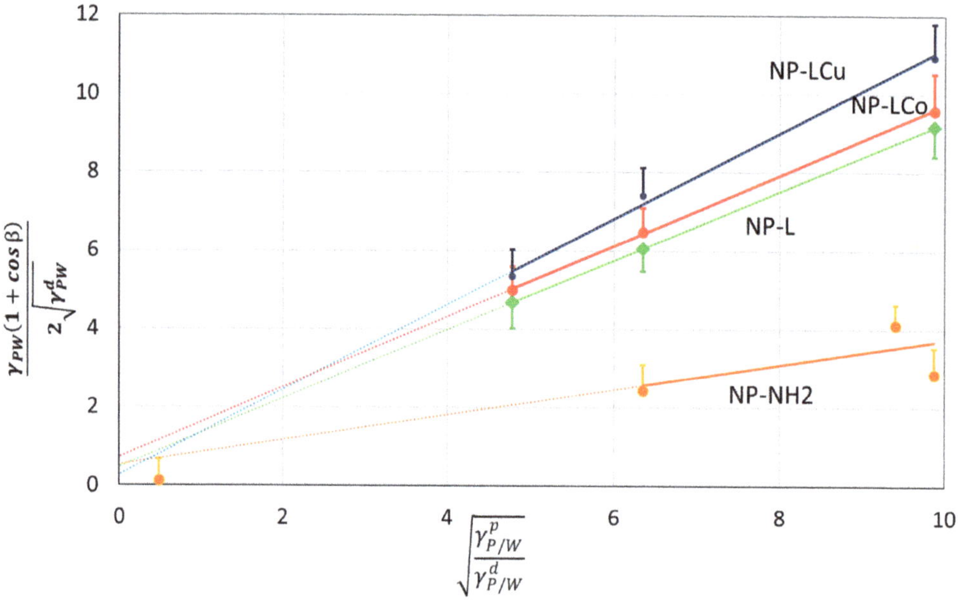

Figure 5. The linear OWRK equation was constructed from the experimental data of the contact angle measured from the circular traces left by the nanoparticles NP-NH_2, NP-L, NP-LCu and NP-LCo. The error bars represent the standard error and are represented only on the positive or negative side of the measured average value.

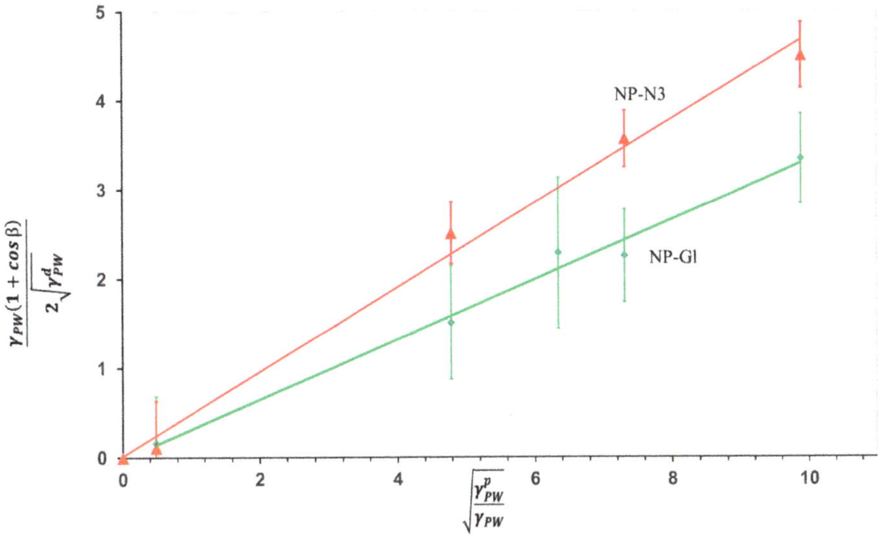

Figure 6. The linear OWRK equation was constructed from the experimental data of the contact angle measured from the circular traces left by the nanoparticles NP-Gly and NP-N_3. The error bars represent the standard error and are represented both on the positive and negative side of the measured average value.

4.2. Monitoring the Change in the Surface Energy of NPs with NanoTraPPED

The surface energies were determined for each of the NPs after each step of surface functionalization, namely the $\gamma^d_{NP/W}$, $\gamma^p_{NP/W}$ and the values are summarized in Table 1.

Table 1. Interfacial energies $\gamma^d_{NP/W}$, $\gamma^p_{NP/W}$ of the NPs adsorbed at the interface calculated with the OWRK equation graphically represented in Figures 5 and 6.

Nanoparticle	$\gamma^p_{NP/water}$ (mN/m)	$\gamma^d_{NP/water}$ (mN/m)	$\gamma_{NP/water}$ (mN/m)
* NP-OH	1.05	0.01	1.06
** NP-Gly	0.66	0.1	0.76
NP-N3	0.21	0.02	0.23
* NP-NH_2	0.13	0.00	0.13
NP-L	0.78	0.22	1.00
NP-LCu	1.16	0.08	1.24
NP-LCo	0.81	0.52	1.33

* values taken from reference [2]. ** values taken from reference [7].

From the data presented in the Table 1, for Reaction 1, we can state that the total interfacial energy of the NPs with water, $\gamma_{NP/water}$, increases in the order NP-NH_2 < NP-L < NP-LCu < NP-LCo. A higher value of the $\gamma_{NP/water}$ indicates a decreasing affinity between the NPs and water. Note that it is generally accepted that a vanishingly small interfacial tension indicates total wettability and adhesion of a liquid on a solid surface, or in the case of a liquid biphasic system, a vanishingly small interfacial tension between the liquids indicates a high level of miscibility to total miscibility. Similarly, here, the NP-NH_2 have the lowest total interfacial energy with water, signaling a good dispersibility of these in water. Although the others are comparable, we predict that the sedimentation of the nanoparticle colloid will be faster (these systems have only kinetic colloid stability) for the NP-LCu and NP-LCo than that of NP-NH_2 due to the observed difference in the interfacial energy. Here we note that the NanoTraPPED method directly yields the interfacial energies of the NPs, because the contact angle is measured directly at the three-phase line, or at the polymer/water interface. Conversely, for Reaction 2, the $\gamma_{NP/water}$ decreases with the substitution reaction, where NP-Gly > NP-N_3. This suggests that the NP-N_3 are better wetted by water.

Furthermore, the interfacial energies $\gamma^d_{NP/W}$, $\gamma^p_{NP/W}$ can easily be converted using the combining rules equation [1] into the surface energy of nanoparticles in air γ^d_{NP}, γ^p_{NP} :

$$\gamma^p_{NP} = \left(\sqrt{\gamma^p_{NP/W}} + \sqrt{\gamma^p_W}\right)^2 \tag{2}$$

Based on the above formula, we have recalculated the values of the surface energies and its disperse and polar components, γ^d_{NP}, γ^p_{NP} that are given in Table 2.

Table 2. Surface energies γ^d_{NP}, γ^p_{NP} of the NPs calculated with the Equation (2).

Nanoparticle	$\gamma^p{}_{NP}$ (mN/m)	$\gamma^d{}_{NP}$ (mN/m)	γ_{NP} (mN/m)
* NP-OH	25.40	33.92	59.32
** NP-Gly	26.02	51.72	77.74
NP-N3	27.43	53.47	80.90
*NP-NH_2	41.96	25.40	67.36
NP-L	59.86	30.83	90.69
NP-LCu	62.96	28.67	91.63
NP-LCo	60.18	33.62	93.80

* values taken from reference [2]. ** values taken from reference [7].

From the values presented in Table 2, a general trend can be observed, for the Reaction 1, namely the surface energy of the NPs, γ_{NP}, increases in the order NP-OH < NP-NH_2 < NP-L $\leq$ NP-LCu < NP-LCo. The steepest increase in the surface energy is first observed in Table 2, from NP-OH upon the attachment of 3-(triethoxysilyl)propylamine with the formation of NP-NH_2 functionality. This increase in the polar component γ^p_{NP} to the almost doubling of its value, although the -NH_2 has a smaller dipole moment than the -OH, see Table S4, can be explained by the introduction of a greater number of these functional groups on the surface of the NPs than that of hydroxyl groups. The second greatest change in surface energy can be observed for NP-L after the surface attachment of the ligand by the reaction between the 2,4-dyhydroxy benzaldehyde and the surface amine groups, see Table 2. Again, the change in the polar surface energy component is the steepest, γ^p_{NP}, and is much larger for NP-L than for NP-NH_2, while the disperse component varies only slightly. This can be explained by the increasing number of polar functional groups; for every ligand attached on the -NH_2, two more polar -OH moieties (Table S4) are introduced, see Scheme 1A.

Upon the exposure of the NP-L to a solution of 10^{-2} M of Cu^{2+} and Co^{2+} ions, the metal ion can be either attached to the surface of the NP-L by chemical reaction, coordination and formation of a chemical complex with the Schiff base, or physical adsorption on the negatively charged NP-L surfaces, see the zeta potential value in Table S1. While the surface attachment of cations via physical adsorption is inevitable, the coordinative binding of the metal ions by the ligand L was also demonstrated from the FTIR spectra. The change in the surface energy NP-L with the physisorption and chemical coordination of the of Co^{2+} with the formation of the NP-LCo complex is minimal, and is indistinguishable within the error limit of the method. Only in the case of Cu^{2+} does the formation of the NP-LCu lead to some more notable change in the total surface energy, mainly due to the increase in the polar component, see Table 2. The increase in the polar component with the coordination of the Cu(II) ions could be interpreted by the formation of highly polar coordinative L- > Cu(II) bonds, but this comes at the expense of the elimination of polar functionalities of imine -CH=N and hydroxyls -OH. However, the overall polarity of these interfacial complexes will also depend on their geometry. The only slight change in the polar γ_{NP}^{p} component with Co(II) and Cu(II) detected suggests that the transformation of the interfacial ligand and the formation of the interfacial complex does not come with a dramatic surface transformation, but it is still detectable with NanoTraPPED.

For Reaction 2, see Table 2, the functionalization from NP-OH to NP-Gly produces a stark change in the disperse component of the surface energy, while the functional group replacement from NP-Gly to NP-N_3 produces only a slight increase in both the polar and disperse energy component such that the total surface energy increases from 77.74 mN/m in the former to 80.9 mN/m in the latter case. Although these observed changes are small, the non-overlapping errors in Figure 6 suggests that these are significant and above the error limit of the method. We further note that in Reaction 2, similarly to Reaction 1, there are subtle changes in the functional groups, as the coordination of metal ions and substitution of glycidyl with azide produces only small changes in the total surface energy or its components. The polar surface energy γ_{NP}^{p} component increases in the order NP-OH > NP-Gly > NP-N_3, which seems to follow the increase in the overall dipole moment of the functional groups, see Table S4. The disperse component due to London dispersive interactions also increases in the same order.

Another interesting aspect to note is that the overall surface energy of the NPs seems to increase with the increase in the complexity of the molecular structure attached to the surface; that is, any surface modification with an increase in the number of bonds, branching or increase in number of atoms leads to an increase in the total surface energy.

5. Conclusions

In this work, we have demonstrated that the NanoTraPPED method can be deployed in monitoring the chemical transformation taking place at the surface of NPs. For this purposed we have designed two surface reactions that change the surface chemistry either by attachment or coupling of a molecular fragment to the surface or by replacement of a surface functionality. Stark changes in the surface energy of NPs are noted when a large molecular fragment is attached to the surface, such as the attachment of 3-(triethoxysilyl)propylamine, or 2,4-dyhydroxy benzaldehyde, and only minor changes in the surface energies are noted when slight changes in the nature of functional groups are performed, such as the substitution of the similar epoxy moiety with azide, or coordination of metal ions. Nonetheless, the NanoTraPPED proves to be a viable method to monitor the surface reactions and to fully characterize the physicochemical state of NPs.

Supplementary Materials: The following are available online at https://www.mdpi.com/article/10.3390/nano13071246/s1, Scheme S1: Surface functionalization reaction scheme, Figure S1: SEM images of the silica NPs before and after functionalization, Figure S2: Cartoon of nanoparticle adsorbed at the oil/water interface, chemical structure of the vinyl bearing monomers, Figure S3: Recipes for Pickering emulsion preparation and polymerization, Figures S4–S7: Microspheres of polymers with the circular traces obtained on their surface after the polymerization of the Pickering emulsion stabilized by different nanoparticles, Figures S8–S10: Diameters of the circular traces on the surface of the microspheres and the contact angles at the interface, Table S1: NP zeta potentials, Table S2: SEM images of microparticles, Table S3: Summary of all hole diameters and contact angles with the polymer, Table S4: calculated values of the electric dipole moments of the functional groups.

Author Contributions: Conceptualization, A.H.; Methodology, A.H. and O.-I.N.; Validation, A.H. and O.-I.N.; Formal analysis, A.H.; Investigation, O.-I.N.; Resources, O.-I.N.; Data curation, A.H.; Writing—original draft preparation, A.H.; Writing—review and editing, A.H. and O.-I.N.; Visualization, O.-I.N.; Supervision, A.H.; Project administration, A.H. and O.-I.N.; Funding acquisition, A.H. All authors have read and agreed to the published version of the manuscript.

Funding: We are very grateful for the generous funding of this work by the Swiss National Science Foundation (SNSF), Grant number 200021_188465: "NanoTraPPED"—Development of a Method for Measuring the Surface Energy of Nanoparticles.

Data Availability Statement: The data generated in this study are publicly available in an open access repository Open Science Framework (OSF) repository at https://doi.org/10.17605/OSF.IO/TYM28.

Conflicts of Interest: The authors declare no conflict of interest.

References

1. Honciuc, A. *Chemistry of Functional Materials Surfaces and Interfaces: Fundamentals and Applications*, 1st ed.; Elsevier: Amsterdam, The Netherlands, 2021; ISBN 978-0-12-821059-8.
2. Honciuc, A.; Negru, O.-I. NanoTraPPED—A New Method for Determining the Surface Energy of Nanoparticles via Pickering Emulsion Polymerization. *Nanomaterials* **2021**, *11*, 3200. [CrossRef] [PubMed]
3. Karde, V.; Ghoroi, C. Fine Powder Flow under Humid Environmental Conditions from the Perspective of Surface Energy. *Int. J. Pharm.* **2015**, *485*, 192–201. [CrossRef] [PubMed]
4. Parent, J.S.; Mrkoci, M.I.; Hennigar, S.L. Silica Agglomeration and Elastomer Reinforcement: Influence of Surface Modifications. *Plast. Rubber Compos.* **2003**, *32*, 114–121. [CrossRef]
5. Shah, U.V.; Olusanmi, D.; Narang, A.S.; Hussain, M.A.; Tobyn, M.J.; Heng, J.Y.Y. Decoupling the Contribution of Dispersive and Acid-Base Components of Surface Energy on the Cohesion of Pharmaceutical Powders. *Int. J. Pharm.* **2014**, *475*, 592–596. [CrossRef] [PubMed]
6. Rudolph, M.; Hartmann, R. Specific Surface Free Energy Component Distributions and Flotabilities of Mineral Microparticles in Flotation—An Inverse Gas Chromatography Study. *Colloids Surf. A Physicochem. Eng. Asp.* **2017**, *513*, 380–388. [CrossRef]
7. Honciuc, A.; Negru, O.-I. Role of Surface Energy of Nanoparticle Stabilizers in the Synthesis of Microspheres via Pickering Emulsion Polymerization. *Nanomaterials* **2022**, *12*, 995. [CrossRef] [PubMed]
8. Owens, D.K.; Wendt, R.C. Estimation of the Surface Free Energy of Polymers. *J. Appl. Polym. Sci.* **1969**, *13*, 1741–1747. [CrossRef]
9. Preuss, M.; Butt, H.-J. Measuring the Contact Angle of Individual Colloidal Particles. *J. Colloid Interface Sci.* **1998**, *208*, 468–477. [CrossRef]
10. Paunov, V.N. Novel Method for Determining the Three-Phase Contact Angle of Colloid Particles Adsorbed at Air–Water and Oil–Water Interfaces. *Langmuir* **2003**, *19*, 7970–7976. [CrossRef]
11. Isa, L.; Lucas, F.; Wepf, R.; Reimhult, E. Measuring Single-Nanoparticle Wetting Properties by Freeze-Fracture Shadow-Casting Cryo-Scanning Electron Microscopy. *Nat. Commun.* **2011**, *2*, 438. [CrossRef] [PubMed]
12. Honciuc, A. Amphiphilic Janus Particles at Interfaces. In *Flowing Matter*; Toschi, F., Sega, M., Eds.; Springer International Publishing: Cham, Switzerland, 2019; pp. 95–136. ISBN 978-3-030-23369-3.
13. Wu, D.; Honciuc, A. Design of Janus Nanoparticles with PH-Triggered Switchable Amphiphilicity for Interfacial Applications. *ACS Appl. Nano Mater.* **2018**, *34*, 1225–1233. [CrossRef]
14. Mihali, V.; Honciuc, A. Evolution of Self-Organized Microcapsules with Variable Conductivities from Self-Assembled Nanoparticles at Interfaces. *ACS Nano* **2019**, *13*, 3483–3491. [CrossRef] [PubMed]
15. Wu, D.; Honciuc, A. Contrasting Mechanisms of Spontaneous Adsorption at Liquid-Liquid Interfaces of Nanoparticles "Constituted of" and "Grafted with" PH-Responsive Polymers. *Langmuir* **2018**, *34*, 6170–6182. [CrossRef] [PubMed]
16. Finkle, P.; Draper, H.D.; Hildebrand, J.H. The Theory of Emulsification. *J. Am. Chem. Soc.* **1923**, *45*, 2780–2788. [CrossRef]
17. Binks, B.P.; Clint, J.H. Solid Wettability from Surface Energy Components: Relevance to Pickering Emulsions. *Langmuir* **2002**, *18*, 1270–1273. [CrossRef]
18. Aveyard, R.; Binks, B.P.; Clint, J.H. Emulsions Stabilised Solely by Colloidal Particles. *Adv. Colloid Interface Sci.* **2003**, *100*, 503–546. [CrossRef]
19. Rabel, W. Einige Aspekte Der Benetzungstheorie Und Ihre Anwendung Auf Die Untersuchung Und Veränderung Der Oberflächeneigenschaften von Polymeren. *Farbe Und Lack* **1971**, *77*, 997–1005.

MDPI
St. Alban-Anlage 66
4052 Basel
Switzerland
www.mdpi.com

Nanomaterials Editorial Office
E-mail: nanomaterials@mdpi.com
www.mdpi.com/journal/nanomaterials

www.ingramcontent.com/pod-product-compliance
Lightning Source LLC
LaVergne TN
LVHW070222100526
838202LV00015B/2073